Lecture Notes in Computer Science

Lecture Notes in Artificial Intelligence 14947

Founding Editor

Jörg Siekmann

Series Editors

Randy Goebel, *University of Alberta, Edmonton, Canada*
Wolfgang Wahlster, *DFKI, Berlin, Germany*
Zhi-Hua Zhou, *Nanjing University, Nanjing, China*

The series Lecture Notes in Artificial Intelligence (LNAI) was established in 1988 as a topical subseries of LNCS devoted to artificial intelligence.

The series publishes state-of-the-art research results at a high level. As with the LNCS mother series, the mission of the series is to serve the international R & D community by providing an invaluable service, mainly focused on the publication of conference and workshop proceedings and postproceedings.

Albert Bifet · Jesse Davis · Tomas Krilavičius ·
Meelis Kull · Eirini Ntoutsi · Indrė Žliobaitė
Editors

Machine Learning and Knowledge Discovery in Databases

Research Track

European Conference, ECML PKDD 2024
Vilnius, Lithuania, September 9–13, 2024
Proceedings, Part VII

 Springer

Editors
Albert Bifet 🆔
LTCI
Télécom Paris
Palaiseau Cedex, France

Tomas Krilavičius 🆔
Faculty of Informatics
Vytautas Magnus University
Akademija, Lithuania

Eirini Ntoutsi 🆔
Department of Computer Science
Bundeswehr University Munich
Munich, Germany

Jesse Davis 🆔
KU Leuven
Leuven, Belgium

Meelis Kull 🆔
Institute of Computer Science
University of Tartu
Tartu, Estonia

Indrė Žliobaitė 🆔
Department of Computer Science
University of Helsinki
Helsinki, Finland

ISSN 0302-9743   ISSN 1611-3349 (electronic)
Lecture Notes in Artificial Intelligence
ISBN 978-3-031-70367-6   ISBN 978-3-031-70368-3 (eBook)
https://doi.org/10.1007/978-3-031-70368-3

LNCS Sublibrary: SL7 – Artificial Intelligence

This Springer imprint is published by the registered company Springer Nature Switzerland AG
The registered company address is: Gewerbestrasse 11, 6330 Cham, Switzerland

If disposing of this product, please recycle the paper.

Preface

The 2024 edition of the European Conference on Machine Learning and Principles and Practice of Knowledge Discovery in Databases (ECML PKDD 2024) was held in Vilnius, Lithuania, from September 9 to 13, 2024.

The annual ECML PKDD conference acts as a world-wide platform showcasing the latest advancements in machine learning and knowledge discovery in databases. Held jointly since 2001, ECML PKDD has established itself as the leading European Machine Learning and Data Mining conference. It offers researchers and practitioners an unparalleled opportunity to exchange knowledge and ideas about the latest technical advancements in these disciplines. Moreover, the conference appreciates the synergy between foundational advances and groundbreaking data science and hence strongly welcomes contributions about how Machine Learning and Data Mining is being employed to solve real-world challenges.

The conference continues to evolve reflecting evolving technological developments and societal needs. For example, in the Research Track this year there has been an increase in submissions on generative AI, especially LLMs, and various aspects of responsible AI.

We received 826 submissions for the Research Track and 224 for the Applied Data Science Track. The Research track accepted 202 papers (out of 826, 24.5%) and the Applied Data Science Track accepted 56 (out of 224, 24.5%). In addition, 31 papers from the Journal Track (accepted out of 65 submissions) and 14 Demo Track papers (accepted out of 30 submissions).

The papers presented over the three main conference days were organized into five distinct tracks:

Research Track: This track featured research and methodology papers spanning all branches within Machine Learning, Knowledge Discovery, and Data Mining.

Applied Data Science Track: Papers in this track focused on novel applications of machine learning, data mining, and knowledge discovery to address real-world challenges, aiming to bridge the gap between theory and practical implementation.

Journal Track: This track included papers that had been published in special issues of the journals *Machine Learning* and *Data Mining and Knowledge Discovery.*

Demo Track: Short papers in this track introduced new prototypes or fully operational systems that leverage data science techniques, demonstrated through working prototypes.

Nectar Track: Concise presentations of recent scientific advances published in related conferences or journals. It aimed to disseminate important research findings to a broader audience within the ECML PKDD community.

The conference featured five keynote talks on diverse topics, reflecting emerging needs like benchmarking and resource-awareness, as well as theoretical understanding and industrial needs.

- Gintarė Karolina Džiugaitė (Google DeepMind): *The Dynamics of Memorization and Unlearning.*
- Moritz Hardt (Max Planck Institute for Intelligent Systems): *The Emerging Science of Benchmarks.*
- Mounia Lalmas-Roelleke (Spotify): *Enhancing User Experience with AI-Powered Search and Recommendations at Spotify.*
- Patrick Lucey (Stats Perform): *How to Utilize (and Generate) Player Tracking Data in Sport.*
- Katharina Morik (TU Dortmund University): *Resource-Aware Machine Learning — a User-Oriented Approach.*

The ECML PKDD 2024 Organizing Committee supported Diversity and Inclusion by awarding some grants that enable early career researchers to attend the conference, present their research activities, and become part of the ECML PKDD community. We provided a total of 3 scholarships of €1000 to individuals that come from the developing countries and/or communities which are underrepresented in science and technology. The scholarships could be used for travel and accommodation. In addition 3 grants covering all of the registration fees were awarded to individuals who belong to underrepresented communities, based on gender and role/position, to attend the conference and present their research activities. The Diversity and Inclusion action also included the Women Networking event and Diversity and Inclusion Panel discussion. The Women Networking event aimed to create a safe and inclusive space for networking and reflecting on the experience of women in science. The event included a structured brainstorm/reflection on the role and experience of women in science and technology, which will be published in the conference newsletter. The Diversity and Inclusion Panel aimed to reach a wider audience and encourage the discussion on the need for diversity in tech, and challenges and solutions in achieving it.

We want to thank the authors, workshop and tutorial organizers, and participants whose scientific contributions make this such an exciting event. Moreover, putting together an outstanding conference program would also not be possible without the dedication and (substantial) time investments of the area chairs, program committee, and organizing committee. The event would not run smoothly without the many volunteers and sessions chairs. Finally, we want to extend a special thanks to all the local organizers – they dealt with all the little details that are needed to make the conference a memorable event.

We want to extend our heartfelt gratitude to our wonderful sponsors for their generous financial support. We also want to thank Springer for their continuous support and Microsoft for allowing us to use their CMT software for conference management and providing help throughout. We very much appreciate the advice and guidance provided by the ECML PKDD Steering Committee over the past two years. Finally, we thank the organizing institution, the Artificial Intelligence Association of Lithuania.

September 2024

Albert Bifet
Tomas Krilavičius
Eirini Ntoutsi
Indrė Žliobaitė
Jesse Davis
Meelis Kull
Ioanna Miliou
Slawomir Nowaczyk

Organization

General Chairs

Albert Bifet   IP Paris, France/University of Waikato, New Zealand

Tomas Krilavičius   Vytautas Magnus University, Lithuania

Research Track Program Chairs

Indrė Žliobaitė   University of Helsinki, Finland
Meelis Kull   University of Tartu, Estonia
Jesse Davis   KU Leuven, Belgium
Eirini Ntoutsi   University of the Bundeswehr Munich, Germany

Applied Data Science Track Program Chairs

Slawomir Nowaczyk   Halmstad University, Sweden
Ioanna Miliou   Stockholm University, Sweden

Journal Track Chairs

Panagiotis Papapetrou   Stockholm University, Sweden
Rita Ribeiro   University of Porto/LIAAD, Portugal
Myra Spiliopoulou   Otto-von-Guericke University Magdeburg, Germany
Šarūnas Girdzijauskas   KTH Royal Institute of Technology, Sweden

Local Chair

Linas Petkevičius   Vilnius University, Lithuania

Workshop and Tutorial Chairs

Mantas Lukoševičius   Kaunas University of Technology, Lithuania
Mykola Pechenizkiy   Technische Universiteit Eindhoven,
                                 the Netherlands

Demo Chairs

Povilas Daniušis   Vytautas Magnus University, Lithuania
Kai Puolamäki   University of Helsinki, Finland

Proceedings Chairs

Wouter Duivesteijn   Technische Universiteit Eindhoven,
                                 the Netherlands
Rianne Schouten   Technische Universiteit Eindhoven,
                                 the Netherlands

PhD Forum Chairs

Virginijus Marcinkevičius   Vilnius University, Lithuania
Simona Ramanauskaitė   Vilnius Tech, Lithuania

Discovery Track Chairs

Peter van der Putten   Universiteit Leiden, the Netherlands
Jan N. van Rijn   Universiteit Leiden, the Netherlands

Workshop Proceedings Chairs

Danguole Kalinauskaite   Vytautas Magnus University, Lithuania
Kristina Šutiene   Kaunas Technology University, Lithuania

Social Media and Web Chairs

Julija Vaitonytė   Tilburg University, the Netherlands
Kamilė Dementavičiūtė   Vilnius University, Lithuania

Sponsorship Chairs

Mariam Barry   BNP Paribas, France
Dalia Breskuvienė   Vilnius University, Lithuania
Daniele Apiletti   Politecnico di Torino, Italy

Diversity and Inclusion Chair

Rūta Binkytė-Sadauskienė   Inria, France

Industry Track Chairs

Pieter Van Hertum   ASML, the Netherlands
Bjoern Bringmann   Deloitte, Germany

Nectar Track Chairs

Heitor Murilo Gomes   Victoria University of Wellington, New Zealand
Jesse Read   École Polytechnique, France

Awards Chairs

Michele Sebag   CNRS, France
João Gama   University of Porto, Portugal

ECML PKDD Steering Committee

Tijl De Bie   Ghent University, Belgium
Francesco Bonchi   ISI Foundation, Italy
Albert Bifet   Télécom ParisTech, France

Andrea Passerini	University of Trento, Italy
Katharina Morik	TU Dortmund, Germany
Arno Siebes	Utrecht University, the Netherlands
Sašo Džeroski	Jožef Stefan Institute, Slovenia
Robert Jan van Wijk	ASML, the Netherlands
Ilaria Bordino	UniCredit, Italy
Siegfried Nijssen	Université catholique de Louvain, Belgium
Albrecht Zimmermann	University of Caen - Normandie, France
Annalisa Appice	University of Bari 'Aldo Moro', Italy
Tania Cerquitelli	Politecnico di Torino, Italy
Alípio Jorge	University of Porto, Portugal
Fernando Perez-Cruz	ETH Zurich, Switzerland
Massih-Reza Amini	University Grenoble Alpes, France
Peggy Cellier	INSA Rennes, IRISA, France
Tias Guns	KU Leuven, Belgium
Grigorios Tsoumakas	Aristote University of Thessaloniki, Greece
Elena Baralis	Politecnico di Torino, Italy
Claudia Plant	Universität Wien, Austria
Manuel Gomez Rodriguez	Max Planck Institute for Software Systems, Germany

Program Committees

Guest Editorial Board, Journal Track

Richard Allmendringer	University of Manchester, UK
Marie Anastacio	Leiden University, the Netherlands
Giuseppina Andresini	Università degli Studi di Bari 'Aldo Moro', Italy
Annalisa Appice	Università degli Studi di Bari 'Aldo Moro', Italy
Jaume Bacardit	Newcastle University, UK
Maria Bampa	Stockholm University, Sweden
Mitra Baratchi	LIACS - University of Leiden, the Netherlands
Szymon Bobek	Jagiellonian University, Poland
Claudio Borile	CENTAI Institute, Italy
Falko Brause	University of Vienna, Austria
Barbara Catania	University of Genoa, Italy
Michelangelo Ceci	University of Bari, Italy
Loïc Cerf	Universidade Federal de Minas Gerais, Brazil
Tianyi Chen	Boston University, USA
Filip Cornell	KTH Royal Institute of Technology, Sweden

Marco Cotogni	University of Pavia, Italy
Claudia Diamantini	Università Politecnica delle Marche, Italy
Sebastien Destercke	UTC, France
César Ferri	Universitat Politécnica Valéncia, Spain
Olga Fink	EPFL, Switzerland
Esther Galbrun	University of Eastern Finland, Finland
Joao Gama	INESC TEC - LIAAD, Portugal
Jose A. Gamez	Universidad de Castilla-La Mancha, Spain
Paolo Garza	Politecnico di Torino, Italy
Carolina Geiersbach	Weierstrass Institute Berlin, Germany
Riccardo Guidotti	University of Pisa, Italy
Francesco Gullo	UniCredit, Italy
Martin Holena	Institute of Computer Science, Czechia
Dino Ienco	INRAE, France
Georgiana Ifrim	University College Dublin, Ireland
Felix Iglesias	Technical University of Vienna, Austria
Angelo Impedovo	University of Bari 'Aldo Moro', Italy
Matthias Jacobs	Technical University Dortmund, Germany
Szymon Jaroszewicz	Polish Academy of Sciences, Poland
Yifei Jin	Ericsson Research/KTH Royal Institute of Technology, Sweden
Panagiotis Karras	University of Copenhagen, Denmark
Mehdi Kaytoue	Infologic R&D, France
Dragi Kocev	Josef Stefan Institute, Slovenia
Helge Langseth	Norwegian Univ of Science and Technology, Norway
Thien Le	MIT, USA
Hsuan-Tien Lin	National Taiwan University, Taiwan
Marco Lippi	University of Modena and Reggio Emilia, Italy
Corrado Loglisci	Università degli Studi di Bari 'Aldo Moro', Italy
Brian Mac Namee	University College Dublin, Ireland
Sindri Magnusson	Stockholm University, Sweden
Giuseppe Manco	ICAR-CNR, Italy
Michael Mathioudakis	University of Helsinki, Finland
Ioanna Miliou	Stockholm University, Sweden
Olof Mogren	RISE Research Institutes, Sweden
Nuno Moniz	University of Notre Dame, France
Anna Monreale	University of Pisa, Italy
Alberto Montresor	University of Trento, Italy
Katharina Morik	Technical University Dortmund, Germany
Lia Morra	Politecnico di Torino, Italy
Amedeo Napoli	LORIA, Nancy, France

Andrea Paudice	University of Milan, Italy
Benjamin Noack	Otto-von-Guericke University Magdeburg, Germany
Slawomir Nowaczyk	Halmstad University, Sweden
Vincenzo Pasquadibisceglie	Università degli Studi di Bari 'Aldo Moro', Italy
Ruggero G. Pensa	University of Turin, Italy
Linas Petkevicius	Vilnius University, Finland
Marc Plantevit	EPITA, France
Kai Puolamäki	University of Helsinki, Finland
Jan Ramon	Inria, France
Matteo Riondato	Amherst College, USA
Isak Samsten	Stockholm University, Sweden
Shinichi Shirakawa	Yokohama National University, Japan
Amira Soliman	Halmstad University, Sweden
Fabian Spaeh	Boston University, USA
Gerasimos Spanakis	Maastricht University, the Netherlands
Mahito Sugiyama	National Institute of Informatics, Japan
Nikolaj Tatti	Helsinki University, Finland
Josephine Thomas	University of Kassel, Germany
Sebastian Stober	Otto-von-Guericke University Magdeburg, Germany
Genoveva Vargas-Solar	CNRS LIRIS, France
Bruno Veloso	University of Porto, Portugal
Pascal Welke	Technical University of Vienna, Austria
Marcel Wever	Ludwig-Maximilian-University Munich, Germany
Ye Zhu	Deakin University, Australia
Albrecht Zimmermann	Université de Caen Normandie, France
Blaz Zupan	University of Ljubljana, Slovenia

Area Chairs, Research Track

Leman Akoglu	CMU, USA
Anthony Bagnall	University of Southampton, UK
Gustavo Batista	UNSW, Australia
Jessa Bekker	KU Leuven, Belgium
Bettina Berendt	TU Berlin, Germany
Hendrik Blockeel	KU Leuven, Belgium
Henrik Bostrom	KTH Royal Institute of Technology, Sweden
Zied Bouraoui	CRIL CNRS & Univ Artois, France
Ulf Brefeld	Leuphana, Germany

Toon Calders	Universiteit Antwerpen, Belgium
Michelangelo Ceci	University of Bari, Italy
Fabrizio Costa	Exeter University, UK
Tijl De Bie	Ghent University, Belgium
Tom Diethe	AstraZeneca, UK
Kurt Driessens	Maastricht University, the Netherlands
Wouter Duivesteijn	TU Eindhoven, the Netherlands
Sebastijan Dumancic	TU Delft, the Netherlands
Tapio Elomaa	Tampere University, Finland
Stefano Ferilli	University of Bari, Italy
Cèsar Ferri	Universitat Politècnica València, Spain
Peter Flach	University of Bristol, UK
Elisa Fromont	Université Rennes 1, IRISA/Inria rba, France
Johannes Fürnkranz	JKU Linz, Austria
Esther Galbrun	University of Eastern Finland, Finland
Joao Gama	INESC TEC - LIAAD, Portugal
Aristides Gionis	KTH Royal Institute of Technology, Sweden
Bart Goethals	Universiteit Antwerpen, Belgium
Chen Gong	Nanjing University of Science and Technology, China
Dimitrios Gunopulos	University of Athens, Greece
Tias Guns	KU Leuven, Belgium
Barbara Hammer	CITEC, Bielefeld University, Germany
José Hernández-Orallo	Universitat Politècnica de València, Spain
Sibylle Hess	TU Eindhoven, the Netherlands
Andreas Hotho	University of Wuerzburg, Germany
Eyke Hüllermeier	University of Munich, Germany
Georgiana Ifrim	University College Dublin, Ireland
Manfred Jaeger	Aalborg University, Denmark
Szymon Jaroszewicz	Polish Academy of Sciences, Poland
George Karypis	University of Minnesota, Twin Cities, USA
Ioannis Katakis	University of Nicosia, Cyprus
Marius Kloft	TU Kaiserslautern, Germany
Dragi Kocev	Jožef Stefan Institute, Slovenia
Parisa Kordjamshidi	Michigan State University, USA
Lars Kotthoff	University of Wyoming, USA
Petra Kralj Novak	Central European University, Austria
Georg Krempl	Utrecht University, the Netherlands
Peer Kröger	Christian-Albrechts-Universität Kiel, Germany
Leo Lahti	University of Turku, Finland
Mark Last	Ben-Gurion University of the Negev, Israel
Jefrey Lijffijt	Ghent University, Belgium

Jessica Lin	George Mason University, USA
Michele Lombardi	University of Bologna, Italy
Donato Malerba	Università degli Studi di Bari 'Aldo Moro', Italy
Fragkiskos Malliaros	CentraleSupelec, France
Giuseppe Marra	KU Leuven, Belgium
Wannes Meert	KU Leuven, Belgium
Ernestina Menasalvas	Universidad Politècnica de Madrid, Spain
Pauli Miettinen	University of Eastern Finland, Finland
Dunja Mladenic	Jozef Stefan Institute, Slovenia
Emmanuel Müller	TU Dortmund, Germany
Siegfried Nijssen	Université catholique de Louvain, Belgium
Symeon Papadopoulos	Information Technologies Institute/Centre for Research & Technology - Hellas, Greece
Evangelos Papalexakis	UC Riverside, USA
Andrea Passerini	University of Trento, Italy
Jaakko Peltonen	Tampere University, Finland
Bernhard Pfahringer	University of Waikato, New Zealand
Claudia Plant	University of Vienna, Austria
Ricardo Prudencio	Universidade Federal de Pernambuco, Brazil
Milos Radovanovic	U. Novi Sad, Serbia
Chedy Raissi	Inria, France
Jesse Read	Ecole Polytechnique, France
Celine Robardet	INSA Lyon, France
Salvatore Ruggieri	University of Pisa, Italy
Steven Schockaert	Cardiff University, Wales, UK
Matthias Schubert	Ludwig-Maximilians-Universität München, Germany
Thomas Seidl	LMU Munich, Germany
Arno Siebes	Universiteit Utrecht, the Netherlands
Fabrizio Silvestri	Sapienza University of Rome, Italy
Jerzy Stefanowski	Poznan University of Technology, Poland
Nikolaj Tatti	Helsinki University, Finland
Evimaria Terzi	Boston University, USA
Grigorios Tsoumakas	Aristotle University of Thessaloniki, Greece
Charalampos Tsourakakis	Boston University, USA
Matthijs van Leeuwen	Leiden University, the Netherlands
Jan Van Rijn	LIACS, Leiden University, the Netherlands
Celine Vens	KU Leuven, Belgium
Jilles Vreeken	CISPA Helmholtz Center for Information Security, Germany
Willem Waegeman	Universiteit Gent, Belgium
Wei Ye	Tongji University, China

Wenbin Zhang	Florida International University, USA
Arthur Zimek	University of Southern Denmark, Denmark
Albrecht Zimmermann	Université de Caen Normandie, France

Area Chairs, Applied Data Science Track

Annalisa Appice	University of Bari 'Aldo Moro', Italy
Sahar Asadi	King (Microsoft), Sweden
Martin Atzmueller	Osnabrück University & DFKI, Germany
Michael R. Berthold	KNIME, Germany
Michelangelo Ceci	University of Bari, Italy
Peggy Cellier	INSA Rennes, IRISA, France
Nicolas Courty	IRISA, Université Bretagne-Sud, France
Bruno Cremilleux	Université de Caen Normandie, France
Tom Diethe	AstraZeneca, UK
Dejing Dou	BCG, USA
Olga Fink	EPFL, Switzerland
Elisa Fromont	Université Rennes 1, IRISA/Inria rba, France
Johannes Fürnkranz	JKU Linz, Austria
Sreenivas Gollapudi	Google, USA
Andreas Hotho	University of Wuerzburg, Germany
Alipio M. G. Jorge	INESC TEC/University of Porto, Portugal
George Karypis	University of Minnesota, Minneapolis, USA
Yun Sing Koh	University of Auckland, New Zealand
Parisa Kordjamshidi	Michigan State University, USA
Niklas Lavesson	Blekinge Institute of Technology, Sweden
Chuan Lei	Amazon, USA
Thomas Liebig	TU Dortmund Artificial Intelligence Unit, Germany
Tony Lindgren	Stockholm University, Sweden
Patrick Loiseau	Inria, France
Giuseppe Manco	ICAR-CNR, Italy
Gabor Melli	PredictionWorks, USA
Ioanna Miliou	Stockholm University, Sweden
Anna Monreale	University of Pisa, Italy
Luis Moreira-Matias	sennder, Germany
Jian Pei	Simon Fraser University, Canada
Fabio Pinelli	IMT Lucca, Italy
Zhiwei (Tony) Qin	Lyft, USA
Visvanathan Ramesh	Independent Researcher, Germany
Fabrizio Silvestri	Sapienza, University of Rome, Italy

Liang Sun   Alibaba Group, China
Jiliang Tang   Michigan State University, USA
Sandeep Tata   Google, USA
Yinglong Xia   Meta, USA
Fuzhen Zhuang   Institute of Artificial Intelligence, Beihang
                                       University, China
Albrecht Zimmermann   Université de Caen Normandie, France

Program Committee Members, Research Track

Zahraa Abdallah   University of Bristol, UK
Ziawasch Abedjan   TU Berlin, Germany
Koren Abitbul   Ben-Gurion University, Israel
Timilehin Aderinola   Insight SFI Research Centre for Data Analytics,
                                       University College Dublin, Ireland
Homayun Afrabandpey   Nokia Technologies, Finland
Reza Akbarinia   Inria, France
Esra Akbas   Georgia State University, USA
Cuneyt Akcora   University of Central Florida, USA
Youhei Akimoto   University of Tsukuba/RIKEN AIP, Japan
Ozge Alacam   University of Bielefeld, Germany
Amr Alkhatib   KTH Royal Institute of Technology, Sweden
Mari-Liis Allikivi   University of Tartu, Estonia
Ranya Almohsen   West Virginia University, USA
Jose Alvarez   Scuola Normale Superiore, Italy
Ehsan Aminian   INESC TEC, Portugal
Christos Anagnostopoulos   University of Glasgow, UK
James Anderson   Columbia University, USA
Thiago Andrade   INESC TEC/University of Porto, Portugal
Jean-Marc Andreoli   Naverlabs Europe, France
Giuseppina Andresini   University of Bari 'Aldo Moro', Italy
Simone Angarano   Politecnico di Torino, Italy
Akash Anil   Cardiff University, Wales, UK
Ekaterina Antonenko   Mines Paris - PSL, France
Alessandro Antonucci   IDSIA, Switzerland
Edward Apeh   Bournemouth University, UK
Nikhilanand Arya   Indian Institute of Technology, Patna, India
Saeed Asadi Bagloee   University of Melbourne, Australia
Ali Ayadi   University of Strasbourg, France
Steve Azzolin   University of Trento, Italy
Lilian Berton   Universidade Federal de Sao Paulo, Brazil

Florian Babl   Universität der Bundeswehr München, Germany
Michael Bain   University of New South Wales, Australia
Chandrajit Bajaj   University of Texas, Austin, USA
Bunil Balabantaray   NIT Meghalaya, India
Federico Baldo   University of Bologna, Italy
Georgia Baltsou   Information Technologies Institute/Centre for
                               Research & Technology - Hellas, Greece

Hubert Baniecki   University of Warsaw, Poland
Mitra Baratchi   LIACS - University of Leiden, the Netherlands
Francesco Bariatti   Univ Rennes, CNRS, IRISA, France
Franka Bause   University of Vienna, Austria
Florian Beck   JKU Linz, Austria
Jacob Beck   LMU Munich, Germany
Rita Beigaite   VTT, Finland
Michael Beigl   Karlsruhe Institute of Technology, Germany
Diana Benavides Prado   University of Auckland, New Zealand
Andreas Bender   LMU Munich, Germany
Idir Benouaret   Epita Research Laboratory, France
Gilberto Bernardes   INESC TEC & University of Porto, Faculty of
                               Engineering, Portugal

Jolita Bernatavičienė   Vilnius University, Lithuania
Cuissart Bertrand   University of Caen, France
Eva Besada-Portas   Universidad Complutense de Madrid, Spain
Jalaj Bhandari   Columbia University, USA
Monowar Bhuyan   Umea University, Sweden
Manuele Bicego   University of Verona, Italy
Przemyslaw Biecek   Warsaw University of Technology, Poland
Albert Bifet   Telecom Paris, France
Livio Bioglio   University of Turin, Italy
Anton Björklund   University of Helsinki, Finland
Szymon Bobek   Jagiellonian University, Poland
Ludovico Boratto   University of Cagliari, Italy
Stefano Bortoli   Huawei Research Center
Annelot Bosman   Universiteit Leiden, the Netherlands
Tassadit Bouadi   Université de Rennes, France
Hamid Bouchachia   Bournemouth University, UK
Jannis Brugger   TU Darmstadt, Germany
Dariusz Brzezinski   Poznan University of Technology, Poland
Maria Sofia Bucarelli   Sapienza University of Rome, Italy
Mirko Bunse   TU Dortmund University, Germany
Tomasz Burzykowski   Hasselt University, Belgium

Sebastian Buschjäger	TU Dortmund Artificial Intelligence Unit, Germany
Maarten Buyl	Ghent University, Belgium
Zaineb Chelly Dagdia	UVSQ, Paris-Saclay, France
Huaming Chen	University of Sydney, Australia
Xiaojun Chen	Institute of Information Engineering, CAS, China
Tobias Callies	Universtiät der Bundeswehr München, Germany
Xiaofeng Cao	University of Technology Sydney, Australia
Cécile Capponi	Aix-Marseille University, France
Lorenzo Cascioli	KU Leuven, Belgium
Guilherme Cassales	University of Waikato, New Zealand
Giovanna Castellano	University of Bari 'Aldo Moro', Italy
Andrea Cavallo	Delft University of Technology, the Netherlands
Remy Cazabet	Lyon, France
Antanas Čenys	Vilnius Gediminas Technical University, Lithuania
Mattia Cerrato	JGU Mainz, Germany
Ricardo Cerri	Federal University of Sao Carlos, Brazil
Prithwish Chakraborty	IBM Corporation
Harry Kai-Ho Chan	University of Sheffield, UK
Laetitia Chapel	IRISA, France
Victor Charpenay	Mines Saint-Etienne, France
Arthur Charpentier	UQAM, Canada
Chunchun Chen	Tongji University, China
Huiping Chen	University of Birmingham, UK
Jin Chen	Hong Kong University of Science and Technology, China
Kuan-Hsun Chen	University of Twente, the Netherlands
Lingwei Chen	Wright State University, USA
Minyu Chen	Shanghai Jiaotong University, China
Xuefeng Chen	Chongqing University, China
Ying Chen	RMIT University, Australia
Zheng Chen	Osaka University, Japan
Zhong Chen	Southern Illinois University, USA
Ziheng Chen	Walmart, USA
Zehua Cheng	University of Oxford, UK
Hua Chu	Xidian University, China
Oana Cocarascu	King's College London, UK
Johanne Cohen	LISN-CNRS, France
Lidia Contreras-Ochando	Universitat Politècnica de València, Spain
Denis Coquenet	IRISA, France
Luca Corbucci	University of Pisa, Italy

Roberto Corizzo	American University, USA
Nathan Cornille	KU Leuven, Belgium
Baris Coskunuzer	University of Texas at Dallas, USA
Andrea Cossu	University of Pisa, Italy
Tiago Cunha	Expedia Group, Portugal
Florence d'Alché-Buc	Télécom Paris, France
Sebastian Dalleiger	KTH Royal Institute of Technology, Sweden
Robertas Damaševičius	Vytautas Magnus University, Lithuania
Xuan-Hong Dang	IBM T.J Watson Research Center, USA
Thi-Bich-Hanh Dao	University of Orleans, France
Paul Davidsson	Malmö University, Sweden
Jasper de Boer	KU Leuven, Belgium
Andre de Carvalho	USP, Brazil
Graziella De Martino	University of Bari 'Aldo Moro', Italy
Lennert De Smet	KU Leuven, Belgium
Marcilio de Souto	LIFO/Univ. Orleans, France
Julien Delaunay	Inria, France
Emanuele Della Valle	Politecnico di Milano, Italy
Pieter Delobelle	KU Leuven, Belgium
Vincent Derkinderen	KU Leuven, Belgium
Guillaume Derval	UCLouvain - ICTEAM, Belgium
Sebastien Destercke	UTC, France
Laurens Devos	KU Leuven, Belgium
Bhaskar Dhariyal	University College Dublin, Ireland
Davide Di Pierro	Università degli Studi di Bari, Italy
Yiqun Diao	National University of Singapore, Singapore
Lucile Dierckx	Université catholique de Louvain, Belgium
Anastasia Dimou	KU Leuven, Belgium
Jingtao Ding	Tsinghua University, China
Zifeng Ding	LMU Munich, Germany
Lamine Diop	EPITA, France
Christos Diou	Harokopio University of Athens, Greece
Alexander Dockhorn	Leibniz University Hannover, Germany
Stephan Doerfel	Kiel University of Applied Sciences, Germany
Hang Dong	University of Oxford, UK
Nanqing Dong	Shanghai Artificial Intelligence Laboratory, China
Emilio Dorigatti	LMU Munich, Germany
Haizhou Du	Shanghai University of Electric Power, China
Stefan Duffner	University of Lyon, France
Inês Dutra	University of Porto, Portugal
Anany Dwivedi	University of Waikato, New Zealand
Sofiane Ennadir	KTH Royal Institute of Technology, Sweden

Mark Eastwood	University of Warwick, UK
Vasilis Efthymiou	Harokopio University of Athens, Greece
Rémi Emonet	Unversité Saint-Etienne, France
Dominik Endres	Philipps-Universität Marburg, Germany
Eshant English	Hasso Plattner Institute, Germany
Bojan Evkoski	Central European University, Austria
Zipei Fan	University of Tokyo, Japan
Hadi Fanaee-T	Halmstad University, Germany
Fabio Fassetti	Universita della Calabria, Italy
Ad Feelders	Universiteit Utrecht, the Netherlands
Wenjie Feng	National University of Singapore, Singapore
Len Feremans	Universiteit Antwerpen, Belgium
Luca Ferragina	University of Calabria, Italy
Carlos Ferreira	INESC TEC, Portugal
Julien Ferry	LAAS-CNRS, France
Michele Fontana	Università di Pisa, Italy
Germain Forestier	University of Haute Alsace, France
Edouard Fouché	Karlsruhe Institute of Technology (KIT), Germany
Matteo Francobaldi	University of Bologna, Italy
Christian Frey	Fraunhofer IIS, Germany
Holger Froening	University of Heidelberg, Germany
Benoît Frénay	University of Namur, Belgium
Fabio Fumarola	Prometeia, Italy
Shanqing Guo	Shandong University, China
Claudio Gallicchio	University of Pisa, Italy
Shengxiang Gao	Kunming University of Science and Technology, China
Yifeng Gao	University of Texas Rio Grande Valley, USA
Manuel Garcia-Piqueras	Universidad de Castilla-La Mancha, Spain
Dario Garigliotti	University of Bergen, Norway
Damien Garreau	Université Côte d'Azur, France
Dominique Gay	Université de La Réunion, France
Alborz Geramifard	Meta, USA
Pierre Geurts	Montefiore Institute, University of Liège, Belgium
Alireza Gharahighehi	KU Leuven, Belgium
Siamak Ghodsi	Leibniz University of Hannover Free University Berlin, Germany
Shreya Ghosh	Penn State, USA
Vasilis Gkolemis	ATHENA RC, Greece
Dorota Glowacka	University of Helsinki, Finland
Heitor Gomes	Victoria University of Wellington, New Zealand

Wenwen Gong   Tsinghua University, China
Adam Goodge   I2R, A*STAR, Singapore
Anastasios Gounaris   Aristotle University of Thessaloniki, Greece
Brandon Gower-Winter   Utrecht University, the Netherlands
Michael Granitzer   University of Passau, Germany
Xinyu Guan   Xian Jiaotong University, China
Massimo Guarascio   ICAR-CNR, Italy
Riccardo Guidotti   University of Pisa, Italy
Dominique Guillot   University of Delaware, USA
Nuwan Gunasekara   AI Institute, University of Waikato, New Zealand
Thomas Guyet   Inria, Centre de Lyon, France
Vanessa Gómez-Verdejo   Universidad Carlos III de Madrid, Spain
Huong Ha   RMIT University, Australia
Benjamin Halstead   University of Auckland, New Zealand
Marwan Hassani   TU Eindhoven, the Netherlands
Yujiang He   University of Kassel, Germany
Edith Heiter   Ghent University, Belgium
Lars Hillebrand   Fraunhofer IAIS and University of Bonn,
                                 Germany
Martin Holena   Institute of Computer Science, Czechia
Mike Holenderski   Eindhoven University of Technology,
                                 the Netherlands
Hongsheng Hu   Data 61, CSIRO, Australia
Chao Huang   University of Hong Kong, China
Denis Huseljic   University of Kassel, Germany
Julian Höllig   University of the Bundeswehr Munich, Germany
Dimitrios Iliadis   UGENT, Belgium
Dino Ienco   INRAE, France
Roberto Interdonato   CIRAD, France
Omid Isfahani Alamdari   University of Pisa, Italy
Elvin Isufi   TU Delft, the Netherlands
Giulio Jacucci   University of Helsinki, Finland
Kuk Jin Jang   University of Pennsylvania, USA
Inigo Jauregi Unanue   University of Technology Sydney, Australia
Renhe Jiang   University of Tokyo, Japan
Pengfei Jiao   Hangzhou Dianzi University, China
Yilun Jin   Hong Kong University of Science and
                                 Technology, China
Rūta Juozaitienė   Vytautas Magnus University, Lithuania
Joonas Jälkö   University of Helsinki, Finland
Mira Jürgens   Ghent University, Belgium

Vana Kalogeraki	Athens University of Economics and Business, Greece
Toshihiro Kamishima	Independent Researcher, Japan
Nikos Kanakaris	University of Southern California, USA
Sevvandi Kandanaarachchi	CSIRO, Australia
Bo Kang	Ghent University, Belgium
Jurgita Kapočiūtė-Dzikienė	Tilde SIA, University of Latvia, Tilde IT, Vytautas Magnus University, Lithuania
Maiju Karjalainen	University of Eastern Finland, Finland
Panagiotis Karras	University of Copenhagen, Denmark
Gjergji Kasneci	TU Munich, Germany
Panagiotis Kasnesis	University of West Attica, Greece
Dimitrios Katsaros	University of Thessaly, Greece
Natthawut Kertkeidkachorn	Japan Advanced Institute of Science and Technology (JAIST), Japan
Stefan Kesselheim	Forschungszentrum Jülich, Germany
Jaleed Khan	University of Oxford, UK
Adem Kikaj	KU Leuven, Belgium
Nadja Klein	University Alliance Ruhr and TU Dortmund, Germany
Tomas Kliegr	University of Economics Prague, Czechia
Astrid Klipfel	CRIL - UMR 8188, France
Simon Koop	Technische Universiteit Eindhoven, the Netherlands
Frederic Koriche	Univ. d'Artois, CRIL CNRS UMR 8188, France
Grazina Korvel	Vilnius University, Lithuania
Ana Kostovska	Jožef Stefan Institute, Slovenia
Stefan Kramer	Johannes Gutenberg University Mainz, Germany
Emmanouil Krasanakis	CERTH, Greece
Anna Krause	Universität Würzburg, Germany
Nils Kriege	University of Vienna, Austria
Ričardas Krikštolaitis	Vytautas Magnus University, Lithuania
Amer Krivosija	TU Dortmund, Germany
Paweł Ksieniewicz	Wrocław University of Science and Technology, Poland
Janne Kujala	University of Turku, Finland
Nitesh Kumar	Cardiff University, UK
Vivek Kumar	Universität der Bundeswehr München, Germany
Olga Kurasova	Vilnius University, Institute of Data Science and Digital Technologies, Lithuania
Marius Köppel	Johannes Gutenberg University Mainz, Germany
Antti Laaksonen	University of Helsinki, Finland
Ville Laitinen	University of Turku, Finland

Carlos Lamuela Orta	University of Helsinki, Finland
Johannes Langguth	Simula Research Laboratory, Norway
Helge Langseth	Norwegian University of Science and Technology, Norway
Martha Larson	Radboud University, the Netherlands
Anton Lautrup	University of Southern Denmark, Denmark
Aonghus Lawlor	University College Dublin, Ireland
Tuan Le	New Mexico State University, USA
Erwan Le Merrer	Inria, France
Thach Le Nguyen	University College Dublin, Ireland
Tai Le Quy	IU International University of Applied Sciences, Germany
Mustapha Lebbah	Paris Saclay University-Versailles, France
Yeon-Chang Lee	Ulsan National Institute of Science and Technology (UNIST), South Korea
Zed Lee	Stockholm University, Sweden
Mathieu Lefort	Univ. Lyon, France
Vincent Lemaire	Orange Innovation
Daniel Lemire	University of Quebec (TELUQ), Canada
Florian Lemmerich	University of Passau, Germany
Daphne Lenders	University of Antwerp, Belgium
Carson Leung	University of Manitoba, Canada
Dan Li	Sun Yat-Sen University, China
Gang Li	Deakin University, Australia
Mark Junjie Li	Shenzhen University, China
Mingxio Li	KU Leuven, Belgium
Nian Li	Tsinghua University, China
Peiyan Li	Ludwig Maximilian University of Munich, Germany
Shuai Li	University of Cambridge, UK and University of Tokyo, Japan and Tsinghua University, China
Tong Li	HKUST, China
Xiang Li	East China Normal University, China
Yinsheng Li	Fudan University, China
Yong Li	Huawei European Research Center, Germany
Zhixin Li	Guangxi Normal University, China
Zhuoqun Li	Louisiana State University, USA
Yuxuan Liang	Hong Kong University of Science and Technology, China
Nick Lim	University of Waikato, New Zealand
Jason Lines	Independent Researcher, UK
Piotr Lipinski	Institute of Computer Science, University of Wroclaw, Poland

Arunas Lipnickas	Kaunas University of Technology, Lithuania
Marco Lippi	University of Florence, Italy
Bin Liu	Chongqing University of Posts and Telecommunications, China
Fenglin Liu	University of Oxford, UK
Junze Liu	University of California, Irvine, USA
Li Liu	Chongqing University, China
Xu Liu	National University of Singapore, Singapore
Zihan Liu	Zhejiang University & Westlake University, China
Corrado Loglisci	Università degli Studi di Bari 'Aldo Moro', Italy
Antonio Longa	University of Trento, Italy
Marco Loog	Radboud University, the Netherlands
Ana Carolina Lorena	ITA, Brazil
Beatriz López	University of Girona, Spain
Tuwe Löfström	Jönköping University, Sweden
Pingchuan Ma	HKUST, China
Ziqiao Ma	University of Michigan, USA
Henryk Maciejewski	Wrocław University of Science and Technology, Poland
Michael Madden	National University of Ireland Galway, Ireland
Sindri Magnusson	Stockholm University, Sweden
Ajay Mahimkar	AT&T, USA
Cedric Malherbe	AstraZeneca, UK
Giuseppe Manco	ICAR-CNR, Italy
Domenico Mandaglio	DIMES Dept., University of Calabria, Italy
Justina Mandravickaitė	Vytautas Magnus University, Lithuania
Silviu Maniu	Université Grenoble Alpes, France
Naresh Manwani	International Institute of Information Technology, Hyderabad, India
Alexandru Mara	Ghent University, Belgium
Virginijus Marcinkevičius	Vilnius University, Lithuania
Timo Martens	KU Leuven, Belgium
Linas Martišauskas	Vytautas Magnus University, Lithuania
Fernando Martínez-Plumed	Universitat Politècnica de València, Spain
Koji Maruhashi	Fujitsu Research, Fujitsu Limited
Rytis Maskeliūnas	Polsl, Poland
Florent Masseglia	Inria, France
Antonio Mastropietro	Università di Pisa, Italy
Sarah Masud	LCS2, IIIT-D, India
Dalius Matuzevicius	Vilnius Gediminas Technical University, Lithuania
Chandresh Maurya	IBM Research, India

Wolfgang Mayer	University of South Australia, Australia
Giacomo Medda	University of Cagliari, Italy
Nida Meddouri	LRE-EPITA, France
Stefano Melacci	University of Siena, Italy
Alessandro Melchiorre	Johannes Kepler University Linz, Austria
Marco Mellia	Politecnico di Torino, Italy
Joao Mendes-Moreira	University of Porto, Portugal
Engelbert Mephu Nguifo	Université Clermont Auvergne, CNRS, LIMOS, France
Fabio Mercorio	University of Milan-Bicocca, Italy
Henning Meyerhenke	Humboldt-Universität zu Berlin, Germany
Matthew Middlehurst	University of Southampton, UK
Jan Mielniczuk	Polish Academy of Sciences, Poland
Paolo Mignone	University of Bari 'Aldo Moro', Italy
Matej Mihelčić	University of Zagreb, Croatia
Tsunenori Mine	Kyushu University, Japan
Pierre Monnin	Université Côte d'Azur, Inria, CNRS, I3S, France
Carlos Monserrat-Aranda	Universitat Politècnica de València, Spain
Raha Moraffah	Arizona State University, USA
Thomas Mortier	Ghent University, Belgium
Frank Mtumbuka	Cardiff University, Wales, UK
Koyel Mukherjee	Adobe Research, India
Mario Andrés Muñoz	University of Melbourne, Australia
Nikolaos Mylonas	Aristotle University of Thessaloniki, Greece
Tommi Mäklin	University of Helsinki, Finland
Felipe Kenji Nakano	KU Leuven, Belgium
Géraldin Nanfack	University of Concordia, Canada
Mirco Nanni	CNR-ISTI Pisa, Italy
Francesca Naretto	University of Pisa, Italy
Fateme Nateghi Haredasht	Stanford University, USA
Benjamin Negrevergne	Université PSL – Paris Dauphine, France
Matti Nelimarkka	University of Helsinki, Finland
Kim Thang Nguyen	LIG, University Grenoble-Alpes, France
Shiwen Ni	Shenzhen Institute of Advanced Technology (SIAT), Chinese Academy of Sciences, China
Mikko Niemi	City of Helsinki, Finland
Nikolaos Nikolaou	University College London, UK
Simona Nisticò	University of Calabria, Italy
Hao Niu	KDDI Research, Inc., Japan
Andreas Nuernberger	Magdeburg University, Germany
Claire Nédellec	INRAE, MaIAGE, France
Barry O'Sullivan	University College Cork, Ireland

Makoto Onizuka	Osaka University, Japan
Jose Oramas	University of Antwerp, IMEC-IDLab, Belgium
Luis Ortega Andrés	Autonomous University of Madrid, Spain
Latifa Oukhellou	IFSTTAR, France
Agne Paulauskaite-Taraseviciene	KTU, Artificial Intelligence Centre, Lithuania
Massimo Piccardi	University of Technology Sydney, Australia
Marc Plantevit	EPITA, France
Andrei Paleyes	University of Cambridge, UK
Emmanouil Panagiotou	Freie Universität Berlin, Germany
George Panagopoulos	University of Luxembourg
Pance Panov	Jozef Stefan Institute, Slovenia
Apostolos Papadopoulos	Aristotle University of Thessaloniki, Greece
Panagiotis Papapetrou	Stockholm University, Sweden
Francesco Parisi	University of Calabria, Italy
Abigail Parker	University of Helsinki, Finland
Antonio Parmezan	University of São Paulo, Brazil
Vincenzo Pasquadibisceglie	University of Bari 'Aldo Moro', Italy
Tatiana Passali	Aristotle University of Thessaloniki, Greece
Eliana Pastor	Politecnico di Torino, Italy
Anand Paul	Louisiana State University HSC, USA
Mykola Pechenizkiy	TU Eindhoven, the Netherlands
Yulong Pei	TU Eindhoven, the Netherlands
Nikos Pelekis	University of Piraeus, Greece
Leonardo Pellegrina	University of Padova, Italy
Charlotte Pelletier	Université de Bretagne du Sud, France
Antonio Pellicani	Università degli Studi di Bari 'Aldo Moro', Italy
Frédéric Pennerath	CentraleSupélec - LORIA, France
Ruggero Pensa	University of Torino, Italy
Lucas Pereira	Interactive Technologies Institute, LARSyS, Técnico Lisboa, Portugal
Pedro Pereira Rodrigues	University of Porto, Portugal
Miquel Perello-Nieto	University of Bristol, UK
Lorenzo Perini	KU Leuven, Belgium
Linas Petkevicius	Vilnius University, Lithuania
Ninh Pham	University of Auckland, New Zealand
Nico Piatkowski	Fraunhofer IAIS, Germany
Francesco Piccialli	Independent Researcher, Italy
Martin Pilát	Charles University, Czechia
Gianvito Pio	University of Bari, Italy
Darius Plonis	Vilnius Gediminas Technical University, Lithuania
Marco Podda	University of Pisa, Italy

Mirko Polato	University of Turin, Italy
Marco Polignano	Università di Bari, Italy
Giovanni Ponti	ENEA, Italy
Alexandru Popa	University of Bucharest, Romania
Fabrice Popineau	CentraleSupélec/LISN, France
Cedric Pradalier	GeorgiaTech Lorraine, France
Paul Prasse	University of Potsdam, Germany
Mahardhika Pratama	University of South Australia, Australia
Bardh Prenkaj	Sapienza University of Rome, Italy
Steven Prestwich	University College Cork, Ireland
Giulia Preti	CENTAI, Italy
Philippe Preux	Inria, France
Danil Provodin	TU Eindhoven, the Netherlands
Chiara Pugliese	ISTI Institute of National Research Council University of Pisa, Italy
Simon Puglisi	University of Helsinki, Finland
Andrea Pugnana	University of Pisa, Italy
Erasmo Purificato	Otto von Guericke University Magdeburg, Germany
Peter van der Putten	Leiden University, the Netherlands
Abdulhakim Qahtan	Utrecht University, the Netherlands
Kun Qian	Amazon, USA
Kallol Roy	University of Tartu, Estonia
Dimitrios Rafailidis	University of Thessaly, Greece
Muhammad Rajabinasab	University of Southern Denmark, Denmark
Chang Rajani	University of Helsinki, Finland
Simona Ramanauskaitė	Vilnius Gediminas Technical University, Lithuania
Jan Ramon	Inria, France
M. José Ramírez-Quintana	Technical University of Valencia, Spain
Rajeev Rastogi	Amazon, USA
Domenico Redavid	University of Bari, Italy
Luis Rei	Jožef Stefan Institute, Slovenia
Christoph Reinders	Leibniz University Hannover, Germany
Qianqian Ren	Heilongjiang University, China
Mina Rezaei	LMU Munich, Germany
Rita Ribeiro	Porto, Portugal
Matteo Riondato	Amherst College, USA
Simon Rittel	University of Vienna, Austria
Giuseppe Rizzo	Niuma s.r.l, Italy
Pieter Robberechts	KU Leuven, Belgium

Christophe Rodrigues	DVRC pôle universitaire Léonard de Vinci, France
Federica Rollo	UNIMORE, Italy
Luca Romeo	University of Macerata, Italy
Nicolas Roque dos Santos	University of São Paulo, Brazil
Céline Rouveirol	LIPN Univ. Sorbonne Paris Nord, France
Arjun Roy	Freie Universität Berlin, Germany
Krzysztof Rudaś	Institute of Computer Science, Polish Academy of Sciences, Poland
Allou Same	Université Gustave Eiffel, France
Oswaldo Solarte-Pabon	Universidad del Valle, Spain
Amal Saadallah	TU Dortmund, Germany
Matthia Sabatelli	University of Groningen, the Netherlands
Chafik Samir	CNRS-UCA, France
Ramses Sanchez	University of Bonn, Germany
Ioannis Sarridis	Information Technologies Institute/Centre for Research & Technology - Hellas, Greece
Milos Savic	University of Novi Sad, Serbia
Nripsuta Saxena	University of Southern California, USA
Alexander Schiendorfer	Technische Hochschule Ingolstadt, Germany
Christian Schlauch	Humboldt-Universität zu Berlin, Germany
Rainer Schlosser	Hasso Plattner Institute, Germany
Johannes Schneider	University of Liechtenstein, Liechtenstein
Rianne Schouten	Technische Universiteit Eindhoven, the Netherlands
Andreas Schwung	Fachhochschule Südwestfalen, Germany
Patrick Schäfer	Humboldt-Universität zu Berlin, Germany
Kristen Scott	KU Leuven, Belgium
Marian Scuturici	LIRIS, France
Raquel Sebastião	ESTGV-IPV & IEETA-UA
Nina Seemann	University of the Bundeswehr, Germany
Artūras Serackis	Vilnius Tech, Lithuania
Giuseppe Serra	Goethe University Frankfurt, Germany
Mattia Setzu	University of Pisa, Italy
Manali Sharma	Samsung, USA
Shubhranshu Shekhar	Brandeis University, USA
Qiang Sheng	Institute of Computing Technology, Chinese Academy of Sciences, China
John Sheppard	Montana State University, USA
Bin Shi	Xi'an Jiaotong University, China
Jimeng Shi	Florida International University, USA
Paula Silva	INESC TEC - LIAAD, Portugal

Telmo Silva Filho	University of Bristol, UK
Esther-Lydia Silva-Ramírez	Universidad de Cádiz, Spain
Raivydas Šimėnas	Vilnius University, Lithuania
Kuldeep Singh	Cerence GmbH, Germany
Andrzej Skowron	University of Warsaw, Poland
Carlos Soares	University of Porto, Portugal
Dennis Soemers	Maastricht University, the Netherlands
Andy Song	RMIT University, Australia
Liyan Song	Harbin Institute of Technology, China
Zixing Song	Chinese University of Hong Kong, China
Sucheta Soundarajan	Syracuse University, USA
Fabian Spaeh	Boston University, USA
Myra Spiliopoulou	Otto-von-Guericke-University Magdeburg, Germany
Dimitri Staufer	TU Berlin, Germany
Kostas Stefanidis	Tampere University, Finland
Pavel Stefanovič	Vilnius Tech, Lithuania
Julian Stier	University of Passau, Germany
Giovanni Stilo	Università of L'Aquila, Italy
Michiel Stock	Ghent University, Belgium
Luca Stradiotti	KU Leuven, Belgium
Lukas Struppek	Technical University of Darmstadt, Germany
Maximilian Stubbemann	University of Hildesheim, Germany
Nikolaos Stylianou	Information Technologies Institute, Greece
Jinyan Su	University of Electronic Science and Technology of China, China
Peijie Sun	Tsinghua University, China
Weiwei Sun	Shandong University, China
Swati Swati	Universität der Bundeswehr München, Germany
Panagiotis Symeonidis	University of the Aegean, Greece
Maryam Tabar	University of Texas at San Antonio, USA
Shazia Tabassum	INESC TEC, Portugal
Andrea Tagarelli	DIMES - UNICAL, Italy
Martin Takac	Mohamed bin Zayed University of Artificial Intelligence, UAE
Acar Tamersoy	NortonLifeLock Research Group, USA
Chang Wei Tan	Monash University, Australia
Xing Tang	Tencent, China
Enzo Tartaglione	Télécom Paris - Institut Polytechnique de Paris, France
Romain Tavenard	Univ. Rennes, LETG/IRISA, France
Gustaf Tegnér	KTH Royal Institute of Technology, Lithuania

Paweł Teisseyre	Warsaw University of Technology, Poland
Alexandre Termier	Université Rennes, France
Stefano Teso	University of Trento, Italy
Surendrabikram Thapa	Virginia Tech, USA
Martin Theobald	University of Luxembourg, Luxembourg
Maximilian Thiessen	TU Wien, Austria
Steffen Thoma	FZI Research Center for Information Technology, Germany
Matteo Tiezzi	University of Siena, Italy
Matteo Tiezzi	SAILab, DIISM, University of Siena, Italy
Gabriele Tolomei	Sapienza University of Rome, Italy
Paulina Tomaszewska	Warsaw University of Technology, Poland
Dinh Tran	King Fahd University of Petroleum & Minerals, Saudi Arabia
Isaac Triguero	Nottingham University, UK
Andre Tättar	University of Tartu, Estonia
Evaldas Vaičiukynas	Kaunas University of Technology, Lithuania
Jente Van Belle	KU Leuven, Belgium
Fabio Vandin	University of Padova, Italy
Aparna S. Varde	Montclair State University, USA
Bruno Veloso	INESC TEC & FEP-UP, Portugal
Dmytro Velychko	University of Oldenburg, Germany
Sreekanth Vempati	Myntra, India
Gabriele Venturato	KU Leuven, Belgium
Michela Venturini	KU Leuven, ITEC, Belgium
Mathias Verbeke	KU Leuven, Belgium
Théo Verhelst	Université libre de Bruxelles, Belgium
Rosana Veroneze	LBiC, UK
Gennaro Vessio	University of Bari 'Aldo Moro', Italy
Paul Viallard	Inria Rennes, France
Herna Viktor	University of Ottawa, Canada
Joao Vinagre	Joint Research Centre - European Commission, Spain
Jean-Noël Vittaut	Sorbonne Université, CNRS, LIP6, France
Maximilian von Zastrow	Southern Denmark University, Denmark
Tomasz Walkowiak	Wrocław University of Science and Technology, Poland
Beilun Wang	Southeast University, China
Huandong Wang	Tsinghua University, China
Hui (Wendy) Wang	Stevens Institute of Technology, USA
Jianwu Wang	University of Maryland, Baltimore County, USA
Jiaqi Wang	Penn State University, USA

Suhang Wang	Pennsylvania State University, USA
Yanhao Wang	East China Normal University, China
Yimu Wang	University of Waterloo, Canada
Yue Wang	Microsoft Research
Zhaonan Wang	University of Illinois Urbana-Champaign, USA
Zichong Wang	Florida International University, USA
Zifu Wang	KU Leuven, Belgium
Zijie J. Wang	Georgia Tech, USA
Roger Wattenhofer	ETH Zurich, Germany
Tonio Weidler	Maastricht University, the Netherlands
Jörg Wicker	University of Auckland, New Zealand
Alicja Wieczorkowska	Polish-Japanese Academy of Information Technology, Poland
Michael Wilbur	Vanderbilt University, USA
David Winkel	LMU Munich, Germany
Moritz Wohlstein	Leuphana Universität Lüneburg, Germany
Szymon Wojciechowski	Wrocław University of Science and Technology, Poland
Bin Wu	Zhengzhou University, China
Chenwang Wu	University of Science and Technology of China, China
Di Wu	Chongqing Institute of Green and Intelligent Technology, Chinese Academy of Sciences, China
Wei Wu	Ben Gurion University of the Negev, Israel
Yongkai Wu	Clemson University, USA
Zhiwen Xiao	Southwest Jiaotong University, China
Cheng Xie	Yunnan University, China
Yaqi Xie	Carnegie Mellon University, USA
Huanlai Xing	Southwest Jiaotong University, China
Xing Xing	Tongji University, China
Ning Xu	Southeast University, China
Weifeng Xu	Weifeng Xu, USA
Ziqi Xu	CSIRO, Australia
Yexiang Xue	Purdue University, USA
Yan Yan	Carleton University, Canada
Yu Yan	School of Information and Cyber Security, People's Public Security University of China, China
Lincen Yang	Leiden University, the Netherlands
Shaofu Yang	Southeast University, China
Muchao Ye	Pennsylvania State University, USA
Kalidas Yeturu	Indian Institute of Technology Tirupati, India

Jaemin Yoo	KAIST, South Korea
Kristina Yordanova	University of Greifswald, Germany
Hang Yu	Shanghai University, China
Jidong Yuan	Beijing Jiaotong University, China
Xiaoyong Yuan	Clemson University, USA
Klim Zaporojets	Aarhus University, Denmark
Claudius Zelenka	Kiel University, Germany
Akka Zemmari	Univ. Bordeaux, France
Guoxi Zhang	Beijing Institute of General Artificial Intelligence, China
Hao Zhang	Fudan University, China
Teng Zhang	Huazhong University of Science and Technology, China
Tianlin Zhang	University of Manchester, UK
Xiang Zhang	National University of Defense Technology, China
Xiao Zhang	Shandong University, China
Xiaoming Zhang	Beihang University, China
Yaqian Zhang	University of Waikato, New Zealand
Yin Zhang	University of Electronic Science and Technology of China
Zhiwen Zhang	University of Tokyo, Japan
Lingxiao Zhao	Carnegie Mellon University, USA
Tongya Zheng	Hangzhou City University, China
Wenhao Zheng	Shopee, Singapore
Yu Zheng	Tsinghua University, China
Yujia Zheng	CMU, USA
Zhengyang Zhou	University of Science and Technology of China, China
Jing Zhu	University of Michigan, Ann Arbor, USA
Ye Zhu	Deakin University, Australia
Yichen Zhu	Midea Group, China
Zirui Zhuang	Beijing University of Posts and Telecommunications, China
Tommaso Zoppi	University of Florence, Italy
Pedro Zuidberg Dos Martires	Örebro University, Sweden
Meiyun Zuo	Renmin University of China, China

Program Committee Members, Applied Data Science Track

Ziawasch Abedjan	TU Berlin, Germany
Shahrooz Abghari	Blekinge Institute of Technology, Sweden
Christian M. Adriano	Hasso-Plattner Institute, Germany
Haluk Akay	KTH Royal Institute of Technology, Lithuania
Fahed Alkhabbas	Malmo University, Sweden
Mohammed Ghaith Altarabichi	Högskolan i Halmstad, Sweden
Evelin Amorim	INESC TEC, Portugal
Giuseppina Andresini	University of Bari 'Aldo Moro', Italy
Sunil Aryal	Deakin University, New Zealand
Awais Ashfaq	Region Halland, Sweden
Asma Atamna	Ruhr-University Bochum, Germany
Berkay Aydin	Georgia State University, USA
Mehdi Bahrami	Fujitsu Research of America, USA
Hareesh Bahuleyan	Zalando, Sweden
Michael Bain	University of New South Wales, Australia
Hubert Baniecki	University of Warsaw, Poland
Enda Barrett	University of Galway, Ireland
Michele Bernardini	Università Politecnica delle Marche, Ancona, Italy
Lilian Berton	Universidade Federal de Sao Paulo, Brazil
Antonio Bevilacqua	Meetecho, Italy
Szymon Bobek	Jagiellonian University, Poland
Veselka Boeva	Blekinge Institute of Technology, Sweden
Martin Boldt	Blekinge Institute of Technology, Sweden
Anton Borg	Blekinge Institute of Technology, Sweden
Cecile Bothorel	IMT Atlantique, France
Mohamed Reda Bouadjenek	Deakin University, New Zealand
Axel Brando	Barcelona Supercomputing Center (BSC) and Universitat de Barcelona (UB), Spain
Stefan Byttner	Halmstad University, Sweden
Ece Calikus	KTH Royal Institute of Technology, Lithuania
Shilei Cao	Tencent, China
Yixuan Cao	Institute of Computing Technology, CAS, China
Hau Chan	University of Nebraska-Lincoln, USA
Chung-Chi Chen	National Taiwan University, Taiwan
Lei Chen	Hong Kong University of Science and Technology, China
Wei-Peng Chen	Fujitsu Research of America, USA
Zhiyu Chen	Amazon, USA
Dawei Cheng	Tongji University, China

Hong Huang	Huazhong University of Science and Technology, China
Yizheng Huang	York University, UK
Yu Huang	University of Florida, USA
Angelo Impedovo	Niuma s.r.l., Italy
Radu Tudor Ionescu	University of Bucharest, Romania
Wei Jin	Emory University, USA
Xiaobo Jin	Xi'an Jiaotong-Liverpool University, China
Xiaolong Jin	Institute of Computing Technology, CAS, China
Pinar Karagoz	Middle East Technical University (METU), Turkey
Saeed Karami Zarandi	Halmstad University, Sweden
Thomas Kober	Zalando, Germany
Elizaveta Kopacheva	LNU, Sweden
Christos Koutras	TU Delft, the Netherlands
Adit Krishnan	University of Illinois at Urbana-Champaign, USA
Rafal Kucharski	Jagiellonian University, Poland
Niraj Kumar	Fujitsu, India
Krzysztof Kutt	Jagiellonian University, Poland
Susana Ladra	University of A Coruña, Spain
Matthieu Latapy	CNRS, France
Niklas Lavesson	Blekinge Institute of Technology, Sweden
Roy Ka-Wei Lee	Singapore University of Technology and Design, Singapore
Alessandro Leite	Inria, France
Daniel Lemire	University of Quebec (TELUQ), Canada
Chang Li	Apple, USA
Daifeng Li	Sun Yat-Sen University, China
Haifang Li	Baidu Inc., China
Junxuan Li	Microsoft, USA
Lei Li	Hong Kong University of Science and Technology, China
Shijun Li	University of Science and Technology of China
Shuai Li	University of Cambridge, UK and University of Tokyo, Japan and Tsinghua University, China
Wei Li	Harbin Engineering University, China
Xiang Lian	Kent State University, USA
Guojun Liang	Halmstad University, Sweden
Zhaohui Liang	National Library of Medicine, NIH, USA
Kwan Hui Lim	Singapore University of Technology and Design, Singapore
Adi Lin	Didi, China

Bang Liu	University of Montreal, Canada
Dugang Liu	Guangdong Laboratory of Artificial Intelligence and Digital Economy (SZ), Shenzhen University, China
Jingjing Liu	MD Anderson Cancer Center, USA
Li Liu	Chongqing University, China
Qing Liu	Zhejiang University, China
Xueyan Liu	Jilin University, China
Yongchao Liu	Ant Group, China
Andreas Lommatzsch	TU Berlin, Germany
Ping Luo	Chinese Academy of Sciences, China
Guixiang Ma	Intel Labs, USA
Zongyang Ma	York University, UK
Saulo Martiello Mastelini	Volt Robotics, Brazil
Elio Masciari	University of Naples, Italy
Nédra Mellouli	LIASD, France
Zoltan Miklos	University of Rennes, France
Mihaela Mitici	Utrecht University, the Netherlands
Martin Mladenov	Google, Brazil
Ahmed K. Mohamed	Meta, USA
Seung-Hoon Na	Jeonbuk National University, South Korea
Sepideh Nahali	York University, UK
Mirco Nanni	CNR-ISTI Pisa, Italy
Richi Nayak	Queensland University of Technology, Brisbane, Australia
Wee Siong Ng	Institute for Infocomm Research, Singapore
Le Nguyen	University of Oulu, Finland
Thanh Thi Nguyen	Monash University, Australia
Slawomir Nowaczyk	Halmstad University, Sweden
Tomas Olsson	RISE SICS, Sweden
Panagiotis Papadakos	FORTH-ICS, Greece
Manos Papagelis	York University, UK
Panagiotis Papapetrou	Stockholm University, Sweden
Luca Pappalardo	ISTI, Italy
Sepideh Pashami	Halmstad University, Sweden
Vincenzo Pasquadibisceglie	University of Bari 'Aldo Moro', Italy
Leonardo Pellegrina	University of Padova, Italy
Pop Petrica	Technical University of Cluj-Napoca, Romania
Pablo Picazo-Sanchez	Halmstad University, Sweden
Srijith PK	IIT, Hyderabad, India
Buyue Qian	Xi'an Jiaotong University, China
Enayat Rajabi	Halmstad University, Sweden

Yanghui Rao	Sun Yat-sen University, China
Salvatore Rinzivillo	KDDLab - ISTI - CNR, Italy
Riccardo Rosati	Università Politecnica delle Marche, Ancona, Italy
Stefan Rueping	Fraunhofer IAIS, Germany
Snehanshu Saha	BITS Pilani Goa Campus, India
Lou Salaün	Nokia Bell Labs, France
Isak Samsten	Stockholm University, Sweden
Eric Sanjuan	Avignon University, France
Johannes Schneider	University of Liechtenstein, Liechtenstein
Wei Shao	Data61, CSIRO, Australia
Nasrullah Sheikh	IBM Research, USA
Jun Shen	University of Wollongong, Australia
Jingwen Shi	Michigan State University, USA
Yue Shi	Meta, USA
Carlos N. Silla	Pontifical Catholic University of Parana (PUCPR), Brazil
Gianmaria Silvello	University of Padova, Italy
Yang Song	Apple, USA
Shafiullah Soomro	Linnaeus University, Sweden
Efstathios Stamatatos	University of the Aegean, Greece
Ting Su	Imperial College London, UK
Gan Sun	South China University of Technology, China
Munira Syed	Procter & Gamble, USA
Zahra Taghiyarrenani	Halmstad University, Sweden
Liang Tang	Google, USA
Xing Tang	Tencent, China
Junichi Tatemura	Google, USA
Joe Tekli	Lebanese American University, Lebanon
Mingfei Teng	Amazon, USA
Sofia Tolmach	Amazon, USA
Gabriele Tolomei	Sapienza University of Rome, Italy
Ismail Hakki Toroslu	METU, Turkey
Md Zia Ullah	Edinburgh Napier University, UK
Maurice Van Keulen	University of Twente, the Netherlands
Ranga Raju Vatsavai	North Carolina State University, USA
Bruno Veloso	INESC TEC & FEP-UP, Portugal
Chang-Dong Wang	Sun Yat-sen University, China
Chengyu Wang	Alibaba Group, China
Kai Wang	Shanghai Jiao Tong University, China
Pengyuan Wang	University of Georgia, USA
Sen Wang	University of Queensland, USA

Senzhang Wang	Central South University, China
Sheng Wang	Wuhan University, China
Wei Wang	Tsinghua University, China
Wentao Wang	Michigan State University, USA
Xiaoli Wang	Xiamen University, China
Yang Wang	University of Science and Technology of China, China
Yu Wang	Vanderbilt University, USA
Zhibo Wang	Zhejiang University, China
Paweł Wawrzyński	IDEAS NCBR, Poland
Hua Wei	Arizona State University, USA
Shi-ting Wen	Ningbo Tech University, China
Zeyi Wen	Hong Kong University of Science and Technology, China
Avani Wildani	Emory University, USA
Fangzhao Wu	MSRA, China
Jun Wu	University of Illinois at Urbana–Champaign, USA
Wentao Wu	Microsoft Research, USA
Xianchao Wu	NVIDIA, Japan
Haoyi Xiong	Baidu, Inc., China
Guandong Xu	University of Technology Sydney, Australia
Yu Yang	City University of Hong Kong, China
Lina Yao	University of New South Wales, Australia
Fanghua Ye	University College London, UK
Dongxiao Yu	Shandong University, China
Haomin Yu	Aalborg University, Denmark
Ran Yu	DSIS Research Group, University of Bonn, Germany
Erik Zeitler	Stream Analyze, Sweden
Chunhui Zhang	Dartmouth College, USA
Denghui Zhang	Rutgers University, USA
Li Zhang	University of Sheffield, UK
Mengxuan Zhang	Australian National University, Australia
Kaiping Zheng	National University of Singapore
Yucheng Zhou	University of Macau, China
Yuanyuan Zhu	Wuhan University, China
Ziwei Zhu	George Mason University, USA
Vasileios Zografos	sennder, Germany

Program Committee Members, Demo Track

Bijaya Adhikari	University of Iowa, USA
Andrius Budrionis	Norwegian Centre for E-health Research, Norway
Luca Cagliero	Politecnico di Torino, Italy
Tania Cerquitelli	Politecnico di Torino, Italy
Gintautas Daunys	Vilnius University, Lithuania
Katharina Dost	University of Auckland, New Zealand
Sourav Dutta	Huawei Research Centre, Ireland
Françoise Fessant	Orange, France
Christelle Godin	CEA, France
Anil Goyal	Amazon, India
Maciej Grzenda	Warsaw University of Technology, Poland
Marius Gudauskis	Institute of Mechatronics, KTU, Lithuania
Thomas Guyet	Inria, Centre de Lyon, France
Andreas Henelius	Independent Researcher, Finland
Rokas Jurevicius	Scandit AG, Lithuania/Switzerland
Pawan Kumar	IIIT, Hyderabad, India
Olga Kurasova	Vilnius University, Institute of Data Science and Digital Technologies, Lithuania
Moreno La Quatra	Kore University of Enna, Italy
Jan Lemeire	Vrije Universiteit Brussel (VUB), Belgium
Martin Luckner	Warsaw University of Technology, Poland
Hoang Phuc Hau Luu	University of Helsinki, Finland
Jarmo Mäkelä	CSC - IT Center for Science Ltd, Finland
Michael Mathioudakis	University of Helsinki, Finland
Darius Miniotas	Vilnius Gediminas Technical University, Lithuania
Michalis Mountantonakis	FORTH-ICS, and CS Department - University of Crete, Greece
Raj Nath Patel	Huawei Ireland Research Center, Ireland
Darius Plikynas	Vilnius Gediminas Technical University, Lithuania
Alexandre Reiffers	IMT Atlantique, France
Marina Reyboz	Univ. Grenoble Alpes, CEA, LIST, France
Yuya Sasaki	Osaka University, Japan
Ines Sousa	Fraunhofer AICOS, Portugal
Jerzy Stefanowski	Poznan University of Technology, Poland
Guoxin Su	University of Wollongong, Australia
Lu-An Tang	NEC Labs America, USA
Michael C. Thrun	Philipps-Universität Marburg, Germany

Yannis Tzitzikas	FORTH-ICS and Computer Science Department, University of Crete, Greece
Aleksandras Voicikas	Vilnius University, Lithuania
Jörg Wicker	University of Auckland, New Zealand
Hao Xue	University of New South Wales, Australia

Sponsors

CENTAI

Invited Talks Abstracts

The Dynamics of Memorization and Unlearning

Gintarė Karolina Džiugaitė

Google DeepMind

Abstract. Deep learning models exhibit a complex interplay between memorization and generalization. This talk will begin by exploring the ubiquitous nature of memorization, drawing on prior work on "data diets", example difficulty, pruning, and other empirical evidence. But is memorization essential for generalization? Our recent theoretical work suggests that eliminating it entirely may not be feasible. Instead, I will discuss strategies to mitigate unwanted memorization by focusing on better data curation and efficient unlearning mechanisms. Additionally, I will examine the potential of pruning techniques to selectively remove memorized examples and explore their impact on factual recall versus in-context learning.

Biography: Gintarė is a senior research scientist at Google DeepMind, based in Toronto, an adjunct professor in the McGill University School of Computer Science, and an associate industry member of Mila, the Quebec AI Institute. Prior to joining Google, Gintarė led the Trustworthy AI program at Element AI/ServiceNow, and obtained her Ph.D. in machine learning from the University of Cambridge, under the supervision of Zoubin Ghahramani. Gintarė was recognized as a Rising Star in Machine Learning by the University of Maryland program in 2019. Her research combines theoretical and empirical approaches to understanding deep learning, with a focus on generalization, memorization, unlearning, and network compression.

The Emerging Science of Benchmarks

Moritz Hardt

Max Planck Institute for Intelligent Systems

Abstract. Benchmarks have played a central role in the progress of machine learning research since the 1980s. Although there's much researchers have done with them, we still know little about how and why benchmarks work. In this talk, I will trace the rudiments of an emerging science of benchmarks through selected empirical and theoretical observations. Looking back at the ImageNet era, I'll discuss what we learned about the validity of model rankings and the role of label errors. Looking ahead, I'll talk about new challenges to benchmarking and evaluation in the era of large language models. The results we'll encounter challenge conventional wisdom and underscore the benefits of developing a science of benchmarks.

Biography: Hardt is a director at the Max Planck Institute for Intelligent Systems, Tübingen. Previously, he was Associate Professor for Electrical Engineering and Computer Sciences at the University of California, Berkeley. His research contributes to the scientific foundations of machine learning and algorithmic decision making with a focus on social questions. He co-authored Fairness and Machine Learning: Limitations and Opportunities (MIT Press) and Patterns, Predictions, and Actions: Foundations of Machine Learning (Princeton University Press).

Enhancing User Experience with AI-Powered Search and Recommendations at Spotify

Mounia Lalmas-Roelleke

Spotify

Abstract. This talk will explore the pivotal role of search and recommendation systems in enhancing the Spotify user experience. These systems serve as the gateway to Spotify's vast audio catalog, helping users navigate millions of music tracks, podcasts, and audiobooks. Effective search functionality allows users to quickly find specific content, whether it is a favorite song, a trending podcast, or an informative audiobook, while also satisfying broader search needs. Meanwhile, recommendation systems suggest new and relevant content that users might not have thought to search for, while ensuring their current needs for familiar content are met. This encourages exploration and discovery of new artists, genres, and shows, enriching the overall listening experience and keeping users engaged with the platform. Achieving this dual objective of precision and discovery requires sophisticated technology. It involves a deep understanding of representation learning, where both content and user preferences are accurately modeled. Advanced AI techniques, including machine learning and generative AI, play a crucial role in this process. These technologies enable the creation of highly personalized recommendations by understanding complex user behaviors and preferences. Generative AI, for instance, allows us to create personalized playlists, thereby enhancing the user experience with innovative features. This presentation is based on the collective research and publications of numerous contributors at Spotify.

Biography: Mounia is a Senior Director of Research at Spotify and the Head of Tech Research in Personalization, where she leads an interdisciplinary team of research scientists. She also holds an honorary professorship at University College London and serves as a Distinguished Research Fellow at the University of Amsterdam. Previously, Mounia was a Director of Research at Yahoo, overseeing a team focused on advertising quality and collaborating on user engagement projects related to news, search, and user-generated content. Before her tenure at Yahoo, Mounia held a Microsoft Research/RAEng Research Chair at the School of Computing Science, University of Glasgow, and before that was a Professor of Information Retrieval at the Department of Computer Science at Queen Mary, University of London. She is a prominent figure in the research community, regularly serving as a senior program committee member at major conferences such as WSDM, KDD, WWW, and SIGIR. She was also a program co-chair for SIGIR 2015, WWW 2018, WSDM 2020, and CIKM 2023. Mounia is widely recognized for her contributions as a speaker and author, with over 250 published papers and appearances on platforms like ACM ByteCast and the AI Business Podcasts series. She was nominated for the VentureBeat Women in AI Awards for Research in both 2022 and 2023.

How to Utilize (and Generate) Player Tracking Data in Sport

Patrick Lucey

Stats Perform

Abstract. Even though player tracking data in sports has been around for 25 years, it still poses as one of the most interesting and challenging datasets in machine learning due to its fine-grained, multi-agent, team-based, and adversarial nature. Despite these challenges, it is also extremely valuable as it is (relatively) low-dimensional, interpretable, and interactive, allowing us to measure performance and answer questions we couldn't objectively address before. In this talk, I will first give a brief history of tracking data in sports, then highlight the challenges associated with utilizing it. I will then show that by obtaining a permutation invariant representation, we can not only measure aspects of sports that couldn't be done before, but also interact with and simulate plays akin to a video game via our "visual search" and "ghosting" technology. Finally, I will show how we can use both tracking and event data to create a multimodal foundation model, which enables us to generate player tracking data at scale and achieve our goal of "digitizing every game of professional sport." Throughout the talk, I will utilize examples from top-tier basketball, soccer, and tennis.

Biography: Patrick Lucey is currently the Chief Scientist at sports data giant Stats Perform, leading the AI team with the goal of maximizing the value of the company's extensive sports data. He has studied and worked in the fields of machine learning and computer vision for the past 20 years, holding research positions at Disney Research and the Robotics Institute at Carnegie Mellon University, as well as spending time at IBM's T.J. Watson Research Center while pursuing his Ph.D. Patrick originally hails from Australia, where he received his BEng(EE) from the University of Southern Queensland and his doctorate from Queensland University of Technology, which focused on multimodal speech modeling. He has authored more than 100 peer-reviewed papers and has been a co-author on papers in the MIT Sloan Sports Analytics Conference Best Research Paper Track for 11 of the last 13 years, winning best paper in 2016 and runner-up in 2017 and 2018. Additionally, he has won best paper awards at INTERSPEECH and WACV international conferences. His main research interests are in artificial intelligence and interactive machine learning in sporting domains, as well as AI education. He has recently piloted a course on "AI in Sport," which aims to give students intuition behind AI methods using the interactive and visual nature of sports data.

Website: www.patricklucey.com

Resource-Aware Machine Learning—A User-Oriented Approach

Katharina Morik

TU Dortmund University

Abstract. Machine Learning (ML) has become integrated into several processes, ranging from medicine, manufacturing, logistics, smart cities, sales, recommendations and advertisements to entertainment and many more business and private processes. The applications together consume a considerable amount of energy and emit CO_2. ML research investigates how to make models smaller and faster through pruning and quantization. Also the use of more energy-efficient hardware is an encouraging field. Research on ML under resource constraints is an active field proposing novel algorithms and scenarios. The aim is that for each application a variety of implementations is offered from which customers and the different types of users may choose the most thrifty one. This, in turn, would push tech providers to focus on the production of economical systems. However, if the customers, users, stakeholders do not know which of the models offers the best tradeoff between performance and energy-efficiency, they cannot select the most frugal one. Hence, testing implementations of learning and inference needs to be developed. They should be easy to use, produce visualizations that are mass-tailored for specific user groups. Automatized testing is difficult due to the diversity of models, computing architectures, training and evaluation data, and the fast rate of changes. The talk will illustrate work on resource-aware ML and advocate to pay more attention to the role of users in the development of scenarios, models, and tests.

Biography: Katharina Morik received her doctorate from the University of Hamburg in 1981 and her habilitation from the TU Berlin in 1988. In 1991, she established the chair of Artificial Intelligence at the TU Dortmund. She retired in 2023. She is a pioneer of bringing machine learning and computing architectures together so that machine learning models may be executed or even trained on resource restricted devices. In 2011, she acquired the Collaborative Research Center CRC 876 "Providing Information by Resource-Constrained Data Analysis" consisting of 12 projects and a graduate school. After the longest possible funding period of 12 years, the CRC ended with the publication of 3 books on Resource-Constrained Machine Learning (De Gruyter). She has participated in numerous European research projects and has been the coordinator of one. She was a founding member and Program Chair of the conference series IEEE International Conference on Data Mining (ICDM) and is a member of the steering committee of ECML PKDD. She is a co-founder of the Lamarr Institute for Machine Learning and Artificial Intelligence. Prof. Morik is a member of the Academy of Technical Sciences and of the North Rhine-Westphalian Academy of Sciences and Arts. She was made a Fellow of the German Society of Computer Science GI e.V. in 2019.

Contents – Part VII

Research Track

Research Track

Data with Density-Based Clusters: A Generator for Systematic Evaluation of Clustering Algorithms

Philipp Jahn[1,2]([⊠]), Christian M. M. Frey[3], Anna Beer[4], Collin Leiber[1,2], and Thomas Seidl[1,2,3]

[1] LMU Munich, Munich, Germany
{jahn,leiber,seidl}@dbs.ifi.lmu.de
[2] Munich Center for Machine Learning (MCML), Munich, Germany
[3] Fraunhofer IIS, Erlangen, Germany
[4] University of Vienna, Vienna, Austria
anna.beer@univie.ac.at

Abstract. Mining data containing density-based clusters is well-established and widespread but faces problems when it comes to systematic and reproducible comparison and evaluation. Although the success of clustering methods hinges on data quality and availability, reproducibly generating suitable data for this setting is not easy, leading to mostly low-dimensional toy datasets being used. To resolve this issue, we propose DENSIRED (**DENSI**ty-based **R**eproducible **E**xperimental **D**ata), a novel data generator for data containing density-based clusters. It is highly flexible w.r.t. a large variety of properties of the data and produces reproducible datasets in a two-step approach. First, skeletons of the clusters are constructed following a random walk. In the second step, these skeletons are enriched with data samples. DENSIRED enables the systematic generation of data for a robust and reliable analysis of methods aimed toward examining data containing density-connected clusters. In extensive experiments, we analyze the impact of user-defined properties on the generated datasets and the intrinsic dimensionalities of synthesized clusters. Our code and novel benchmark datasets are publicly available at: https://github.com/PhilJahn/DENSIRED.

Keywords: Clustering · Evaluation · Data Generator · Density-based

1 Introduction

In order to properly develop and evaluate new clustering methods, access to suitable data is essential. However, real-world data is often limited in terms of availability, expensive to annotate, or even biased in certain ways [1]. By generating synthetic data, machine learning engineers and researchers can evaluate

Supplementary Information The online version contains supplementary material available at https://doi.org/10.1007/978-3-031-70368-3_1.

their novel methods more thoroughly and in a controlled, reproducible, and flexible environment. The manual creation of such data is not easy for use cases like density-based clustering, where no suitable data generators exist. Existing generators are insufficient w.r.t. generating arbitrarily shaped clusters, guarantees of the cluster's density-separability, or the capability to generate similar, yet distinct datasets. State-of-the-art data generators for synthetic data are limited in terms of complexity, especially in higher-dimensional spaces: e.g., they cannot produce non-convex, high-dimensional clusters. Hence, we develop DENSIRED (**DENSI**ty-based **R**eproducible **E**xperimental **D**ata), a tool to synthesize data containing density-connected structures in high-dimensional spaces in a controlled, reproducible, and customizable way. Users can adjust a large set of potentially relevant properties of a dataset, e.g., its size, dimensionality, number of clusters, density of clusters, or noise ratio. DENSIRED can create density-connected clusters as known from DBSCAN [2] as well as other non-convex clusters of arbitrary shape, varying density, non-linearly separable clusters, and clusters with several branches. This set of properties is of special interest for evaluating advantages of density-based clustering methods. Using a two-step approach, DENSIRED separates the generation of the underlying structures ("skeletons") of clusters and the instantiation of the actual data points. This allows the synthetization of various datasets with the same overall structure that are equivalently hard to cluster. We define and show the importance of this property. The process is depicted in Fig. 1 along with some examples for the two steps.

We showcase various benefits of our data generator by performing a comprehensive and broad benchmarking study. In this study, we compare 14 algorithms based on diverse heuristics for clustering on various datasets. As DENSIRED is able to maintain the general structure with increasing dimensionality, interesting conclusions can be drawn about the behavior of the considered methods. The generated data exhibits the intended characteristics: density-based algorithms deliver substantially better results than, e.g., centroid-based methods. Our contributions are as follows:

- We introduce the novel data generator DENSIRED for synthesizing data that consists of density-connected structures in high-dimensional spaces
- DENSIRED supports the reproducible generation of clusters of arbitrary shapes, varying intrinsic dimensionality, different densities, and diverse geometric characteristics
- We perform a broad benchmark study on exemplary generated datasets to demonstrate the evaluation possibilities for novel clustering approaches

2   Related Work

Benchmark Data
There are numerous well-established benchmark datasets covering various domains for evaluating data mining methods.

1) Real-world data. Real-world datasets are, e.g., used for the evaluation of clustering approaches [2–5]. The inclusion of real-world datasets is essential for a

Table 1. Data properties in the evaluation of density-connectivity-based clustering methods (sorted by year of publication)

Method	synthetic 2d	synthetic >2d	real
DBSCAN [2]	✔	✗	✔
GDBSCAN [6]	✔	✗	✔
OPTICS [7]	✔	✗	✔
MDD [9]	✔	✗	✗
HDBSCAN [11]	✗	✗	✔
DBSCAN Revisited [14]	✔	✔	✔
AnyDBC [15]	✔	✔	✔
DSets-DBSCAN [16]	✔	✗	✔
RNN-DBSCAN [17]	✔	(✔)	✔
AA-DBSCAN [18]	✔	✔	✔
DDC [19]	✔	✗	✔
BLOCK-DBSCAN [20]	✔	✗	✔
HDBSCAN-MR [21]	✗	✔	✔
AMD-DBSCAN [22]	✔	✗	✔
GriT-DBSCAN [23]	✔	✔	✔
K-DBSCAN [24]	✔	✗	✔
MDBSCAN [25]	✔	✗	✔

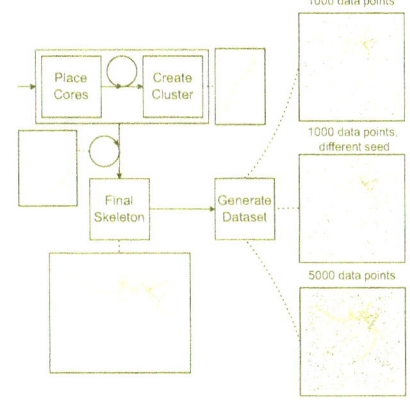

Fig. 1. Process of DENSIRED: Multiple clusters, each consisting of a set of cores, are created to produce a final skeleton. This can then be used to generate datasets of the same shape but with different data points (top and middle) and data numbers (bottom). The setting favors elongated datasets branching at the start point.

proper evaluation. Still, a systematic evaluation often requires specific tuning of data parameters, which can not easily be achieved with real-world datasets.

2) Synthetic data. To procure datasets of varying and sufficient complexity, synthetic datasets have been established as benchmarks. Widespread (but dated) datasets [37] allow examining methods' performances w.r.t. varying cardinality and number of clusters but are, for the most part, only 2-dimensional. Many are based on Gaussian distributions (including the few higher dimensional datasets), which do not necessarily conform with the density-based setting.

Evaluation of Density-Based Clustering
While many methods assume certain properties for density-based clusterings, they are often described only cursory, if at all. Often, such density-based clusters are described as sets of points that are separated from other sets of points by an area of low density. The most frequently used definition of density-based clusters describes them as a maximal set of density-connected points [2].

The concept of *density-connectivity* became famous with the introduction of DBSCAN [2] and builds upon the concept of *direct density-reachability*. Two points p and q are *directly density-reachable* iff they are both *core* points, i.e., they have at least $minPts$ many points within their ε-range. For density-connected points, it is sufficient if there exists a chain of directly density-reachable core points between them. There are plenty clustering methods that aim to detect density-connected clusters, e.g.: DBSCAN [2], GDBSCAN [6], Multi-density DBSCAN (MDD) [9].

Table 2. Properties of selected data generators for data with >2 dimensions

Name	Arbitrarily shaped underlying model?	Not linearly separable?	Guaranteed density separability?	Ability to generate equivalent datasets? (see Sect. 3)
Milligan's [26]	✗	✗	✔	✗
Improved Milligan's [27]	✗	✗	✔	✗
MixSim [28]	✗	✔	✗	✗
swiss-roll [29]	✗	✔	✗	✔
blobs [29]	✗	✔	✗	✔
Clugen [30]	✗	✔	✗	✔
OCLUS [31]	✗	✔	✔	✗
HAWKS [32]	✗	✔	✔	✗
MDCGEN [33]	✗	✔	✔	✗
SynDECA (Irregular 2) [34]	✔	✗	✔	✗
SynDECA (Irregular 1) [34]	✔	✔	✗	✗
Seed Spreader [14,35]	✔	✔	✗	✗
DataGen [36]	✔	✔	✗	✔
DENSIRED	✔	✔	✔	✔

Both, real-world and synthetic data, are used to evaluate these methods, as shown in Table 1. High-dimensional synthetic data is used least commonly and the way these datasets are obtained also varies, e.g., making use of specific benchmarking datasets [18,37], creating their own data [15], or using data generators [14,17,21,23,38]. Seed Spreader [14,35], as a generator for specifically density-based data, was used in [14,23]. RNN-DBSCAN [17] was evaluated using 3-dimensional data produced through scikit-learn[1] [29] (swiss-roll (3d) and 3d blobs). MixSim [28] was used for HDBSCAN-MR [21]. Still, many evaluations rely on 2d synthetic data for benchmarking. While evaluating on real-world data is essential, not all real-world datasets are equally suitable for density-connectivity-based clustering, as pointed out in [39]. Certain relevant properties, like high variability in density [22], cannot be easily evaluated in a systematic manner solely based on real-world data. As most properties of real-world data can not be tuned, their distinct effects on the quality of novel methods are hard to observe. Thus, there is a need for high-dimensional density-based synthetic data.

Data Generators
Data-generating functions have been established as a useful tool for a controlled evaluation of procedures in a plethora of domains ranging from healthcare [12],

---

[1] https://scikit-learn.org, last accessed: Oct 5th, 2023.

routing problems [10], flow networks [8], to manufacturing [13]. However, in contrast to data-driven generators (data simulators) that aim at recreating observed data, we regard process-driven data generators that are based on computational models [40]. This allows a generalized generative system that can produce a large variety of datasets containing density-based structures and is not limited to a specific use case. However, many existing generators cannot synthesize the required high-dimensional data (see Table 1). E.g., Pei's generator [41] offers arbitrarily shaped clusters but is restricted to two dimensions.

Table 2 shows which generators can create data with the denoted important properties for evaluating density-based clustering:

1) Arbitrarily shaped underlying model. In contrast to k-Means [42], density-based clustering methods are capable of dealing with arbitrarily shaped clusters [43,44]. To ensure that this ability is properly evaluated when investigating density-based methods, the generation of clusters with arbitrary shapes, e.g., by using random walks, is important. However, many generators [26–28,30–33] rely solely on using distributions with a singular midpoint for each component. Relying on a single distribution per dimension is not sufficient to generate arbitrary cluster shapes.

2) Not linearly separable. We want to generate data with clusters that are not pairwise linearly separable to enhance differences to centroid-based clustering methods. Clusters that are pairwise linearly separable can be detected easily by traditional, non-density-based clustering algorithms like k-Means [42]. The way to prevent overlap employed by some methods [26,27,34] causes them to always generate data that can be linearly separated.

3) Guaranteed density separability. Many algorithms do not have sufficient overlap handling [14,28,30] to guarantee that data from different clusters is not density-connected, leading to a flawed ground truth. The goal is to evaluate methods that use density-based assumptions. As a result, it leads to a lowered usability of the produced datasets and misleading results if these do not apply.

4) Ability to generate equivalent datasets. For a robust analysis, creating datasets that are equivalently hard to cluster is important, see Sect. 3. This requires either restating certain properties of the dataset [29] or storing and re-using the underlying model [30,36]. Some data generators, like MixSim [28], come close to fulfilling this requirement but cannot guarantee it. The property makes it possible to generate new datasets that share the same underlying structure while containing different points. This allows for, e.g., the expansion of existing datasets with additional points without generating the data points from new distributions or by excluding some data points initially and adding them again later on.

SynDECA [34] can generate "irregularly" [34] shaped clusters. However, they are either surrounded by hyperrectangular bounding boxes for overlap prevention (Irregular 2), making them easily separable, or potentially overlap (Irregular 1) due to violation of those bounding boxes. Datagen [36] fulfills desired properties but tends towards generating disconnected subclusters separated by other clusters. Seed Spreader [14,35] aims at creating high-dimensional synthetic data for density-connectivity-based clustering methods. It is, thus, both our inspiration and main competitor. Like DENSIRED, Seed Spreader uses a random walk and generates its data points uniformly distributed within hyperspheres along this path. However, Seed Spreader generally offers very little control over the generation process and can lead to clusters that are not density-separable. We describe its deficits, which we overcome with DENSIRED, in detail in Sect. 3.4.

3   A Reliable Data Generator for Density-Based Clusters

For a robust analysis of clustering models, we want to examine datasets that are similar, but not equal. This alleviates effects of outliers or random events.

DENSIRED creates datasets containing density-connected clusters as known from, e.g., DBSCAN [2]. Note that this differs from mode-seeking density-based concepts as, e.g., used in Mean Shift [45]. The ε-neighborhood $N_\varepsilon(p)$ of a point p in a dataset X is $N_\varepsilon(p) = \{q \in X : dist(p, q) \le \varepsilon\}$ for $\varepsilon > 0$ and a distance metric $dist(\cdot, \cdot)$. A point is a *core* point w.r.t. $\varepsilon \in \mathbb{R}^+$ and $\mu \in \mathbb{N}$ iff it has at least μ points in its ε-neighborhood. A point q is *directly density-reachable* from a point p for some $\varepsilon > 0$, $\mu \in \mathbb{N}$ iff p is in the ε-neighborhood of q and q is a core point. They are *density-reachable* iff there is a chain of directly density-connected core points between them and they are *density-connected* iff there is a point o such that p and q are both density-reachable from o. A (density-based) cluster $C^{\varepsilon,\mu} \subseteq X$ is a "maximal set of density-connected objects" [15].

For analyzing centroid-based clustering algorithms, clusters are usually generated by drawing from a random distribution. The resulting datasets are similar, but not equal. We define equivalence between two density-based clusterings:

Definition (Equivalence of density-based clusterings) Two datasets X, Y are equivalent regarding their density-connectivity if, for every ε and μ, there exist δ, ν, such that the ε, μ- clustering of X yields corresponding clusters to the δ, ν-clustering of Y. I.e., there is a bijective mapping between the density-connected components of both datasets.

For an unbiased evaluation of a novel method, it is important to eliminate unwanted random effects, like the single-link effect, and prevent overoptimism [46]. Thus, we need to carefully model the data in order to create equivalent datasets. For this reason, generating equivalent density-based clusters is a challenging problem because of its discrete, non-continuous aspect expressed by the parameter $\mu \in \mathbb{N}$.

3.1   Main Concept of DENSIRED

We construct equivalent but not equal density-connected clusters by following a two-step approach: 1) Defining the data distribution, i.e., the "skeleton" of the data, and 2) Instantiating the data points within this distribution. By separating the generation of the underlying dense areas from the actual instantiation of the data objects, we allow the synthesis of datasets with similar properties (e.g., shape and distance between clusters) but different instantiations. The goal is to produce arbitrarily shaped clusters that are density-connected even for higher dimensionalities. Preventing overlap is needed in order to ensure that the ground truth corresponds to the generated data, but the risk of the dataset being trivially separable due to the clusters being too far apart needs to be addressed.

As density-connected clusters do not follow any predefined shape or correlation, a random walk in high-dimensional space is a suitable model for generating the cluster's expansion [14]. In step one, we create a skeleton for each cluster with a random walk of a given step size that outputs core points. These are not necessarily the only core points of a cluster, but rather a subset that is sufficient to define a cluster's shape (cf. Section 3.2). In step two, we instantiate the data by filling the hypersphere of radius ε around each core point (cf. Section 3.3). This ensures that each core point has enough points in its ε-range. Each instantiation produces different datasets because of the randomness within each hypersphere. Clusters of different instantiations are equivalent and have the same shape in high-dimensional space but have deviations that are sufficient to avoid unwanted random effects. We offer users a large variety of optional parameters to customize data's properties, which are explained in the next subsections. A simplified pseudo code of DENSIRED is given in Algorithm 1, though lacks, e.g., restarting and noise adjustment.

3.2   Generation of Skeletons

In the first step, the local hyperspheres, also referred to as cores, are used as a skeleton to define the data distribution without sampling the actual data. The creation of the skeleton is based on a random walk assumption.

- **Number of clusters and cores** The number of clusters and cores can be defined by users. They can either specify the number of cores per cluster, or each cluster receives a random ratio of the overall core number. The starting positions of each cluster random walk can be randomly assigned or placed between two different clusters' random cores to increase the closeness of clusters. In case the second option fails because of, e.g., overlapping clusters, a random starting position is chosen for this cluster. If no position allows to start a new cluster, the value range ds in which a cluster can start is increased.
- **Momentum factor** ω Users can bias the random walk's direction of each step with a momentum factor $\omega \in [0, 1]$ individually per cluster. The default value is 0.5. Lower values lead to a rather randomly chosen direction for each step, while higher values keep the momentum and the direction v_{prev} of previous steps. For this, each step's direction is given by the weighted sum of a random vector v_{rand} and the previous step's direction vector, i.e., $v_{new} = (1 - \omega)v_{rand} + \omega v_{prev}$.
- **Different Densities** DENSIRED allows to create clusters of varying density: scale factors for every cluster adjust the ε-radius of the cores and the distance between cores of the same cluster. Users can set these factors for every cluster separately, as a fixed value across all clusters, assign them randomly, or disable them to maintain the same density factor (default: 1) for all clusters.

- **Branching β and Stars $\varkappa$** The restart probability at every step of the random walk is determined by the branch factor β (default: 0.05). Restarts continue the random walk starting from an earlier step of the current walk. A star factor $\varkappa$ (default: 0) can be used to give the initial core of the walk a higher probability than other cores, else any of the cores is chosen randomly. Restarts are also performed whenever the random walk comes too close to another cluster's skeleton such that they keep a determined separability and do not overlap.
- **Overlapping Clusters** The separability between two clusters can be determined by a factor o (default: 1.1). Values less than 1 can lead to overlapping core spheres of different clusters. If a user-given amount of random walk step attempts failed due to coming too close to a core of another cluster, the random walk continues from a different core. If this fails, the cluster is fully restarted.

While the previous steps ensure the creation of clusters that fulfill the notion of density-based clusters, DENSIRED also allows to include properties that make the clustering task more challenging:

- **Step Sizes** The step size δ between consecutive cores can be set by users. Additionally, DENSIRED offers a flag to vary it slightly by choosing values from a truncated Gaussian distribution around the original step size δ^o, where $\frac{2}{3} \cdot \delta^o \leq \delta \leq 1.5 \cdot \delta^o$. Per default, this deviation is turned off.
- **Adding Chains.** DENSIRED can add linkages between clusters in the form of low-density paths between random points from different clusters. To create this path, the direction of each step of the random walk is randomly chosen, but the chain-specific version of the momentum factor ω is used to maintain the direction between the two clusters. To prevent a deadlock (due to, e.g., other clusters being in the way), any connection that fails too many times is dropped.

3.3   Instantiating Data Points

In the second step, data points within a skeleton's hypersphere are instantiated.

Users can define the approximate data's distribution among different clusters and noise. For each cluster, the number of points per core is computed based on this distribution and the size of the full dataset. For each core, the points are generated uniformly at random within the cluster-specific radius. Points at the center of the hypersphere to ensure density-connectivity for a given ε and $minPts$ are guaranteed by setting the flag *center*. Using the flag *equal* can ensure the same number of points per core of a cluster. Similar to the notion of [2], we define noise as additional random data points that are not density-connected to any cluster. We sample noise uniformly within the maximal bounds in high-dimensional space that is spanned by the radii of the hyperspheres of the cores. Additionally, noise points are not within a user-defined, cluster-dependent range of any cluster core.

Algorithm 1: Simplified Pseudo-Code for DENSIRED

---

1 $cores \leftarrow \{\}$;
2 **for** $cluid \in range(cluster\ number)$ **do**
3   Add core at starting position pos to $cores[cluid]$;
4       $pos_{old} \leftarrow pos$; $v_{old} \leftarrow norm(v_{rand} - 0.5 \cdot \mathbf{1})$
5   **for** $coreid \in range(core\ number\ for\ cluid\ -\ 1)$ **do**
6           $v \leftarrow norm((1-\omega) \cdot norm(v_{rand} - 0.5 \cdot \mathbf{1}) + \omega \cdot v_{old})$; $pos \leftarrow pos_{old} + \delta \cdot v$;
7   **if** *pos too close to other core in $cores[\neg cluid]$ based on ε and o, or due to β* **then**
8   Select pos_{old} based on $\varkappa$; $v \leftarrow norm(v_{rand} - 0.5 \cdot \mathbf{1})$; Go to 5;
9   **else**
10   add core at pos to $cores[cluid]$; $pos_{old} \leftarrow pos$; $v_{old} \leftarrow v$
11
12   **end for**
13 **end for**
14 Generate connections (3 to 12 with v_{old} pointing to a randomly chosen core of the target cluster);
15 Generate $data\ amount \cdot (1 - noise\ ratio)$ data points based on cluster cores in $cores$ and ε of the cluster;
16 Generate $data\ amount \cdot noise\ ratio$ data points uniformly at random (while replacing those too close to cores);

3.4 Delimitations

Applying random walks with restarts and local hyperspheres within which data points are uniformly generated are based on ideas proposed for Seed Spreader [14]. However, there are crucial differences: Seed Spreader does not track individual hyperspheres and completely randomly expands the clusters, potentially leading to overlapping clusters, especially in lower dimensional space. This also means that Seed Spreader does not offer any control over the clusters' spatial expansion. Moreover, Seed Spreader only maintains a list of data points, limiting the ability to increase the number of data points generated from the same underlying data distributions, as further generation of data points will only randomly expand the latest existing clusters or add new clusters. The final number of clusters can not be set, but is instead based on a restart chance. Thus, any aspects on an individual cluster level can not be properly user-controlled. While density variance has been introduced in [35], the setting is limited to a fixed scaling factor on the size without affecting the number of points within a hypersphere.

We overcome these limitations as follows: DENSIRED follows a two-step approach, where the creation of the skeletons are decoupled from their instantiations. By tracking core points, DENSIRED controls the number of clusters and prevents cluster overlaps. Additional degrees of freedom are given by the parameters momentum ω, branching β, and the star factor $\varkappa$, which in combination impact the shape of (individual) clusters. Different cluster densities are achieved through the relative ratio of data points and the size of hyperspheres. Lastly, the instantiation relies solely on the underlying distribution rather than changing a cluster's skeleton. This enables the generation of equivalent but not equal cluster structures.

3.5 Analysis Intrinsic Dimensionality

The intrinsic dimensionality (id) of a dataset [47] refers to the topological dimension of (nonlinear) manifolds a high-dimensional dataset can be mapped into. Given a dataset $X := \{x_i\}_{i=1}^N \subseteq \mathbb{R}^d$ in d-dimensional space, i.e., $dim(X) = d$, the aim of projective id estimators is to compute a mapping $\phi : x_i \to \tilde{x}_i \in \mathcal{M} \subseteq \mathbb{R}^l$ with $l \ll d$. The minimal number of vectors that linearly span the subspace $\mathcal{M}$ can be estimated by applying PCA. Assuming a data space $X \sim \mathcal{N}(\mu, \sigma)$ where samples are generated by a multivariate Gaussian distribution with covariance matrix $\Sigma \in \mathbb{R}^{p \times p}$ for which all pairs of components are uncorrelated and each component having the same variance. In the asymptotic case of $d \to \infty$, the empirical variances $\sigma_k, \forall k = 1, \ldots, p$, are converging to a unit value σ. In the case of zero correlation amongst the components the eigenvalues are equal to the respective standard deviation, i.e., the diagonal entries of the covariance matrix Σ. Moreover, the eigenvector matrix becomes an identity matrix leading to a permutation-invariant result, where each principal component contributes equally to the explained variance of X. Note that for a multivariate Gaussian distribution $X \sim \mathcal{N}(\mu, \sigma)$ with m components, and diagonal covariance matrix $\sigma \in \mathbb{R}^{p \times p}$ with variance $\sigma = \sigma_1 = \ldots = \sigma_m$, the $id(X) \propto dim(X)$.

Figure 2 shows the explained variance computed by a PCA w.r.t. the dimensions for a single cluster in $\mathbb{R}^{100}$ with 1000 cores. The datasets are scaled to a range between 0 and 1 based on their maximal values. The straight blue line corresponds to equally distributed principal components as described for the Gaussian distribution. *DENSIRED* shows the default setting of our approach, whereas for the other variants, all parameters are fixed besides the one indicated by the respective superscript. The impact of our generator's parameters on a cluster's id is as follows:

Increasing the momentum factor $\omega \to 1$ shown for *DENSIRED*$^\omega$ ($\omega = 1$, default: 0.5) diminishes the amount of uncertainty in the direction vector and, thus, emphasizes the most representative eigenvectors. The id(X) of a cluster X is reduced as a higher explained variance is already reached with a lower amount of dimensions being taken into account. Increasing the star factor $\varkappa \to 1$ for *DENSIRED*$^\varkappa$ ($\varkappa = 1$; default: 0) increases the probability that new core points are set in the ϵ-neighborhood of a cluster's initial core point. It results in star-like topologies where progressively, i.e., with a higher amount of the star factor, more eigenvectors need to be considered to explain the variance. Thus, it increases id(X) of a cluster X. Increasing the branching factor $\beta \to 1$ for *DENSIRED*$^\beta$ ($\beta = 1.0$; default: 0.05) increases the probability that new core points are set in the ε-neighborhood of a cluster's existing core points, where more eigenvectors are required to reach a higher amount of explained variance. It increases the id(X) of a cluster X. Lastly, increasing the step size δ in the set of core points

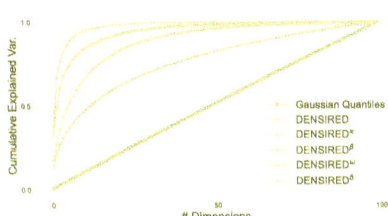

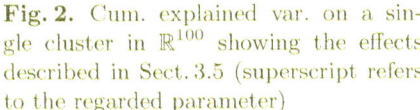

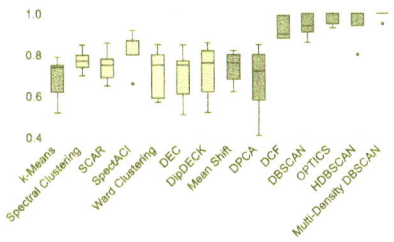

Fig. 2. Cum. explained var. on a single cluster in $\mathbb{R}^{100}$ showing the effects described in Sect. 3.5 (superscript refers to the regarded parameter)

Fig. 3. NMI across all examined dimensionalities. Color indicates clustering concept as described in Sect. 4.2. (Color figure online)

that define the skeleton disperses data points farther apart along the respective direction vectors. This primarily leads to a reduction of the id(X) of a cluster X, especially for smaller core numbers where singular movements have a bigger impact. For larger core numbers, the id(X) remains similar to the default settings as shown with $DENSIRED^\delta$ ($\delta = 2$; default: 1), though it is still reduced.

4   Experiments

4.1   Discussion of the Data Generator

Suitability for High Dimensionality
One of the effects of the *curse of dimensionality* is that pairwise distances become successively similar with increasing dimensionality [48]. Thus, a good data generator for density-based clusters should yield low pairwise intra-cluster distances for varying dimensionalities. In Fig. 4, we evaluate pairwise distances for points generated by various data generators with increasing dimensionality. The results are shown for three metrics: cosine, Euclidean, and density-connectivity distance (*dc-dist*) [49]. The effect is visualized for two widely used generating functions, *Blobs* and *Gaussian Quantiles* from scikit-learn, and for the generators MDCGen [33] and Seed Spreader [14]. Our method is shown with four different settings: i) *DENSIRED*: all parameters are set to their default values; ii) $DENSIRED^\varkappa$: star parameter set to $\varkappa = 1$; iii) $DENSIRED^\beta$: the branch parameter set to $\beta = 1$; iv) $DENSIRED^\omega$: the momentum factor ω set to 1; when varying one parameter, all other parameters are fixed to their default values. For the cosine metric, the expected behavior for Gaussian distributed clusters is that the average distance reaches 90 degrees when $n \to \infty$. With increasing dimensionality, our method preserves lower angles between the data points due to densely connected structures. The effect can also be observed for the Euclidean metric, for which, e.g., MDCGen, Blobs, and Gaussian Quantiles show degenerating effects on preserving low distances with increasing dimensionality, whereas the variants of the DENSIRED preserve lower distances in higher dimensional space. By construction, our method enriches data points in the skeleton such that data

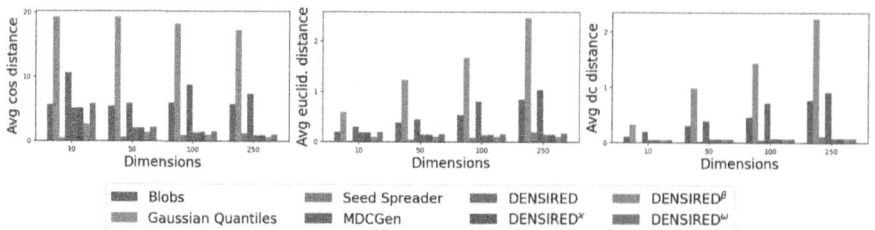

Fig. 4. Avg. pairwise distances of one cluster without noise for the metrics *cos*, *euclidean*, *dcdist* [49] for varying dimensionalities on datasets synthesized by *Blobs*, *Gaussian Quantiles* (sklearn), *MDCGen* [33], *Seed Spreader* [14] and variants of our generator: i) *DENSIRED*: default; ii) $DENSIRED^{\varkappa}$: $\varkappa = 1$; iii) $DENSIRED^{\beta}$: $\beta = 1$; iv) $DENSIRED^{\omega}$: $\omega = 1$; (all other parameters are fixed in each case). All datasets are scaled to a range between 0 and 1 based on their maximal values.

points are less likely extrema in higher dimensional space that would lead to a higher average Euclidean distance. By examining the structures with the dc-dist, the density-preserving characteristic, i.e., low intra-cluster distance values, is expected to be captured best, as both the metric and our data-generating function rely on the basic assumptions of density-connected core points. Our method can keep the lowest average dc-dist for a high dimensionality.

Impact of Parameters
The effect on a cluster's intrinsic dimensionality (id) on scaling the parameters *momentum* ω, *branch* β, *star* $\varkappa$, and *step size* δ is evaluated and summarized in Fig. 5. The evaluation shows the average results on 10 various seeds on a 100-dimensional dataset and we scale the number of generated cores in the range of $[0.1k, 0.5k, 1k]$. When scaling a specific parameter, the other ones are fixed to their default values. The y-axis shows the id with an explained variance of 0.99 of the input data. As shown on the left of Fig. 5, the id decreases with an increase of the momentum factor. Preserving the previous direction with a high momentum value for the random walk results in elongated clusters with low id. An increased number of cores increases the number of possibilities to elongate clusters in various directions, resulting in a slightly increased id for lower values for the momentum before it decreases as expected. As shown in the second part of Fig. 5, high values for β increase the id as they increase the restart probability during the random walk, making elongated clusters rather unlikely. As the *star* parameter $\varkappa$ indicates the behavior when restarting, it only has a marginal influence on the id for a low restart chance. Lastly, we evaluate the influence of the *step size* δ between a cluster's consecutive core points. As expected, when increasing the distance between the cores, the effect of the momentum's direction vector increases, which results in a lower id.

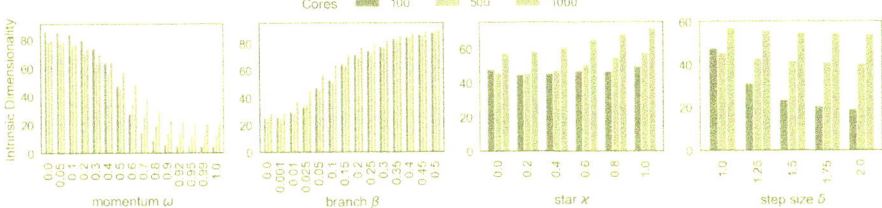

Fig. 5. Intrinsic dimensionality with a datasets's explained variance of 0.99 in $\mathbb{R}^{100}$ on scaling the parameters $\omega \in [0.0, 1.0]$, $\beta \in [0.0, 0.5]$, $\varkappa \in [0.0, 1.0]$, $\delta \in [1.0, 2.0]$ with $cores \in [0.1k, 0.5k, 1k]$ on 10 runs. When evaluating one parameter, others are set to default values: $\omega = 0.5$, $\beta = 0.05$, $\varkappa = 0.0$, $\delta = 1.0$. All datasets are scaled to a range between 0 and 1 based on their maximal values.

4.2 Benchmarking

Setup - Comptetitors + Dataset
For our benchmark study, we evaluated the following 14 clustering algorithms:

Centroid-Based Clustering. k-Means [42] minimizes the distances of points to their respective cluster centroids.

Spectral Clustering. Spectral Clustering [50] is based on the eigendecomposition of a similarity graph. SCAR [5] is an accelerated spectral clustering method. SpectACl [4] combines spectral and density-based clustering.

Hierarchical Clustering. Agglomerative clustering methods, like Ward's method [51], produce nested clusters in a bottom-up manner.

Deep Clustering. DEC [52] combines a clustering objective with neural networks. Unlike DEC, DipDECK [53] is capable of estimating the number of clusters within the latent space of an autoencoder.

Density-Based Clustering. Mean Shift [45] aims at finding modes of an underlying density function. The Density Peak Clustering Algorithm (DPCA) [54] finds clusters that are centered around a local density maximum. DCF [3] determines cluster cores for instances with highest densities based on kNN.

Density-Connectivity-Based Clustering. DBSCAN [2] finds density-connected clusters by connecting *core* points. OPTICS [7] is based on the concepts of DBSCAN and computes an augmented cluster-ordering. HDBSCAN [11] is an extension of DBSCAN that makes use of concepts from hierarchical clustering. MDD [9] can also identify clusters of different densities.

We used grid search to find the best hyperparameters for all algorithms. Where required, the number of clusters was set to number of synthesized ground truth clusters: $k{=}10$. The implementations for k-Means, Mean Shift, Spectral Clustering, Ward Clustering, DBSCAN and OPTICS stem from scikit-learn. The implementations for DEC, DipDeck and MDD [9] originate from ClustPy[2] [55].

---

[2] https://github.com/collinleiber/ClustPy, last accessed: Nov 15th, 2023.

Table 3. Clustering results for different dimensionalities

dimensionality	2-dim		5-dim		10-dim		50-dim		100-dim	
Metric	ARI	NMI	ARI	NMI	ARI	NMI	ARI	NMI	ARI	NMI
k-Means	0.50	0.62	0.32	0.52	0.65	0.74	0.61	0.75	0.61	0.79
Spectral Clust.	0.57	0.74	0.43	0.70	0.64	0.80	0.62	0.77	0.78	0.85
SCAR	0.60	0.69	0.47	0.65	0.58	0.75	0.71	0.78	0.84	0.86
SpectACl	0.60	0.66	0.54	0.80	0.69	0.87	0.78	0.87	0.85	0.92
Ward Clust.	0.43	0.59	0.35	0.57	0.61	0.75	0.66	0.80	0.72	0.85
DEC	0.36	0.51	0.41	0.61	0.72	0.85	0.59	0.77	0.52	0.75
DipDECK	0.52	0.62	0.34	0.52	0.66	0.76	0.75	0.82	0.77	0.86
Mean Shift	0.64	0.68	0.44	0.62	0.73	0.76	0.83	0.80	0.87	0.82
DPCA	0.43	0.58	0.19	0.41	0.61	0.72	0.77	0.80	0.82	0.85
DCF	0.91	0.90	1.00	0.99	1.00	0.99	0.95	0.88	0.95	0.88
DBSCAN	0.87	0.86	0.93	0.91	0.92	0.94	1.00	1.00	1.00	1.00
OPTICS	0.89	0.93	0.91	0.95	1.00	1.00	1.00	1.00	1.00	1.00
HDBSCAN	0.81	0.80	0.95	0.94	1.00	1.00	1.00	1.00	1.00	1.00
MDD	0.96	0.95	1.00	1.00	1.00	1.00	1.00	1.00	1.00	1.00

The implementation of HDBSCAN stems from the hdbscan python package[3]. DCF[4], SCAR[5], SpectACl[6], and DPCA[7] originate from the linked repositories.

The dataset contains branching data with high momentum and larger step size between consecutive cores[8], yielding high intrinsic dimensionalities and an elongated shape for each cluster. The two-dimensional dataset can be seen in Fig. 1 in the bottom-right. We generate datasets of varying dimensionality $d \in [2, 5, 10, 50, 100]$ scaled to a range between 0 and 1 based on their respective maximal values with 5000 data points each. We provide the datasets as novel benchmark datasets in our repository. We also include further evaluations on more datasets with different settings in our supplementary materials, which can be found in our GitHub repository.

Evaluation Metrics
We evaluate clustering results based on the ground truth labels with the commonly used Adjusted Rand Index (ARI) [56] and Normalized Mutual Information (NMI) [57]. ARI can take values $\in [-1, 1]$ and NMI lies in $[0, 1]$.

---

[3] https://github.com/scikit-learn-contrib/hdbscan, last accessed: Oct 5th, 2023.

[4] https://github.com/tobinjo96/DCFcluster, last accessed: Oct 5th, 2023.

[5] https://github.com/SpectralClusteringAcceleratedRobust/SCAR, last accessed: Nov 4th, 2023.

[6] https://bitbucket.org/Sibylse/spectacl, last accessed: Oct 11th, 2023.

[7] https://github.com/colinwke/dpca, last accessed: Oct 31st, 2023.

[8] Exact values: $\omega = 0.8$, $\delta = 1.5$, $\beta = 0.1$, $\varkappa = 1$, 200 cores.

Benchmarking Results

Table 3 summarizes the results and Fig. 3 showcases the general trend among the different clustering algorithms. Algorithms that support the usage of seeds were performed with 10 different seeds while keeping hyperparameter settings. k-Means cannot capture arbitrarily shaped clusters or noise. Ward Clustering is not robust against noise, and tends to fuse multiple clusters, which results in low performance scores. Spectral Clustering is generally sensitive to high noise levels in a dataset, as the decomposition of the affinity matrix does not yield distinct indicator vectors for various clusters. Hence, vanilla Spectral Clustering lacks in identifying the dense structures. SCAR has similar issues but manages to outperform Spectral Clustering. SpectACl faces similar difficulties for elongated clusters but is better at detecting dense structures in high-dimensional settings. DEC targets a latent representation optimizing the distributions of cluster assignments, which is not density-specific. DipDECK performs better as its micro-clusters are partially able to describe non-convex clusters in the latent space. Mean Shift often splits apart widespread clusters. Furthermore, the sensitivity of the bandwidth parameter in Mean Shift tends to misassign noisy data samples. DPCA tends to underestimate the size of the clusters and assigns large parts of clusters to noise (named cluster halo in [54]). This effect is more prominent for lower dimensionalities and for elongated clusters. Still, the method is robust against noise points. DCF performs reasonably well, but in areas of lower densities, where fewer core points overlap, DCF identifies subclusters. Noise in the dataset often produces individual density peaks, resulting in a lower clustering quality. DBSCAN introduced the concept of density-connectivity and succeeds at detecting the clusters. However, differing density levels exacerbate the choice of the right hyperparameters. Less dense clusters are often split into smaller clusters or are assigned to noise. Likewise to DBSCAN, OPTICS is highly dependent on the correct choice of a minimal number of samples to identify core points. Small values lead to significant parts of the clusters being considered as noise or subsumed into separate clusters, especially in lower dimensionalities. HDBSCAN subdivides less dense clusters and assigns noise to multiple smaller clusters. This effect is reduced in higher dimensionalities, inducing higher distances between noisy parts. MDD performs best out of all the evaluated algorithms. Notably, random noise is assigned to its own cluster for lower dimensionalities, whilst noisy points between clusters are correctly classified as noise.

In summary, due to the concept of density-connectivity, the class of density-based methods performs in general superior to other methods. While DBSCAN-related algorithms perform the best on the synthesized data, not all other density-based heuristics succeed in identifying the ground truth. Our benchmarking uncovers the necessity of proper data generators to enhance the modeling and evaluation of density-based clustering methods in a variety of settings.

5 Conclusion

A fair and reproducible comparison and evaluation of new algorithms is crucial for good research. While systematic testing is easy for data containing Gaussian clusters, evaluation of density-based clustering algorithms was rather empirical in the past, as synthetic datasets were usually non-reproducible. In this paper, we present DENSIRED, a flexible and reliable data generator for data containing density-based clusters. It enables a systematic, reproducible, and controlled evaluation of all kinds of algorithms working on data that contain arbitrary density-connected shapes that are not generated by Gaussian processes. As the data generation process is separated into two steps, the model generation and the instantiation, it produces similar yet different datasets where users can define a multitude of properties for their specific use case. DENSIRED builds the basis for reproducible, fair, non-arbitrary comparison, evaluation, and benchmarking of data mining algorithms. The easy-to-use code as well as a set of benchmark datasets created with DENSIRED is publicly available in our repository.

References

1. Tommasi, T., Patricia, N., Caputo, B., Tuytelaars, T.: A deeper look at dataset bias. In: Csurka, G. (ed.) Domain Adaptation in Computer Vision Applications. ACVPR, pp. 37–55. Springer, Cham (2017). https://doi.org/10.1007/978-3-319-58347-1_2
2. Ester, M., Kriegel, H., Sander, J., Xu, X.: A density-based algorithm for discovering clusters in large spatial databases with noise. In: KDD, AAAI Press, pp. 226–231 (1996)
3. Tobin, J., Zhang, M.: DCF: an efficient and robust density-based clustering method. In: ICDM, pp. 629–638. IEEE (2021)
4. Hess, S., Duivesteijn, W., Honysz, P., Morik, K.: The SpectACl of nonconvex clustering: a spectral approach to density-based clustering. In: AAAI, AAAI Press, pp. 3788–3795 (2019)
5. Hohma, E., Frey, C.M.M., Beer, A., Seidl, T.: SCAR - spectral clustering accelerated and robustified. Proc. VLDB Endow. **15**(11), 3031–3044 (2022)
6. Sander, J., Ester, M., Kriegel, H., Xu, X.: Density-based clustering in spatial databases: the algorithm GDBSCAN and its applications. Data Min. Knowl. Discov. **2**(2), 169–194 (1998)
7. Ankerst, M., Breunig, M.M., Kriegel, H., Sander, J.: OPTICS: ordering points to identify the clustering structure, pp. 49–60 (1999)
8. Frey, C., Züfle, A., Emrich, T., Renz, M.: Efficient information flow maximization in probabilistic graphs. IEEE Trans. Knowl. Data Eng. **30**(5), 880–894 (2018)
9. Ashour, W., Sunoallah, S.: Multi density DBSCAN. In: Yin, H., Wang, W., Rayward-Smith, V. (eds.) IDEAL 2011. LNCS, vol. 6936, pp. 446–453. Springer, Heidelberg (2011). https://doi.org/10.1007/978-3-642-23878-9_53
10. Frey, C.M., Jungwirth, A., Frey, M., Kolisch, R.: The vehicle routing problem with time windows and flexible delivery locations. Eur. J. Oper. Res. **308**(3), 1142–1159 (2023). ISSN 0377-2217
11. Campello, R.J.G.B., Moulavi, D., Sander, J.: Density-based clustering based on hierarchical density estimates. In: Pei, J., Tseng, V.S., Cao, L., Motoda, H., Xu, G. (eds.) PAKDD 2013. LNCS (LNAI), vol. 7819, pp. 160–172. Springer, Heidelberg (2013). https://doi.org/10.1007/978-3-642-37456-2_14
12. Yale, A., Dash, S., Dutta, R., Guyon, I., Pavao, A., Bennett, K.P.: Generation and evaluation of privacy preserving synthetic health data. Neurocomputing **416**, 244–255 (2020). ISSN 0925-2312

13. Libes, D., Lechevalier, D., Jain, S.: Issues in synthetic data generation for advanced manufacturing. In: 2017 IEEE International Conference on Big Data (Big Data), Boston, MA, USA, pp. 1746-1754 (2017)
14. Gan, J., Tao, Y.: DBSCAN revisited: Mis-claim, un-fixability, and approximation. In: SIGMOD Conference, pp. 519–530. ACM (2015)
15. Mai, S.T., Assent, I., Storgaard, M.: AnyDBC: an efficient anytime density-based clustering algorithm for very large complex datasets. In: KDD, pp. 1025–1034. ACM (2016)
16. Hou, J., Gao, H., Li, X.: DSets-DBSCAN: a parameter-free clustering algorithm. IEEE Trans. Image Process. **25**(7), 3182–3193 (2016)
17. Bryant, A., Cios, K.J.: RNN-DBSCAN: a density-based clustering algorithm using reverse nearest neighbor density estimates. IEEE Trans. Knowl. Data Eng. **30**(6), 1109–1121 (2018)
18. Kim, J., Choi, J., Yoo, K., Nasridinov, A.: AA-DBSCAN: an approximate adaptive DBSCAN for finding clusters with varying densities. J. Supercomput. **75**(1), 142–169 (2019)
19. Ren, Y., Wang, N., Li, M., Xu, Z.: Deep density-based image clustering. Knowl. Based Syst. **197**, 105841 (2020)
20. Chen, Y., Zhou, L., Bouguila, N., Wang, C., Chen, Y., Du, J.: BLOCK-DBSCAN: fast clustering for large scale data. Pattern Recognit. **109**, 107624 (2021)
21. dos Santos, J.A., Iqbal, S.T., Naldi, M.C., Campello, R.J.G.B., Sander, J.: Hierarchical density-based clustering using MapReduce. IEEE Trans. Big Data **7**(1), 102–114 (2021)
22. Wang, Z., et al.: AMD-DBSCAN: an adaptive multi-density DBSCAN for datasets of extremely variable density. In: DSAA, pp. 1–10. IEEE (2022)
23. Huang, X., Ma, T., Liu, C., Liu, S.: GriT-DBSCAN: a spatial clustering algorithm for very large databases. Pattern Recognit. **142**, 109658 (2023)
24. Ma, B., Yang, C., Li, A., Chi, Y., Chen, L.: A faster dbscan algorithm based on self-adaptive determination of parameters. Procedia Comput. Sci. **221**, 113–120 (2023). (ITQM 2023)
25. Qian, J., Zhou, Y., Han, X., Wang, Y.: MDBSCAN: a multi-density dbscan based on relative density. Neurocomputing **576**, 127329 (2024)
26. Milligan, G.W.: An algorithm for generating artificial test clusters. Psychometrika **50**, 123–127 (1985)
27. Qiu, W., Joe, H.: Generation of random clusters with specified degree of separation. J. Classif. **23**(2), 315–334 (2006)
28. Melnykov, V., Chen, W.-C., Maitra, R.: MixSim: an r package for simulating data to study performance of clustering algorithms. J. Stat. Softw. **51**, 1–25 (2012)
29. Pedregosa, F., et al.: Scikit-learn: machine learning in Python. J. Mach. Learn. Res. **12**, 2825–2830 (2011)
30. Fachada, N., de Andrade, D.: Generating multidimensional clusters with support lines. Knowl. Based Syst. **277**, 110836 (2023)
31. Steinley, D.L., Henson, R.: OCLUS: an analytic method for generating clusters with known overlap. J. Classif. **22**(2), 221–250 (2005)
32. Shand, C., Allmendinger, R., Handl, J., Webb, A.M., Keane, J.: HAWKS: evolving challenging benchmark sets for cluster analysis. IEEE Trans. Evol. Comput. **26**(6), 1206–1220 (2022)
33. Iglesias, F., Zseby, T., Ferreira, D.C., Zimek, A.: MDCGen: multidimensional dataset generator for clustering. J. Classif. **36**(3), 599–618 (2019)

34. Vennam, J.R., Vadapalli, S.: SynDECA: a tool to generate synthetic datasets for evaluation of clustering algorithms. In: COMAD, Computer Society of India, pp. 27–36 (2005)
35. Gan, J., Tao, Y.: On the hardness and approximation of euclidean DBSCAN. ACM Trans. Database Syst. **42**(3), 14:1–14:45 (2017)
36. Rachkovskij, D.A., Kussul, E.M.: DataGen: a generator of datasets for evaluation of classification algorithms. Pattern Recognit. Lett. **19**(7), 537–544 (1998)
37. Fränti, P., Sieranoja, S.: K-means properties on six clustering benchmark datasets, pp. 4743–4759 (2018). http://cs.uef.fi/sipu/datasets/
38. Beer, A., Schüler, N.S., Seidl, T.: A generator for subspace clusters. In: LWDA, ser. CEUR Workshop Proceedings, vol. 2454, pp. 69–73 (2019). CEUR-WS.org
39. Schubert, E., Sander, J., Ester, M., Kriegel, H., Xu, X.: DBSCAN revisited, revisited: why and how you should (still) use DBSCAN. ACM Trans. Database Syst. **42**(3), 19:1–19:21 (2017)
40. Goncalves, A., Ray, P., Soper, B., Stevens, J., Coyle, L., Sales, A.P.: Generation and evaluation of synthetic patient data. BMC Med. Res. Methodol. **20**(1), 1–40 (2020)
41. Pei, Y., Zaiane, O.R.: A synthetic data generator for clustering and outlier analysis (2006)
42. Lloyd, S.P.: Least squares quantization in PCM. IEEE Trans. Inf. Theory **28**(2), 129–136 (1982)
43. Georgoulas, G.K., Konstantaras, A., Katsifarakis, E., Stylios, C.D., Maravelakis, E., Vachtsevanos, G.J.: "Seismic-mass" density-based algorithm for spatio-temporal clustering. Expert Syst. Appl. **40**(10), 4183–4189 (2013)
44. Jain, A.K.: Data clustering: 50 years beyond k-means. Pattern Recognit. Lett. **31**(8), 651–666 (2010)
45. Comaniciu, D., Meer, P.: Mean Shift: a robust approach toward feature space analysis. IEEE Trans. Pattern Anal. Mach. Intell. **24**(5), 603–619 (2002)
46. Ullmann, T., Beer, A., Hünemörder, M., Seidl, T., Boulesteix, A.: Over-optimistic evaluation and reporting of novel cluster algorithms: an illustrative study. Adv. Data Anal. Classif. **17**(1), 211–238 (2023)
47. Levina, E., Bickel, P.: Maximum likelihood estimation of intrinsic dimension. In: Saul, L., Weiss, Y., Bottou, L. (eds.) NIPS, vol. 17. MIT Press (2004)
48. Beyer, K., Goldstein, J., Ramakrishnan, R., Shaft, U.: When Is "Nearest Neighbor" meaningful? In: Beeri, C., Buneman, P. (eds.) ICDT 1999. LNCS, vol. 1540, pp. 217–235. Springer, Heidelberg (1999). https://doi.org/10.1007/3-540-49257-7_15
49. Beer, A., Draganov, A., Hohma, E., Jahn, P., Frey, C.M., Assent, I.: Connecting the dots - density-connectivity distance unifies dbscan, k-center and spectral clustering. In: KDD, pp. 80–92. ACM (2023)
50. von Luxburg, U.: A tutorial on spectral clustering. Stat. Comput. **17**(4), 395–416 (2007)
51. Ward, J.H., Jr.: Hierarchical grouping to optimize an objective function. J. Am. Stat. Assoc. **58**(301), 236–244 (1963)
52. Xie, J., Girshick, R.B., Farhadi, A.: Unsupervised deep embedding for clustering analysis. In: ICML, ser. JMLR Workshop and Conference Proceedings, vol. 48, pp. 478–487 (2016). JMLR.org
53. Leiber, C., Bauer, L.G.M., Schelling, B., Böhm, C., Plant, C.: Dip-based deep embedded clustering with k-estimation. In: KDD, pp. 903–913. ACM (2021)
54. Rodriguez, A., Laio, A.: Clustering by fast search and find of density peaks. Science **344**(6191), 1492–1496 (2014)

55. Leiber, C., Miklautz, L., Plant, C., Böhm, C.: Benchmarking deep clustering algorithms with clustpy. In: ICDM (Workshops), pp. 625–632. IEEE (2023)
56. Hubert, L., Arabie, P.: Comparing partitions. J. Classif. **2**, 193–218 (1985)
57. Strehl, A., Ghosh, J.: Cluster ensembles – a knowledge reuse framework for combining multiple partitions. J. Mach. Learn. Res. **3**, 583–617 (2002)

Model-Based Reinforcement Learning with Multi-task Offline Pretraining

Minting Pan, Yitao Zheng, Yunbo Wang$^{(\boxtimes)}$, and Xiaokang Yang

MoE Key Lab of Artificial Intelligence, AI Institute, Shanghai Jiao Tong University,
Shanghai, China
{panmt53,iorisou0826,yunbow,xkyang}@sjtu.edu.cn

Abstract. Pretraining reinforcement learning (RL) models on offline datasets is a promising way to improve their training efficiency in online tasks, but challenging due to the inherent mismatch in dynamics and behaviors across various tasks. We present a model-based RL method that learns to transfer potentially useful dynamics and action demonstrations from offline data to a novel task. The main idea is to use the world models not only as simulators for behavior learning but also as tools to measure the task relevance for both dynamics representation transfer and policy transfer. We build a time-varying, domain-selective distillation loss to generate a set of offline-to-online similarity weights. These weights serve two purposes: (i) adaptively transferring the task-agnostic knowledge of physical dynamics to facilitate world model training, and (ii) learning to replay relevant source actions to guide the target policy. We demonstrate the advantages of our approach compared with the state-of-the-art methods in Meta-World and DeepMind Control Suite.

1 Introduction

Reinforcement learning (RL) approaches have made significant advancements in solving a wide range of sequential control problems [7,20,26]. In the realm of visual RL, agents need to not only conduct representation learning from raw image inputs but also perform behavior learning in the learned state space, which requires a large number of interactions with an online environment and limits the applications in the real world. Recently, model-based RL algorithms have greatly improved sample efficiency by concurrently learning a differentiable simulator of the environment (*i.e.*, the world model), and using imagined rollouts generated by the world model for policy optimization [10,17]. Nevertheless, the process of training an effective world model from scratch remains a time-consuming and challenging pursuit, often yielding less generalizable representations.

To address this problem, many recent approaches [29–31,34] adopt the *pre-training and finetuning* paradigm to pre-learn representation models on off-the-shelf offline datasets and transfer the learned prior knowledge to a novel online

---

M. Pan and Y. Zheng—Equal contribution. Code available at https://github.com/
panmt/Vid2Act.

© The Author(s), under exclusive license to Springer Nature Switzerland AG 2024
A. Bifet et al. (Eds.): ECML PKDD 2024, LNAI 14947, pp. 22–39, 2024.
https://doi.org/10.1007/978-3-031-70368-3_2

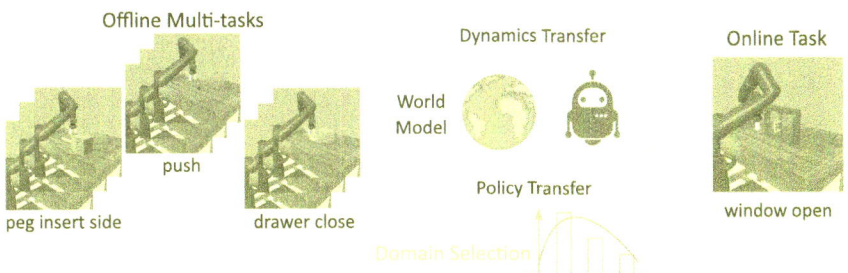

Fig. 1. We aim to build an offline-to-online transfer RL agent for visual control problems, which is challenging due to the discrepancies between the target task and the source tasks from which the offline datasets are collected. The key idea of our approach is to leverage the world models to enable positive knowledge transfer through *domain-selective dynamics distillation* and *behavior guidance.*

RL domain. For example, SMART [30] exploits a Transformer model to learn generalizable visual representations from reward-free, offline interaction data under a control-centric pretraining objective. Similarly, our focus lies in leveraging multi-task offline data without reward to improve the visual RL performance in a novel online task. However, it is crucial to recognize that, despite the effectiveness of the pretraining method, a straightforward finetuning method may still suffer from the potential discrepancy in visual observations, physical dynamics, or even action spaces across task domains. Unlike SMART, our method aims to:

1. *Adaptively identify the relevance between offline and online tasks in an unsupervised manner, allowing for positive domain transfer even when some offline data may seem unrelated.*
2. *Exploit relevant actions from the offline datasets to effectively guide and enhance the policy optimization process for the new task.*

As shown in Fig. 1, we propose a new *domain-selective* transfer RL approach called Vid2Act to reduce the potential discrepancies between the pretraining stage and the transferring stage. In the pretraining stage, we exploit reward-free offline trajectories of image-action pairs to train a mixture world model, which learns the task-specific observation-to-state mapping functions and state-to-state transition functions based on task index for different source tasks. In the transferring stage, instead of performing direct finetuning, we leverage the mixture world model as the teacher model to provide flexible regularization to the representation learning process of the target-domain agent. This is achieved through a domain-selective distillation loss, where we learn a set of importance weights over the teacher model with different label indexes to adaptively transfer the prior knowledge of physical dynamics gathered from the offline data to the target world model.

In addition to their impact on representation learning, the importance weights also directly contribute to the policy optimization process conducted over the imaginations of the target world model. Specifically, Vid2Act incorporates a *"generative action replay"* module. During behavior learning, it serves to reproduce source-domain actions based on the target-domain states, which have been aligned to the corresponding source-domain state spaces by the distillation loss. By reusing the importance weights, we can dynamically select the most relevant source task at different time steps and replay its source expert behaviors to provide effective guidance for target policy improvement.

In summary, the main technical contributions of Vid2Act are as follows:

- Our work introduces a novel pretraining and finetuning pipeline for visual model-based RL. It transfers the dynamics from multiple source tasks with a set of importance weights learned by the world model.
- Vid2Act presents a novel domain-selective behavior learning scheme that identifies potentially valuable source actions and employs them as exemplar guidance for the target policy.

We evaluate Vid2Act on the Meta-World benchmark [40] and the DeepMind Control Suite [32]. Our approach shows remarkable performance improvements over both the vanilla model-based RL baselines like DreamerV2 [12] and existing unsupervised pretraining methods for transfer RL, such as APV [27] and SMART [30]. Importantly, our experimental results consistently demonstrate that Vid2Act can achieve positive domain transfer, even when the available source offline data seems less relevant to the target task.

2   Related Work

Visual RL. In visual control tasks, the agent needs to learn policy from high-dimensional and complex observations. Learning generalized representation by either unsupervised [9,19,29,38] or self-supervised manners [4,35,41], is a natural way to learn an auxiliary encoder of images for visual control tasks. Prior approaches consist of model-based methods to optimize latent dynamics model [10–12,24], and model-free methods to utilize data augmentation [3,20] and contrastive representation learning [1,19,21,23]. Similar to our work, several methods pretrain RL models on offline datasets and then finetune them on the online target task [6,19,25,27,29]. Except for not bridging the domain gap between pretraining source data and RL tasks, they have shown attractive performance on vision-based RL tasks. In our framework, we do not directly finetune the parameters of the pretrained models, but rather learn more useful world models by distillation technique.

Transfer RL. Previous experiences across a diverse range of tasks can be beneficial in solving online control tasks, even when encountering them for the first time. To quickly leverage the past information to the new environments, many transfer learning approaches [15,16,22,36,37] are proposed to bridge the gap across different tasks or domains. APV [27] employs action-free videos of multiple domains to pretrain an action-free recurrent state-space model (RSSM), which focuses on learning visual representation from offline datasets. XTRA [34] proposes a framework based on EfficientZero [39] to use multiple offline tasks with rewards both in pretraining and finetuning stages for cross-task transfer. Recently, some methods leveraging Transformer have been proposed to facilitate transfer learning in control tasks [30,33]. SMART [30] designs a control-centric pretraining objective for Decision Transformers [2] to capture the common essential information relevant to short-term control and long-term control across tasks. A work closely related to our approach is Knowledge Flow [22], which involves training multiple teacher models and distilling knowledge from their layers to a student model. In our work, we propose a domain-selective distillation strategy to fully utilize both the dynamics and action information from the source tasks. It introduces a more flexible way to adaptively transfer useful knowledge to help downstream tasks.

3   Problem Formulation

In the visual control task, the agent learns the behavior policy directly from high-dimensional observations, which is formulated as a partially observable Markov decision process (POMDP) with a tuple $(\mathcal{S}, \mathcal{A}, \mathcal{O}, \mathcal{T}, \mathcal{R})$. Here, $\mathcal{S}$ is the state space, $\mathcal{A}$ is the action space, $\mathcal{O}$ is the observation space, $\mathcal{R}(s_t, a_t)$ is the reward function, and $\mathcal{T}(s_{t+1} \mid s_t, a_t)$ is the state-transition distribution. In this setting, the agent cannot access the true states in $\mathcal{S}$. At each timestep $t \in [1;T]$, the agent takes an action $a_t \in A$ to interact with the environment and receives a reward $r_t = \mathcal{R}(s_t, a_t)$. The objective is to learn a policy that maximizes the expected cumulative reward $\mathbb{E}_p[\sum_{\tau=1}^{T} r_\tau]$.

To improve policy learning and sample efficiency of visual RL, we aim to transfer previous knowledge from multiple offline tasks. The offline datasets are reward-free and exclusively consist of image-action pairs $\{(o_t, a_t)\}$. It is important to note that there might be substantial distribution shifts in observations $(\mathcal{O})$, state transition functions $(\mathcal{T})$, and behaviors $(\mathcal{A})$ across task domains, which pose significant challenges in transfer learning, providing strong motivation for the development of a dynamic domain-selective transfer RL approach. The primary goal of our approach is to efficiently bridge the gap between tasks in terms of state representations, physical dynamics, and action behaviors.

4   Method

In this section, we present a comprehensive overview of the pretraining process in the source datasets and the subsequent transfer learning process in the target task. The transfer learning process consists of two stages, *i.e.*, *domain-selective dynamics transfer* and *behavior learning with generative action replay*, as shown in Fig. 2 and described in detail in Algorithm 1.

4.1   Why Model-Based RL for Domain Transfer?

Our overall pipeline is built upon model-based RL, which involves learning the underlying dynamics from a buffer of past experiences, optimizing the control policy through future rollouts of compact model states, and executing actions in the environment to append the experience buffer. More precisely, we introduce a transfer RL approach based on the model-based DreamerV2 method [12]. Unlike previous work, the world model in our approach serves not only as a simulator for policy learning but also provides a measure of task relevance for both dynamics representation transfer and behavior transfer discussed in the following sections. Additionally, after pretraining the source world model, subsequent algorithms can rely on the fixed parameters of this model, making it more universal in real-world scenarios and decoupled from the source data.

4.2   Multi-task Offline Pretraining

Mixture World Model as the Teacher Model. As illustrated in Fig. 2, we consider multiple reward-free, action-conditioned datasets denoted as $\mathcal{D}$. These datasets comprise expert data that has been previously collected from N tasks and is readily available for our use. Initially, we pretrain an action-conditioned video prediction model, denoted as F_ϕ, with the explicit task label $k \in \{1, \ldots, N\}$. In contrast to APV [27], an existing model-based pretraining-finetuning transfer RL method, our approach incorporates actions during the pretraining phase, which is reasonable in learning the consequences of state transitions. The pretrained models consist of three main components as follows:

$$
\begin{aligned}
\text{Representation model:} &\quad q(s_t \mid s_{t-1}, a_{t-1}^k, o_t^k, k) \\
\text{Dynamics model:} &\quad p(\hat{s}_t \mid s_{t-1}, a_{t-1}^k, k) \\
\text{Decoder model:} &\quad p(\hat{o}_t \mid s_t, k).
\end{aligned}
\tag{1}
$$

The representation model extracts posterior latent states s_t from observations o_t, previous states s_{t-1}, previous actions a_{t-1} and task label k. The dynamics model follows the Recurrent State Space Model (RSSM) architecture from PlaNet [11] to predict the prior latent states $\hat{s}_t$ without access to the corresponding o_t. The decoder reconstructs $\hat{o}_t$ given the latent states. For task T_k, all components are optimized jointly using the following loss function:

$$
\begin{aligned}
\mathcal{L}_{\text{source}} = \mathbb{E} \Big\{ \sum_{t=1}^{T} \underbrace{-\ln p(\hat{o}_t \mid s_t, k)}_{\text{Image reconstruction}} \\
+ \underbrace{\beta\, \text{KL}[q(s_t \mid s_{t-1}, a_{t-1}^k, o_t^k, k) \parallel p(\hat{s}_t \mid s_{t-1}, a_{t-1}^k, k)]}_{\text{KL divergence}} \Big\},
\end{aligned}
\tag{2}
$$

where β is a hyperparameter of the Kullback-Leibler (KL) divergence that regularizes the approximate posterior learned from the representation model toward the prior learned from the dynamics model.

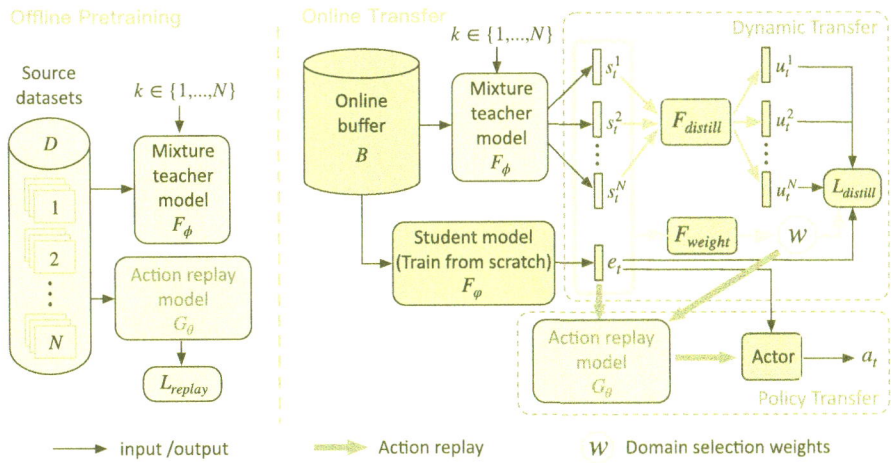

Fig. 2. Left: We employ multiple offline domains to train a mixture world model (F_ϕ), whose parameters are frozen during the subsequent transfer learning process. **Right:** In the target domain, we use F_ϕ as the teacher model and dynamically distill prior knowledge from it with a set of domain-similarity weights $\mathcal{W}$. These weights are further used to reproduce the most relevant source actions to guide the target policy.

Behavior Replay. We simultaneously utilize the offline source datasets to learn an action replay model to guide subsequent target behavior learning. Inspired by BCQ [8], which is an offline RL method, we design G_θ using a state-conditioned variational auto-encoder (VAE) [18,28]. The action replay model G_θ consists of an encoder $E_{\theta 1}$ and a decoder $D_{\theta 2}$. The encoder takes a state-action pair and a task label k, and outputs a Gaussian distribution $\mathcal{N}(\mu, \sigma)$. The state s, along with a latent vector z sampled from the Gaussian distribution and a task label k, is passed to the decoder $D_{\theta 2}$ which outputs an action:

$$\mu, \sigma = E_{\theta 1}(s, a, k), \quad \hat{a} = D_{\theta 2}(s, z, k), \quad z \sim \mathcal{N}(\mu, \sigma). \tag{3}$$

The action replay model G_θ is optimized by

$$\mathcal{L}_{\text{replay}} = \mathbb{E} \left[\sum_{(s,a) \in D} (a - \hat{a})^2 + \text{KL}\left(\mathcal{N}(\mu, \sigma) \parallel \mathcal{N}(0, 1)\right) \right]. \tag{4}$$

4.3   Domain-Selective Dynamics Transfer

It is important to note that, even though the pretraining method is effective, a simple finetuning approach may encounter challenges due to the potential discrepancy in visual observations, physical dynamics, or even action spaces across task domains. Therefore, when a novel target task emerges, we initialize a student world model F_φ from scratch, while freezing the parameters of the teacher model F_ϕ to transfer the dynamics representations from the source domains (see

---

Algorithm 1: Vid2Act with improved dynamics learning, behavior learning & policy deployment

---

1 **Hyperparameters:** H: Imagination horizon
2 Initialize the online replay buffer $\mathcal{B}$ with random episodes.
3 **while** *not converged* **do**
4   **for** *update step* $c = 1 \ldots C$ **do**
5   Draw data sequences $\{(o_t, a_t, r_t)\}_{t=1}^T \sim \mathcal{B}$.
6   `// Dynamics learning`
7   Compute distillation loss using Equation (6) and update world model parameters using Equation (7)
8   `// Behavior learning`
9   **for** *time step* $i = t \ldots t + H$ **do**
10   Select the task label k with highest confidence in Equation (5)
11   Imagine an action $a_i \sim \pi(a_i \mid e_i, G_\theta(e_i, k))$
12   Predict rewards $r_i \sim p(\hat{r}_i \mid e_i)$ and values $v_\psi(e_i)$
13   **end**
14   Update the actor and value models in Equation (8) using estimated rewards and values.
15   **end**
16   `// Environment interaction`
17     $o_1 \leftarrow$ `env.reset()`
18   **for** *time step* $t = 1 \ldots T$ **do**
19   Calculate the posterior state $e_t \sim q(e_t \mid e_{t-1}, a_{t-1}, o_t; \varphi)$ from history.
20   Use the teacher model to obtain $\{\hat{s}_t^i \sim p(e_{t-1}, a_{t-1}, i; \phi) \mid i \in [1, N]\}$ and determine the task label k with highest confidence in Equation (5)
21   Compute $a_t \sim \pi(a_t \mid e_t, G_\theta(e_t, k))$
22         $r_t, o_{t+1} \leftarrow$ `env.step`(a_t)
23   **end**
24   Add experience to the online replay buffer $\mathcal{B} \leftarrow \mathcal{B} \cup \{(o_t, a_t, r_t)_{t=1}^T\}$.
25 **end**

Fig. 2). In addition to the model components outlined in Eq. (1), F_φ also incorporates a reward model represented as $\hat{r}_t \sim p(\hat{r}_t \mid s_t)$. To avoid confusion of notations, we use s_t^k to denote the state obtained from the teacher model with task label k, and e_t to denote the state of the target student model. Given a latent state denoted by e_{t-1} and a corresponding action a_{t-1}, we first transit this state to the next time step individually using the teacher model and the student model, obtaining $\{s_t^k \sim p(e_{t-1}, a_{t-1}; \phi)\}_{k=1}^N$ and $e_t \sim p(e_{t-1}, a_{t-1}; \varphi)$.

To close the distance between the marginal distributions of state transitions produced by the student world model and the dynamics estimated by the teacher model, we incorporate a distillation network in F_φ, denoted as F_{distill}, which takes the form of a multilayer perceptron (MLP). The role of this module is to extract transferable features from the predicted states of the teacher model. In other words, it transforms the states s_t^k predicted by the teacher model into a set of transferable features $\{u_t^k = F_{\text{distill}}(s_t^k)\}_{k=1}^N$. These features are then used in the knowledge distillation loss.

Intuitively, each source task may hold varying impacts on the dynamics learning of the target visual control task. We introduce the concept of domain-similarity weights and propose to optimize these weights through the knowledge distillation loss. By learning this set of weights, we can dynamically transfer knowledge in an adaptive manner based on offline-online task relevance. To compute the similarity weight $\mathcal{W}$, we concatenate the predicted state s_t^k of teacher model and the predicted state e_t of the student model. This concatenated representation is then fed into a fully-connected layer F_{weight}, followed by a softmax activation function:

$$\text{Domain selection:} \quad \mathcal{W} = \{w_k\}_{k=1}^N = \text{Softmax}(\{F_{\text{weight}}(s_t^k * e_t)\}_{k=1}^N), \quad (5)$$

where $*$ denotes the operation of concatenation. In order to avoid the collapse of domain-specific weights, wherein $w_i = 1$ when $i = c$ and $w_i = 0$ for $i \neq c$, with c denoting the offline task most akin to the present online task, we establish a minimum threshold of 0.1 for the weights. We then minimize the Euclidean distance between pairs of states as follows, taking into account the corresponding domain-similarity weights:

$$\mathcal{L}_{\text{distill}} = \sum_{k=1}^N \sum_{t=1}^T w_k \cdot \| e_t - u_t^k \|_2^2. \quad (6)$$

The overall objective of the student model can be written as follows, where α is a hyperparameter:

$$\mathcal{L}_{\text{target}} = \mathbb{E}\left[\left[\sum_{t=1}^T \underbrace{\beta \, \text{KL}\big[q(e_t \mid e_{t-1}, a_{t-1}, o_t) \parallel p(\hat{e}_t \mid e_{t-1}, a_{t-1})\big]}_{\text{KL divergence}}\right.\right.$$
$$\left.\left.\underbrace{-\ln p(\hat{o}_t \mid e_t)}_{\text{Image reconstruction}} \quad \underbrace{-\ln p(\hat{r}_t \mid e_t)}_{\text{Reward prediction}}\right] + \alpha \, \mathcal{L}_{\text{distill}}\right]. \quad (7)$$

Equation (6) is the fundamental basis for Vid2Act. When the dynamics of the source domain are similar to the target task, the latter term of this loss naturally becomes smaller. On the other hand, for source tasks with significantly different dynamics from the target task, the model will minimize the weight term to minimize this loss. The *domain-selective* distillation loss enables the student model to adaptively learn from the teacher model, acquiring significant prior knowledge regarding intricate physical dynamics from the most relevant source tasks. By selectively distilling knowledge from these source tasks, the student model can adapt and incorporate valuable information to enhance its overall learning capabilities.

4.4   Domain-Selective Behavior Transfer

We utilize an actor-critic algorithm to learn the policy over the predicted future state and reward trajectories. As shown in Fig. 2, we use the action replay model

G_θ to promote policy learning, which *1) provides an efficient indication when a strong correlation exists between the source and target tasks,* and *2) expends exploration of action space when there is little correlation between them.* The parameters in G_θ are frozen at this stage. Reusing the similarity weights learned in dynamics transfer, we can dynamically select task label k with the highest confidence to generate action guidance. We exclusively employ the decoder $D_{\theta 2}$ of action generation model G_θ to replay source-domain actions, which takes the state e_t of the student model and the selected task label k with highest confidence as inputs. We modify the actor model and the value model as follows:

$$
\begin{aligned}
\text{Actor model:} \quad & a_t \sim \pi(a_t \mid e_t, G_\theta(e_t, k)), \\
\text{Value model:} \quad & v_\psi(e_t) \approx \mathbb{E}_{\pi(\cdot \mid e_t, G_\theta(e_t, k))} \sum_{t'=t}^{t+H} \gamma^{t'-t} r_k,
\end{aligned}
\tag{8}
$$

where H is the imagination time horizon and γ is the reward discount. The actor model is optimized to maximize the value estimation, while the value model is optimized to approximate the expected imagined rewards. The training target for the value model is:

$$
V_t = r_t + \gamma \begin{cases} (1-\lambda)v_\psi(e_{t+1}) + \lambda V_{t+1} & \text{if} \quad t < H, \\ v_\psi(e_H) & \text{if} \quad t = H, \end{cases}
\tag{9}
$$

where λ equals to 0.95. Similar to the process of behavior learning, we also utilize the action replay model G_θ to draw action from the actor model during policy deployment. As shown in Lines 20–21 in Algorithm 1, the action guidance is dependent on current states e_i and the source task label with the highest domain-similarity weights, which may evolve over time.

5 Experiments

5.1 Experimental Setup

Benchmarks. We evaluate Vid2Act on three visual RL environments in an offline-to-online domain transfer setup:

- **Meta-World [40]:** It simulates 50 manipulation tasks, all involving the same robotic arm. We collect 6 offline datasets using expert experiences from *button press topdown, door open, drawer close, peg insert side, pick place,* and *push.* Each of them contains 10 demonstrations.
- **DeepMind Control Suite [32]:** It is a standard benchmark for visual-based RL that contains a diverse set of continuous control tasks. We collect offline datasets from 4 tasks, *i.e., cheetah run, hopper stand, walker walk,* and *walker run.* Each task contains 50 trajectories of expert experiences.
- **CARLA [5]:** It is an open-source simulator that provides more intricate and lifelike visual observations for research in autonomous driving. The objective of the agent is to maximize its driving distance within 1000 time steps while

avoiding collisions with 30 other moving vehicles or barriers. As a result, the episode length is 1000 steps with the action repeat of 4. To encourage highway progression and penalise collisions, the reward is formulated as: $r_t = v_{ego}^T \hat{u}_h \cdot \Delta t - \xi_1 \cdot \mathbb{I} - \xi_2 \cdot |steer|$, where v_{ego} represents the velocity vector of the ego-vehicle, projected onto the highway's unit vector $\hat{u}_h$, and multiplied by time discretization $\Delta t = 0.05$ to measure highway progression in meters. The impulse $\mathbb{I} \in \mathbb{R}^+$ indicates the impact caused by collisions, and a steering penalty $steer \in [-1, 1]$ aids in maintaining lane position. The visualization samples of four towns utilized in our experiments are shown in Fig. 3.

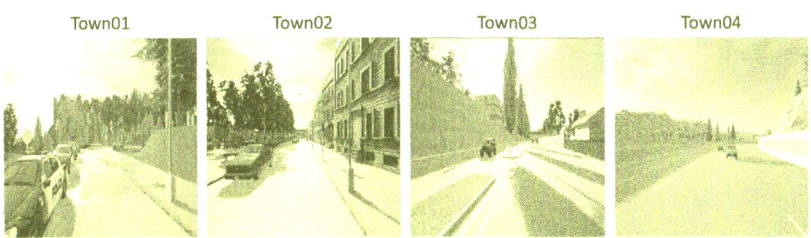

Fig. 3. Showcases of selected towns in CARLA environment.

Compared Methods. We compare Vid2Act with the following approaches:

- **DreamerV2** [12]: A model-based RL method that learns the policy directly from latent states in the world model. The latent representation enables agents to imagine thousands of trajectories simultaneously.
- **APV** [27]: A model-based RL method that stacks an action-conditional RSSM model on top of the pretrained action-free RSSM model. We train this model by following its two-step training setting.
- **Iso-Dream** [24]: A strong baseline for visual RL that learns different dynamics based on controllability. It rolls out noncontrollable states into the future and performs policy optimization based on the decoupled latent imaginations.
- **SMART** [30]: A generic multi-task pretraining framework that designs a Control Transformer coupled with a control-centric pretraining objective in a self-supervised manner.
- **TD-MPC2** [14]: A model-based RL method that primarily uses state information to learn task-oriented latent dynamics model purely from rewards, ignoring nuances unnecessary for the task at hand.

It is reasonable to compare our method with APV and Iso-Dream, as they are also built upon DreamerV2. Furthermore, our proposed transfer RL techniques can also be seamlessly integrated with DreamerV3 [13], enhancing its overall performance. In this paper, our method is based on DreamerV2 unless otherwise specified.

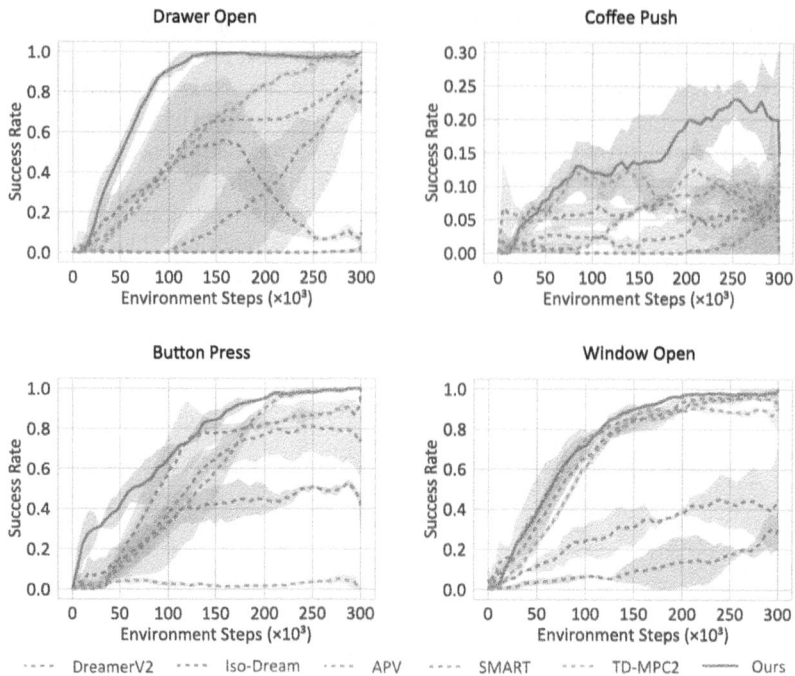

Fig. 4. Performance comparison with the state-of-the-art methods on Meta-World as measured on the success rate. Vid2Act outperforms the compared models.

5.2   Main Results

Meta-World. We first pretrain the teacher model of the action-conditioned video prediction model by minimizing the objective in Eq. (2) for $200K$ gradient steps. The hyperparameters β and α are set to 1 in Eq. (7). Our model is evaluated in 4 tasks, *i.e.*, *drawer open, coffee push, button press*, and *window open*. In all tasks, the episode length is 500 steps without any action repeat. The number of environment steps is limited to $300K$. We run all tasks with 3 seeds and report the mean success rate and standard deviations of 10 episodes. As shown in Fig. 4, our Vid2Act generally outperforms other methods on four tasks. Specifically, we improve DreamerV2 20% in *drawer open* and 90% in *coffee push*. TD-MPC2, despite its ability to handle state information effectively, exhibits weaker performance than our model when processing visual image inputs.

DeepMind Control Suite. In this environment, the episode length is 1,000 steps with the action repeat of 2, and the reward ranges from 0 to 1. For the online target tasks, we train our method for $200K$ iterations, which results in $400K$ environment steps. We evaluate Vid2Act with baselines on the mean episode rewards and standard deviations. The results of *quadruped walk* and *quadruped run* are illustrated in Fig. 5. Our framework achieves significant improvements compared with existing model-based RL approaches. For example, Vid2Act

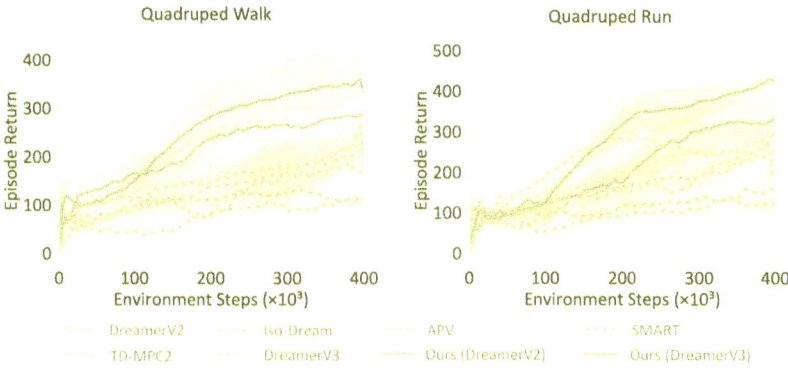

Fig. 5. Performance comparison on two tasks from DeepMind Control Suite as measured on the episode rewards. Our Vid2Act with dynamic knowledge distillation achieves significant improvements compared with existing model-based RL approaches.

performs nearly 100 higher performance on the task of *quadruped walk* and *quadruped run* than DreamerV2 after 400k steps environment interactions. Iso-Dream, which serves as a robust baseline for addressing visual control tasks through isolated state transition branches, exhibits limitations in handling these two tasks. Compared with APV, which only uses the pretrained action-free world model as initialization to train downstream tasks, Vid2Act is encouraged to learn more precise state transitions based on action input and more useful source dynamics based on domain-selective knowledge distillation. Moreover, the learned domain selection weights help the agent adaptively transfer potentially useful action demonstrations from offline datasets. In addition, we utilize DreamerV3 as the network backbone and observe that our proposed techniques can be seamlessly integrated with DreamerV2/V3 and consistently enhance their performance.

5.3   Ablation Studies

We conduct ablation studies to confirm the validity of learning a set of time-varying domain selection weights and behavior learning with action replay on two tasks, as shown in Fig. 6. Without the process of learning the importance weights (orange) to measure the similarity between source and target tasks, the performance of our model has decreased by about 25% in *button press*, and it requires more timesteps to improve the behavior policy in *drawer open*. It demonstrates that information in different source tasks has different impacts on the target task, and a domain-selective knowledge distillation loss with importance weights encourages the student model to adaptively find useful prior knowledge and transfer it to help the dynamics learning in downstream tasks. Moreover, we evaluate Vid2Act without action replay model for behavior learning (green). The result shows that our proposed domain-selective behavior learning strategy

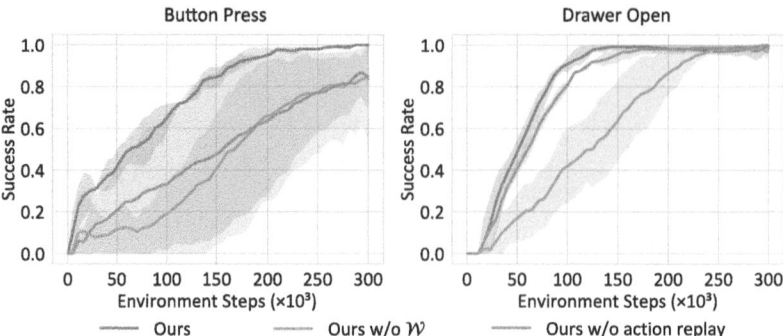

Fig. 6. Ablations of Vid2Act that illustrate the impact of learning time-varying domain selection weights and optimizing behavior learning with action replay. (Color figure online)

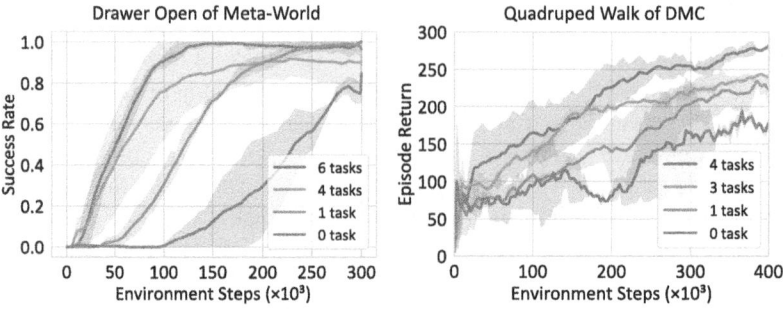

Fig. 7. Analyses on the impact of different source task configurations. Compared with DreamerV2 (0 source task) which is exclusively trained on the target task, our method consistently achieves positive offline-to-online transfer even when it has access to only one task with the lowest importance weights (1 task). (Color figure online)

can identify potentially valuable source actions and employ them as exemplar guidance for the target policy.

5.4   Analyses of Task Relations

Impact of Fewer or Less-relevant Source Tasks. To analyze the robustness of our approach to different source domain configurations, we sequentially decrease the number of source domain tasks according to the learned importance weights, *i.e.*, gradually removing the task with the highest importance weight. The results are shown in Fig. 7. We have two observations in this figure. First, compared with the baseline model that is solely trained on the target task, our approach consistently achieves positive offline-to-online transfer even when it can only access parts of the source datasets with lower importance weights. Second,

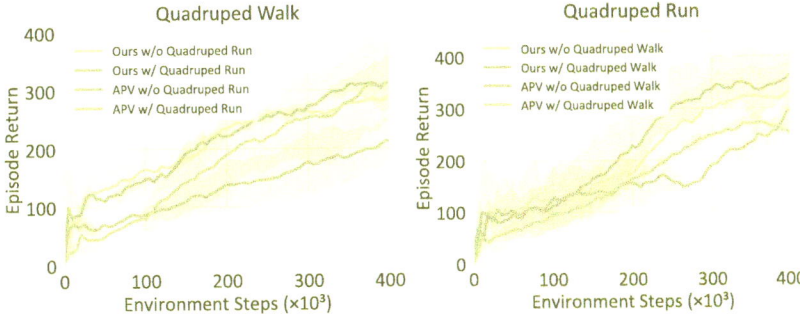

Fig. 8. Analyses on the impact of same dynamics between offline source tasks and target task. Our model shows more stable performance compared with APV, eliminating the reliance on similar physical dynamics across domains.

as the number of the source tasks grows, the performance of Vid2Act improves as well, demonstrating its effectiveness in identifying task similarity and improving the target policy with the expanded offline datasets.

Impact of Various Dynamics between Source/Target Tasks. Furthermore, we use a setup where the underlying dynamics of the target task are already seen in the source domain, but the task is different from the source tasks, *i.e.*, the reward functions are different. Specifically, we add the task of *quadruped walk* (*quadruped run*) to the offline dataset and then transfer the knowledge to the task of *quadruped run* (*quadruped walk*). In Fig. 8, our model shows superior performance, regardless of the presence of similar dynamics between the source and target domains. In contrast, APV is unstable and depends heavily on the similarity of physical dynamics across domains, such as *quadruped walk*.

Changes in Domain Selection Weights. In Fig. 9, we show the weights of different source tasks during the training phase. For example, in the online *button press* task, as the training progresses, the weight of *button press todown* in source tasks increases and then becomes dominant. This shows that our model can dynamically transfer knowledge in an adaptive manner.

5.5   Results on CARLA Environment

We also demonstrate the performance of Vid2Act in CARLA. In our experiments, we use the expert datasets collected from three distinct maps, *i.e.*, "Town01", "Town02", and "Town03", and evaluate our model in a first-person highway driving task in "Town04". The visualization samples can be found in the supplementary materials. We employ Iso-Dream, a model demonstrated effective in the CARLA environment, instead of Dreamerv2 as our network backbone. Our method is trained for $75K$ iterations, resulting in $300K$ environment steps. The results are shown in Fig. 10. Compared with APV, which also uses offline datasets to pretrain, our model presents a remarkable advantage. Comparing the

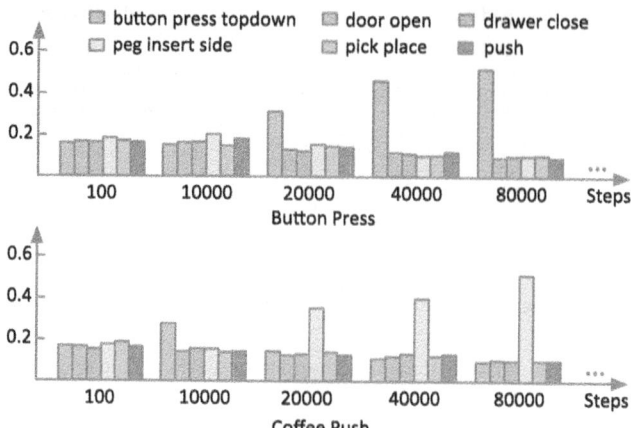

Fig. 9. Weight distribution of different source tasks during the training phase.

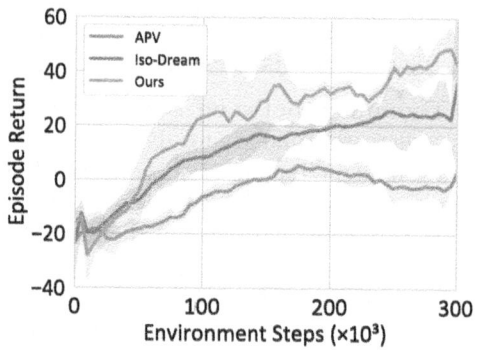

Fig. 10. Performance comparison in CARLA environment as measured on the episode rewards. Our Vid2Act can improve the performance of Iso-Dream.

orange and blue curves, we see that our framework can improve the performance of Iso-Dream.

5.6   Results with Medium Offline Data

In this section, we further use the "medium" datasets as the source domains to verify the generalization and robustness of our Vid2Act. The medium datasets are generated by first training a policy online using DreamerV2, early-stopping the training, and collecting 200 episodes from this partially-trained policy for each task. We collect 6 offline datasets in *door close, faucet open, handle press, plate slide, reach wall,* and *window close.* Our model is compared with the methods that also use offline datasets for pretraining the model, and the results are shown in Table 1. We can observe that our Vid2Act presents a remarkable advantage against other methods in terms of both success rate and episode return.

Table 1. Performance comparison, measured by the success rate and episode return, with baselines on the Meta-World environment using "medium" datasets as source domains.

Methods	DreamerV2	APV	SMART	Ours
Success Rate				
Coffee Push	0.20 ± 0.19	0.18 ± 0.15	0 ± 0	$\mathbf{0.35 \pm 0.15}$
Drawer Open	0.62 ± 0.44	0.63 ± 0.37	0 ± 0	$\mathbf{0.95 \pm 0.08}$
Episode Return				
Coffee Push	328 ± 370	253 ± 789	14 ± 7	$\mathbf{561 \pm 261}$
Drawer Open	3726 ± 715	3942 ± 762	664 ± 116	$\mathbf{4276 \pm 501}$

In *coffee push*, Vid2Act improves DreamerV2 by around **75%** ($0.20 \to 0.35$) in success rate and by over **70%** ($328 \to 561$) in episode return. The results demonstrate that our model is not sensitive to the quality of source domain data, as it can achieve impressive performance even with medium datasets.

6 Conclusion

In this paper, we proposed a new domain-selective transfer learning framework called Vid2Act that improves visual RL with offline datasets with multiple tasks. Vid2Act has two contributions. First, it provides a novel model-based pretraining and transfer learning pipeline for visual RL. Unlike APV [27], it transfers action-conditioned dynamics from multiple source tasks with a set of importance weights learned by the world models. Second, it provides a novel domain-selective behavior learning strategy that identifies potentially valuable source actions and employs them as exemplar guidance for the target policy. Experiments in the Meta-World, DeepMind Control and CARLA environments demonstrated that Vid2Act significantly outperforms existing visual RL approaches.

Our Vid2Act has a limitation in the time consumption during offline pretraining, as the utilization of a mixture world model to simultaneously learn the dynamics of multiple tasks requires a significant amount of time for the model to converge.

Acknowledgment. This work was supported by the National Natural Science Foundation of China (Grant No. 62250062, 62106144), the Shanghai Municipal Science and Technology Major Project (Grant No. 2021SHZDZX0102), the Fundamental Research Funds for the Central Universities, and the CCF-Tencent Rhino-Bird Open Research Fund.

Disclosure of Interests. The authors have no competing interests to declare that are relevant to the content of this article.

References

1. Anand, A., Racah, E., Ozair, S., Bengio, Y., Côté, M.A., Hjelm, R.D.: Unsupervised state representation learning in Atari. In: NeurIPS, vol. 32 (2019)
2. Chen, L., et al.: Decision transformer: reinforcement learning via sequence modeling. In: NeurIPS, vol. 34, pp. 15084–15097 (2021)
3. Cho, D., Shim, D., Kim, H.J.: S2p: state-conditioned image synthesis for data augmentation in offline reinforcement learning. In: NeurIPS (2022)
4. Choudhary, R., Walambe, R., Kotecha, K.: Spatial and temporal features unified self-supervised representation learning networks. Robot. Auton. Syst. **157**, 104256 (2022)
5. Dosovitskiy, A., Ros, G., Codevilla, F., López, A.M., Koltun, V.: CARLA: an open urban driving simulator. In: CoRL, vol. 78, pp. 1–16 (2017)
6. Dwibedi, D., Tompson, J., Lynch, C., Sermanet, P.: Learning actionable representations from visual observations. In: IROS, pp. 1577–1584. IEEE (2018)
7. Ebert, F., Finn, C., Dasari, S., Xie, A., Lee, A., Levine, S.: Visual foresight: model-based deep reinforcement learning for vision-based robotic control. arXiv preprint arXiv:1812.00568 (2018)
8. Fujimoto, S., Meger, D., Precup, D.: Off-policy deep reinforcement learning without exploration. In: ICML, pp. 2052–2062. PMLR (2019)
9. Gelada, C., Kumar, S., Buckman, J., Nachum, O., Bellemare, M.G.: DeepMDP: learning continuous latent space models for representation learning. In: ICML, pp. 2170–2179. PMLR (2019)
10. Hafner, D., Lillicrap, T., Ba, J., Norouzi, M.: Dream to control: learning behaviors by latent imagination. In: ICLR (2020)
11. Hafner, D., et al.: Learning latent dynamics for planning from pixels. In: ICML, pp. 2555–2565. PMLR (2019)
12. Hafner, D., Lillicrap, T., Norouzi, M., Ba, J.: Mastering Atari with discrete world models. In: ICLR (2021)
13. Hafner, D., Pasukonis, J., Ba, J., Lillicrap, T.: Mastering diverse domains through world models. arXiv preprint arXiv:2301.04104 (2023)
14. Hansen, N., Su, H., Wang, X.: Td-mpc2: scalable, robust world models for continuous control. arXiv preprint arXiv:2310.16828 (2023)
15. Hester, T., et al.: Deep q-learning from demonstrations. In: AAAI (2018)
16. Kadokawa, Y., Zhu, L., Tsurumine, Y., Matsubara, T.: Cyclic policy distillation: sample-efficient sim-to-real reinforcement learning with domain randomization. Robot. Auton. Syst. **165**, 104425 (2023)
17. Kaiser, L., et al.: Model-based reinforcement learning for Atari. In: ICLR (2019)
18. Kingma, D.P., Welling, M.: Auto-encoding variational bayes. arXiv preprint arXiv:1312.6114 (2013)
19. Laskin, M., Srinivas, A., Abbeel, P.: CURL: contrastive unsupervised representations for reinforcement learning. In: ICML, vol. 119, pp. 5639–5650. PMLR (2020)
20. Laskin, M., Lee, K., Stooke, A., Pinto, L., Abbeel, P., Srinivas, A.: Reinforcement learning with augmented data. In: NeurIPS, vol. 33, pp. 19884–19895 (2020)
21. Li, D., Wang, S., Chen, K., Li, B.: Contrastive inductive bias controlling networks for reinforcement learning. In: ACML, pp. 563–578. PMLR (2023)
22. Liu, I.J., Peng, J., Schwing, A.G.: Knowledge flow: improve upon your teachers. In: ICLR (2019)
23. Nair, S., Rajeswaran, A., Kumar, V., Finn, C., Gupta, A.: R3m: a universal visual representation for robot manipulation. arXiv preprint arXiv:2203.12601 (2022)

24. Pan, M., Zhu, X., Wang, Y., Yang, X.: Iso-dream: isolating and leveraging noncontrollable visual dynamics in world models. In: NeurIPS, vol. 35, pp. 23178–23191 (2022)
25. Schwarzer, M., et al.: Pretraining representations for data-efficient reinforcement learning. In: NeurIPS, vol. 34, pp. 12686–12699 (2021)
26. Sekar, R., Rybkin, O., Daniilidis, K., Abbeel, P., Hafner, D., Pathak, D.: Planning to explore via self-supervised world models. In: ICML, pp. 8583–8592 (2020)
27. Seo, Y., Lee, K., James, S.L., Abbeel, P.: Reinforcement learning with action-free pre-training from videos. In: ICML, pp. 19561–19579. PMLR (2022)
28. Sohn, K., Lee, H., Yan, X.: Learning structured output representation using deep conditional generative models. In: NeurIPS, vol. 28 (2015)
29. Stooke, A., Lee, K., Abbeel, P., Laskin, M.: Decoupling representation learning from reinforcement learning. In: ICML, pp. 9870–9879. PMLR (2021)
30. Sun, Y., Ma, S., Madaan, R., Bonatti, R., Huang, F., Kapoor, A.: Smart: self-supervised multi-task pretraining with control transformers. In: ICLR (2023)
31. Taiga, A.A., Agarwal, R., Farebrother, J., Courville, A., Bellemare, M.G.: Investigating multi-task pretraining and generalization in reinforcement learning. In: ICLR (2023)
32. Tassa, Y., et al.: Deepmind control suite. arXiv preprint arXiv:1801.00690 (2018)
33. Xie, Z., Lin, Z., Ye, D., Fu, Q., Wei, Y., Li, S.: Future-conditioned unsupervised pretraining for decision transformer. In: ICML, pp. 38187–38203. PMLR (2023)
34. Xu, Y., et al.: On the feasibility of cross-task transfer with model-based reinforcement learning. In: ICLR (2023)
35. Yang, H., et al.: Self-supervised representations for multi-view reinforcement learning. In: UAI (2022)
36. Yang, M., Nachum, O.: Representation matters: offline pretraining for sequential decision making. In: ICML, pp. 11784–11794. PMLR (2021)
37. Yao, Z., Wang, Y., Long, M., Wang, J.: Unsupervised transfer learning for spatiotemporal predictive networks. In: ICML, pp. 10778–10788. PMLR (2020)
38. Yarats, D., Zhang, A., Kostrikov, I., Amos, B., Pineau, J., Fergus, R.: Improving sample efficiency in model-free reinforcement learning from images. In: AAAI, pp. 10674–10681 (2021)
39. Ye, W., Liu, S., Kurutach, T., Abbeel, P., Gao, Y.: Mastering Atari games with limited data. In: NeurIPS (2021)
40. Yu, T., et al.: Meta-world: a benchmark and evaluation for multi-task and meta reinforcement learning. In: CoRL, pp. 1094–1100. PMLR (2020)
41. Ze, Y., Hansen, N., Chen, Y., Jain, M., Wang, X.: Visual reinforcement learning with self-supervised 3d representations. IEEE Robot. Autom. Lett. **8**(5), 2890–2897 (2023)

Advancing Graph Counterfactual Fairness Through Fair Representation Learning

Zichong Wang[1], Zhibo Chu[1], Ronald Blanco[1], Zhong Chen[2], Shu-Ching Chen[3], and Wenbin Zhang[1(✉)]

[1] Florida International University, Miami, USA
{ziwang,wenbin.zhang}@fiu.edu
[2] Southern Illinois University, Carbondale, USA
[3] University of Missouri-Kansas City, Kansas City, USA

Abstract. Graph neural networks (GNNs) have shown remarkable success in various domains. Nonetheless, studies have shown that GNNs may inherit and amplify societal bias, which critically hinders their application in high-stakes scenarios. Although efforts have been exerted to enhance the fairness of GNNs, most of them rely on the statistical fairness notion, which assumes that biases arise solely from sensitive attributes, neglecting the pervasive issue of labeling bias prevalent in real-world scenarios. To this end, recent works extend counterfactual fairness in graph data to address label bias, but they neglect the graph structure bias, where nodes sharing sensitive attributes tend to connect more closely. To bridge these gaps, we propose a novel GNN framework, Fair Disentangled GNN (FDGNN), designed to mitigate multi-sources biases to enhance the fairness of GNNs while preserving task-related information via fair node representation learning. Specifically, FDGNN initiates by mitigating graph structure bias by ensuring consistent representation of different subgroups. Subsequently, to achieve fair node representation, identified counterfactual instances are utilized as guides for disentangling a node's representation and eliminating sensitive attribute-related information via a de-identifiable sensitive attribute mechanism. Extensive experiments on multiple real-world graph datasets demonstrate the superiority of FDGNN in graph fairness compared to other state-of-the-art methods while achieving comparable utility performance.

Keywords: GNNs · Counterfactual fairness · Fair representation

1 Introduction

Graph neural networks (GNNs) have emerged as a powerful tool for learning node representation from graph-structured data, which are employed in various domains such as recommendation systems [10], social network analysis [13], and online advertisement [34]. Generally, GNNs adopt a message-passing mechanism (MP) [32], aggregating local neighborhood information for every node in each

A. Bifet et al. (Eds.): ECML PKDD 2024, LNAI 14947, pp. 40–58, 2024.
https://doi.org/10.1007/978-3-031-70368-3_3

layer. This aggregation process effectively renders the distinction between similar and dissimilar nodes while preserving node attributes and graph structure information, thereby enhancing the performance of downstream graph tasks [1]. Despite these successes, GNNs may make discriminatory predictions for subgroups defined by *sensitive attributes* (*e.g.*, gender or race) due to biases inherited from training data and further amplified by their message-passing mechanism. Such biased predictions give rise to ethical and societal concerns, which severely limits the adoption of GNNs in high-stake decision-making scenarios, such as job screening [22], healthcare [40] and criminal prediction [16]. For instance, a bank's loan decision-making process is influenced by the race information of the applicant and their close contacts, constituting a serious ethical problem [21,42,43].

To this end, many efforts have been taken towards fair GNNs [28]. Among them, most existing fairness work utilizes statistical fairness notions to evaluate and address bias in node representation learning on graphs, which highlights algorithmic decisions should equally treat subgroups or individuals, with these methods primarily focusing on *sensitive attributes* (*e.g.*, race or gender) as the only source of bias [22]. However, these strategies cannot quantify and mitigate labeling bias which arises when societal biases, prejudices, or discriminatory practices skew the data collection process [23]. This distortion introduces systemic biases into the training dataset, which GNNs may then learn and perpetuate, exacerbating the bias against the *deprived subgroups* (*e.g.*, female) [21].

To this end, recent research has incorporated counterfactual fairness into graph learning, aiming to address the model's bias from a causal perspective [27]. Typically, these approaches fall into two categories: generation of counterfactual instances based on real sample distributions or identification of potential counterfactual instances within the dataset. For example, GEAR [21] employs GraphVAE [24] to generate counterfactuals aimed at minimizing the disparity between original and counterfactual node representation to eliminate the impact of sensitive attributes. On the other hand, RFCGNN [27] aims to identify corresponding counterfactual instances directly from the representation space and learn disentangled representations, thereby removing sensitive attribute-related information to enhance fairness. A significant limitation of these approaches is neglecting the intricate interplay between sensitive attribute-related information and task-related information. Specifically, they aim to eliminate the sensitive attribute information to force GNNs to make decisions independent of the sensitive attribute, which inadvertently leads to the unintentional removal of the task-related information due to its correlations with the sensitive attribute.

Furthermore, these methods often overlook the graph structure bias present in the graph data, where nodes sharing the same sensitive attributes are likely to be connected [30]. Specifically, GNNs aggregate each node's neighboring node information and its own features to obtain a final node representation. However, the disparity in the distribution of neighboring nodes of the target node can lead to an over-association of node representation with sensitive attributes. This results in the obtained counterfactual instances being too tightly connected to neighboring nodes with the same sensitive attributes, resulting in inaccurate counterfactual scenarios.

In this paper, we investigate counterfactual fairness to mitigate the root causes of bias, focusing on the potential causal interactions between each node and its neighboring nodes. While great progress has been made in the field, the application of counterfactual fairness to graphs faces distinctive challenges due to fundamental obstacles as follows. **1) Complexity of Counterfactual Graph Data Structures:** Unlike tabular data, graph-structured data contains node features and graph structure information. Thus, given the complexity of these relationships, in counterfactual scenarios, it is imperative to consider the implications of sensitive attribute flipping not only on the target node features but also on its connectivity with neighboring nodes. **2) Mitigating Bias in Node Representations:** To achieve fairness in GNNs, it is essential to mitigate bias while preserving model performance, which requires reasonably handling task-related information that is also associated with the sensitive attribute. This involves disentangling node representations to isolate sensitive attributes related information effectively, thereby ensuring the retention of valuable task-related information. **3) Obtaining Accurate Counterfactual Scenarios:** The essence of counterfactual fairness hinges on accurate counterfactual scenarios. Existing fairness works often overlook the graph structure bias, leading to the derivation of inaccurate counterfactual instances. An effective strategy is thus required to mitigate the association of learned representations with sensitive attributes while maintaining important information.

In order to address all the above-mentioned challenges, this paper proposes a novel framework named *Fair Disentangled Graph neural networks* (FDGNN), which aims to learn fair node representation while preserving task-related information. *To the best of our knowledge, this is the first work that utilizes authentic counterfactual samples to learn disentangled node representation to mitigate the multi-source biases from sensitive attributes, graph structure, and the labeling process collectively.* Specifically, we conduct a comprehensive causal analysis of both original and counterfactual instances, establishing a set of constraints that foster the learning of disentangled representations. This strategy effectively diminishes the associations between sensitive attributes and unrelated representation dimensions. Moreover, by imposing fairness constraints on components associated with sensitive attributes, FDGNN minimizes the influence of the sensitive attribute-related information on other representation channels. This approach prevents unnecessary task-related information loss, leading to a more balanced and effective model. The main contributions are as follows:

- **A novel graph causal model.** We introduce a novel causal formulation that paves the way for understanding the generation process of graph structures and the fair learning task of node representation.
- **A novel framework for mitigating graph-structured data bias via counterfactual instance.** We propose FDGNN, a fair graph representation learning framework that utilizes accurate counterfactual instances to mitigate multi-source biases, including sensitive attributes, graph structure, and

the labeling process. In addition, FDGNN preserves task-relevant information associated with sensitive attributes by effectively disentangling sensitive attributes. This approach enables our model to enhance fairness without compromising performance.

– **Extensive experiments are conducted to evaluate our proposed approach.** We conduct extensive experiments on three real-world datasets and five evaluative metrics, the results show that FDGNN acquires superior performance and significantly enhances fairness compared with baselines.

The organization of this paper is as follows: An overview of the relevant literature is provided in Sect. 2. Notations are presented in Sect. 3. Our proposed method is detailed in Sect. 4. Section 5 describes the experimental framework and discusses the experiment results. Lastly, Sect. 6 concludes the paper.

2 Related Work

2.1 Graph Neural Networks

Graph Neural Networks have shown great ability in representation learning on graph-structured data and have been used in a variety of tasks such as node classification [18], graph classification [25], and link prediction [32]. Their notable success across these diverse tasks has propelled GNNs into the forefront of research and application, extending their utility into critical decision-making systems [27]. For instance, financial institutions increasingly rely on GNNs to evaluate credit card applications or make loan approval decisions [29]. However, the application in critical decision-making systems places higher demand for GNNs to not only be effective but also fair and interpretable [38]. In this context, there is a trend for the research community to design fairer GNNs to mitigate biases and ensure fair outcomes in graph-based tasks [28].

2.2 Fairness in Graph

Fairness in the graph has received intensive attention [4,5,8,31,37,39]. Most existing fair graph learning works are based on statistical fairness notation, including individual fairness [17,26,41] and group fairness [6,7,40], aiming to ensure fair GNN predictions. While these approaches have achieved notable success, their focus on correlation metrics often renders them ineffective at addressing biases introduced by statistical anomalies. To address this limitation, counterfactual fairness [19] leverages the causal perspective to measure and eliminate the root bias. For example, NIFTY [1] generates counterfactual instances by directly flipping the sensitive attributes of nodes to enhance the consistency between original and counterfactual representations. Similarly, GEAR [21] employs GraphVAE [24] to generate counterfactuals, focusing on minimizing the difference between representations derived from the original and counterfactuals. Furthermore, RFCGNN [27] identifies counterfactuals within the existing representation space to learn fair representation.

Our work is distinct from existing works in that it: i) employs disentangled representation learning to preserve essential task-relevant information, minimizing performance loss; and ii) pays attention to fundamental yet neglected graph structural bias.

3   Notations

Let $\mathcal{G} = \{\mathcal{V}, \mathcal{E}, \mathbf{X}\}$ denote an undirected attributed graph, comprised of a set of $\mathcal{V} = \{v_1, v_2, \ldots, v_n\}$ nodes and a set of $\mathcal{E} \subseteq \mathcal{V} \times \mathcal{V}$ edges. $\mathbf{X} \in \mathbb{R}^{n \times d}$ represents the node feature matrix with the i-th row of $\mathbf{X}$, i.e., $\mathbf{X}_{i,:}$ as node feature of v_i with d being the dimension of node features. The adjacency matrix $\mathbf{A} \in \{0,1\}^{n \times n}$ encapsulates the graph structure information, where $\mathbf{A}_{i,j} = 1$ indicates that there exists edge $e_{ij} \in \mathcal{E}$ between the node v_i and v_j, and $\mathbf{A}_{i,j} = 0$ otherwise. Meanwhile, in this work, we focus on binary sensitive attributes and binary node classification tasks. Each node v_i has a sensitive attribute $s_i \in \{0,1\}$, where $s_i = 0$ indicates that node v_i belongs to the deprived group $S_0 = \{\forall\, v_i : v_i \in \mathcal{V} \wedge s_i = 0\}$; if $s_i = 1$, v_i belongs to the *favored group* $S_1 = \{\forall\, v_i : v_i \in \mathcal{V} \wedge s_i = 1\}$. It is important to note that the sensitive attribute s_i is incorporated within the feature vector $\mathbf{X}_{i,:}$ of each node. In addition, we let $\mathcal{L}$ denote the set of labeled vertices, and let Y denote the corresponding set of ground-truth labels.

4   Methodology

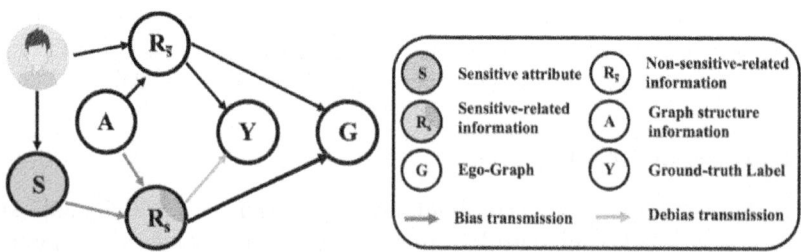

Fig. 1. The causal model of FDGNN with the red color denoting sensitive related information and white color representing non-sensitive related information, while green color is task-related information that is also related to the sensitive attribute.

4.1   Causal Model

This section introduces the proposed causal model, which is pivotal for examining counterfactual scenarios, i.e., querying outcomes in a counterfactual world under certain conditions were altered. To address multi-source biases, a scenario that exposes the limitations of the fairness notions solely based on statistics, a Structural Causal Model (SCM) is constructed from the observed graph, as depicted in Fig. 1. Specifically, SCM encapsulates causal relationships among five key variables: sensitive attribute (S), ground-truth label (Y), the graph structure (A), ego-graph (G), and information-related (R_S) or unrelated $(R_{\overline{S}})$ to sensitive attribute (S). In SCM, every connection denotes a deterministic causal link between variables, with the reasoning and explanations outlined as follows:

- $S \to R_S$: This link denotes that the node representation learned by the GNNs is influenced by sensitive attribute (S), thereby introducing bias into the final node representation. To enhance model fairness, we need to accurately identify R_S in the node representation, thus paving the way for mitigating bias in subsequent processes.
- $R_S \leftarrow A \to R_{\overline{S}}$: A impacts R_S and $R_{\overline{S}}$. For example, the connection between two nodes might stem from sensitive attribute-influenced interactions (*e.g.*, two individuals sharing the same neighborhood) or from non-sensitive factors (*e.g.*, common interests in activities such as soccer).
- $R_S \perp\!\!\!\perp R_{\overline{S}}$: To effectively address biases while minimizing the impact on performance, it's imperative to disentangle and isolate R_S from $R_{\overline{S}}$. This separation ensures that only R_S is adjusted to ensure fairness without unnecessarily compromising the information crucial for predicting Y.
- $R_S \to Y \leftarrow R_{\overline{S}}$: This model structure guarantees that both R_S and $R_{\overline{S}}$ influence the prediction of Y. The objective is to carefully modulate the impact of R_S to mitigate bias, *i.e.*, minimizing the sensitive information represented by red in R_S, while concurrently maintaining task-related information encapsulated within both R_S (*e.g.*, represented by   ) and $R_{\overline{S}}$.
- $R_S \to G \leftarrow R_{\overline{S}}$: Same as the above substructure, but from a graph structure perspective, both R_S and $R_{\overline{S}}$ have direct causal effects on G, ensuring that it an accurate reconstruction of the ego-graph.

4.2   Framework Overview

Building upon the proposed causal model, a novel framework is designed to enhance the fairness of GNNs. This framework initially identifies accurate counterfactual instances from the existing samples. Subsequently, it utilizes de-identifiable sensitive attribute mechanisms to preserve task-relevant information while eliminating biased information from node representation. Figure 2 presents the overview of FDGNN, which incorporates three major phases. First, the Fair Ego-graph Generation Module aims to generate a subgraph for each node that contains important neighboring nodes while ensuring a fair and consistent representation of different subgroups. Second, the Counterfactual Data Augmentation Module finds accurate counterfactual instances to facilitate subsequent disentanglement learning. Last, the Fair Disentangled Representation Learning Module aims to perform sensitive information decomposition in node representation to keep task-relevant information while removing biased information through de-identifiable sensitive attributes. Each of these components will be introduced in the following sections.

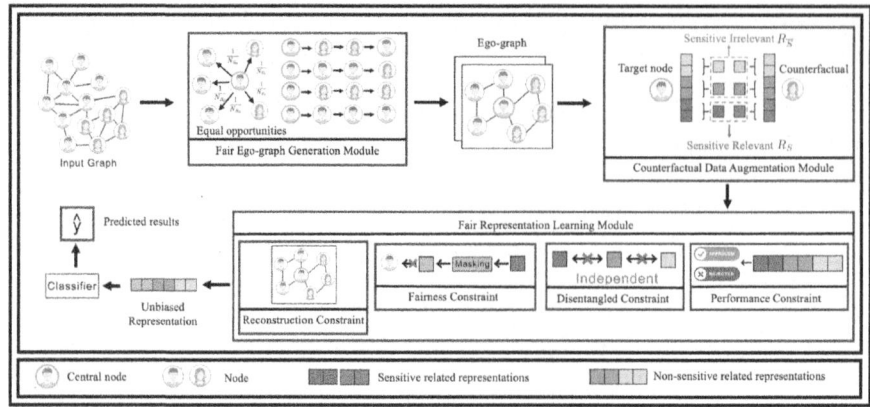

Fig. 2. Overview of the proposed FDGNN framework.

4.3 Fair Ego-Graph Generation Module

The inherent complexity of graph data poses a computational challenge to directly constructing causal models, especially for large-scale networks such as social networks. To this end, most existing methods aim to extract an ego graph for each node. This strategy is based on the local dependency assumption, *i.e.*, a node is primarily influenced by its nearest neighbors [15]. Despite the efficiency, it overlooks the critical aspect of local fairness within each node's ego graph. Specifically, the existing work may result in a biased ego graph, where an ego graph disproportionately consists of nodes sharing the same sensitive attribute. Such disparity in neighbor node distribution can result in over-association of the learned representation with sensitive attributes. In response, a Fair Ego-graph Generation Module is introduced to foster equitable representation across different subgroups within each ego-graph ($\mathcal{G}_{v_i}$) while avoiding limiting the distance of neighboring nodes, which can lead to the loss of important neighboring nodes. To achieve this, the concept of a *Related Score* (RS) for each node pair is introduced, inspired by PageRank [14], to quantify the relevance of node v_j to node v_i. Mathematically, it is represented as:

$$RS = \xi(I - (1 - \xi)\widetilde{A}) \tag{1}$$

where $\xi \in [0, 1]$ represents a parameter that controls the probability of a random walk restarting at the central node, while I is the identity matrix, and $\widetilde{A} = \mathbf{A}D^{-1}$ represents the transfer probability with D being the diagonal matrix where $D_{i,j} = \sum_j A_{i,j}$. Each entry of this matrix, denoted as $IS_{i,j}$, measures the relevance of node v_j to node v_i. Moreover, $IS_{i,:}$ denotes the vector of importance scores for node v_i.

However, PageRank, designed to assign uniform transfer probabilities to each neighboring node, can lead to biases, particularly in networks where nodes sharing the same sensitive attribute tend to form stronger connections, thereby skewing transitions toward these neighbors. To address this, a fairness constraint is introduced to adjust these probabilities, promoting equitable representation among nodes from different subgroups. This fairness constraint categorizes neighbors based on their sensitive attributes and then adjusts the selection probabilities to balance the representation of each group during node transitions. As illustrated in Fig. 2, this adjustment grants male and female nodes probabilities of $\frac{1}{N_{S_0}}$ and $\frac{1}{N_{S_1}}$, respectively, ensuring an even representation of both subgroups in the sampling outcome. Mathematically, it is represented as:

$$\sum (P_{v_j} | \overline{A}_{i,j} = 1, s_j \in S_1) = \sum (P_{v_u} | \overline{A}_{i,j} = 1, s_u \in S_0) \tag{2}$$

where P_{v_j} and P_{v_u} represent the transition probabilities to neighboring nodes belonging to the deprived and favored groups, respectively.

4.4   Counterfactual Data Augmentation Module

With the learned fair ego-graph, central to the proposed causal framework SCM (*c.f.*, Sect. 4.1) is the distinction between sensitive attribute-related node representation (R_S, illustrated as red or blue squares within the red box in this module in Fig. 2) and sensitive information irrelevant node representation ($R_{\overline{S}}$, depicted as green or yellow squares within the gray box in this module in Fig. 2). To ensure accurate dissociation of these representations, identifying accurate counterfactual instances is essential. Specifically, consider a node v_i characterized by a factual sensitive attribute s_i and a corresponding label y_i. When flipping its sensitive attribute to $1 - s_i$, the representation independent of sensitive attributes, $R_{\overline{S}}$, should remain consistent, while the representation associated with sensitive attributes, R_S, should adapt to reflect this change. This forms the counterfactual subgraph $\mathcal{G}_{v_i}^C$, expressed mathematically as:

$$\mathcal{G}_{v_i}^C = \min \sum_{m=1}^{M} \left(d(\mathcal{G}_{v_i}, \mathcal{G}_{v_j}^m) | y_i = y_j, s_i \neq s_j \right) \tag{3}$$

where $\mathcal{G} = \{\mathcal{G}_{v_i} | v_i \in \mathcal{V}\}$, v_j denotes the corresponding counterfactuals of v_i, and $d(\cdot)$ measures the distance between pairs of ego-graphs.

Existing methods for generating graph counterfactual samples, as discussed in Sect. 1, may obtain inaccurate counterfactual samples. Therefore, we aim to find potential candidate counterfactual instances with the observed factual graphs. This strategy avoids making assumptions about how graphs that include sensitive attributes are generated while obviating the necessity for additional supervised signals to select counterfactuals. However, computing pairwise distances between ego-graphs becomes highly inefficient and impractical Given the complexity of graph structures and the vast search space of graph data. To address this issue, we aim to measure distances in the representation space, leveraging the captured graph structure and node attribute information to enhance computational efficiency. The task in Eq. 3 in thus reformulated as:

$$\mathcal{G}_{v_i}^C = \min_{h_j \in H} \sum_{m=1}^{M} \left(\|h_i - h_j\|_2^2 | y_i = y_j, s_i \neq s_j \right) \tag{4}$$

where $H = \{h_i | v_i \in \mathcal{V}\}$ is learned representation matrix and the L2 distance is employed to calculate the distance between h_i and h_j. Note that for each v_i, a set of counterfactual samples is obtained instead of one sample. Consequently, the counterfactual $\mathcal{G}_{v_i}^{c_i}$ can naturally extend to a set of counterfactuals consisting of M samples $\{\mathcal{G}_{v_i}^{C_i} | i = 1, \ldots, M\}$, where M is a constant number.

4.5   Fair Disentangled Representation Learning Module

FDGNN is now prepared to disentangle $R_{\overline{S}}$ and R_S within the node representation space, guided by the identified counterfactual instances. Besides, given that both $R_{\overline{S}}$ and R_S contain critical information for downstream tasks, our strategy aims to obtain informative yet sensitive-irrelevant node representation. To this end, we aim to segregate sensitive related information into a distinct component of the node representation and subsequently dissociate the sensitive related information within that component. This methodology prevents unnecessary performance degradation linked to enforcing fairness constraints on sensitive-relevant components, thereby minimizing performance loss while enhancing fairness. To effectively implement this disentanglement, the following four specific constraints are introduced:

1) Disentangled Constraint ($\mathcal{L}_D$). This constraint ensures the independence of $R_{\overline{S}}$ and R_S, preventing information leakage between them. To achieve this, we disentanglement the node representation into c distinct channels, with each channel influenced by a unique latent factor K, ensuring that they operate independently. Notably, only one of these factors, K_i, is associated with the sensitive attributes S, thus effectively segregating sensitive attribute-related information from the overall node representation. To assess the impact of different node neighbors on these partitioned representations, we employ an adaptive encoder configured as a multilayer perceptron (MLP). Specifically, for any pair of nodes v_i and v_j, their attributes x_i and x_j are input into the adaptive encoder $(\boldsymbol{\rho}_{v_i,v_j} = F_\rho([x_i, x_j]))$ to evaluate the relevance of connection e_{ij} across c latent factors, where $\boldsymbol{\rho}_{v_i,v_j}$ is the vector of score indicates importance for e_{ij}, with $\rho_{v_i,v_j}^c \in \boldsymbol{\rho}_{v_i,v_j}$ representing scores for each latent factor c, and $F_\rho(\cdot)$ denoting the adaptive encoder operation. This score is normalized via a Softmax function to derive connection weights ω_{v_i,v_j}^c, as follows:

$$\omega_{v_i,v_j}^c = \text{Softmax}(\rho_{v_i,v_j}^c) \tag{5}$$

where ω_{v_i,v_j}^c represents the weight from node v_i to node v_j for channel c, indicating the likelihood that the connection is influenced by a latent factor c, with N_c reflecting the total number of channels.

Building on this, we further employ disentangled layers for graph convolution across multiple channels. Each disentangled layer comprises c channels of graph convolution, all sharing the same network architecture, with each channel dedicated to a specific latent factor. Initially, we reduce the dimensionality of the original node attributes by projecting these attributes into different subspaces, each corresponding to a latent factor. For any given node representation R_{v_i}, a linear layer is employed for dimensionality reduction, transforming the representation from R_{v_i}-dimensional to N_c-dimensional space. This reduction operation $F_R(\cdot)$ is independently applied N_c times to generate N_c reduced node attributes, corresponding to the different latent factors K_i. Consequently, the disentangled node representation of a node v_i at the lth layer, denoted by $h_{v_i}^l$, is formed by concatenating the reduced representations across all channels: $h_{v_i}^l = [r_{v_i,1}^l, r_{v_i,2}^l, \ldots, r_{v_i,N_c}^l]$. Extending this to all nodes, R^c represents the disentangled representations of all nodes within the cth channel. Thus, h^l symbolizes the aggregated disentangled representations across all channels at layer l: $h^l = [R_1^l, R_2^l, \ldots, R_{N_c}^l]$.

However, the above process primarily addresses disentanglement at the sample level, neglecting the independence among latent factors, especially mutual independence across different channels. As depicted in Fig. 2, the goal is to achieve zero correlation between distinct channel representations, such as the blue, green, and yellow squares, to truly enhance the disentanglement process. To achieve this, we propose the Independence Constraint, mathematically formulated as:

$$\mathcal{L}_I = \sum_{c_1=1}^{N_c} \sum_{c_2 \neq c_1}^{N_c} \frac{D(K_{c_1}, K_{c_2})}{Norm(K_{c_1}, K_{c_2})} \tag{6}$$

where $D(\cdot)$ denotes the distance covariance, and $Norm(\cdot)$ represents the normalization function.

With fully disentangled channels, the next step is to pinpoint the latent factors that correlate with sensitive information. Counterfactual instances serve as a pivotal guide in this process. Specifically, for a given counterfactual (CI_i), its non-sensitive representation $(R_{\overline{S}}^{CI_i})$ is similar to the target sample, while its sensitive representation $(R_S^{CI_i})$ is distinct. By leveraging counterfactuals, facilitates a strategy aimed at minimizing the similarity between channels unrelated to sensitive attributes and maximizing it between channels that are related to sensitive attributes. Consequently, the Sensitive Identification Constraints are proposed:

$$\mathcal{L}_C = \frac{1}{|\mathcal{V}| \times M} \sum_{v_i \in \mathcal{V}} \sum_{m=1}^{M} \left[d(R_{\overline{S}}^{v_i}, R_{\overline{S}}^{CI_m}) - d(R_S^{v_i}, R_S^{CI_m}) \right] \tag{7}$$

where $d(\cdot)$ is a distance metric, with M indicating the size of the counterfactual sample set. By amalgamating $\mathcal{L}_I$ and $\mathcal{L}_S$, the Disentangled Constraint is formally defined as:

$$\mathcal{L}_D = \mathcal{L}_I + \mathcal{L}_C \tag{8}$$

2) Fairness Constraint ($\mathcal{L}_M$). The objective of this constraint is to disassociate the component associated with sensitive attributes. As demonstrated in this module in Fig. 2, it ensures that the gender information of the node from the purple square cannot be inferred, thereby preventing bias from impacting downstream tasks. To achieve this, we employ a de-identifiable sensitive attribute technique, which utilizes a learnable vector ($\mathbf{W}$) to remove the identifiability of sensitive attributes from node representations. This process transforms R_S into an unbiased representation, denoted as $\overline{R_S} = R_S \odot \mathbf{W}$, making it indiscernible whether a node belongs to any specific subgroup. This unbiased representation $\overline{R_S}$ and $R_{\overline{S}}$ are subsequently used for predicting instance labels. Further, we use covariance as a constraint to ensure the effective removal of sensitive information, aiming to minimize the absolute covariance between the sensitive attribute and the label predictions. Mathematically, this is expressed as:

$$\mathcal{L}_F = \sum_{i=1}^{d_{R_{v_i}}} Abs(\mathbb{E}\left[(S_{v_i} - \mathbb{E}(S_{v_i}))(\overline{R}_{v_i} - \mathbb{E}(\overline{R}_{v_i}))\right]), \tag{9}$$

where $\overline{R}_{v_i}$ denotes the unbiased node representation for node v_i. In addition, we let $\mathbb{E}(\cdot)$ indicate the expectation operation, and $Abs(\cdot)$ is the absolute value function. This constraint ensures that the predictions are unbiased by sensitive attributes, thereby enhancing model fairness.

3) Performance Constraint ($\mathcal{L}_P$). Ensuring that the representations $R_{\overline{S}}$ and $\overline{R_S}$ for each node v_i incorporate vital node attributes and neighborhood information is essential to uphold their utility for downstream tasks, thereby aiding accurate label predictions. Thus, the Performance Constraint is established to enforce alignment between the prediction $\hat{y}_i$ and the ground truth y_i:

$$\mathcal{L}_P = \frac{1}{|\mathcal{V}_L|} \sum_{v_i \in \mathcal{V}_L} -(y_i \log(\hat{y}_i) + (1 - y_i) \log(1 - \hat{y}_i)) \tag{10}$$

where classifier takes $R_{\overline{S}}$ and $\overline{R_S}$ as input and $\hat{y}_i$ is prediction results for node v_i.

4) Reconstruction Constraint ($\mathcal{L}_R$). For each node v_i, the learned representations $R_{\overline{S}}$ and $\overline{R_S}$ should be sufficient to reconstruct the observed egograph G_{v_i}, transforming into an adjacency matrix reconstruction task. The effectiveness of node representation is thus evaluated by the discrepancies between the reconstructed adjacency matrices and the original graph structure. In addition, considering the sparsity of positive edges, FDGNN also incorporates negative sampling to address the distribution disparity between existent (positive) and non-existent (negative) edges. Specifically, for each positive edge $\{A(v_i, v_j) = 1 \;\forall\; i, j\}$, we counterpart this with a randomly selected non-existent edge $\{A(v_i, v_k) = 0 \;\forall\; i, k\}$, thereby forming a set of negative samples, M^-. Lastly, the Reconstruction Constraint, $\mathcal{L}_R$, is defined mathematically as:

$$\mathcal{L}_R = \sum_{A(v_i,v_j)\in M^+, A(v_i,v_k)\in M^-} \|\hat{A}(v_i, v_j) - A(v_i, v_j)\|_F^2 + \|\hat{A}(v_i, v_k) - A(v_i, v_k)\|_F^2 \tag{11}$$

where $\hat{A}$ and A are the predicted and observed adjacency matrices of input graph $\mathcal{G}$.

4.6   Final Optimization Objectives

The final objective function of FDGNN, as presented in Eq. 12, brings together the above three modules. Specifically, this function consists of four parts and is governed by the tunable hyperparameters α, β, and γ to balance the contributions of various elements: i) $\mathcal{L}_P$ aims to minimize the prediction loss, ii) $\mathcal{L}_I$ encourages the decomposition of learned representations into different independent channels and distinguishes between sensitive relevant and irrelevant representations, iii) $\mathcal{L}_F$ aims to mitigate sensitive-related information in node representation thereby improving the fairness of the model, and iv) $\mathcal{L}_R$ works to minimize the reconstruction loss for the node representations.

$$\min \mathcal{L}_{total} = \mathcal{L}_P + \alpha \mathcal{L}_D + \beta \mathcal{L}_F + \gamma \mathcal{L}_R \tag{12}$$

5   Experiment

5.1   Datasets

Experiments are conducted on three real-world graph datasets: i) The **German** dataset [2] contains credit information from clients at a German bank. Each node in this dataset represents a client, with edges reflecting the similarity between clients' credit profiles. The sensitive attribute is the clients' gender, and the classification task focuses on distinguishing clients into good versus bad credit risks. ii) The **Credit** dataset [36] consists of default payment records for individuals, where each node denotes an individual and edges indicate similarities in their expenditure and payment behaviors. The age of the individuals serves as the sensitive attribute, and the predictive task aims to determine whether an individual is likely to default on their credit card payments. iii) The **Bail** dataset [1] presents data related to defendants granted bail in U.S. state courts. Nodes represent defendants, and edges between nodes denote similarities in criminal records and demographic information. The sensitive attribute in this dataset is the race of the defendants, with the classification objective being to identify defendants as either suitable or unsuitable for bail (Table 1).

5.2   Evaluation Metrics

To effectively evaluate our proposed model, we measured our model performance from two perspectives: classification performance and fairness. For classification performance, we adopt Accuracy, F1-Score, and AUROC to evaluate the performance on node classification tasks. All three performance metrics close to 1 indicate better classification performance. To evaluate fairness, we use two commonly used fairness metrics, *i.e.*, Statistical Parity Difference (SPD) [20] and Equal Opportunity Differences (EOD) [12]. For both fairness metrics, values closer to 0 are indicative of greater model fairness.

Table 1. Summary of the datasets used in the experiments.

Dataset	German	Credit	Bail
Vertices	1,000	30,000	18,876
Edges	21,742	137,377	311,870
Feature dimension	27	13	18
Sensitive Attribute	Gender	Age	Race

5.3 Baselines

The proposed FDGNN is compared against seven state-of-the-art methods across three categories to evaluate its effectiveness. These include vanilla models like GCN [18], which leverages spatial graph convolutions for neighbor representation aggregation; GraphSAGE [11], which addresses GCN's scalability by training on node mini-batches; and GIN [33] enhancing node representation learning through MLP. Additionally, the fair node classification method FairGNN [7], which employs adversarial training to achieve group fairness by obscuring deprived group identities from discriminators, is also considered, in addition to graph counterfactual fairness methods like NIFTY [1], GEAR [21], and RFCGNN [27], detailed in Sect. 2.

5.4 Experiment Results

For thorough evaluation, the following research questions are addressed:

RQ1: How well does FDGNN performance compared to the state-of-the-art bias mitigation algorithms?

To answer RQ1, we experiment on three datasets with the comparison to the baselines on the node classification task. Each experiment is conducted 10 times, the results are shown in Table 2. As we can see, FDGNN outperforms all baseline methods across all evaluation metrics in most cases. Specifically, FDGNN demonstrates superior fairness performance, as evidenced by the significant margin overall baseline methods across all datasets. The enhancement of fairness is attributed to FDGNN accurately identifying sensitive attribute-related information via counterfactual instances and disentangling it into an independent component. It then mitigates its influence on prediction outcomes via a de-identifiable sensitive attribute mechanism. Simultaneously, FDGNN showcases commendable utility performance, surpassing other methods in most cases, which is indicative of FDGNN's capability to maintain important task-relevant information. This is because FDGNN avoids directly enforcing the fairness constraints by disentangling node representation, which facilitates the retention of task information related to sensitive attributes. Overall, the experimental results demonstrate the effectiveness of FDGNN in improving fairness while achieving comparable performance.

Table 2. Results on performance and fairness for FDGNN and baselines. The darkest cells indicate the top rank, while lighter cells represent the second rank.

Dataset	Methods Metrics	SPD (↓)	EOD (↓)	Accuracy (↑)	F1-Score (↑)	AUROC (↑)
German	GCN	0.364	0.312	0.684	0.786	0.654
	GraphSAGE	0.231	0.157	0.746	0.817	0.781
	GIN	0.148	0.091	0.720	0.812	0.734
	FairGNN	0.086	0.054	0.653	0.817	0.671
	NIFTY	0.077	0.049	0.674	0.792	0.736
	GEAR	0.085	0.046	0.681	0.780	0.722
	RFCGNN	0.067	0.041	0.721	0.823	0.747
	FDGNN	0.058	0.024	0.727	0.837	0.781
Credit	GCN	0.108	0.096	0.689	0.835	0.707
	GraphSAGE	0.113	0.124	0.739	0.859	0.767
	GIN	0.132	0.127	0.724	0.823	0.729
	FairGNN	0.126	0.104	0.674	0.812	0.711
	NIFTY	0.094	0.113	0.703	0.806	0.727
	GEAR	0.097	0.084	0.734	0.817	0.738
	RFCGNN	0.074	0.064	0.735	0.849	0.743
	FDGNN	0.056	0.047	0.736	0.861	0.747
Bail	GCN	0.093	0.044	0.828	0.784	0.871
	GraphSAGE	0.086	0.041	0.847	0.793	0.894
	GIN	0.072	0.043	0.728	0.658	0.768
	FairGNN	0.067	0.044	0.815	0.776	0.872
	NIFTY	0.035	0.028	0.753	0.671	0.796
	GEAR	0.047	0.024	0.823	0.783	0.786
	RFCGNN	0.031	0.024	0.861	0.802	0.747
	FDGNN	0.025	0.020	0.854	0.785	0.896

RQ2: What is the impact on FDGNN's performance when individual components are ablated?

To answer RQ2, we conduct ablation studies to gain insights into the effect of each module of FDGNN on improving fairness. Initially, our first analysis examined the significance of the fair ego-graph generation module. By substituting this module with the FDGNN-NFG variant, which employs an extractor to capture 2-hop neighboring nodes as ego-graphs for each node. Figure 3 presents ablation results on German, Credit, and Bail datasets. We observe that the fairness of FDGNN-NFG noticeably decreases. This reduction in fairness is ascribed to the FDGNN-NFG variant's inability to equitably represent diverse

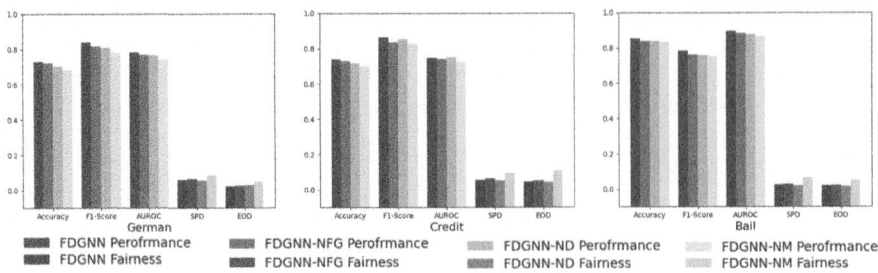

Fig. 3. Ablation study results for FDGNN, FDGNN-NFG, FDGNN-ND, and FDGNN-NM.

subgroups within subgraphs, leading to oversight of information from neighboring nodes with different sensitive attributes and, consequently, introducing graph structural bias. Next, we assessed the impact of the disentangled constraint by introducing the FDGNN-ND variant, which eschews this constraint by setting $N_c = 1$ and excluding $\mathcal{L}_D$. The results, depicted in Fig. 3, revealed a decline in both fairness and overall performance. This downturn can be attributed to the direct application of fairness constraints across the entire representation space in the absence of disentanglement, inevitably removing some task-related information. Lastly, we evaluate the effectiveness of our fairness constraint by creating the FDGNN-NF variant, removing $\mathcal{L}_F$. Compared to the FDGNN, there is a marked degradation in fairness, demonstrating the critical role of the fairness constraint in removing sensitive attribute information from node representation. To sum up, experimental results demonstrate the indispensability and efficacy of each component within the FDGNN framework.

> **RQ3: What the effect of Different N_c Values on Fairness and Predictive Performance?**

To answer RQ3, we conducted experiments with a variety of values for N_c as $\{1, 2, 4, 8, 16\}$, keeping all other training factors the same. We compare FDGNN's predictive performance and fairness under different settings. We observe that (Fig. 4): i) As N_c increases, the FDGNN achieves better fairness, demonstrating better disentanglement of information related to sensitive attributes. ii) When N_c is a modest value, the model fairness is hardly affected or even increases. The FDGNN mostly strikes the right balance between maintaining model utility and fostering fairness with proper choices of N_c in here. iii) When N_c is significant, , a noticeable decline in fairness is observed. This is attributed to the model's inability to isolate sensitive attribute information within a singular channel, resulting in cross-channel correlations that retain sensitive attribute data within node representation. In essence, increasing N_c allows for finer disentanglement and recognition of sensitive attribute-related information up to a point. Beyond this threshold, however, the decomposition into an excessive number of chan-

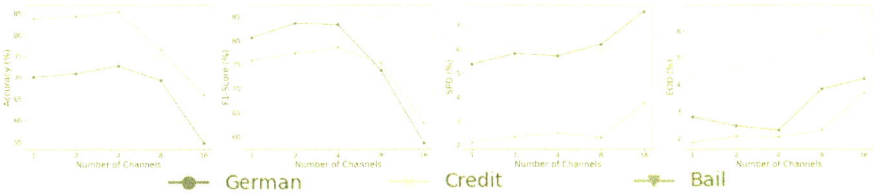

Fig. 4. Study the choice of N_c-value on German, Credit and Bail datasets.

nels introduces interference among them. This complexity hampers the model's ability to ensure channel independence, resulting in a decrease in model fairness.

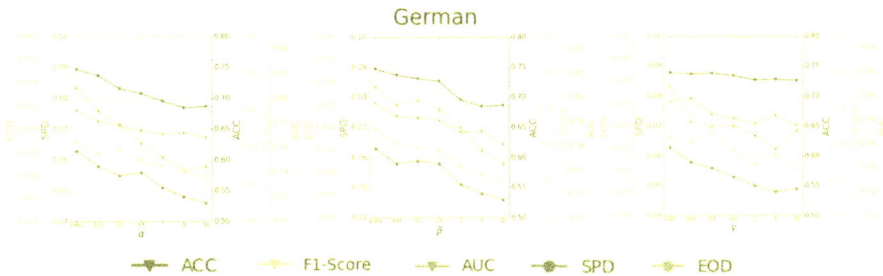

Fig. 5. Exploring hyperparameters study results in the German dataset.

RQ4: How do hyperparameters affect the performance and fairness of FDGNN?

To answer RQ4, we delve into the effects of three critical hyperparameters, i.e., α, β, and γ, which respectively modulate the influence of disentanglement, decorrelation, and the model's reconstruction performance within FDGNN. For this analysis, we individually varied each hyperparameter through a range from 0.001 to 10, keeping all other training factors the same. Figure 5 presents the relevant findings from the German dataset. Specifically, an increase in α and β will increase model fairness but at the cost of some predictive performance degradation. This phenomenon occurs as the increased weightage of these parameters strengthens the model's ability to disentangle node representation and mitigate the correlation with sensitive attributes. Consequently, this diminishes the influence of sensitive attribute information on node representation, thereby advancing model fairness. As for γ, its increment initially bolsters fairness up to a certain point, beyond which fairness begins to decrease, though without significantly affecting performance. This is because a higher γ value improves the model's fidelity in reconstructing the graph structure, thereby avoiding the introduction of noise into the node representation and improving the model's ability to capture the underlying factors behind the data.

6 Conclusion

In this work, we study the problem of learning fair graph representation within GNNs. Inspired by causal theory, we introduce the Fair Disentangled Graph Neural Network (FDGNN) framework, which aims to achieve counterfactual fairness in graph-based representations while preserving important task-related information. FDGNN conducts a causal analysis of both original and counterfactual samples, effectively disentangling sensitive attributes into distinct components and subsequently mitigating their undue influence on the learned representations. This strategy allows FDGNN to enhance fairness without compromising the utility of the node representation for downstream tasks. Empirical evaluations on three real-world datasets validate the effectiveness of our framework with respect to both prediction performance and fairness.

Acknowledgement. This work was supported in part by the National Science Foundation (NSF) under Grant No. 2245895.

References

1. Agarwal, C., Lakkaraju, H., Zitnik, M.: Towards a unified framework for fair and stable graph representation learning. In: Uncertainty in Artificial Intelligence, pp. 2114–2124. PMLR (2021)
2. Asuncion, A., Newman, D.: UCI machine learning repository (2007)
3. Chinta, S.V., et al.: Optimization and improvement of fake news detection using voting technique for societal benefit. In: 2023 IEEE International Conference on Data Mining Workshops (ICDMW), pp. 1565–1574. IEEE (2023)
4. Chu, Z., et al.: History, development, and principles of large language models-an introductory survey. arXiv preprint arXiv:2402.06853 (2024)
5. Chu, Z., Wang, Z., Zhang, W.: Fairness in large language models: a taxonomic survey. arXiv preprint arXiv:2404.01349 (2024)
6. Creager, E., et al.: Flexibly fair representation learning by disentanglement. In: International Conference on Machine Learning, pp. 1436–1445. PMLR (2019)
7. Dai, E., Wang, S.: Say no to the discrimination: learning fair graph neural networks with limited sensitive attribute information. In: Proceedings of the 14th ACM International Conference on Web Search and Data Mining, pp. 680–688 (2021)
8. Doan, T.V., Chu, Z., Wang, Z., Zhang, W.: Fairness definitions in language models explained (2024)
9. Dzuong, J., Wang, Z., Zhang, W.: Uncertain boundaries: multidisciplinary approaches to copyright issues in generative AI. arXiv preprint arXiv:2404.08221 (2024)
10. Gao, C., Wang, X., He, X., Li, Y.: Graph neural networks for recommender system. In: Proceedings of the Fifteenth ACM International Conference on Web Search and Data Mining, pp. 1623–1625 (2022)
11. Hamilton, W., Ying, Z., Leskovec, J.: Inductive representation learning on large graphs. In: Advances in Neural Information Processing Systems, vol. 30 (2017)
12. Hardt, M., Price, E., Srebro, N.: Equality of opportunity in supervised learning. In: Advances in Neural Information Processing Systems, vol. 29 (2016)

13. He, X., Deng, K., Wang, X., Li, Y., Zhang, Y., Wang, M.: LightGCN: simplifying and powering graph convolution network for recommendation. In: Proceedings of the 43rd International ACM SIGIR Conference on Research and Development in Information Retrieval, pp. 639–648 (2020)
14. Jeh, G., Widom, J.: Scaling personalized web search. In: Proceedings of the 12th International Conference on World Wide Web, pp. 271–279 (2003)
15. Jiao, Y., Xiong, Y., Zhang, J., Zhang, Y., Zhang, T., Zhu, Y.: Sub-graph contrast for scalable self-supervised graph representation learning. In: 2020 IEEE International Conference on Data Mining (ICDM), pp. 222–231. IEEE (2020)
16. Jin, G., Wang, Q., Zhu, C., Feng, Y., Huang, J., Zhou, J.: Addressing crime situation forecasting task with temporal graph convolutional neural network approach. In: 2020 12th International Conference on Measuring Technology and Mechatronics Automation (ICMTMA), pp. 474–478. IEEE (2020)
17. Kang, J., He, J., Maciejewski, R., Tong, H.: Inform: Individual fairness on graph mining. In: Proceedings of the 26th ACM SIGKDD International Conference on Knowledge Discovery & Data Mining, pp. 379–389 (2020)
18. Kipf, T.N., Welling, M.: Semi-supervised classification with graph convolutional networks. arXiv preprint arXiv:1609.02907 (2016)
19. Kusner, M.J., Loftus, J., Russell, C., Silva, R.: Counterfactual fairness. In: Advances in Neural Information Processing Systems, vol. 30 (2017)
20. Le Quy, T., Roy, A., Iosifidis, V., Zhang, W., Ntoutsi, E.: A survey on datasets for fairness-aware machine learning. Wiley Interdisc. Rev. Data Min. Knowl. Discov. **12**(3), e1452 (2022)
21. Ma, J., Guo, R., Wan, M., Yang, L., Zhang, A., Li, J.: Learning fair node representations with graph counterfactual fairness. In: Proceedings of the Fifteenth ACM International Conference on Web Search and Data Mining, pp. 695–703 (2022)
22. Mehrabi, N., Morstatter, F., Saxena, N., Lerman, K., Galstyan, A.: A survey on bias and fairness in machine learning. ACM Comput. Surv. (CSUR) **54**(6), 1–35 (2021)
23. Olteanu, A., Castillo, C., Diaz, F., Kıcıman, E.: Social data: Biases, methodological pitfalls, and ethical boundaries. Front. Big Data **2**, 13 (2019)
24. Simonovsky, M., Komodakis, N.: GraphVAE: towards generation of small graphs using variational autoencoders. In: Kůrková, V., Manolopoulos, Y., Hammer, B., Iliadis, L., Maglogiannis, I. (eds.) ICANN 2018. LNCS, vol. 11139, pp. 412–422. Springer, Cham (2018). https://doi.org/10.1007/978-3-030-01418-6_41
25. Sui, Y., Wang, X., Wu, J., Lin, M., He, X., Chua, T.S.: Causal attention for interpretable and generalizable graph classification. In: Proceedings of the 28th ACM SIGKDD Conference on Knowledge Discovery and Data Mining, pp. 1696–1705 (2022)
26. Wang, Z., et al.: Individual fairness with group awareness under uncertainty. In: Joint European Conference on Machine Learning and Knowledge Discovery in Databases. Springer Nature Switzerland (2024)
27. Wang, Z., Narasimhan, G., Yao, X., Zhang, W.: Mitigating multisource biases in graph neural networks via real counterfactual samples. In: 2023 IEEE International Conference on Data Mining (ICDM), pp. 638–647. IEEE (2023)
28. Wang, Z., Qiu, M., Chen, M., Salem, M.B., Yao, X., Zhang, W.: Towards fair graph neural networks via real counterfactual samples. Knowl. Inf. Syst. (2024). https://doi.org/10.1007/s10115-024-02161-z
29. Wang, Z., Saxena, N., et al.: Preventing discriminatory decision-making in evolving data streams. In: Proceedings of the 2023 ACM Conference on Fairness, Accountability, and Transparency (FAccT) (2023)

30. Wang, Z., Wallace, C., Bifet, A., Yao, X., Zhang, W.: FG^2AN: fairness-aware graph generative adversarial networks. In: Koutra, D., Plant, C., Gomez Rodriguez, M., Baralis, E., Bonchi, F. (eds.) Machine Learning and Knowledge Discovery in Databases: Research Track, ECML PKDD 2023, LNCS, vol. 14170, pp. 259–275. Springer, Cham (2023). https://doi.org/10.1007/978-3-031-43415-0_16

31. Wang, Z., et al.: Towards fair machine learning software: Understanding and addressing model bias through counterfactual thinking. arXiv preprint arXiv:2302.08018 (2023)

32. Wu, Z., Pan, S., Chen, F., Long, G., Zhang, C., Philip, S.Y.: A comprehensive survey on graph neural networks. IEEE Trans. Neural Netw. Learn. Syst. **32**(1), 4–24 (2020)

33. Xu, K., Hu, W., Leskovec, J., Jegelka, S.: How powerful are graph neural networks? arXiv preprint arXiv:1810.00826 (2018)

34. Yang, Z., Pei, W., Chen, M., Yue, C.: Wtagraph: web tracking and advertising detection using graph neural networks. In: 2022 IEEE Symposium on Security and Privacy (SP), pp. 1540–1557. IEEE (2022)

35. Yazdani, S., Saxena, N., Wang, Z., Wu, Y., Zhang, W.: A comprehensive survey of image and video generative AI: recent advances, variants, and applications (2024)

36. Yeh, I.C., Lien, C.H.: The comparisons of data mining techniques for the predictive accuracy of probability of default of credit card clients. Expert Syst. Appl. **36**(2), 2473–2480 (2009)

37. Yin, Z., Wang, Z., Zhang, W.: Improving fairness in machine learning software via counterfactual fairness thinking. In: Proceedings of the 2024 IEEE/ACM 46th International Conference on Software Engineering: Companion Proceedings, pp. 420–421 (2024)

38. Yuan, H., Yu, H., Gui, S., Ji, S.: Explainability in graph neural networks: a taxonomic survey. IEEE Trans. Pattern Anal. Mach. Intell. **45**(5), 5782–5799 (2022)

39. Zhang, W., Wang, Z., Kim, J., Cheng, C., Oommen, T., Ravikumar, P., Weiss, J.: Individual fairness under uncertainty. In: 26th European Conference on Artificial Intelligence, pp. 3042–3049 (2023)

40. Zhang, W., Weiss, J.C.: Fair decision-making under uncertainty. In: 2021 IEEE International Conference on Data Mining (ICDM), pp. 886–895. IEEE (2021)

41. Zhang, W., Weiss, J.C.: Longitudinal fairness with censorship. In: Proceedings of the AAAI Conference on Artificial Intelligence, vol. 36, pp. 12235–12243 (2022)

42. Zhang, W., Weiss, J.C., Zhou, S., Walsh, T.: Fairness amidst non-iid graph data: a literature review. arXiv preprint arXiv:2202.07170 (2022)

43. Zhang, W., Zhang, L., Pfoser, D., Zhao, L.: Disentangled dynamic graph deep generation. In: Proceedings of the 2021 SIAM International Conference on Data Mining (SDM), pp. 738–746. SIAM (2021)

Continuously Deep Recurrent Neural Networks

Andrea Ceni[1(✉)] 🆔, Peter Ford Dominey[2] 🆔, Claudio Gallicchio[1] 🆔,
Alessio Micheli[1] 🆔, Luca Pedrelli[1] 🆔, and Domenico Tortorella[1] 🆔

[1] Department of Computer Science, University of Pisa, Largo Bruno Pontecorvo 3,
56127 Pisa, Italy
{andrea.ceni,micheli,luca.pedrelli}@di.unipi.it,
claudio.gallicchio@unipi.it, domenico.tortorella@phd.unipi.it
[2] INSERM UMR1093-CAPS, Université de Bourgogne, UFR des Sciences du Sport,
Dijon, France
Peter-Ford.Dominey@u-bourgogne.fr

Abstract. The architecture of multi-layer dynamic neural systems traditionally contains recurrent neurons organized in successive well-defined
layers. In this paper, we introduce a new class of recurrent neural models
based on a fundamentally different type of topological organization than
the conventionally used deep recurrent networks, and directly inspired
by the way cortical networks in the brain process information at multiple
temporal scales. We explore the novel paradigm from the perspective of
Reservoir Computing (RC), a popular approach to designing efficiently
trainable recurrent neural networks, and introduce the Continuously-
Deep Echo State Network (C-DESN). The proposed C-DESN architecture comprises a reservoir layer of untrained recurrent neurons connected
in a biologically inspired exponentially decaying pattern based on distance. The depth of the resulting neural information processing system
is modulated by a single depth hyperparameter that controls the extent
of local connectivity. Mathematically, we analyze the dynamical stability properties of the continuously deep reservoir, providing a rigorous
bound on the resulting eigenspectrum. Empirically, we show that the
novel recurrent architecture is biased toward tunable temporal resolution processing in the same way as conventional deep recurrent neural
networks. Additionally, our experiments on short-term memory capacity and real-world time-series reconstruction demonstrate how the depth
hyperparameter of C-DESN can effectively regulate the temporal scale
in the reservoir's dynamic behavior.

Keywords: Reservoir Computing · Echo State Networks · Recurrent
Neural Networks

1 Introduction

Recurrent Neural Networks (RNNs) are machine learning models designed to
process sequential data. Unlike feedforward neural networks, RNNs spatially

A. Bifet et al. (Eds.): ECML PKDD 2024, LNAI 14947, pp. 59–73, 2024.
https://doi.org/10.1007/978-3-031-70368-3_4

encode temporal features of the input in their hidden state allowing for context-dependent computation, a key feature of prefrontal cortex microcircuits [17]. Cortical networks in the brain exhibit hierarchies of temporal processing, with long integrative timescales in the frontal cortex and rapid transient responses in input-driven areas [3,5]. This temporal hierarchy is likely due to the structural organization of the cortex, which is dominated by neural connection probability diminishing exponentially with distance [3]. This motivates the study of RNNs whose network connectivity topology is similarly constrained by such exponential distance rules. It has been shown that such connectivity patterns increase temporal integration timescales in areas far from the input [7], and one would predict that these integrative areas would have higher memory capacity than earlier areas, but the prediction that this flow topology will impact memory capacity remains to be tested. In this paper, we explore this route from the perspective of Reservoir Computing (RC) [18,21]. RC is a paradigm for designing and training RNNs that leverages fixed randomized recurrent connections while adapting only a readout layer. This approach, other than making the model amenable to neuromorphic implementations [19], has the benefit of isolating the effects of this recurrent architecture, keeping aside the influence of the specific training algorithm used to adapt the recurrent weights.

In this paper, adopting the mathematical formalism of Echo State Networks (ESNs) [16], we introduce a novel class of RC-based architectures called *Continuously-Deep ESN* (C-DESN). Our proposal allows the design of biologically plausible ESNs characterized by local connections among neurons based on an exponentially decaying rule while modulating the depth of the internal information processing via a single hyperparameter of the C-DESN model. Unlike the deep architectures previously explored in the RC literature, and fundamentally based on the concept of a pool of "discrete" recurrent layers nested one inside the other [11], the recurrent architecture of the model proposed in this paper presents a more nuanced and "continuous" concept of layering. Here, by investigating the intrinsic multiple temporal scale processing capability of C-DESN, our analysis aims to shed light on the architectural bias induced by a hierarchically continuous organization of recurrent neurons in RNNs. From an RC-oriented perspective, we mathematically analyze the behavior of the resulting continuously deep reservoir system. This provides a characterization of its dynamical behavior from the stability standpoint. Furthermore, we perform comparative experiments with conventional shallow and deep RC models and show that the proposed class of RC models excels in short-term memorization [15] and time-series reconstruction capabilities. Additionally, the introduced continuously defined depth concept simplifies the construction of the hierarchy of dynamics in the reservoir.

The rest of this paper is organized as follows. In Sect. 2 we introduce the fundamental concepts of RC neural networks, and the utilized mathematical formalism to describe the temporal computation performed by an ESN-like model. We introduce our proposed C-DESN model in Sect. 3. In Sect. 4 we study the intrinsic properties of continuously deep recurrent neural systems in terms of their temporal response. Then, in Sect. 5 we delve into the mathematical analysis of the resulting dynamical reservoir. We provide our experimental analysis in Sect. 6. Finally, Sect. 7 concludes the paper.

2 Shallow and Deep Echo State Networks

Echo State Networks (ESNs) [16] are a type of discrete-time neural network based on RC. ESNs have a vanilla RNN structure, where the recurrent hidden layer is called the *reservoir*. The reservoir is initialized under asymptotic stability conditions and left untrained. The output layer, called the *readout* layer, is the only part of the network that is trained on the specific task at hand. ESNs are widely used in applications due to their exceptional trade-off between efficiency and effectiveness, which makes them well-suited for pervasive AI applications in both edge-AI scenarios [1,6] and neuromorphic hardware [19].

At each timestep, the reservoir updates its internal state, denoted as $\mathbf{h}[t] \in \mathbb{R}^{N_h}$, based on the following state transition function:

$$\mathbf{h}[t+1] = \sigma\left(\mathbf{W}_h\,\mathbf{h}[t] + \mathbf{W}_x\,\mathbf{x}[t+1]\right), \tag{1}$$

where $\mathbf{x}[t] \in \mathbb{R}^{N_x}$ is the input at timestep t, σ is the activation function (we use tanh, the hyperbolic tangent, as nonlinearity), $\mathbf{W}_h \in \mathbb{R}^{N_h \times N_h}$ is the matrix of the recurrent weights, and $\mathbf{W}_x \in \mathbb{R}^{N_h \times N_x}$ is the matrix of the input weights. Driven by the external input dynamics, the reservoir system is started in the null state, i.e. $\mathbf{h}[0] = \mathbf{0} \in \mathbb{R}^{N_h}$, and evolved according to the above Eq. (1). The output of the network is computed by a linear readout layer, i.e., $\mathbf{y}[t] = \mathbf{W}_y\mathbf{h}[t]$, where $\mathbf{y}[t] \in \mathbb{R}^{N_y}$ is the output of the network and $\mathbf{W}_y \in \mathbb{R}^{N_y \times N_x}$ is the matrix of the output weights. Typically, connections pointing to the readout are trained in closed form, e.g. by pseudo-inverse or ridge regression.

Crucially, the parameters of the reservoir, i.e., the weight values in $\mathbf{W}_h$ and $\mathbf{W}_x$ can be left untrained provided that the reservoir has a dynamical behavior in agreement with the Echo State Property (ESP) [16,23], a form of global asymptotic stability condition imposed to the input-driven reservoir dynamics. This property is strictly linked to the contractive characterization of Lipschitz continuous iterated maps, and a sufficient condition for the ESP is traditionally given in terms of the Euclidean norm of the recurrent weight matrix, i.e., $\|\mathbf{W}_h\| < 1$. While of interest for mathematical analysis, this is a too restrictive condition in most practical cases and, in applications, a necessary condition is often used instead, which consists of controlling the eigenspectrum of $\mathbf{W}_h$. In particular, the weight values in $\mathbf{W}_h$ can be randomly drawn, e.g., from a uniform distribution on $(-1, 1)$, and then re-scaled such that the resulting spectral radius[1] has a value below 1, i.e., $\rho(\mathbf{W}_h) < 1$. Similarly, the input weight matrix $\mathbf{W}_x$ is usually randomly initialized with values drawn from a uniform distribution in $(-\omega, \omega)$, where ω denotes an input-scaling coefficient (that we set to 1 in this paper).

---

[1] The largest modulus of an eigenvalue of the matrix.

More recently, a class of networks based on hierarchically organized reservoirs has become progressively more popular, and known as *deep reservoirs*. The reference model in this class is the Deep Echo State Network (DeepESN) [11], which generalizes conventional shallow ESN models presenting a dynamical component that is organized in a hierarchy of layers. In particular, the architecture comprises a discrete number of recurrent reservoir layers L. While the first reservoir layer is driven by the external input (as in shallow ESNs), the dynamical evolution of each successive reservoir layer is driven by the state of the preceding one. In formulas:

$$\mathbf{h}^{(1)}[t+1] = \sigma \left(\mathbf{W}_h^{(1)} \, \mathbf{h}^{(1)}[t] + \mathbf{W}_x^{(1)} \, \mathbf{x}[t+1] \right), \tag{2}$$

for the first layer, and

$$\mathbf{h}^{(i)}[t+1] = \sigma \left(\mathbf{W}_h^{(i)} \, \mathbf{h}^{(i)}[t] + \mathbf{W}_x^{(i)} \, \mathbf{h}^{(i-1)}[t+1] \right), \tag{3}$$

for layers $i > 1$, where we use the superscript (i) to refer to the i-th layer in the architecture. The deep reservoir constructed in this way is left untrained after initialization under a generalization of the ESP conditions [9], which essentially prescribe control of the spectral values of the $\mathbf{W}_h^{(i)}$ matrices. The readout is fed by the concatenation of the reservoir states computed at every layer and is trained in closed form, as for the shallow case. DeepESNs, despite having the same efficiency advantage as conventional shallow ESNs, are capable of processing sequential information at multiple temporal resolutions in a very effective way. This results in a significantly advantageous architectural setup when dealing with temporal tasks that require multiple time-scale processing [10].

3   Continuously Deep Echo State Networks

In order to exploit the advantages of a deep architecture while effectively keeping a single-layer recurrent structure, we introduce Continuously Deep ESNs (C-DESN). C-DESN consists of a hierarchically connected untrained recurrent reservoir layer that captures the temporal evolution of the input sequence. Different from typical deep RNNs, in which the layers are completely separated, C-DESNs are composed of a hierarchy of connectivity in which the number of recurrent layers is not discrete and not architecturally fixed. Additionally, as in conventional ESN models, a trainable readout layer computes the output starting from the internal state representation developed by the reservoir. An example of C-DESN architecture is shown in Fig. 1.

The network state is computed as in the following equation:

$$\mathbf{h}[t+1] = \sigma \left(\widetilde{\mathbf{W}}_h(\beta) \, \mathbf{h}[t] + \widetilde{\mathbf{W}}_x(\beta) \, \mathbf{x}[t+1] \right), \tag{4}$$

where $\widetilde{\mathbf{W}}_h(\beta) = \exp[-\beta \mathbf{D}_h] \odot \mathbf{W}_h$, and $\widetilde{\mathbf{W}}_x(\beta) = \exp[-\beta \mathbf{D}_x] \odot \mathbf{W}_x$, and $\odot$ represents the component-wise multiplication of matrices (also called Hadamard

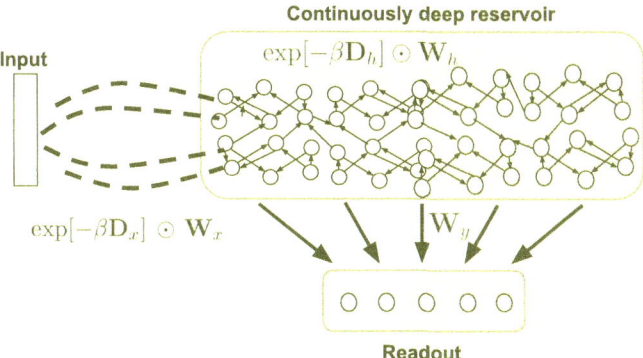

Fig. 1. An example of C-DESN architecture. The external input drives the dynamics of the internal reservoir. Along the recurrent architecture, the distance between the hidden neurons and the external input source increases continuously. Only the readout connections are trained.

product), i.e. $(\mathbf{A} \odot \mathbf{B})_{ij} = (\mathbf{A})_{ij}(\mathbf{B})_{ij}$. Here, $\exp[\mathbf{M}]$ denotes the component-wise exponentiation of all the entries of the matrix $\mathbf{M}$, i.e. $(\exp[\mathbf{M}])_{ij} = e^{(\mathbf{M})_{ij}}$, not to be confused with the matrix exponential. As introduced in Sect. 2, the vector $\mathbf{h}[t] \in \mathbb{R}^{N_h}$ is the state of the network at timestep t, $\mathbf{x}[t] \in \mathbb{R}^{N_x}$ is the input at timestep t, σ is the activation function (we use tanh, the hyperbolic tangent, as nonlinearity), $\mathbf{W}_h \in \mathbb{R}^{N_h \times N_h}$ is the matrix of the recurrent weights, $\mathbf{W}_x \in \mathbb{R}^{N_h \times N_x}$ is the matrix of the input weights. Different from standard ESN-like models in Sect. 2, here we introduce two additional matrices that are used to parametrize a local connectivity pattern among the neurons. Specifically, $\mathbf{D}_h \in \mathbb{R}^{N_h \times N_h}$ is the hidden mask that represents the hierarchical structure of the architecture (defined in the following) and $\mathbf{D}_x \in \mathbb{R}^{N_h \times N_x}$ is the mask which determines the structure of the input-to-hidden connections. We define the hidden mask as $\exp[-\beta \mathbf{D}_h]$, where $\beta > 0$, and

$$\mathbf{D}_h = \frac{1}{N_h - 1} \begin{bmatrix} 0 & 1 & 2 & \dots N_h - 1 \\ 1 & 0 & 1 & \dots N_h - 2 \\ 2 & 1 & 0 & \dots N_h - 3 \\ \vdots & \vdots & \vdots & \dots & \vdots \\ N_h - 1 & N_h - 2 & N_h - 3 & \dots & 0 \end{bmatrix}. \tag{5}$$

The entries $(\mathbf{D}_h)_{ij}$ of the matrix in Eq. (5) represent the discrete distance between the i-th neuron and the j-th neuron, normalized by the total number of neurons. Therefore, the mask $\exp[-\beta \mathbf{D}_h]$ applied on the matrix $\mathbf{W}_h$ has the effect of reducing the off-diagonal elements, giving values approaching 0 when pushing $\beta \to +\infty$. Large values of β force an exponentially weak coupling between i-th and j-th neurons if $|i - j|$ is large. This promotes the emergence of an effective depth in the recurrent architecture as a function of β. For this reason, we dub the hyperparameter β the *continuous depth* (or just *depth*) of the model. Similarly, we use the mask $\exp[-\beta \mathbf{D}_x]$ for the input-to-hidden matrix $\mathbf{W}_x$, with $\mathbf{D}_x \in \mathbb{R}^{N_h \times N_x}$ defined as follows:

$$
\mathbf{D}_x = \frac{1}{N_h - 1}
\begin{bmatrix}
0 & 0 & \dots & 0 \\
1 & 1 & \dots & 1 \\
\vdots & \vdots & \dots & \vdots \\
N_h - 1 & N_h - 1 & \dots & N_h - 1
\end{bmatrix}.
\tag{6}
$$

Note that we use the same hyperparameter β for masking both the hidden-to-hidden and the input-to-hidden matrices. Finally, the output of the network is computed by a linear readout layer, i.e., $\mathbf{y}[t] = \mathbf{W}_y \mathbf{h}[t]$, where $\mathbf{y}[t] \in \mathbb{R}^{N_y}$ is the output of the network and $\mathbf{W}_y \in \mathbb{R}^{N_y \times N_x}$ is the matrix of the output weights. The reservoir of the C-DESN is initialized similarly to the conventional ESN systems, i.e., by controlling the spectral radius ρ of the recurrent weight matrix $\mathbf{W}_h$ and the scaling of the weights in the input matrix $\mathbf{W}_x$. The masking matrices $\mathbf{D}_h$ and $\mathbf{D}_x$ are fixed (given the number of reservoir neurons N_h), and the value of the continuous depth β is treated as a hyperparameter. The readout is trained in closed form as in conventional RC.

Overall, it is relevant to note that the continuously deep reservoir introduced for C-DESN networks, presents an architecture in which the reservoir units are located at progressively greater distances from the point where the input signal is injected. As we will see in the following, this enables the processing of multiple time scales and longer short-term memorization than a shallow reservoir, similar to what can be achieved with a DeepESN. Crucially, unlike the case of Deep-ESNs, where the number of layers must be predetermined during architecture construction, in C-DESNs, deep temporal signal processing is controlled by a single non-architectural hyper-parameter (β). As a result, the network's depth can be adjusted to solve a specific task simply by fine-tuning one value without changing the underlying architecture.

4   Analysis of Deep Dynamics

We investigate the properties of the architectural bias of our proposed C-DESN model as a general case of a recurrent neural network, characterizing it as a dynamical system in terms of its response to unit impulse inputs. In a deep layered RNN, we expect such response to be faster in shallower units and slower in deep units due to the propagation delay between layers, as a consequence enhancing the ability of the model to hold long short-term information concerning input history in the sequence embeddings [9,11,20]. We expect this behavior to hold for an RNN with continuous depth connectivity (Fig. 2a) as in an RNN with discrete layer connectivity (Fig. 2b). In the latter, the diagonal blocks correspond to recurrent matrices of layers, and the lower diagonal blocks to input matrices of each layer from the previous one. To experimentally validate this claim, we measure the peak response time in the deepest unit to a unit impulse input in

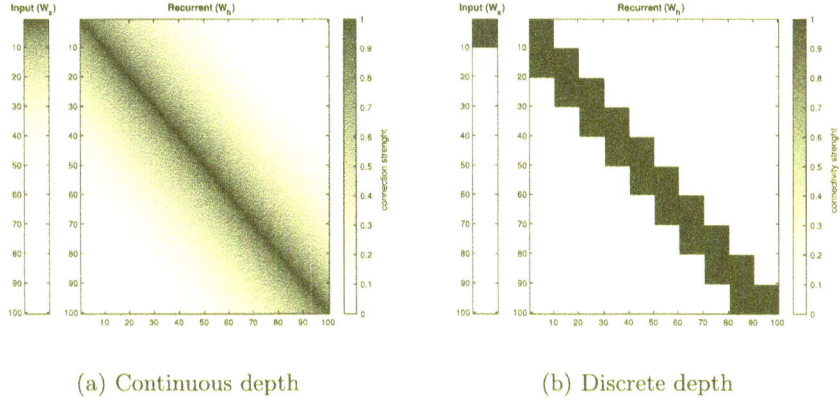

(a) Continuous depth   (b) Discrete depth

Fig. 2. Connectivity patterns in a 100-units recurrent neural network: (a) Continuously deep RNN with $\beta = 4$; (b) Discretely deep RNN with 10 layers.

a 100-unit RNN with continuous connectivity for different values of $\beta \in [0.2, 20]$ and with discrete connectivity with different number of layers in $[1, 50]$. For simplicity, we assume the RNN to have linear activation as $\sigma(\cdot)$ and as recurrent matrices $\mathbf{W}_h$ the connectivity masks rescaled to unit norm, i.e. $\|\mathbf{W}_h\| = 1$. We set input scaling to $\omega = 1$. Figure 3 shows that the response times increase linearly with the number of layers in a discretely deep RNN. This behavior similarly holds for continuously deep RNNs, with β acting as the continuous equivalent of the number of layers. As the two systems share a common qualitative behavior, we expect the same advantages of deep layering in DeepESNs to translate in C-DESNs.

5   Mathematical Analysis

In this section, we analyze the behavior of the masked matrix $\widetilde{\mathbf{W}}_h(\beta)$ when varying β in the continuum $(0, +\infty)$, providing a characterization of the eigenvalues distribution of $\widetilde{\mathbf{W}}_h(\beta)$, and an estimation of the spectral radius of $\widetilde{\mathbf{W}}_h(\beta)$ for large values of the depth hyperparameter β.

In the following, we use the notation $A \preceq B$ to express the component-wise relation $(A)_{ij} \leq (B)_{ij}$, for all i, j. We start noticing that the monotonic property of the exponential function implies that $\mathbf{I} \preceq \exp[-\beta_1 \mathbf{D}_h] \preceq \exp[-\beta_2 \mathbf{D}_h] \preceq \mathbf{J}$, for all $0 \leq \beta_2 \leq \beta_1$, where $\mathbf{I}$ is the identity matrix of dimension N_h, and $\mathbf{J}$ is the $N_h \times N_h$ matrix with all entries equal to 1. These component-wise bounds imply that $\mathbf{I} \odot \mathbf{W}_h \preceq \exp[-\beta_1 \mathbf{D}_h] \odot \mathbf{W}_h \preceq \exp[-\beta_2 \mathbf{D}_h] \odot \mathbf{W}_h \preceq \mathbf{J} \odot \mathbf{W}_h$, which can be equivalently written, for all $0 \leq \beta_2 \leq \beta_1$, as:

$$\operatorname{diag}(\mathbf{W}_h) \preceq \exp[-\beta_1 \mathbf{D}_h] \odot \mathbf{W}_h \preceq \exp[-\beta_2 \mathbf{D}_h] \odot \mathbf{W}_h \preceq \mathbf{W}_h, \qquad (7)$$

where $\operatorname{diag}(\mathbf{W}_h)$ denotes the diagonal matrix with diagonal entries of $\mathbf{W}_h$. The matrix $\widetilde{\mathbf{W}}_h(\beta)$ draws a continuous trajectory in the space of real-valued squared

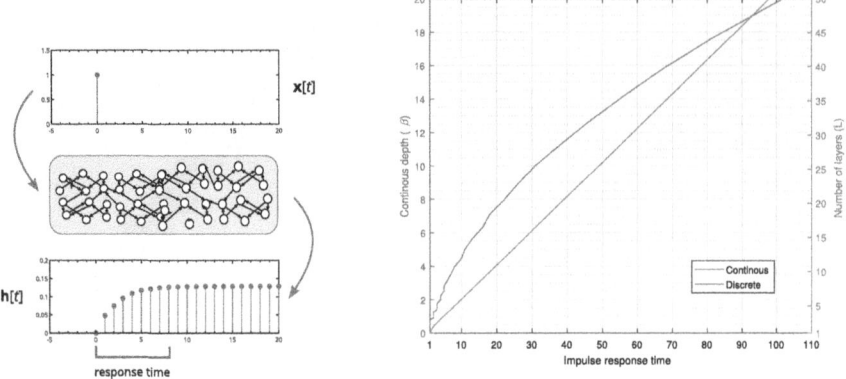

Fig. 3. Response time to an unit impulse at $t = 0$ in the deepest unit $\mathbf{h}_{100}$ of a recurrent neural network with 100 units and linear activation.

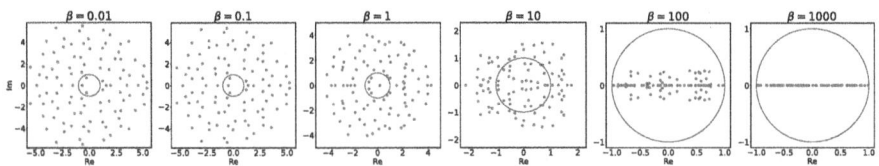

Fig. 4. Spectrum of eigenvalues of $\widetilde{\mathbf{W}}_h(\beta) = \exp[-\beta \mathbf{D}_h] \odot \mathbf{W}_h$ for $\beta = 0.01, 0.1, 1, 10, 100, 1000$. We generate the matrix $\mathbf{W}_h$ with random uniformly sampled i.i.d entries in $(-1, 1)$, and dimension $N_h = 100$. The blue circle is the complex unitary circle. (Color figure online)

matrices from $\mathbf{W}_h$ to $\text{diag}(\mathbf{W}_h)$ as the parameter β goes from 0 to $+\infty$. We summarize these findings in the following proposition.

Proposition 1. *For any given square matrix $\mathbf{W}_h$ of dimension $N_h \times N_h$ the following component-wise limits hold*

$$\lim_{\beta \to 0} \widetilde{\mathbf{W}}_h(\beta) = \mathbf{W}_h; \tag{8}$$

$$\lim_{\beta \to +\infty} \widetilde{\mathbf{W}}_h(\beta) = \text{diag}(\mathbf{W}_h). \tag{9}$$

Since the spectrum of a diagonal matrix is the set of its entries on the diagonal, we deduce from Eq. (9) that the eigenvalues distribution of $\widetilde{\mathbf{W}}_h(\beta)$ tends toward the set of diagonal values of $\mathbf{W}_h$ for large values of β. We provide below a more precise characterization of the eigenvalues distribution of $\widetilde{\mathbf{W}}_h(\beta)$ when varying the hyperparameter β.

Theorem 1. *For each eigenvalue λ of the matrix $\widetilde{\mathbf{W}}_h(\beta)$, there exists an element $(\mathbf{W}_h)_{ii}$ in the diagonal of $\mathbf{W}_h$, for some $i \in \{1, \dots, N_h\}$, such that*

$$|\lambda - (\mathbf{W}_h)_{ii}| \leq e^{-\beta/(N_h-1)} \|\mathbf{W}_h\|. \tag{10}$$

Proof. We decompose $\widetilde{\mathbf{W}}_h(\beta)$ as follows

$$
\begin{aligned}
\widetilde{\mathbf{W}}_h(\beta) &= \exp[-\beta\mathbf{D}_h] \odot \mathbf{W}_h \\
&= \Big(\mathbf{I} + (\exp[-\beta\mathbf{D}_h] - \mathbf{I})\Big)\odot\mathbf{W}_h \\
&= \mathbf{I} \odot \mathbf{W}_h + (\exp[-\beta\mathbf{D}_h] - \mathbf{I}) \odot \mathbf{W}_h \\
&= \mathrm{diag}(\mathbf{W}_h) + (\exp[-\beta\mathbf{D}_h] - \mathbf{I}) \odot \mathbf{W}_h.
\end{aligned}
$$

We know that the spectrum of $\mathrm{diag}(\mathbf{W}_h)$ is exactly the set of entries on the diagonal, and we consider the term $(\exp[-\beta\mathbf{D}_h] - \mathbf{I}) \odot \mathbf{W}_h$ as a perturbative term from the diagonal form. Now Bauer-Fike's theorem [2] implies that each eigenvalue of $\widetilde{\mathbf{W}}_h(\beta)$ lies at maximum distance $\|(\exp[-\beta\mathbf{D}_h] - \mathbf{I}) \odot \mathbf{W}_h\|$ from the eigenvalues of $\mathrm{diag}(\mathbf{W}_h)$. The norm of the perturbative term $(\exp[-\beta\mathbf{D}_h] - \mathbf{I}) \odot \mathbf{W}_h$ can be upper-bounded as follows

$$
\begin{aligned}
\|(\exp[-\beta\mathbf{D}_h] - \mathbf{I}) \odot \mathbf{W}_h\| &\le \max_{i,j}\Big|(\exp[-\beta\mathbf{D}_h] - \mathbf{I})_{ij}\Big|\|\mathbf{W}_h\| \\
&= e^{-\beta/(N_h-1)}\|\mathbf{W}_h\|,
\end{aligned}
$$

from which it follows the thesis. $\square$

We provide in Fig. 4 some visual examples of the spectrum of $\widetilde{\mathbf{W}}_h(\beta)$ varying the value of β in the range $[0.01, 1000]$. When $\beta \gg 1$, the masked recurrent matrix $\widetilde{\mathbf{W}}_h(\beta)$ pushes its eigenvalues along the real axis.

We conclude this theoretical section deriving from (10) the following bound on the spectral radius of $\widetilde{\mathbf{W}}_h(\beta)$, which we denote as $\rho(\widetilde{\mathbf{W}}_h(\beta))$:

$$
\Big|\rho(\widetilde{\mathbf{W}}_h(\beta)) - \max_{i=1,\dots,N_h} |(\mathbf{W}_h)_{ii}|\Big| \le e^{-\beta/(N_h-1)}\|\mathbf{W}_h\|. \tag{11}
$$

The inequality in Eq. (11) informs us that, for large values of β, the spectral radius of $\widetilde{\mathbf{W}}_h(\beta)$ converges to the maximum value on the diagonal of $\mathbf{W}_h$. We deduce that randomly generating $\mathbf{W}_h$ with i.i.d. entries distributed uniformly in $(-\rho_{\mathrm{des}}, \rho_{\mathrm{des}})$ gives, for sufficiently large values of β, a spectral radius $\rho(\widetilde{\mathbf{W}}_h(\beta)) \approx \rho_{\mathrm{des}}$. Notice also that as a corollary, the imaginary part of the spectrum tends to zero as β increases, as can be observed in the leftmost plots of Fig. 4, thus affecting the quality of the RNN dynamics.

6   Experiments

We evaluate our proposed approach in two experimental settings. First, in Sect. 6.1, we consider the memory capacity task, and then, in Sect. 6.2, we test it on time-series reconstruction.

6.1   Memory Capacity

In the Memory Capacity (MC) task, we investigate the role of continuous depth parameter β on the ability of C-DESN to preserve information on past inputs. In particular, we compare the β role in C-DESN architecture with the role of the number of layers in DeepESN. We, therefore, measure the short-term memory capacity [8,14] of our models, that is, how well they are capable of recalling delayed versions of the input sequence based on the current hidden state. In MC tasks, we consider 10000 scalar input sequences $\mathbf{x}[t]$ with $t = [1, 2, ..., T]$. We allocate 5000 sequences for training, 2500 sequences for validation, and test the model on the remaining 2500 sequences. A linear readout is trained via pseudo-inverse to find $\mathbf{W}_y$ that minimize the distance between the output network $\mathbf{y}[T] = \mathbf{W}_y\mathbf{h}[T]$ and the target $\hat{\mathbf{y}}$ where $\hat{\mathbf{y}}_\tau = \mathbf{x}[T - \tau]$ with $\tau = [1, 2, ..., \Delta]$ and $\Delta = 200$. The memory capacity score for sequence p is defined as:

$$\mathrm{MC}_\Delta^{(p)} = \sum_{\tau=1}^{\Delta} \mathrm{corr}^2\left(\mathbf{y}_\tau^{(p)}, \hat{\mathbf{y}}_\tau^{(p)}\right). \tag{12}$$

We recall [14] that the best achievable MC for a N_h-units reservoir is MC $= N_h$.

Settings. We consider two tasks, in the first (called in this work *Random_MC* task) we generate 10000 scalar input sequences $\mathbf{x}[t]$ with $t = [1, 2, ..., 300]$ where each $\mathbf{x}[t]$ value is sampled i.i.d. from the uniform distribution in the range $[-0.8, 0.8]$. On second task (called in this work *MNIST_MC* task), we consider 10000 sequences from MNIST dataset $\mathbf{x}[t]$ with $t = [1, 2, ..., 500]$. In order to have a similar setting as in Random_MC task we rescaled the values of the MNIST dataset to the range $[-0.8, 0.8]$. For these tasks, we consider a grid search with 10 values in a logarithmic scale of $\rho \in [0.1, 10]$, a total number of recurrent units 100, and input scaling $\omega = 0.1$. In the case of C-DESN we consider 10 values in the logarithmic scale of $\beta \in [10^{-2}, 10^3]$ while in the case of DeepESN we consider a number of layers L from 1 to 10 (all layers sharing the same values of hyperparameters). It is worth mentioning that a C-DESN with high β value is very similar to a DeepESN with $L = 1$, which in turn is equivalent to a simple ESN.

Results. We present the results obtained on MC tasks comparing C-DESN and DeepESN considering different β and L for C-DESN and DeepESN, respectively. Figure 5a shows the results obtained by C-DESN by varying the β value on Random_MC task. The best result obtained by C-DESN is 24.37 MC with a β of 5.99. In the shallow case (with $\beta = 0.01$), the model achieved 21.15 MC. As depicted in Fig. 5a, optimizing the β value up to 5.99 leads to the best results. Beyond this point, with higher β values, the model's MC starts to significantly decrease.

Figure 5b shows the results obtained by DeepESN with $L = [1, 2, ..., 10]$ in the Random_MC task. The best result obtained by DeepESN is 26.28 MC with $L = 4$. Notably, this result is very similar to the C-DESN case. In the shallow

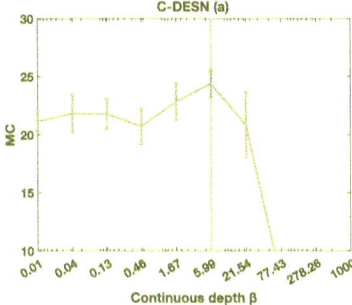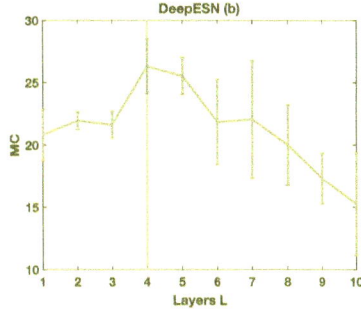

Fig. 5. MC results obtained on Random_MC task by C-DESN (a) and DeepESN (b) by varying the continuous depth β and the discrete depth L.

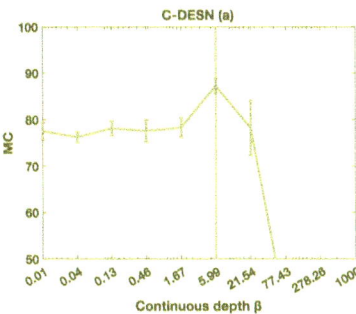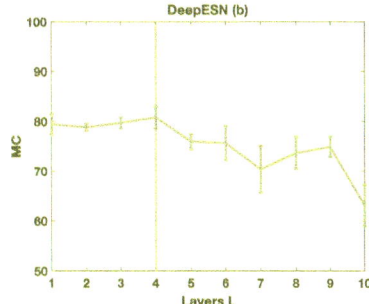

Fig. 6. MC results obtained on MNIST_MC task by C-DESN (a) and DeepESN (b) by varying the continuous depth β and the discrete depth L.

case (with $L = 1$), the model achieved 20.82. Similar to C-DESN, it is observed that beyond a certain depth value ($L = 4$ in this case), the MC achieved by DeepESN tends to deteriorate.

Now we compare the models on the MNIST_MC task. Figure 6a shows the results obtained by C-DESN by varying the β value on MNIST_MC task. The best result obtained by C-DESN is 87.27 MC with a β of 5.99. In the shallow case (with $\beta = 0.01$), the model achieved 77.56 MC. As illustrated in Fig. 6a, it can be observed that by increasing β up to 5.99, optimal results are reached. Beyond this point, with higher β values, the model's MC begins to significantly decrease.

Figure 6b shows the results obtained by DeepESN with $L = [1, 2, ..., 10]$ on MNIST_MC task. The best result achieved by DeepESN is 80.79 MC with $L = 4$, whereas in the shallow case (with $L = 1$), the model obtained 79.42.

Interestingly, we can observe that for both models, varying β or L yields very similar results in both the shallow case (with $\beta = 0.01$ and $L = 1$) and the deep case ($\beta = 5.99$ and $L = 4$). In Random_MC task, both models achieved significantly better results in the deep case compared to the shallow case. While in MNIST_MC task, the optimization of beta allows C-DESN to significantly

Table 1. Statistics of real-world time-series

Name	Sequences	Length	Input dimension
Arabic	6599	93	13
Blink	500	510	4
Epilepsy	137	206	3
Phoneme	214	1024	1

improve the MC while the optimization of the number of layers allows DeepESN to obtain a small improvement than C-DESN. Furthermore, in both cases, if we have too low β or too high L value, the models exhibit a decrease in memory MC. This provides empirical evidence that a C-DESN has MC that resembles those of hierarchical recurrent architectures. Moreover, the β value plays a similar role as the number of layers in a deep recurrent neural network for the MC of the model.

6.2 Time-Series Reconstruction

In this set of experiments, we consider four real-world collections of sequences whose statistics are reported in Table 1. Arabic [12] is a collection of audio sequences spoken in Arabic by different speakers. Blink [4] is a collection of 4-channels EEGs recorded on different subjects during eye blinking. Epilepsy [22] is a collection of accelerometer data on the dominant wrist collected on different subjects during normal activities and seizures. Finally, Phoneme [13] is a collection of segmented audio phonemes by different speakers.

Settings. We evaluate the ability of C-DESNs to reconstruct previous inputs $\mathbf{x}[T-\tau]$ at different delays $\tau > 0$ from the final state $\mathbf{h}[T]$. The datasets are split into 60:20:20 training/validation/test sets. As in the memory capacity experiments, we consider reservoirs of 100 units and select values of $\beta \in [10^{-2}, 10^3]$ and $\rho \in [10^{-1}, 10^1]$ according to the performance on the validation set. We set input scaling to $\omega = 1$. The readout is trained by ridge regression, and we select the corresponding regularization in the range $[10^{-12}, 10^{-1}]$.

Results. In Fig. 7 we report curves of the test set mean squared error (MSE) of the reconstructed inputs for different delays and different values of β. In all four tasks, we notice a valley around the optimal value of β, evidenced by the red line in the plots; the corresponding connectivity patterns are reported on the right. This behavior is most pronounced on the Blink task, where reconstruction delays up to 50 are considered, and the optimal β is ≈ 21.55. In the remaining tasks in which we considered delays up to 8–12, the valley around the optimum is less pronounced, and the optimal value is a slightly smaller $\beta \approx 6.00$. These choices of β correspond to a trade-off between depth and width in the connectivity patterns of C-DESNs: (a) for larger values of β the units become poorly connected to

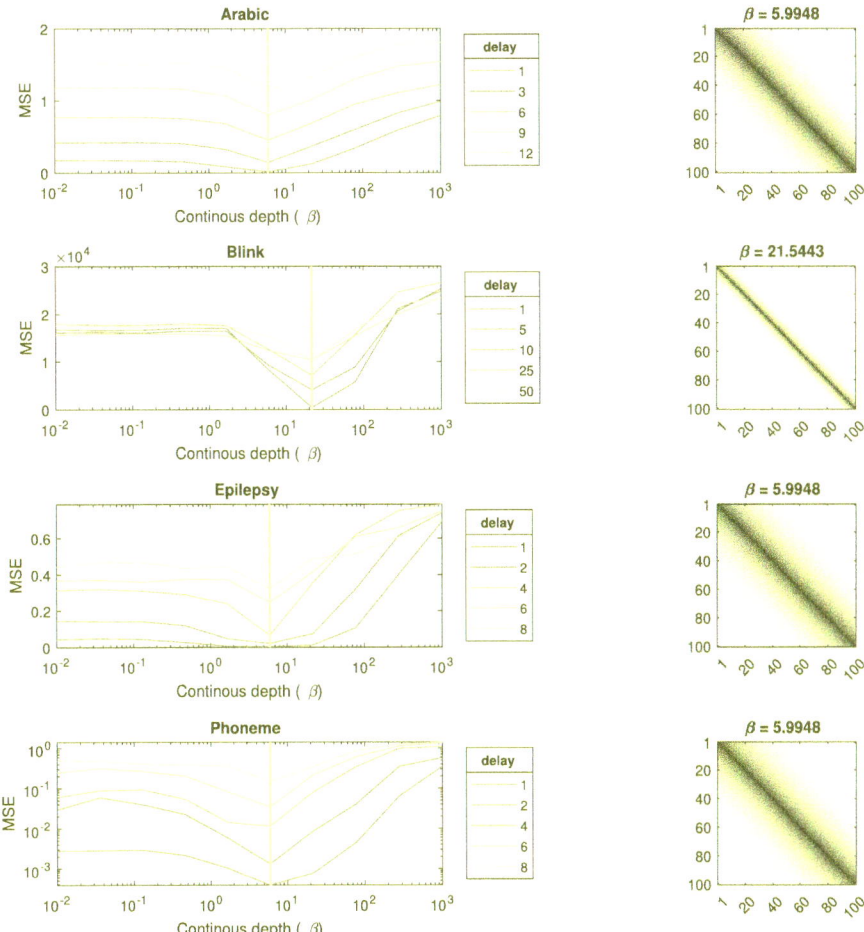

Fig. 7. Reconstruction of delayed inputs on four real-world tasks by C-DESN with 100 units. The connectivity of the best β highlighted with the red bar on the left plots is displayed on the right of each task.

each other, and in the limit completely disconnected for $\beta \to \infty$ as the recurrent matrix $\mathbf{W}_h$ becomes diagonal; while (b) for smaller values of β the connectivity loses depth, and thus the ability to hold longer history on the input sequence as observed in previous sections.

7   Conclusions

In this paper, we have introduced a novel class of recurrent neural models under the name of Continuously-Deep Echo State Network (C-DESN). The proposed recurrent architecture is based on a brain-like local connectivity pattern among neurons and is framed in the context of Reservoir Computing (RC). In the paper, we have presented both analytical and experimental evidence that the resulting dynamics of the recurrent hidden layer in C-DESN have the flexibility to process temporal data latching information on past inputs over multiple time steps and establishing longer-term memorization similarly to conventional deep recurrent networks. Crucially, the effective depth of the developed temporal processing is determined by a single non-architectural hyperparameter β. As a result, the depth of the network can be tuned to solve a specific task simply by fine-tuning a single value, without changing the underlying architecture, providing a clear design advantage over conventional deep recurrent systems.

In our future work, we plan to expand the potential of the presented C-DESN architecture into the broader landscape of fully trainable systems, while investigating possible outcomes in terms of neuromorphic hardware implementations.

Acknowledgments. This work has been supported by EU-EIC EMERGE (Grant No. 101070918), by NEURONE, a project funded by the Italian Ministry of University and Research (PRIN 20229JRTZA), by PNRR - M4C2 - Investimento 1.3, Partenariato Esteso PE00000013 - "FAIR - Future Artificial Intelligence Research" - Spoke 1 "Human-centered AI", funded by the European Commission under the NextGeneration EU programme, and by the project BrAID under the Bando Ricerca Salute 2018 - Regional public call for research and development projects aimed at supporting clinical and organisational innovation processes of the Regional Health Service - Regione Toscana.

References

1. Bacciu, D., et al.: Teaching-trustworthy autonomous cyber-physical applications through human-centred intelligence. In: 2021 IEEE International Conference on Omni-Layer Intelligent Systems (COINS), pp. 1–6. IEEE (2021)
2. Bauer, F.L., Fike, C.T.: Norms and exclusion theorems. Numer. Math. **2**(1), 137–141 (1960)
3. Chaudhuri, R., Bernacchia, A., Wang, X.J.: A diversity of localized timescales in network activity. Elife **3**, e01239 (2014)
4. Chicaiza, K.O., Benalcázar, M.E.: A brain-computer interface for controlling IoT devices using EEG signals. In: 2021 IEEE Fifth Ecuador Technical Chapters Meeting (ETCM), pp. 1–6 (2021). https://doi.org/10.1109/ETCM53643.2021.9590711
5. Chien, H.Y.S., Honey, C.J.: Constructing and forgetting temporal context in the human cerebral cortex. Neuron **106**(4), 675–686 (2020)
6. De Caro, V., Gallicchio, C., Bacciu, D.: Continual adaptation of federated reservoirs in pervasive environments. Neurocomputing **556**, 126638 (2023)
7. Dominey, P.F., Ellmore, T.M., Ventre-Dominey, J.: Effects of connectivity on narrative temporal processing in structured reservoir computing. In: 2022 International Joint Conference on Neural Networks (IJCNN), pp. 1–8. IEEE (2022)
8. Gallicchio, C.: Short-term memory of deep RNN. In: Proceedings of the 26th European Symposium on Artificial Neural Networks, Computational Intelligence and Machine Learning (ESANN 2018), pp. 633–638 (2018)
9. Gallicchio, C., Micheli, A.: Echo state property of deep reservoir computing networks. Cogn. Comput. **9**(3), 337–350 (2017). https://doi.org/10.1007/s12559-017-9461-9

10. Gallicchio, C., Micheli, A.: Why layering in recurrent neural networks? A deep-esn survey. In: 2018 International Joint Conference on Neural Networks (IJCNN), pp. 1–8. IEEE (2018)
11. Gallicchio, C., Micheli, A., Pedrelli, L.: Deep reservoir computing: a critical experimental analysis. Neurocomputing **268**, 87–99 (2017). https://doi.org/10.1016/j.neucom.2016.12.089
12. Hammami, N., Sellam, M.: Tree distribution classifier for automatic spoken Arabic digit recognition. In: 2009 International Conference for Internet Technology and Secured Transactions (ICITST), pp. 1–4 (2009). https://doi.org/10.1109/ICITST.2009.5402575
13. Hamooni, H., Mueen, A.: Dual-domain hierarchical classification of phonetic time series. In: 2014 IEEE International Conference on Data Mining, pp. 160–169 (2014). https://doi.org/10.1109/ICDM.2014.92
14. Jaeger, H.: Short term memory in echo state networks. Technical report 152, German National Research Institute for Computer Science (2002)
15. Jaeger, H.: Short term memory in echo state networks. gmd-report 152. In: GMD-German National Research Institute for Computer Science (2002). http://www.faculty.jacobs-university.de/hjaeger/pubs/STMEchoStatesTechRep.pdf.Citeseer
16. Jaeger, H., Haas, H.: Harnessing nonlinearity: predicting chaotic systems and saving energy in wireless communication. Science **304**(5667), 78–80 (2004)
17. Mante, V., Sussillo, D., Shenoy, K.V., Newsome, W.T.: Context-dependent computation by recurrent dynamics in prefrontal cortex. Nature **503**(7474), 78–84 (2013)
18. Nakajima, K., Fischer, I.: Reservoir Computing. Springer, Singapore (2021). https://doi.org/10.1007/978-981-13-1687-6
19. Tanaka, G., et al.: Recent advances in physical reservoir computing: a review. Neural Netw. **115**, 100–123 (2019)
20. Tortorella, D., Gallicchio, C., Micheli, A.: Hierarchical dynamics in deep echo state networks. In: Pimenidis, E., Angelov, P., Jayne, C., Papaleonidas, A., Aydin, M. (eds.) ICANN 2022. LNCS, vol. 13531, pp. 668–679. Springer, Cham (2022). https://doi.org/10.1007/978-3-031-15934-3_55
21. Verstraeten, D., Schrauwen, B., d'Haene, M., Stroobandt, D.: An experimental unification of reservoir computing methods. Neural Netw. **20**(3), 391–403 (2007)
22. Villar, J.R., Vergara, P., Menéndez, M., de la Cal, E., González, V.M., Sedano, J.: Generalized models for the classification of abnormal movements in daily life and its applicability to epilepsy convulsion recognition. Int. J. Neural Syst. **26**(06), 1650037 (2016). https://doi.org/10.1142/S0129065716500374
23. Yildiz, I.B., Jaeger, H., Kiebel, S.J.: Re-visiting the echo state property. Neural Netw. **35**, 1–9 (2012)

Dynamics Adaptive Safe Reinforcement Learning with a Misspecified Simulator

Ruiqi Xue[1,2], Ziqian Zhang[1,2], Lihe Li[1,2], Feng Chen[1,2], Yi-Chen Li[1,2], Yang Yu[1,2,3], and Lei Yuan[1,2,3(✉)]

[1] National Key Laboratory for Novel Software Technology, Nanjing University, Nanjing, China
yuanl@lamda.nju.edu.cn
[2] School of Artificial Intelligence, Nanjing University, Nanjing, China
ruiqixue@smail.nju.edu.cn
[3] Polixir Technologies, Nanjing, China

Abstract. Sim-to-real reinforcement learning offers the advantage of learning safe policies within simulators, circumventing the need for costly trial-and-error in the real world. Traditional approaches often rest on the assumption of consistent state-action transition between the simulator and the real-world environment. However, this assumption can be violated due to the poor fidelity of simulators, leading to a constrained trust region for effective policy learning. The limitation can be more pronounced when safety issues are considered, potentially resulting in threatening policies if no safe samples exist in the trust region. To overcome these challenges, we propose **D**ynamics **A**daptive **S**afe **R**einforcement Learning with a Misspecified Simulator (DASaR). Our approach begins by relaxing the assumption to expand the trust region and theoretically demonstrate the unbounded performance gap inherent in traditional methods. Subsequently, DASaR aligns the estimated value functions in the simulator and the real-world environment via inverse dynamics-based relabeling of reward and cost signals. Furthermore, to deal with the underestimation of cost value functions, DASaR employs uncertainty estimation to improve its conservatism, ensuring the safety of the learned policy. Experiments in various complex environments thoroughly demonstrate DASaR's outstanding ability to balance safety satisfaction and reward maximization across diverse dynamics gaps.

Keywords: Safe Reinforcement Learning · Domain Adaptation · Sim-to-real Reinforcement Learning

1 Introduction

Learning optimal policies through reinforcement learning (RL) has demonstrated outstanding performance in various fields, including game playing [35], recom-

---

Supplementary Information The online version contains supplementary material available at https://doi.org/10.1007/978-3-031-70368-3_5.

mendation systems [2], and robotic control [19]. Nonetheless, the application of RL in real-world settings, which often come with multiple constraints, presents the critical challenge of ensuring policy safety. For instance, while an RL agent can maximize rewards by driving autonomous vehicles at high speeds, it must also comply with speed limits to prevent collisions [20]. Safe RL [13], which balances reward maximization and constraint satisfaction, can effectively improve the safety of the learned policies. However, the real-world environment is not amenable to such trial-and-error learning due to numerous costly errors caused during policy exploration [25]. To reduce the risk and cost in the real-world environment, the strategy of training policies within a high-fidelity simulator before deployment has emerged as a viable and efficient paradigm for safe RL.

However, it is challenging to construct high-fidelity simulators of systems with complex physical laws, leading to significant dynamics gaps between the simulator and the real-world environment [32]. Directly deploying policies learned within such simulators can result in substantial performance declines [18]. To tackle the problem, sim-to-real RL enhances the robustness of the policies against different dynamics [37]. Domain randomization creates augmented environments with different physical parameters to enhance policy generalization [28,32], but it necessitates expert knowledge of simulators and the training complexity scales with the number of variations. Alternatively, domain adaptation presumes access to a limited set of offline real-world samples [46]. It strives for better usage of online simulator explorations by training classifiers to differentiate between the simulator and the real-world transitions. Such classifiers then aid in providing intrinsic rewards or filtering data, steering the policy towards regions where the simulator and real-world dynamics align closely [10,30].

Despite the effective adaptation to dynamics changes of past domain adaptation approaches, they all assume that every possible transition in the real world will also have a non-zero probability in the simulator [11]. This enables the estimation of the policy's value function within the simulated environment as a bound for real-world returns. However, it can be violated due to poor fidelity of the simulator, leading to unbounded performance gaps. In this case, only transitions in the intersection between offline data and the support set of the dynamics transition function in the simulator, referred to as the trust region, can be fully leveraged for policy optimization. Furthermore, the lack of safe transitions within this trust region could lead to significant safety issues.

Addressing the mentioned challenges, this paper presents the approach of learning safe policies in misspecified simulators using a limited amount of offline data from the real-world environment. Specifically, we propose a novel algorithmic framework named **D**ynamics **A**daptive **S**afe **R**einforcement Learning with a Misspecified Simulator (DASaR). We first relax the assumption to the state-transition level, not only allowing the method to be applied to a broader range of simulators but also serving as a foundation for the extension of the trust region. Afterward, to avoid the unbounded performance gap, we align the estimated value functions in the simulator and the real-world environment via inverse dynamics-based relabeling. Finally, as the inaccuracy of inverse dynamics models leads to the underestimation of cost critics, we strengthen the conservatism via uncertainty estimation, thus ensuring the safety of the learned policy. Through extensive experiments on safe tasks within MuJoCo environments [40], we demonstrate that our approach can achieve superior performance in balancing safety satisfaction and reward maximization under various dynamics gaps.

2 Related Work

2.1 Safe Reinforcement Learning

Safety has been one of the major roadblocks in the way of deploying RL policies to the real world [13,14]. To solve the problem, Safe RL typically models the environment as a Constrained Markov Decision Process (CMDP) [3] and employs constrained optimization methods for policy learning [5,13]. Lagrangian-based methods are traditional approaches to solving constrained optimization problems and are widely used in Safe RL. These methods utilize a learnable multiplier to penalize violations of constraints [7,36,38]. Additionally, trust region methods have been proposed to maintain policies within a safe trust region via low-order Taylor expansions [1,15,43] or variational inference [24,44]. Despite their advantages, both Lagrangian-based and trust region methods encounter difficulties in preventing unsafe interactions during the trial-and-error phase. An emerging approach is offline safe RL, which aims to learn policies ensuring safety exclusively from offline data, thus avoiding unsafe explorations [22,23,25,42]. However, the efficacy of offline safe RL heavily depends on the quantity and quality of offline data, restricting the application in limited offline data scenarios. Our previous work [16] attempts to solve the challenge of cost function changes in offline safe RL, but it also struggles when changes in dynamics appear.

2.2 Sim-to-Real Reinforcement Learning

To deal with the dynamics gap between the simulator and the real-world environment, sim-to-real RL studies the problems in two contexts based on the accessibility of real-world data. Without samples from real-world environments, zero-shot sim-to-real RL unleashes the potential of simulators to ensure the robustness of polices [28]. As a widely-used solution, domain randomization necessitates expert knowledge of the simulator and creates augmented environments by randomizing the physical parameters of the simulator, thus training policies robust to dynamics changes [27–29,39]. When real-world samples can be obtained, different methods are developed in few-shot sim-to-real RL. Among them, simulator calibration methods improve the accuracy of simulators by adjusting physical parameters based on offline data [6,9,12,34]. However, the requirement of expert knowledge is not always feasible. Meanwhile, some approaches align the dynamics between the simulator and the real-world environment via inverse dynamics model learned from offline data [8], but the generalization error emerges as a critical challenge. Alternatively, domain adaptation methods have gained widespread attention in recent years. DARC [11] employs two classifiers to distinguish transitions between the simulator and the real-world environment, providing intrinsic rewards to guide the policy. H2O [30] combines the conservative regularization term of CQL [21] with the classifiers, filtering simulator samples with a small dynamics gap for policy optimization. However, the assumption of transition dynamics limits their applicability and neither of the mentioned methods takes the safety issue into consideration.

3   Problem Formulation

We model the standard safe RL problem as a Constrained Markov Decision Process (CMDP) denoted as $\mathcal{M} = (S, A, P_{\mathcal{M}}, R, C, \rho, \gamma, b)$, where S and A represent the state space and the action space, $R : S \times A \times S \rightarrow [R_{\min}, R_{\max}]$ and $C : S \times A \times S \rightarrow \{0, 1\}$ denote the reward and cost functions respectively. $P_{\mathcal{M}} : S \times A \rightarrow \Delta S$ is the transition dynamics function. ρ represents the initial state distribution, $\gamma \in (0, 1)$ is the discount factor, and b represents the safety constraint limit. Specifically, for a deterministic CMDP $\mathcal{M}$ where there is no uncertainty on the next state s' given s and a, we can describe the transition dynamics with a deterministic transition function $f_{\mathcal{M}} : S \times A \rightarrow S$. For the given policy $\pi : S \mapsto \Delta(A)$ which specifies the action distribution on state s, the expected reward return and cost return within the CMDP $\mathcal{M}$ can be expressed as $J_{\mathcal{M}}^R(\pi) = \mathbb{E}_{s_0 \sim \rho, a_t \sim \pi(\cdot|s_t), s_{t+1} \sim P_{\mathcal{M}}(\cdot|s_t, a_t)}[\sum_{t=0}^{\infty} \gamma^t R(s_t, a_t, s_{t+1})]$ and $J_{\mathcal{M}}^C(\pi) = \mathbb{E}_{s_0 \sim \rho, a_t \sim \pi(\cdot|s_t), s_{t+1} \sim P_{\mathcal{M}}(\cdot|s_t, a_t)}[\sum_{t=0}^{\infty} \gamma^t C(s_t, a_t, s_{t+1})]$, respectively.

The safe sim-to-real RL problem involves two CMDPs: $\mathcal{M}_S$ representing the source domain (simulator) and $\mathcal{M}_T$ representing the target domain (real world). The main difference between $\mathcal{M}_S$ and $\mathcal{M}_T$ lies in their transition dynamics functions $P_{\mathcal{M}_S}$ and $P_{\mathcal{M}_T}$, which we abbreviate as P_S and P_T. The objective is using interactions in $\mathcal{M}_S$ together with a small number of offline data $\mathcal{D}_T$ sampled from $\mathcal{M}_T$, under the given reward function R and cost function C, to acquire a policy π that serves as the optimal solution to the following constrained optimization problem:

$$\max_{\pi} J_{\mathcal{M}_T}^R(\pi),$$
$$s.t. \ J_{\mathcal{M}_T}^C(\pi) \leq b. \tag{1}$$

To facilitate later analysis, we introduce the *discounted stationary state transition occupancy* $d_{P_{\mathcal{M}}}^{\pi}(s, s') = (1 - \gamma)\mathbb{E}_{\rho, \pi, P_{\mathcal{M}}}[\sum_{t=0}^{\infty} \gamma^t \mathbb{P}(s_t = s, s_{t+1} = s')]$, and the *discounted stationary state-action transition occupancy* $\mu_{P_{\mathcal{M}}}^{\pi}(s, a, s') = (1 - \gamma)\mathbb{E}_{\rho, \pi, P_{\mathcal{M}}}[\sum_{t=0}^{\infty} \gamma^t \mathbb{P}(s_t = s, a_t = a, s_{t+1} = s')]$. Intuitively, they measure the overall frequency of visiting a specific state(-action) transition. For simplicity, we will omit "discounted stationary" throughout.

As is introduced in Sect. 2.2, domain adaptation methods make the common assumption that every possible state-action transition (s, a, s') in the target domain also has non-zero probability in the source domain. However, this could be violated due to the poor fidelity of the simulator. In this paper, we loosen it and make the following assumption:

Assumption 1. *For every possible state transition in the deterministic target domain, it has a non-zero probability in the deterministic source domain.*

$$\forall s, s' \in S, (\exists a \in A, P_T(s'|s,a) > 0 \Rightarrow \exists a' \in A, P_S(s'|s,a') > 0). \quad (2)$$

Intuitively, by changing the restriction from state-action transition space to state transition space, the assumption can be satisfied more easily, allowing for broader application. We here only study the problem of deterministic CMDPs for that it is hard to capture the probabilistic information with a small offline dataset, further studies in non-deterministic settings are encouraged.

4   Method

In this section, we will provide a detailed description of our proposed DASaR. This algorithm is designed to learn safe and high-performance policies within simulators having dynamics gaps. All proofs, additional theoretical results and implementation details are provided in our full paper at the following link: http://www.lamda.nju.edu.cn/lilh/file/dasar.pdf.

4.1   Theoretical Motivation

In safe sim-to-real RL, we aim to learn the policy with high performance and safety guarantees in the real-world with only an offline dataset and a simulator. While direct performance evaluations may not be feasible, we can evaluate the performance gap between policies in the simulator and the real-world environment, as demonstrated in Proposition 1.

Proposition 1. *For any two policies π_1, π_2 and two CMDPs $\mathcal{M}_S, \mathcal{M}_T$, the following hold for expected reward and cost returns of policies within CMDPs:*

$$J^R_{\mathcal{M}_T}(\pi_2) \geq J^R_{\mathcal{M}_S}(\pi_1) - \frac{\sqrt{2}R_{max}}{1-\gamma}\sqrt{\mathbb{D}_{KL}(\mu^{\pi_1}_{P_S}(s,a,s')||\mu^{\pi_2}_{P_T}(s,a,s'))}, \quad (3)$$

$$J^C_{\mathcal{M}_T}(\pi_2) \leq J^C_{\mathcal{M}_S}(\pi_1) + \frac{\sqrt{2}}{1-\gamma}\sqrt{\mathbb{D}_{KL}(\mu^{\pi_1}_{P_S}(s,a,s')||\mu^{\pi_2}_{P_T}(s,a,s'))}, \quad (4)$$

where $\mathbb{D}_{KL}(\cdot||\cdot)$ is the KL divergence, $\mu^{\pi_1}_{P_S}(s,a,s')$ and $\mu^{\pi_2}_{P_T}(s,a,s')$) are state-action transition occupancies of π_1 and π_2 within $\mathcal{M}_S$ and $\mathcal{M}_P$, respectively.

Intuitively, with a well-performed policy π_1 in the simulator, we could enhance the performance of π_2 in the real world by minimizing the KL divergence between two state-action transition occupancies. The cost return of π_2 can be minimized in the same way. Since the traditional domain adaptation methods make the assumption that every possible state-action transition in the real world also has a non-zero probability in the simulator, the KL divergence term can be effectively minimized within a broad trust region. However, such traditional approaches will fail under the loosened Assumption 1. To facilitate further analysis, we introduce the concept of inverse dynamics probability first:

Definition 1. *Inverse Dynamics Probability.* *The inverse dynamics probability $\rho_{\mathcal{M}}^{\pi}(a|s, s')$ given the policy π and CMDP $\mathcal{M}$ is defined as:*

$$\rho_{\mathcal{M}}^{\pi}(a|s, s') := \frac{P_{\mathcal{M}}(s'|s, a)\pi(a|s)}{\int_A P_{\mathcal{M}}(s'|s, \bar{a})\pi(\bar{a}|s)d\bar{a}}. \tag{5}$$

As $\rho_{\mathcal{M}}^{\pi}$ is injective and independent of π when $\mathcal{M}$ is deterministic, we denote it as $\rho_{\mathcal{M}}$ for brevity. Meanwhile, we can further define the **deterministic inverse dynamics function** *$g_{\mathcal{M}}(s, s')$ given the inverse dynamics probability $\rho_{\mathcal{M}}$:*

$$\forall s, a, s', (g_{\mathcal{M}}(s, s') = a \;\Leftrightarrow\; \rho_{\mathcal{M}}(a|s, s') = \delta(0)), \tag{6}$$

where δ is the Dirac delta function [4].

With the definition of inverse dynamics probability function, we can further decompose the KL divergence term as is shown in Theorem 1:

Theorem 1. *For any two policies π_1 and π_2, and any two CMDPs $\mathcal{M}_S$ and $\mathcal{M}_T$, the following holds:*

$$\mathbb{D}_{KL}(\mu_{P_S}^{\pi_1}(s, a, s')||\mu_{P_T}^{\pi_2}(s, a, s')) = \underbrace{\mathbb{D}_{KL}(d_{P_S}^{\pi_1}(s, s')||d_{P_T}^{\pi_2}(s, s'))}_{term(a)}$$
$$+ \underbrace{\mathbb{E}_{(s, s')\sim d_{P_S}^{\pi_1}}[\mathbb{D}_{KL}(\rho_{\mathcal{M}_S}(a|s, s')||\rho_{\mathcal{M}_T}(a|s, s'))]}_{term(b)}. \tag{7}$$

Theorem 1 decomposes the KL divergence term between the state-action transition occupancy into terms (a) and (b). Although term (a) can be optimized to 0 under Assumption 1, term (b) is an uncontrollable factor. This implies that all past domain adaptation approaches that estimate $J_{\mathcal{M}_S}^R(\pi)$ and $J_{\mathcal{M}_S}^C(\pi)$ in the simulator and minimize the KL divergence term $\mathbb{D}_{KL}(\mu_{P_S}^{\pi_1}(s, a, s')||\mu_{P_T}^{\pi_2}(s, a, s'))$ will fail to bound the performance gap.

4.2 Value Estimation Alignment with an Inverse Dynamics Model

DASaR is a framework-agnostic module designed for learning safe policies, any off-policy safe RL methods can be combined with it. We here choose a widely-used safe RL algorithm SAC_Lagrange [33] for its simplicity and efficiency. Specifically, it learns the reward and cost state-action value functions via the maximum entropy Bellman operator and standard Bellman operator respectively:

$$\hat{Q}^R(s, a) = R(s, a, s') + \gamma\mathbb{E}_{a'\sim\pi(\cdot|s')}[\hat{Q}^R(s', a') - \xi\log(\pi(a'|s'))], \tag{8}$$

$$\hat{Q}^C(s, a) = C(s, a, s') + \gamma\mathbb{E}_{a'\sim\pi(\cdot|s')}[\hat{Q}^C(s', a')], \tag{9}$$

where ξ is a hyperparameter. Then $\mathbb{E}_{s\sim\rho, a\sim\pi(\cdot|s)}[\hat{Q}^R(s, a)]$, $\mathbb{E}_{s\sim\rho, a\sim\pi(\cdot|s)}[\hat{Q}^C(s, a)]$ are used as approximations of $J_{\mathcal{M}}^R(\pi)$ and $J_{\mathcal{M}}^C(\pi)$ to guide the policy update. As is discussed in Sect. 4.1, directly applying SAC_Lagrange in the simulator may lead to an unbounded performance gap in both reward and cost return under Assumption 1. However, if we can have access to the critic values within the target domain, the policy can be optimized to directly maximize the reward return while restricting its cost value when deployed in real-world environments. The difficulty lies in the lack of reward and cost signals from the real world. As the reward and cost functions are available in simulators, we can relabel the signals of transition (s, a, s') into $R(s, a', s')$ and $C(s, a', s')$ if given the appropriate action $a' = g_{\mathcal{M}_T}(s, s')$, which implies that (s, a', s') is possible in real-world environments. Accordingly, we first learn the deterministic inverse dynamics model g_I to mimic $g_{\mathcal{M}_T}$ using samples from offline datasets D_T:

$$\min_{g_I} \mathbb{E}_{(s,a,s') \sim D_T}[||g_I(s, s') - a||_2^2]. \tag{10}$$

With the well-trained inverse dynamics model g_I, we can now relabel the reward and cost signals of transition (s, a, s'), which is sampled from the simulator, by defining the corrected reward and cost functions as follows:

$$r' = R'(s, a, s') = R(s, g_I(s, s'), s'), \tag{11}$$
$$c' = C'(s, a, s') = C(s, g_I(s, s'), s'). \tag{12}$$

The relabeled transition (s, a, r', c', s') will then be used for the updates of state-action value functions $\hat{Q}^R(s, a)$ and $\hat{Q}^C(s, a)$ as introduced in Eqs. (8) and (9), thus achieving the alignment of value estimation.

Although the relabeled transition (s, a, r', c', s') helps align the value functions $\hat{Q}^R(s, a)$ and $\hat{Q}^C(s, a)$, the action $a \sim \pi(\cdot|s)$ cannot lead to the transition of s' in real-world environments due to the dynamics gap. Instead, we utilize the learned inverse dynamics model to derive the desired action via the composition of two models $\pi \circ g_I$, which is defined as:

$$\pi \circ g_I(a|s) = \int_A \pi(a'|s)\delta(g_I(s, f_S(s, a')) - a)da', \tag{13}$$

where f_S is the deterministic state transition function of $\mathcal{M}_S$. Intuitively, we first roll out the policy π in the simulator to get $s' = f_S(s, a')|_{a' \sim \pi(\cdot|s)}$ and derive the executable action in the real world via the inverse dynamics model $a = g_I(s, s')$. In other words, the state-action value functions are equivalently used to estimate the performance of the policy $\pi \circ g_I$ in real-world environments. We then theoretically prove that both the reward and cost returns of such policies can be bounded and further justify the approach. To promote the analysis, we define the Inverse Dynamics Induced CMDP as follows:

Definition 2. *Inverse Dynamics Induced CMDP*. *Given a deterministic inverse dynamics model $g(s, s')$, it can induce a CMDP with deterministic transition dynamics function f satisfying the following property:*

$$\forall s, a, s', (g(s, s') = a \iff f(s, a) = s'). \tag{14}$$

The reward and cost functions of such inverse dynamics induced CMDP can be designated the same as those in the simulator or the real-world environment.

It can be shown that, with the inverse dynamics model g_I that mimics $g_{\mathcal{M}_T}$ perfectly, the CMDP $\mathcal{M}_I$ induced by g_I is exactly the same as the target domain $\mathcal{M}_T$. Furthermore, we illustrate that the optimization of policies under the aligned value functions is consistent with optimization in the induced CMDP $\mathcal{M}_I$ in Theorem 2.

Theorem 2. *For a given inverse dynamics model g_I and its induced CMDP $\mathcal{M}_I$ with transition dynamics function P_I, the following equations hold: $d_{P_S}^{\pi}(s,s') = d_{P_I}^{\pi \circ g_I}(s,s')$, $J_{\mathcal{M}_S}^{R'}(\pi) = J_{\mathcal{M}_I}^{R}(\pi \circ g_I)$, $J_{\mathcal{M}_S}^{C'}(\pi) = J_{\mathcal{M}_I}^{C}(\pi \circ g_I)$, where $J_{\mathcal{M}_S}^{R'}(\pi)$ and $J_{\mathcal{M}_S}^{C'}(\pi)$ are the expected return of π under the simulator $\mathcal{M}_S$ evaluated by the corrected reward and cost functions R' and C', defined in Eq. 11.*

Although Theorem 2 proves that we can optimize the policy $\pi \circ g_I$ under the induced CMDP $\mathcal{M}_I$ through updates in simulators, the differences between $\mathcal{M}_I$ and the target domain $\mathcal{M}_T$ will hinder the performance improvement in the real world. The differences result from the inaccuracy and generalization ability of g_I, and the coverage of offline samples. To deal with the challenge, we show that the cost and reward returns can both be bounded by $\mathbb{D}_{\mathrm{KL}}(d_{P_I}^{\pi \circ g_I}(s,s')||d_{P_I}^{\pi_B}(s,s'))$ with the adequate assumption, which is equivalent to $\mathbb{D}_{\mathrm{KL}}(d_{P_S}^{\pi}(s,s')||d_{P_I}^{\pi_B}(s,s'))$ through Theorem 2. Here, π^B is the behavioral policy for collecting offline samples, corresponding theoretical results can be found in our full paper. Intuitively, this result comes from that g_I has higher accuracy on its training samples. Leveraging advancements in imitation from observation [41], we introduce a binary discriminator $K(s,s')$ which is optimized through:

$$\max_K \ \mathbb{E}_{d_{P_S}^{\pi}}[\ln K(s,s')] + \mathbb{E}_{d_{P_T}^{\pi_B}}[\ln(1 - K(s,s'))]. \tag{15}$$

Following the paradigm widely used in imitation learning, we introduce the additional reward signal $\Delta r = \ln \frac{1-K(s,s')}{K(s,s')}$ and optimize $\hat{Q}^R$ with $r' + \alpha \Delta r$ instead, where α is a hyperparameter. In this way, we successfully expand past methods' trust region based on state-action transitions to a new trust region based on state transitions, covering the whole offline dataset under Assumption 1.

4.3   Conservative Cost Critic Learning via Uncertainty Estimation

Although the value estimation alignment via inverse dynamics-based relabeling can effectively bound the performance gap between the simulator and the real-world environment under Assumption 1, the safety issue could still emerge as a challenge. On one hand, the inherent approximation errors in neural networks will affect the estimation of the cost critic. On the other hand, the inaccuracy of the inverse dynamics model may lead to misleading cost relabel. These might result in underestimation of the cost critic, making the policy execute unsafe actions with low cost critic values during deployment. To alleviate the problem, we introduce uncertainty estimation using model ensemble approaches.

To deal with the approximation errors of the cost critic's network, we train E_C base models, denoted as $\{\hat{Q}^{C,i}(s,a)\}_{i=1}^{E_c}$. Then, we use the upper confidence bound (UCB) [31] as the estimation of the expected cost return under (s,a):

$$\hat{Q}^{C,\text{UCB}}(s,a) = \mathbb{E}_{i\in\{1,...,E_C\}}[\hat{Q}^{C,i}(s,a)] + \beta_C \cdot \sqrt{\text{Var}_{i\in\{1,...,E_C\}}[\hat{Q}^{C,i}(s,a)]}, \quad (16)$$

where β_C is a hyperparameter.

As for the inaccuracy of the inverse dynamics model, we first analyze how it leads to the underestimation of the cost critic. We update the cost critic with relabeled cost value $C(s, g_I(s, s'), s')$, with the hope that $f_T(s, g_I(s, s')) = s'$ when g_I perfectly mimics $g_{\mathcal{M}_T}$ within the real world $\mathcal{M}_T$. However, the inaccuracy of g_I might lead to undesired state transition to $f_T(s, g_I(s, s')) = s' + e$ where e is the state transition error. Meanwhile, the aligned cost critic evaluates the expected cost return of policy $\pi \circ g_I$ under state-action pair $(s, g_I(s, s'))$ in the real world. Accordingly, the sample supposed to be used to update the critic is $(s, g_I(s, s'), s' + e)$ instead. As the lack of transition dynamics f_T of the real-world environment prevents us from attaining the state error e, we introduce the proxy error by model ensemble and pessimistic estimation.

First of all, we can approximate the action error $e^A = a' - g_I(s, s')$ with regards to the desired action $a' = g_{\mathcal{M}_T}(s, s')$ via uncertainty estimation. Specifically, we apply model ensemble by training E_I inverse dynamics models $\{g_I^i\}_{i=1}^{E_I}$. Without loss of generality, we assume the existence of β_T such that for any s, s', the following holds:

$$e^A \in [-\beta_T \cdot \sqrt{\text{Var}_{i\in\{1,...,E_I\}}[g_I^i(s,s')]}, \beta_T \cdot \sqrt{\text{Var}_{i\in\{1,...,E_I\}}[g_I^i(s,s')]}\,]. \quad (17)$$

After determining the range of the action error e^A, we can then decide the range of state error e based on the assumption on the continuity of f_T and f_S as follows:

Assumption 2. *Given two CMDPs $\mathcal{M}_S, \mathcal{M}_T$, together with their deterministic inverse dynamics functions g_S, g_T and deterministic state transition functions f_S, f_T, and the range of action error e_T^A in $\mathcal{M}_T$, there exists $\beta_S > 0$ so that the range of state error in $\mathcal{M}_T$ can be included by that in $\mathcal{M}_S$:*

$$\{f_T(s, g_T(s, s') + e_T^A)|e_T^A \in [-l, l]\} \subseteq \{f_S(s, g_S(s, s') + e_S)|e_S^A \in [-\beta_S \cdot l, \beta_S \cdot l]\}, \quad (18)$$

where l is the range of action errors in $\mathcal{M}_T$.

Intuitively, Assumption 2 tells that we can get all possible state errors within the real world by rollouts in the simulator, it is actually a direct result of Assumption 1. When f_T and f_S are continuous, the parameter β_S will not need to be a large number. Combining Eq. 17 and Assumption 2, we derive the range of action errors under simulators: $E_S^A = [-\beta_I \cdot \sqrt{\text{Var}_{i\in\{1,...,E_I\}}[g_I^i(s,s')]}, \beta_I \cdot \sqrt{\text{Var}_{i\in\{1,...,E_I\}}[g_I^i(s,s')]}\,]$, where $\beta_I = \beta_T \times \beta_S$ and we set it as a hyperparameter. Since it is impossible to give a certain value of action error, we make the pessimistic estimation and update the cost critic based on the term with the highest cost:

$$\mathcal{B}^e \hat{Q}^{C,i}(s,a) = \max_{e^A \in E_S^A} (C(s, g_I^E(s,s'), s^{e^A}) + \gamma \mathbb{E}_{a' \sim \pi(\cdot|s^{e'})}[\hat{Q}^{C,i}(s^{e^A}, a')]), \quad (19)$$

where $g_I^E(s,s') = \mathbb{E}_{i \subset \{1,\dots,E_I\}}[g_I^i(s,s')]$ is the expectation over all the inverse dynamics models, and $s^{e^A} = f_S(s, g_I^E(s,s') + e^A)$ is the simulated state under action with the error. By improving the conservatism of the cost critics, the safety of the learned policy can be further guaranteed.

5   Experiments

In this section, we present our experimental analysis conducted in four MuJoCo environments, each featuring three different simulators. The experiments are designed to address the following critical questions: (1) Can DASaR learn the policy that is both safe and high-performing in the real world via trial-and-error in different simulators (Sect. 5.2)? (2) What contributions do different components of DASaR make and how does the quantity of offline data impact its performance (Sect. 5.3)? (3) Does the policy learned by DASaR exhibit remarkable sim-to-real adaptation capabilities compared with other baselines (Sect. 5.4)? (4) How do different choices of hyperparameters in DASaR affect its performance (Sect. 5.5)?

For a thorough evaluation, we compare DASaR against multiple baselines. All experimental results within a simulator are averaged across ten evaluation episodes, five random seeds, and three different parameter values accompanied by standard deviation information. Detailed experimental information and additional results will be provided in our full paper.

5.1   Baselines and Environments

To thoroughly assess the performance of DASaR, we compare it against a range of baselines: (1) **SAC_Lagrange(sim)** [33] employs the Lagrangian version of SAC [17] to train a policy in the simulator and deploys it directly to the real world. (2) **SAC_Lagrange(id)** [8] also utilizes SAC_Lagrange for policy training in simulators but employs an inverse dynamics model learned from the offline dataset during real-world deployment. (3) **DARC_Lagrange** is the Lagrangian version of the sim-to-real approach, DARC [11], which utilizes two classifiers to differentiate transitions from the simulator and the real-world environment, thus constraining the policy learning to trust region with small dynamics gaps. (4) **H2O_Lagrange** also combines Lagrangian methods with H2O [30], which adds conservative regularization of CQL [21] into DARC's classifiers, filtering samples for policy updates. (5) **CPQ** [42] is a pure offline safe algorithm without explorations in simulators, emphasizing safety and conservatism via regularization.

For our evaluation, we chose four Gym MuJoCo environments [40]: Ant, HalfCheetah, Hopper, and Walker, and modified them for safety considerations. The offline dataset of each environment includes 100 trajectories, where 20 of them are collected from a safe policy and others are generated by diverse behavioral policies violating safety constraints. For each environment, we create three simulators, each differs from the real world in specific physical parameters, including gravity, friction, and density. The experiments are conducted within each simulator for three parameter values: 2.0, 1.5, and 0.5, indicating how much it deviates from the real-world parameter value 1.0.

Table 1. Average test return ± std across various environments. Rewards and costs are normalized via the offline dataset and safety constraint limit $b = 5$, respectively. If the normalized cost exceeds 1.0, the method with a lower cost is preferred, otherwise, the method with a higher reward is better. The method with the best performance in each simulator is emphasized in blue. Letters 'g', 'f', and 'd' represent simulators with different gravity, friction, and density parameter values compared with the real world.

Env		DASaR reward↑	DASaR cost↓	SAC_Lagrange(id) reward↑	SAC_Lagrange(id) cost↓	SAC_Lagrange(sim) reward↑	SAC_Lagrange(sim) cost↓
Ant	g	72.8 ± 14.2	1.3 ± 1.2	48.2 ± 16.5	3.4 ± 3.0	40.9 ± 29.3	5.9 ± 7.0
	f	78.8 ± 8.8	1.1 ± 1.4	52.8 ± 16.1	1.8 ± 1.4	51.6 ± 10.5	3.0 ± 2.2
	d	97.0 ± 4.7	0.6 ± 0.5	62.7 ± 16.5	1.1 ± 0.9	67.7 ± 20.9	2.5 ± 1.3
Cheetah	g	63.2 ± 8.4	1.7 ± 1.6	−184.9 ± 124.3	2.0 ± 2.6	28.0 ± 63.6	1.2 ± 2.0
	f	86.2 ± 8.5	1.2 ± 1.2	-106.0 ± 171.6	0.8 ± 1.4	65.1 ± 20.3	1.7 ± 1.9
	d	83.3 ± 6.0	0.5 ± 0.5	−215.9 ± 151.5	1.5 ± 2.7	61.4 ± 45.5	0.5 ± 0.8
Hopper	g	44.5 ± 25.3	3.0 ± 2.9	38.4 ± 19.0	5.2 ± 4.0	45.5 ± 19.1	7.5 ± 4.6
	f	40.1 ± 26.7	4.6 ± 4.7	35.5 ± 29.4	5.6 ± 6.2	49.5 ± 29.5	6.7 ± 4.1
	d	59.2 ± 29.5	3.2 ± 3.2	45.1 ± 20.2	4.8 ± 4.9	51.4 ± 18.0	6.5 ± 3.4
Walker	g	62.5 ± 7.7	3.0 ± 4.1	30.7 ± 29.6	7.0 ± 4.9	48.7 ± 18.9	5.3 ± 6.8
	f	58.8 ± 23.8	2.5 ± 3.4	21.6 ± 30.7	6.7 ± 5.3	73.1 ± 7.2	5.3 ± 4.3
	d	65.4 ± 5.6	1.5 ± 1.8	14.3 ± 30.5	6.3 ± 4.2	57.3 ± 21.0	8.1 ± 6.1
Overall		67.7	2.0	−13.1	3.9	53.4	4.5

Env		DARC_Lagrange reward	DARC_Lagrange cost	H2O_Lagrange reward	H2O_Lagrange cost	CPQ reward	CPQ cost
Ant	g	76.9 ± 33.5	3.6 ± 2.7	31.9 ± 28.3	13.7 ± 8.8		
	f	97.2 ± 8.1	5.0 ± 2.6	56.3 ± 20.9	9.4 ± 6.6	49.6 ± 6.0	9.7 ± 2.7
	d	111.2 ± 5.2	3.0 ± 1.7	27.4 ± 22.9	16.2 ± 8.1		
Cheetah	g	−21.3 ± 73.7	11.3 ± 6.7	−116.5 ± 228.1	46.5 ± 15.7		
	f	2.5 ± 76.4	4.2 ± 9.2	−329.3 ± 0.3	60.0 ± 0.0	79.6 ± 3.7	8.3 ± 1.3
	d	9.5 ± 72.4	1.5 ± 2.1	−301.0 ± 104.9	56.4 ± 13.4		
Hopper	g	63.5 ± 25.0	11.3 ± 10.9	22.3 ± 33.1	8.5 ± 10.8		
	f	58.3 ± 37.5	11.1 ± 8.2	33.1 ± 52.0	11.6 ± 12.5	98.0 ± 3.5	11.3 ± 2.6
	d	71.1 ± 32.0	11.8 ± 10.0	49.7 ± 42.0	13.1 ± 9.3		
Walker	g	65.5 ± 9.6	9.7 ± 11.3	45.5 ± 18.8	12.0 ± 4.5		
	f	73.1 ± 3.1	5.5 ± 4.8	65.6 ± 16.5	16.8 ± 5.0	72.7 ± 9.2	8.1 ± 3.1
	d	68.5 ± 7.3	12.5 ± 7.2	48.3 ± 17.1	17.0 ± 4.3		
Overall		56.3	7.5	−30.6	23.4	75.0	9.4

5.2   Overall Performance Comparison

The detailed experimental results are shown in Table 1. Firstly, it is observed that in scenarios with limited and low-quality offline data, the pure offline safe algorithm CPQ severely violates safety constraints. This underscores the necessity to introduce online trial-and-error for data augmentation and policy learning. When a certain dynamics gap exists in the simulator, SAC_Lagrange(sim) improves safety compared to CPQ but comes with a decrease in reward. Meanwhile, as it lacks theoretical guarantees, the performance of learned policies is highly dependent on the fidelity of simulators, emphasizing the importance of utilizing offline data to mitigate dynamics gaps. SAC_Lagrange(id) further incorporates an inverse dynamics model for adaptation in the real world, enhancing safety but causing a substantial drop in reward. This highlights the substantial negative impact of the inverse dynamics model's empirical and generalization errors on the policy.

Domain adaptation methods, DARC_Lagrange and H2O_Lagrange, fail to perform well when considering safety issues. While DARC_Lagrange achieves higher rewards in some settings, it significantly violates safety constraints. This validates the conflict between a trust region based on similar dynamics and safe data, with methods leaning towards the trust region as the primary constraint, tending to overlook safety constraints. H2O_Lagrange performs poorly in both reward and cost, diverging significantly compared to its outstanding performance in traditional environments without safety constraints. The issue arises from H2O's application of conservative regularization, which assigns transition samples with larger dynamics gaps lower rewards to avoid taking corresponding actions. However, in the safe scenario, the situation arises where samples with larger dynamics gaps, despite having lower rewards, also have lower estimated costs. Such actions are preferred compared to those with high costs in safe RL. Accordingly, these samples with larger dynamic gaps and low costs are used to update policy, leading to a significant performance gap.

Our method, DASaR, stands out by achieving the safest performance on most simulators while maintaining high reward returns. Compared to the safest baseline, DASaR improves safety performance by approximately 48%, and reward performance by approximately 80%. As for the approach with the highest reward, DASaR experiences a less than 10% decrease while achieving a remarkable 78% improvement in safety.

5.3   Ablation Studies and Data Sensitivity Study

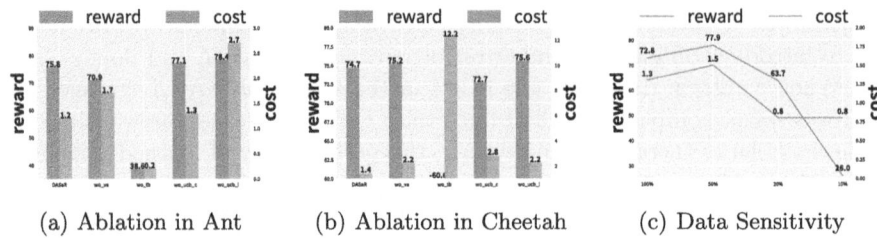

(a) Ablation in Ant   (b) Ablation in Cheetah   (c) Data Sensitivity

Fig. 1. Ablation studies in two environments and data sensitivity study.

To illustrate the impact of different modules of DASaR on its policy performance, we conducted ablation studies in two environments: Ant and HalfCheetah. Results from two simulators parameterized with different gravity and friction are averaged for each environment. Detailed statistical data along with standard deviation information are provided in the full paper. We here present four different baselines: (1) **wo_va** does not use the inverse dynamics-based relabling. (2) **wo_tb** dismisses the discriminator reward signal. (3) **wo_ucb_c** uses the original cost critic value instead of UCB estimation. (4) **wo_ucb_i** applies the traditional Bellman operator into cost critic updating. The experimental results are shown in Fig. 1(a) and 1(b). It can be observed that wo_va, wo_ucb_c, and wo_ucb_i do not show significant differences in reward compared to DASaR, but they all exhibit an increase in cost, indicating the necessity of these modules for guaranteeing the safety of the learned policy. On the other hand, the performance of wo_tb varies between two environments. In Ant, while its reward decreases significantly, it achieves the lowest cost. However, in HalfCheetah, its reward drops while the cost increases significantly. This highlights the importance of constraining the policy within the trust region of higher accuracy inverse dynamics models for performance gap bounding.

Next, to verify the sensitivity of DASaR to the quantity of offline data, we conducted a data sensitivity study in the Ant environment's gravity simulator. We designated the original 100 trajectories as 100%, then reduced the data volume to 50%, 20%, and 10% while maintaining the same quality proportion, ensuring that only 20% of the data was sampled by the safe policy. As shown in Fig. 1(c), when the data volume is reduced from 100 trajectories to 50 or 20 trajectories, no significant undulation is observed. However, a significant performance decline happens when it is reduced to 10 trajectories, with only 2 of them safe. Despite the low cost, the failure in reward indicates the unbounded performance gap. This illustrates that DASaR exhibits adequate tolerance to reduced data volume, learning safe and high-performing policies under more limited offline data is promising in future work.

5.4   Visualization Analysis

To assess whether the policy learned in the simulator exhibits remarkable sim-to-real adaptation in the real world, we compare the state distribution induced by the policy in the simulator and the real-world environment. The state distribution of DASaR and other baselines in the Ant environment are visualized using t-SNE [26], as is shown in Fig. 2.

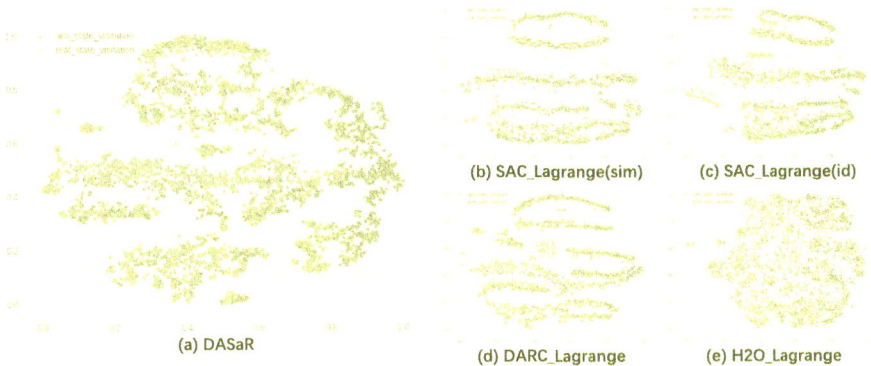

Fig. 2. Visualization of the state distributions in both simulator and the real.

First of all, SAC_Lagrange(sim) displays notable discrepancies in state distribution between the simulator and the real world, indicating failure in sim-to-real adaptation. In contrast, SAC_Lagrange(id) shows increased similarity in state distribution compared to SAC_Lagrange(sim), despite that the discrepancy still exists. This underscores the necessity of employing sim-to-real techniques to enhance adaptation in the real-world environment. While H2O_Lagrange utilizes domain adaptation techniques, substantial disparities persist between simulated and real state distributions. This suggests the need for further adjustments in safe RL. Both DASaR and DARC_Lagrange exhibit notably enhanced similarity in state distribution compared to other algorithms. This showcases the remarkable adaptation capabilities of these methods, which is in accordance with their higher performance in primary experiments. However, DARC_Lagrange's adaptation comes at the cost of compromising policy safety to some extent. Only DASaR achieves remarkable adaptation capabilities while maintaining the relative safety of the learned policy.

5.5   Parameter Sensitivity Studies

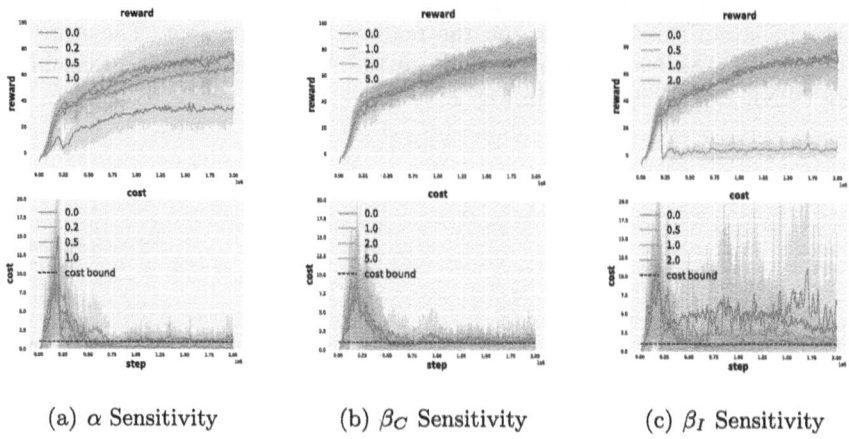

(a) α Sensitivity   (b) β_C Sensitivity   (c) β_I Sensitivity

Fig. 3. Visualization of the curves depicting the changes in the reward and costs during the training process of different parameter values.

To assess the sensitivity of DASaR to different hyperparameter selections, we conduct hyperparameter sensitivity studies in the Ant gravity simulator for three hyperparameters: α, β_C, and β_I. Specifically, α is varied between 0.0, 0.2, 0.5, and 1.0; β_c is varied between 0.0, 1.0, 2.0, and 5.0; and β_I is varied between 0.0, 0.5, 1.0, and 2.0. Results are shown in Fig. 3.

Firstly, concerning the hyperparameter α, it is evident that when α is 0, the policy yields a significantly lower reward return, indicating a considerable performance gap. As α increases to 0.2 and 0.5, the reward return experiences a substantial increase while maintaining safety performance within an acceptable range. However, as α further increases to 1.0, both the reward and safety performance of the policy decrease. This underscores the importance of both reducing the performance gap and expanding the optimization search space of the policy to achieve better performance in real-world environments.

Moving on to the hyperparameter β_C, it is observed that in the Ant gravity environment, the values of β_C do not have a significant impact on the performance of the policy. This suggests that in this environment, the learning of the cost critic is relatively accurate, thus the upper confidence bound of the cost critic will not undergo significant changes regardless of the value of β_C.

Finally, regarding the hyperparameter β_I, it is observed that as β_I gradually increases from 0 to 1.0, the safety performance of the policy also gradually improves. However, when β_I further increases to 2.0, the policy learning collapses, with the reward essentially dropping to near 0 and safety no longer guaranteed. This phenomenon primarily occurs because an excessively large β_I makes the cost critic overly conservative, potentially preventing the discovery of a policy that satisfies the cost critic's estimation being less than $b = 5$. Consequently, the Lagrange multiplier continues to increase, leading to the collapse of policy learning. This underscores the importance of selecting β_l appropriately.

6   Final Remarks

In this work, we present a novel algorithm, DASaR, designed to tackle the challenge of training a policy in a simulator with a dynamics gap while ensuring safe performance in the real. DASaR aligns the value estimation from the simulator with the real by incorporating an inverse dynamics model, resulting in a lower performance gap across the entire offline data distribution where the inverse dynamics model exhibits high accuracy. Additionally, DASaR employs uncertainty estimation to robustly model the conservative cost critic, addressing neural network approximation errors and further enhancing safety performance. Extensive experimental results demonstrate DASaR's remarkable adaptation capabilities across various simulators. However, communication with the simulator during deployment introduces additional overhead, future work can further apply a behavior cloning process to solve this problem. Additionally, the applicability of DASaR is somewhat constrained by the presence of Assumption 1, and expanding its scope remains a promising direction for further exploration. Extending this work to the multi-agent setting is also a good choice for solving the open environment reinforcement learning challenges [45].

Acknowledgements. This work is supported by National Science Foundation of China (61921006). We would like to express our gratitude to the anonymous reviewers for their kind reviews and constructive feedback.

References

1. Achiam, J., Held, D., Tamar, A., Abbeel, P.: Constrained policy optimization. In: ICML, pp. 22–31 (2017)
2. Afsar, M.M., Crump, T., Far, B.: Reinforcement learning based recommender systems: a survey. ACM Comput. Surv. **55**(7), 1–38 (2022)
3. Altman, E.: Constrained Markov Decision Processes. Routledge, London (2021)
4. Arfken, G.B., Weber, H.J., Harris, F.E.: Mathematical Methods for Physicists: A Comprehensive Guide. Academic Press, Cambridge (2011)
5. Brunke, L., et al.: Safe learning in robotics: from learning-based control to safe reinforcement learning. Annu. Rev. Control Robot. Auton. Syst. **5**, 411–444 (2022)
6. Chebotar, Y., et al.: Closing the sim-to-real loop: adapting simulation randomization with real world experience. In: ICRA, pp. 8973–8979 (2019)
7. Chow, Y., Ghavamzadeh, M., Janson, L., Pavone, M.: Risk-constrained reinforcement learning with percentile risk criteria. J. Mach. Learn. Res. **18**(167), 1–51 (2018)
8. Christiano, P., et al.: Transfer from simulation to real world through learning deep inverse dynamics model. arXiv preprint arXiv:1610.03518 (2016)
9. Collins, J., Brown, R., Leitner, J., Howard, D.: Traversing the reality gap via simulator tuning. In: ACRA, pp. 1–10 (2021)

10. Desai, S., Durugkar, I., Karnan, H., Warnell, G., Hanna, J., Stone, P.: An imitation from observation approach to transfer learning with dynamics mismatch. In: NeurIPS, pp. 3917–3929 (2020)
11. Eysenbach, B., Asawa, S., Chaudhari, S., Levine, S., Salakhutdinov, R.: Off-dynamics reinforcement learning: training for transfer with domain classifiers. arXiv preprint arXiv:2006.13916 (2020)
12. Farchy, A., Barrett, S., MacAlpine, P., Stone, P.: Humanoid robots learning to walk faster: from the real world to simulation and back. In: AAMAS, pp. 39–46 (2013)
13. García, J., Fernández, F.: A comprehensive survey on safe reinforcement learning. J. Mach. Learn. Res. **16**(1), 1437–1480 (2015)
14. Gu, S., et al.: Safe multi-agent reinforcement learning for multi-robot control. Artif. Intell. **319**, 103905 (2023)
15. Gu, S., et al.: Multi-agent constrained policy optimisation. arXiv preprint arXiv:2110.02793 (2021)
16. Guan, C., et al.: Cost-aware offline safe meta reinforcement learning with robust in-distribution online task adaptation. In: Proceedings of the 23rd International Conference on Autonomous Agents and Multiagent Systems, pp. 743–751 (2024)
17. Haarnoja, T., Zhou, A., Abbeel, P., Levine, S.: Soft actor-critic: off-policy maximum entropy deep reinforcement learning with a stochastic actor. In: ICML, pp. 1861–1870 (2018)
18. Höfer, S., et al.: Sim2real in robotics and automation: applications and challenges. IEEE Trans. Autom. Sci. Eng. **18**(2), 398–400 (2021)
19. Ibarz, J., Tan, J., Finn, C., Kalakrishnan, M., Pastor, P., Levine, S.: How to train your robot with deep reinforcement learning: lessons we have learned. Int. J. Robot. Res. **40**(4–5), 698–721 (2021)
20. Kiran, B.R., et al.: Deep reinforcement learning for autonomous driving: a survey. IEEE Trans. Intell. Transp. Syst. **23**(6), 4909–4926 (2021)
21. Kumar, A., Zhou, A., Tucker, G., Levine, S.: Conservative Q-learning for offline reinforcement learning. In: NeurIPS, pp. 1179–1191 (2020)
22. Le, H., Voloshin, C., Yue, Y.: Batch policy learning under constraints. In: ICML, pp. 3703–3712 (2019)
23. Lee, J., et al.: Coptidice: offline constrained reinforcement learning via stationary distribution correction estimation. arXiv preprint arXiv:2204.08957 (2022)
24. Liu, Z., et al.: Constrained variational policy optimization for safe reinforcement learning. In: ICML, pp. 13644–13668 (2022)
25. Liu, Z., et al.: Constrained decision transformer for offline safe reinforcement learning. arXiv preprint arXiv:2302.07351 (2023)
26. Van der Maaten, L., Hinton, G.: Visualizing data using t-SNE. J. Mach. Learn. Res. **9**(11) (2008)
27. Mehta, B., Diaz, M., Golemo, F., Pal, C.J., Paull, L.: Active domain randomization. In: CoRL, pp. 1162–1176. PMLR (2020)
28. Mordatch, I., Lowrey, K., Todorov, E.: Ensemble-CIO: full-body dynamic motion planning that transfers to physical humanoids. In: IROS, pp. 5307–5314 (2015)
29. Nagabandi, A., et al.: Learning to adapt in dynamic, real-world environments through meta-reinforcement learning. In: ICLR (2018)
30. Niu, H., et al.: When to trust your simulator: dynamics-aware hybrid offline-and-online reinforcement learning. In: NeurIPS, pp. 36599–36612 (2022)
31. Osband, I., Blundell, C., Pritzel, A., Van Roy, B.: Deep exploration via bootstrapped DQN. In: NeurIPS, pp. 4026–4034 (2016)

32. Peng, X.B., Andrychowicz, M., Zaremba, W., Abbeel, P.: Sim-to-real transfer of robotic control with dynamics randomization. In: ICRA, pp. 3803–3810 (2018)
33. Ray, A., Achiam, J., Amodei, D.: Benchmarking safe exploration in deep reinforcement learning. arXiv preprint arXiv:1910.01708 (2019)
34. Ren, A.Z., Dai, H., Burchfiel, B., Majumdar, A.: Adaptsim: task-driven simulation adaptation for sim-to-real transfer. arXiv preprint arXiv:2302.04903 (2023)
35. Silver, D., et al.: Mastering the game of go without human knowledge. Nature **550**(7676), 354–359 (2017)
36. Stooke, A., Achiam, J., Abbeel, P.: Responsive safety in reinforcement learning by PID Lagrangian methods. In: ICML, pp. 9133–9143 (2020)
37. Tan, J., et al.: Sim-to-real: learning agile locomotion for quadruped robots. arXiv preprint arXiv:1804.10332 (2018)
38. Tessler, C., Mankowitz, D.J., Mannor, S.: Reward constrained policy optimization. arXiv preprint arXiv:1805.11074 (2018)
39. Tobin, J., Fong, R., Ray, A., Schneider, J., Zaremba, W., Abbeel, P.: Domain randomization for transferring deep neural networks from simulation to the real world. In: IROS, pp. 23–30 (2017)
40. Todorov, E., Erez, T., Tassa, Y.: Mujoco: a physics engine for model-based control. In: 2012 IEEE/RSJ International Conference on Intelligent Robots and Systems, pp. 5026–5033. IEEE (2012)
41. Torabi, F., Warnell, G., Stone, P.: Generative adversarial imitation from observation. arXiv preprint arXiv:1807.06158 (2018)
42. Xu, H., Zhan, X., Zhu, X.: Constraints penalized q-learning for safe offline reinforcement learning. In: AAAI, pp. 8753–8760 (2022)
43. Yang, T.Y., Rosca, J., Narasimhan, K., Ramadge, P.J.: Projection-based constrained policy optimization. In: ICLR (2019)
44. Yao, Y., et al.: Constraint-conditioned policy optimization for versatile safe reinforcement learning. In: NeurIPS, vol. 36 (2024)
45. Yuan, L., Zhang, Z., Li, L., Guan, C., Yu, Y.: A survey of progress on cooperative multi-agent reinforcement learning in open environment. arXiv preprint arXiv:2312.01058 (2023)
46. Zhou, K., Liu, Z., Qiao, Y., Xiang, T., Loy, C.C.: Domain generalization: a survey. IEEE Trans. Pattern Anal. Mach. Intell. **45**(4), 4396–4415 (2022)

CRISPert: A Transformer-Based Model for CRISPR-Cas Off-Target Prediction

William Jobson Pargeter[1], Rolf Backofen[1,2], and Van Dinh Tran[3,4](✉)

[1] Bioinformatics Group, Department of Computer Science, University of Freiburg, Freiburg im Breisgau, Germany
{pargeter,backofen}@informatik.uni-freiburg.de
[2] Signalling Research Centre CIBSS, University of Freiburg, Freiburg im Breisgau, Germany
[3] Department of Information and Computer Science, King Fahd University of Petroleum and Minerals, Dhahran, Saudi Arabia
[4] SDAIA-KFUPM Joint Research Center for Artificial Intelligence, Dhahran, Saudi Arabia
vandinh.tran@kfupm.edu.sa

Abstract. CRISPR-Cas9 has emerged as a popular gene-editing technique due to its flexibility, precision, and ease of use. It is a complex that consists of a Cas9 protein and a designed, synthetic single guide-RNA (sgRNA) that guides the Cas9 protein to its intended genomic target site, where it induces editing of the DNA through cleavage. Despite its popularity, the potential side effects caused by unintended cleavage of CRISPR-Cas9 have been a critical issue that hinders its development and clinical applications. Therefore, predicting the potential off-target sites will help evaluate the safety of a designed CRISPR-Cas9 system. Many methods have been proposed for off-target site prediction. However, they only obtain moderate results. This is partly due to the high imbalance of data, the choice of network architecture, and the neglect of additional useful information. Here, we introduce CRISPert, a transformer-based model that overcomes these issues. Empirical results from various experimental settings show that our proposed method outperforms many compared methods and confirms its potential for practical use.

Keywords: Off-target prediction · CRISPR-Cas · Transformer-based model

1 Introduction

CRISPR-Cas systems are an adaptive viral defence mechanism used by bacteria and archaea cells. Copies of DNA from viruses that have previously attacked the cell are stored within a sequence of repeats. This stored viral DNA acts like a genetic memory of past infections, allowing cells to recognise and respond to future attacks. During the defence, the copy of viral DNA is transcribed to RNA molecules. These molecules guide an enzyme or enzyme complex to the DNA

A. Bifet et al. (Eds.): ECML PKDD 2024, LNAI 14947, pp. 92–104, 2024.
https://doi.org/10.1007/978-3-031-70368-3_6

of the attacking virus. When it binds to the viral DNA, the cleavage activity will neutralise the virus. One of the most straightforward functioning systems is CRISPR-Cas9, which consists of the Cas9 protein, a tracer RNA, and a copy of the viral DNA. The Cas9 protein acts as a molecular scissor that cuts DNA at a specific location. The role of the tracer RNA is to bind the copy of the viral DNA to the Cas9 protein. An interesting fact is that CRISPR-Cas9 can also function in other cell types. Scientists [7] found that the tracer RNA and viral DNA pair can be replaced by a synthetically created single guide RNA (sgRNA). Therefore, the sgRNA can be designed to target the exact location on the genome at which insertion or deletion is desired. This means that scientists can program the CRISPR-Cas9 system to make precise changes to the DNA of any organism. Its flexibility and ease of implementation have made CRISPR-Cas9 a popular gene editing technique.

Despite having promising results, the effectiveness of CRISPR-Cas9 gene editing is of concern. A sound Cas9 system (complex) should bind and cleave the desired location (on-target) and have minimal off-target effects [27], occurring when Cas9 cleaves unintended sites (off-targets). A typical sgRNA is 20 nucleotides long and complementary to the target strand, followed by a Protospacer Adjacent Motif (PAM). Binding and cleavage can happen at sites that have as many as 6 mismatches compared to the sgRNA. Off-target effects might lead to adverse outcomes, e.g., a loss of gene function [27]. These side effects hinder the application of CRISPR-Cas9 in clinical and therapeutic settings, especially when it concerns humans. Therefore, evaluating the potential of off-target sites for a given sgRNA is vital to designing a safe CRISPR-Cas9 system for gene editing. Classically, wet-lab methods are employed. These include in-vitro techniques, e.g., CIRCLE-seq [21] and SITE-seq [2], as well as cell-based techniques, e.g., BLESS [19] and GUIDE-seq [22]. Despite being the most accurate, such methods are costly and time-consuming. Thus, there has been a call for proposals for high-performance off-target site prediction methods.

2   Computational Methods for Off-Target Prediction

Several off-target prediction methods have been proposed in the literature. We can essentially divide them into three groups, namely alignment-based, hypothesis-driven, and learning-based methods. Alignment-based methods, of which Cas-OFFinder [1] is a typical example, search the genome for sequences that differ from the sgRNA by a fixed number of mismatches. A drawback of these methods is their high false positive rate. This means many sites are predicted as potential off-target, but only a small portion has been confirmed as true off-targets. To address this issue, hypothesis-driven methods were developed using predefined rules and/or epigenetic features as additional information to provide more accurate results. CROP-IT [20] is an example. It considers the chromatin state information at the off-target site. A downside of such methods is that their performances cannot be increased even when more available samples are given. In the learning-based group, models are trained on known data to generalise to unseen data. Thus, they tend to outperform methods in other groups.

Many machine learning-based models have been introduced for off-target prediction. An early proposed method is DeepCRISPR [4], which adopts a Convolutional Neural Network (CNN) architecture to build the model. First, an Autoencoder-based pre-training is performed to learn a generic data representation for each input. Then, a model fine-tuning is carried out using limited labeled data to distinguish between the true and false off-targets. CnnCrispr [10] employs learned GloVe [17] embedding for encoding along with a CNN and a bi-directional Long Short-Term Memory network (LSTM). CRISPR-OFFT [26] is also a CNN-based model. It uses pairwise encoding for each pair of a sgRNA and an off-target, and Word2Vec [15] for embedding. In addition, it makes use of an attention block between the convolutional layers. CRISPR-IP [28] architecture comprises a CNN, followed by a bi-directional LSTM and an attention layer. In this method, one-hot encoding is used to encode sequences. Due to the success of Transformers [23], several methods based on this architecture have been introduced. DNA-BERT [3] uses a BERT [5] encoder pre-trained on unlabelled off-target candidates to encode the candidate and sgRNA pair. These are combined with additional non-sequence information and used as input for a gradient-boosting model. CRISPR-BERT [12] encodes the sgRNA and off-target candidate pair using one-hot encoding and pair-encoding. The network includes two modules. The first contains a CNN and a bidirectional Gated Recurrent Unit (GRU) that takes one hot encoded sequence as input while the second has a BERT encoder pre-trained on text that uses the pair-encoded sequence as input. It is followed by a bidirectional GRU.

Although the current machine learning models have shown encouraging results, their performances are still far from perfect. The reason is that these methods do not effectively address the common problems when constructing models for off-target prediction. Firstly, we can only access a limited amount of lab-verified off-target site data (labeled data), which can be insufficient for model training. Secondly, while candidate off-target sites can be generated using alignment methods, these vastly outnumber the lab-verified sites (true off-targets). This leads to data imbalance issues. When trained with such data, the model tends to be biased toward the primary class. Lastly, the sequence similarity between the true off-targets and their corresponding candidates is high, making it hard for the model to distinguish between the two. In this paper, we propose CRISPert, which aims to overcome the shortcomings of existing methods and the common issues mentioned above. In particular, our model is based on BERT with an adapted input encoding, allowing it to integrate sequence and additional features effectively. Like other BERT-based models, CRISPert can be pre-trained on massive unlabelled data to learn abstract representations. The key advantage of our model comes from its utilisation of CasKas features (see Sect. 4), which are profiles of CRISPR-Cas binding sites.

3   Method

In this section, we first formalise the problem of CRISPR-Cas off-target prediction. Then we introduce the architecture and the features of our proposed method, CRISPert in detail.

3.1   Problem Formalisation

We notate G, and S as a set of sgRNAs, and off-target candidates, respectively. We consider a dataset of all examples as $D = \{(g, s, c) \mid g \in G, s \in S, c \in \{0, 1\}\}$, where c equals to 1 if s is a true off-target site corresponding to the sgRNA, g, otherwise 0. The off-target prediction problem can be defined as the following function.

$$f : G \times S \longrightarrow \{0, 1\}$$

Note that the function f is an unknown. Therefore, given a training set (set of observations) $D_{tr} \subset D$, a machine learning model aims at learning from D an estimation of the function f to predict for unseen data, $(D - D_{tr})$.

3.2   Model Architecture

An overview of CRISPert's architecture is illustrated in Fig. 1. As the model is based on BERT [5], a transformer-based model, its architecture consists of three main parts: input encoding, encoders, and classifiers. The input encoding converts input data into token numerical representations, while the encoders aim to learn an abstract representation for each token by utilising the self-attention mechanism. The classifiers include two fully connected networks used for off-target classification and pre-training. In the following, we will describe each component in more detail.

Input Encoding

The input for the model consists of a pair of 23-nucleotide long sequences (sgRNA and off-target site) and point-wise concentration features (see Fig. 2). Regarding the sequence input, we first encode every pair of sequences as a single sequence composed of 16 distinct letters, each corresponding to a unique pair of nucleotides. Then, we tokenise this obtained sequence and encode each token by two numeric vectors, i.e., token embedding and positional embedding. A procedure similar to the sequence input is performed for each type of concentration feature. In particular, after tokenising, we take an average of features in each token and embed it in the same dimensional space as the token embedding and positional embedding. Finally, we sum up all embeddings corresponding to each token to form its final representation.

Encoders

The encoders follow the same structure as that of BERT and DNABERT [6], which use the encoder initially proposed in the Transformer architecture. This consists of 12 stacked identical blocks containing a multi-head self-attention mechanism and a position-wise multilayer perceptron. The encoders take the representations resulting from the input encoding as input and learn more abstract representations for the tokens.

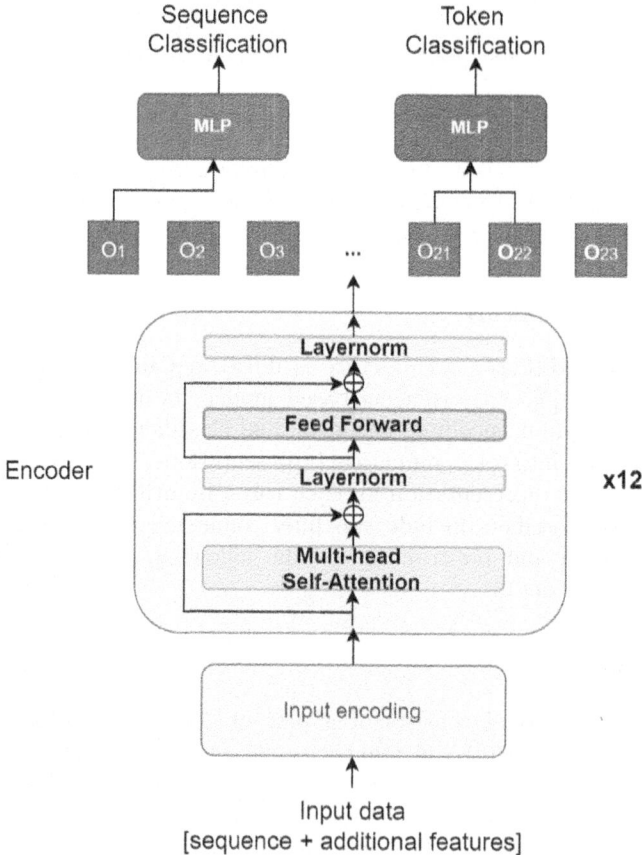

Fig. 1. An overview of CRISPert model architecture

Classifiers

The classifiers comprise two multilayer perceptrons (fully connected feed-forward networks), each employed for either pre-training or training. Pre-training aims to learn generic representations for the input data. It is done in a self-supervised learning manner in which some tokens are masked and used as labels (see [5,6] for more detail). The network weights obtained by the pre-training are used as initialisation for network weights in the training phase. Then, the training performs fine-tuning using input labels for supervision.

3.3 CRISPR-Cas Binding Concentration Features

The proposed model accepts additional features in addition to sequence input. Although additional features, such as epigenetic features, have been employed in earlier models like DeepCRISPR, no clear evidence has been shown that they help improve models' performance. We assume that models can benefit from using the CRISPR-Cas binding concentration profile, in which a score associated with each DNA locus shows the likelihood of a CRISPR-Cas binding being involved in this locus. An experiment is conducted (see Sect. 4.1 for more detail) utilising a CRISPR-Cas system with a certain sgRNA to obtain such a profile.

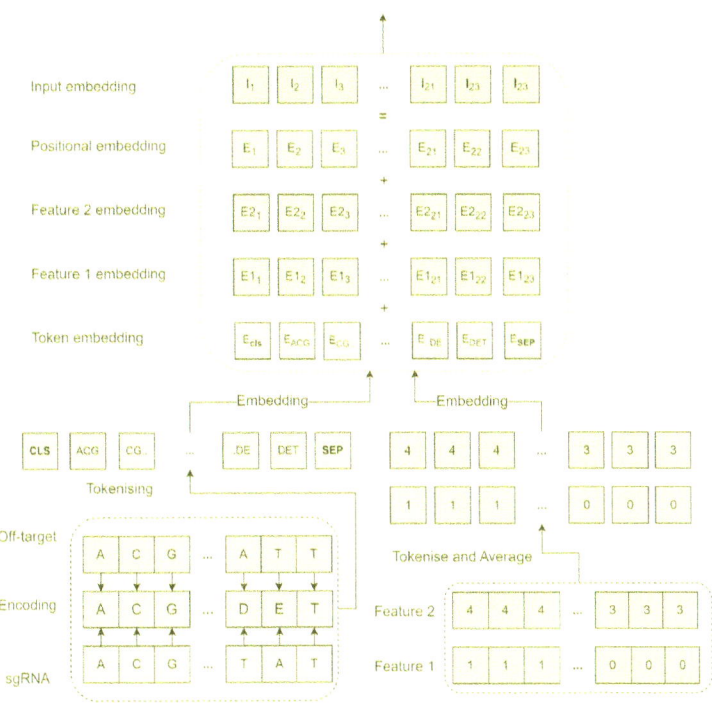

Fig. 2. CRISPert input encoding

Then, the recorded binding results are used to form the binding profile. For each candidate site, we extract corresponding sequences of concentration scores. The number of concentration score sequences depends on the number of available profiles obtained from CasKas experiments. Later, these score sequences and the common input sequence are employed to provide more information for the model in the training and prediction phase (see Fig. 2).

3.4   Data Imbalance Handling

It is common knowledge that machine learning models work best when trained on balanced data, i.e., a similar number of examples in each class. In imbalanced data, we observe a skewed class distribution due to the domination of the majority class. In such cases, the model tends to be biased toward the majority class and does not generalise well. Unfortunately, CRISPR-Cas off-target datasets are highly imbalanced. This is because only a limited number of true off-targets have been collected by lab experiments, whereas an abundance of potential off-targets can be obtained from the genome. Therefore, employing techniques to overcome data imbalance issues is necessary when constructing machine learning models for off-target prediction. Different methods have been proposed to solve this problem, such as over-sampling, under-sampling, and class weights; each has its advantages. Here, we choose a combination of over-sampling and under-sampling. In particular, we construct balanced data for training by over-sampling the minority class and under-sampling the majority one.

3.5   Model Implementation

CRISPert is implemented in Python, using PyTorch [16] and the Hugging-face transformer library [24]. A modified version of the DNA tokeniser from DNABERT [6] is used to tokenise the input sequences. We adopt DNABERT [6] and adapt the input encoding part. We use AdamW [11] as the optimiser, Cross Entropy as the loss function for model training, and Ray tune [9] for hyperparameter optimisation.

The code and supplementary information for CRISPert can be found at the following Github link

4   Experimental Setting

This section shows how different experiments are conducted using two separate settings to evaluate CRISPert's performance and compare it with state-of-the-art methods for off-target prediction.

4.1   Data

DeepCRISPR Dataset: This dataset is downloaded from [4]. It is the most commonly used one for benchmarking off-target prediction models and contains results from various studies in two cell types, namely Hek293t and K562. In particular, Hek293t comprises 18 sgRNAs with 515 true off-targets and 132,399 potential off-targets (candidates), while K562 consists of 11 sgRNAs with 118 true off-targets and 20,201 potential off-targets. The candidates are genome sequences adjacent to "NGG" PAM and have at most 6 mismatches compared to the respected sgRNAs. Since both cell-type data contain duplicates, a duplicate removal is performed. Ultimately, we get 57,245 potential off-targets in Hek295 and 18,434 in K562. Notice that this dataset is highly imbalanced, with an average imbalance ratio of around 1:100.

CasKas Data: CasKas [14] uses Kas-seq [13], a genome-wide method for sequencing of single-stranded DNA and shows it can be applied to CRISPR-Cas data. Since unwound single-stranded DNA breaks are created when the CRISPR-Cas complex binds to the gnome, off-target binding sites can be recorded. The authors perform this method to form genome-wide concentration binding profiles for various sgRNAs. It is important to note that CasKas is a fast and inexpensive method for accurately mapping CRISPR-Cas binding sites for a given sgRNA. Although most of these are irrelevant to our study, we are interested in the two sgRNAs that target the EMX1 and VEGF genes in the Hek293t cell. Hence, our model uses two concentration profiles. To this end, we employ Bed tools coverage [18] to extract two concentration score sequences corresponding to each candidate site from the Hek239t cell in DeepCRISPR dataset.

4.2 Test Scenarios

We conduct experiments in two separate test scenarios for the model evaluation and comparison. Each scenario is described below.

- **Test scenario 1:** The objective of the first scenario is to compare the performance of CRISPert-base, i.e., our model that uses input sequence only, with state-of-the-art off-target predictive methods. For this, we perform experiments using the DeepCRISPR dataset, which contains data regarding 29 sgRNAs and two cell types. We use the leave-(one-sgRNA)-out to evaluate the performance. Precisely, we iteratively perform experiments. Each time, data related to a sgRNA is left out to use as a test set, and the remaining data concerning the other 28 sgRNAs is used for the model training. For model selection, 20% of the train set is taken out and used as the validation set. As the data has a high imbalance ratio, we use AUC-PR (Area under precision and recall curve) as the evaluation metric. We choose six state-of-the-art models for model comparison: CnnCRISPR [10], DeepCRISPR [4], CRISPR-OFFT [26], DNA-BERT [3], CRISPR-BERT [12], CRISPR-IP [28].
- **Test scenario 2:** In the second test scenario, we would like to test our assumption that models can benefit from using concentration profiles as additional input data. Thus, we employ the CasKas data. As mentioned earlier, we have access to CasKas data for only two sgRNAs on Hek293t. We use this data to form additional features for the other 16 sgRNAs in Hek293t. Similar to the first scenario, the leave-(one-sgRNA)-out evaluation method

is adopted. It is worth noticing that we disregard data from the two sgR-NAs that were used to generate the CasKas data to avoid bias. AUC-PR is also used as the evaluation metric in this scenario. Here, we compare the CRISPert-base model with the CRISPert version that uses CasKas features to check our assumption. We also compare our model with the state-of-the-art methods used in test scenario 1. However, DeepCRISPR and DNA-BERT are excluded from the comparison as the source code of their methods is not available.

4.3 Hyper-parameter Optimisation

Our model has several hyperparameters, e.g. learning rate, drop-out, and weight-decay. Hence, we optimise the hyperparameter using grid search with ASHA (Asynchronous Successive Halving) [8] to find the optimal hyperparameter tuple for model training. Please see the supplementary for the complete list of hyper-parameters used and their defined value ranges.

4.4 Pre-training

The data without labels from the DeepCRISPR dataset is employed for our model pre-training. Remember that for pre-training, the MLP classifier, i.e., top right of Fig. 1, is used to perform token-level classification. The self-supervised learning is done using masked-language modeling, as shown in [6].

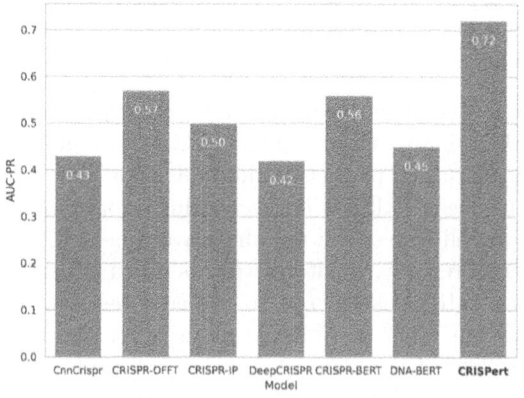

Fig. 3. The performances in AUC-PR of different models in test scenario 1

5   Results and Analysis

Figures 3, 4 show the results from both test scenarios in AUC-PR. As can be seen from the figures, CRISPert outperforms the compared methods in both test scenarios by a high margin.

In test scenario 1, we compare different methods for off-target prediction on 29 sgRNAs from two cell types. CRISPert-base, our model using input sequence only, gains the best performance with 0.72, while CRISPR-OFFT and CRISPR-BERT are placed second with 0.57 and 0.56, respectively. The worst-performing models are CnnCrispr, DeepCRISPR, and DNA-BERT, having results that are less than or equal to 0.45. It is noted that CRISPert performs at least 15% higher in comparison to other methods (see Fig. 3). While the main aim of test scenario 1 is to compare CRISPert with state-of-the-art methods, that of test scenario 2 is to test our assumption on the benefit of models when using CasKas features. In Fig. 4, we can observe that all CRISPert models that integrate CasKas features illustrate higher performances than CRISPert-base with an improvement from 3 to 9%. This implies the clear benefit of using CasKas as an additional feature for our model. Comparing models that employ CasKas features, CRISPert-EMX1+VEGF, which combines both CasKas features, gets higher results than using only a single one. Regarding the comparison with other methods, we also notice that the CRISPert models show superior results with a gap to the second best of around 10%. It is noticeable that there is a decline in the performance of each method in this test scenario. This can be explained by the use of a smaller data size for model training, i.e., data regarding 15 gRNAs instead of 28 sgRNAs in test scenario 1.

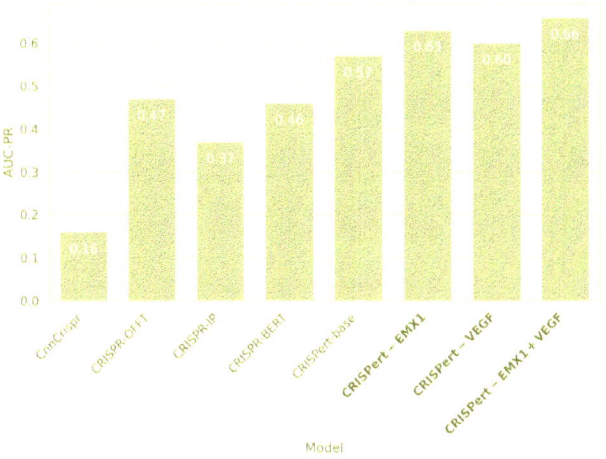

Fig. 4. The performances in AUC-PR of different models in test scenario 2

6   Conclusion

CRISPR-Cas9 is a powerful tool for gene editing, but off-target effects limit its effectiveness and potential clinical use. These can be mitigated using an effective sgRNA design tool employing an off-target prediction model to suggest safe sgRNAs. In this work, we have proposed CRISPert, a BERT-based model for off-target prediction. The proposed model has advantages, from the proposed input encoding and the integration of CasKas features. Experimental results show that it achieves state-of-the-art results and suggests utilising CasKas features when constructing off-target predictive models.

In general, the lack of true off-target data is one of the critical problems for off-target prediction methods. Therefore, we will investigate the possibility of using synthetic data generated by generative models such as SeqGAN [25]. Although we confirm the advantage of using CasKas in our model, we will study how it impacts the other methods.

Acknowledgments. The author would like to acknowledge the support received from Saudi Data and AI Authority (SDAIA) and King Fahd University of Petroleum and Minerals (KFUPM) under SDAIA-KFUPM Joint Research Center for Artificial Intelligence Grant no. JRC-AI-RFP-18.

This research is also funded by Deutsche Forschungsgemeinschaft (DFG, German Research Foundation) - Project-ID 499552394 - SFB 1597 and Project ID 390939984 - CIBSS - EXC-2189.

Disclosure of Interests. The authors have no competing interests to declare relevant to this article's content.

References

1. Bae, S., Park, J., Kim, J.S.: Cas-OFFinder: a fast and versatile algorithm that searches for potential off-target sites of Cas9 RNA-guided endonucleases. Bioinformatics **30**(10), 1473–1475 (2014). https://doi.org/10.1093/bioinformatics/btu048
2. Cameron, P., et al.: Mapping the genomic landscape of CRISPR-Cas9 cleavage. Nat. Methods **14**(6), 600–606 (2017). https://doi.org/10.1038/nmeth.4284
3. Chen, D., Shu, W., Peng, S.: Predicting CRISPR-Cas9 off-target with self-supervised neural networks. In: 2020 IEEE International Conference on Bioinformatics and Biomedicine (BIBM), pp. 245–250. IEEE (2020)
4. Chuai, G., et al.: DeepCRISPR: optimized CRISPR guide RNA design by deep learning. Genome Biol. **19**(1), 80 (2018). https://doi.org/10.1186/s13059-018-1459-4
5. Devlin, J., Chang, M.W., Lee, K., Toutanova, K.: BERT: pre-training of Deep Bidirectional Transformers for Language Understanding (2019). https://doi.org/10.48550/arXiv.1810.04805. arXiv:1810.04805
6. Ji, Y., Zhou, Z., Liu, H., Davuluri, R.V.: DNABERT: pre-trained bidirectional encoder representations from transformers model for DNA-language in genome. Bioinformatics **37**(15), 2112–2120 (2021). https://doi.org/10.1093/bioinformatics/btab083
7. Jinek, M., Chylinski, K., Fonfara, I., Hauer, M., Doudna, J.A., Charpentier, E.: A programmable dual-RNA-guided DNA endonuclease in adaptive bacterial immunity. Science **337**(6096), 816–821 (2012). https://doi.org/10.1126/science.1225829
8. Li, L., et al.: A system for massively parallel hyperparameter tuning. Proc. Mach. Learn. Syst. **2**, 230–246 (2020)

9. Liaw, R., Liang, E., Nishihara, R., Moritz, P., Gonzalez, J.E., Stoica, I.: Tune: a research platform for distributed model selection and training. arXiv preprint arXiv:1807.05118 (2018)

10. Liu, Q., Cheng, X., Liu, G., Li, B., Liu, X.: Deep learning improves the ability of sgRNA off-target propensity prediction. BMC Bioinform. **21**(1), 51 (2020). https://doi.org/10.1186/s12859-020-3395-z

11. Loshchilov, I., Hutter, F.: Decoupled weight decay regularization. arXiv preprint arXiv:1711.05101 (2017)

12. Luo, Y., Chen, Y., Xie, H., Zhu, W., Zhang, G.: Interpretable CRISPR/Cas9 off-target activities with mismatches and indels prediction using BERT. Comput. Biol. Med. **169**, 107932 (2024)

13. Lyu, R., Wu, T., Zhu, A.C., West-Szymanski, D.C., Weng, X., Chen, M., He, C.: KAS-seq: genome-wide sequencing of single-stranded DNA by N3-kethoxal-assisted labeling. Nat. Protoc. **17**(2), 402–420 (2022)

14. Marinov, G.K., et al.: CasKAS: direct profiling of genome-wide dCas9 and Cas9 specificity using ssDNA mapping. Genome Biol. **24**(1), 85 (2023). https://doi.org/10.1186/s13059-023-02930-z

15. Mikolov, T., Chen, K., Corrado, G., Dean, J.: Efficient Estimation of Word Representations in Vector Space (2013). http://arxiv.org/abs/1301.3781. arXiv:1301.3781

16. Paszke, A., et al.: Pytorch: an imperative style, high-performance deep learning library. In: Advances in Neural Information Processing Systems, vol. 32 (2019)

17. Pennington, J., Socher, R., Manning, C.D.: Glove: global vectors for word representation. In: Empirical Methods in Natural Language Processing (EMNLP), pp. 1532–1543 (2014). http://www.aclweb.org/anthology/D14-1162

18. Quinlan, A.R., Hall, I.M.: Bedtools: a flexible suite of utilities for comparing genomic features. Bioinformatics **26**(6), 841–842 (2010)

19. Ran, F.A., et al.: In vivo genome editing using Staphylococcus aureus Cas9. Nature **520**(7546), 186–191 (2015). https://doi.org/10.1038/nature14299

20. Singh, R., Kuscu, C., Quinlan, A., Qi, Y., Adli, M.: Cas9-chromatin binding information enables more accurate CRISPR off-target prediction. Nucleic Acids Res. **43**(18), e118 (2015). https://doi.org/10.1093/nar/gkv575

21. Tsai, S.Q., Nguyen, N.T., Malagon-Lopez, J., Topkar, V.V., Aryee, M.J., Joung, J.K.: CIRCLE-seq: a highly sensitive in vitro screen for genome-wide CRISPR-Cas9 nuclease off-targets. Nat. Methods **14**(6), 607–614 (2017). https://doi.org/10.1038/nmeth.4278

22. Tsai, S.Q., et al.: GUIDE-seq enables genome-wide profiling of off-target cleavage by CRISPR-Cas nucleases. Nat. Biotechnol. **33**(2), 187–197 (2015). https://doi.org/10.1038/nbt.3117

23. Vaswani, A., et al.: Attention is all you need. In: Advances in Neural Information Processing Systems, vol. 30 (2017)

24. Wolf, T., et al.: Huggingface's transformers: state-of-the-art natural language processing. arXiv preprint arXiv:1910.03771 (2019)

25. Yu, L., Zhang, W., Wang, J., Yu, Y.S.: Sequence generative adversarial nets with policy gradient. 492 in. In: AAAI Conference on Artificial Intelligence, vol. 493 (2017)

26. Zhang, G., Zeng, T., Dai, Z., Dai, X.: Prediction of CRISPR/Cas9 single guide RNA cleavage efficiency and specificity by attention-based convolutional neural networks. Comput. Struct. Biotechnol. J. **19**, 1445–1457 (2021). https://doi.org/10.1016/j.csbj.2021.03.001

27. Zhang, X.H., Tee, L.Y., Wang, X.G., Huang, Q.S., Yang, S.H.: Off-target effects in CRISPR/Cas9-mediated genome engineering. Mol. Ther. Nucleic Acids **4**, e264 (2015). https://doi.org/10.1038/mtna.2015.37
28. Zhang, Z.R., Jiang, Z.R.: Effective use of sequence information to predict CRISPR-Cas9 off-target. Comput. Struct. Biotechnol. J. **20**, 650–661 (2022). https://doi.org/10.1016/j.csbj.2022.01.006. https://www.ncbi.nlm.nih.gov/pmc/articles/PMC8804193/

Improved Topology Features for Node Classification on Heterophilic Graphs

Yurui Lai[1], Taiyan Zhang[1,2], and Rui Fan[1(⊠)]

[1] ShanghaiTech University, Shanghai 201210, China
{laiyr,zhangty2022,fanrui}@shanghaitech.edu.cn
[2] Shanghai Innovation Center for Processor Technologies, Pudong,
Shanghai 201210, China

Abstract. Graph neural networks (GNNs) are one of the most effective techniques for node classification tasks. However, standard GNNs
strongly depend on the graph homophily assumption, and their accuracy can degrade substantially on heterophilic graphs. A number of
recent works have found that using graph topology as an explicit feature can improve classification performance. However, we observe that
the sparsity of graphs often limits the amount of information first order
connectivity can provide. We propose an embedding method which uses
higher-order connectivity information to further improve accuracy, while
limiting the amount of extra computational overhead. We further observe
that standard features-based GNNs and newer topology-based models
each have their own strengths and weaknesses on the same graph, and we
introduce a technique to combine information from both types of GNNs
to achieve higher accuracy than either type alone. We conduct extensive
experiments on graphs with a range of sizes and heterophily levels and
show that our proposed GNN architecture achieves state of the art accuracy, especially on highly heterophilic graphs. We also conduct further
experiments and ablations to validate the observations underlying our
GNN's design and analyze the importance of different components. Our
code is available in https://github.com/lannester666/BoPE-GNN.

Keywords: Node classification · Heterophily · Graph Neural Network

1 Introduction

Graphs are a powerful model for complex relationships, and many graphs exhibit
a key property of *homophily*, i.e. a tendency for same-class nodes to be connected.
For example, academic citations often occur within the same field [3], and people
with similar interests tend to connect on social media [18]. Many graph neural
networks (GNNs) [4,7,9,17,20] make crucial use of the homophily assumption
when performing node classification. However, there are also important classes
of graphs that have less structured, *heterophilic* edge relationships. For example,
chemical compounds frequently contain connections between different types of

© The Author(s), under exclusive license to Springer Nature Switzerland AG 2024
A. Bifet et al. (Eds.): ECML PKDD 2024, LNAI 14947, pp. 105–123, 2024.
https://doi.org/10.1007/978-3-031-70368-3_7

atoms [21], while some citation networks define classes based on year of publication and hence mostly contain edges between different classes [11]. Heterophilic graphs pose a challenge to traditional GNNs, and straightforward application of these architectures leads to significantly reduced accuracy [2,13,25]. This has led to a number of recent studies on GNNs designed for heterophilic graphs. One of the most effective techniques in these works [10,11,13] is to explicitly use the connectivity pattern, i.e. the set of neighbors that a node is connected to, as a feature. However, this method also has drawbacks. For example, it leads to neural networks whose parameter count is at least linear in the number of nodes, which increases computational cost and may cause training instability and overfitting. Furthermore, single hop connectivities may not be sufficiently informative, especially in sparse graphs.

To deal with these issues, we propose to use higher-order, multi-hop connectivity information as features. These features are informative both because they are denser, and also because nodes exhibit greater homophily within their l-hop neighborhoods [25], for moderate values of l, as compared to their immediate neighbors. Higher homophily is helpful for GNNs based on homophily assumption,e.g. GCN [9] and GAT [17]. However, direct use of higher-order connectivity would lead to exponentially increasing memory usage and compute time, and also would not reduce the number of parameters. To this end, we propose a simple embedding method which reduces the size of each feature down to a constant, as compared to being linear in the number of nodes when encoding full adjacency information. We show that this approach is highly effective and leads to substantially increased accuracy. However, for certain types of graphs, the basic embedding method fails to sufficiently distinguish between different nodes, so we also propose a modified embedding which adds positional information about each node to further increase the informativeness of the topological features. We call our method Bin of Paths Embedding (BoPE).

Another observation we made is that in many graphs, GNNs and neural networks using topological features (which for ease of exposition we term TNNs) have different accuracies on different classes, with GNNs being significantly better on some classes and TNNs being much better on others. This motivated us to design an architecture that can take advantage of both GNNs and TNNs and be equally or more accurate than both models on each class. To this end, we propose a method which mixes results from GNNs and TNNs, using a per node mixing parameter which is computed using the training accuracy of each model on the node and the node's predicted class. We call this method Confidence and Class-wise training Accuracy Weighting (CCAW).

Our overall proposed architecture, which we call BoPE-GNN, combines BoPE with CCAW. BoPE-GNN is flexible, as the BoPE component can be combined with any other existing node classifier via CCAW. We performed extensive experiments of our model on a variety of graphs with a range of sizes and homophily levels, and show that it significantly improves over the existing state of the art, with especially high performance gains on the most challenging heterophilic graphs. We also show that in most cases BoPE-GNN does indeed improve on the accuracy of the underlying GNN and TNN on a per class basis, validating the design of CCAW. We conducted a number of ablations on the importance of various components of BoPE-GNN, and also studied the best selection of different parameters. Finally, we show that BoPE-GNN is resource efficient, and can be trained with low additional time and memory overhead.

Our contributions can be summarized as follows:

1. We propose a simple but performant method called BoPE which makes use of higher-order connectivity information to improve classification accuracy.
2. We give a method to integrate results from standard GNNs and topology based neural networks which combines the advantages of both models and typically exceeds both models in accuracy.
3. We thoroughly validate our model and show that it achieves significantly higher accuracy than existing models, especially when heterophily is high. We also analyzed different components of the model to determine their effectiveness.

2   Related Work

Graph homophily measures the tendency of nodes to connect with those from the same class. It is usually quantified as a number from 0 to 1, with higher values indicating a greater likelihood of same-class connections. Graphs with low homophily levels are termed *heterophilic* and can exhibit complex correlations between classes of connected nodes. Heterophilic graphs are a broad and diverse family which poses significant challenges for current graph-based learning methods.

A number of existing GNN architectures leverage the high homophily of certain graphs to achieve high classification accuracy. Graph Convolutional Networks (GCN) [9] implement graph convolution using a normalized adjacency matrix combined with a non-linear transformation. Graph Attention Networks (GAT) [17] introduced masked attention to dynamically adjust the weights in the normalized adjacency matrix for adaptive aggregation. To mitigate over-smoothing, GCNII [4] deepens GNN layers using initial residual connections and identity mappings. These GNNs often perform poorly on heterophilic graphs, leading to the proposal of new architectures [2,5,13,15,25] with features to deal with heterophily.

In [25], a comprehensive definition of graph heterophily is given and a model called H$_2$GCN is proposed, which aggregates representations from one and two hop neighborhoods and concatenates them with those of the ego node. An effective technique in heterophilic graphs is to use signed aggregation weights [5,10,13,22]. The aggregation weights of standard GNNs are positive, and thus have a tendency to bring the representations of neighboring nodes together, leading them to be classified as the same class. By allowing negative aggregation weights, neighboring representations can instead be pushed apart, thus producing heterophilic class predictions. Another technique which can substantially improve accuracy on heterophilic graphs is to explicitly use graph topology as a feature. LINK [24] compiled an extensive collection of large-scale heterophilic graphs to establish more generalized benchmarks. The model uses MLPs to encode node attributes and a graph's adjacency matrix independently, achieving both high accuracy and efficiency across numerous datasets. GloGNN [10] creates node representations by combining node attributes with topology features using MLPs, then aggregates information from global nodes in the graph using a learned coefficient matrix. These GNNs directly use the graph's adjacency matrix for features without modifying or embedding it, resulting in a large number of model parameters and possible overfitting.

A number of techniques exist to deal with the high dimensionality and sparsity of adjacency matrices. Graph embeddings [6] represent graph topology using representations of a specified size. Decomposition-based graph embeddings such as PRONE [23] perform SVD or t-SVD on graph adjacency matrices or related matrices and propagate the embeddings in a spectrally modulated space. Matrix sketching-based embeddings such as Random Projection [1] use a binary random sketching matrix to effectively reduce dimensionality. Other methods use neural networks to generate graph embeddings efficiently. Deep-walk [16] does random walks on a graph to generate node sequences, then uses word2vec [14] to obtain a node embedding.

3   Methodology

In this section, we describe the design of our BoPE-GNN architecture and its implementation. We first discuss some observations motivating its design and then formally define the model.

3.1   Notation

We first introduce some notation. Given a graph $\mathcal{G} = (\mathcal{V}, \mathcal{E})$, let $N = |\mathcal{V}|$ and $E = |\mathcal{E}|$ be the number of nodes and edges, respectively. Also, let the adjacency matrix of $\mathcal{G}$ be $\mathbf{A} \in \{0, 1\}^{N \times N}$. Each node has an F-dimensional attribute vector to help with its classification, and we write the attributes matrix for all the nodes as $\mathbf{X} \in \mathbb{R}^{N \times F}$. Each node also has a label from one of C classes, and we let $\mathbf{Y} \in \{0, 1\}^{N \times C}$ be the one-hot encoding of the labels of all the nodes. The degree matrix $\mathbf{D} \in \mathbb{R}^{N \times N}$ is a diagonal matrix whose nonzero elements are the row sums of $\mathbf{A}$.

3.2   Motivations

Higher-Order Connectivity. As discussed in Sect. 2, standard GNNs which make use of only node attributes often perform poorly on heterophilic graphs, while neural networks such as LINK [24], which makes explicit use of the adjacency information of each node, e.g. using the adjacency matrix $\mathbf{A}$ as an additional feature, can be substantially more accurate on a range of graph datasets. However, one problem with directly using $\mathbf{A}$ is that it is typically very sparse, which can lead to insufficient feature information being provided to each node. To increase the amount of information available to each node, we consider using higher-order connectivities, in particular the matrices $\mathbf{A}, \mathbf{A}^2, \ldots$. Note that for $l \geq 1$ and $i, j \in [N]$, $\mathbf{A}^l_{i,j}$ gives the number of paths of length l between nodes i and j, so that $\mathbf{A}^l$ is typically much denser than $\mathbf{A}$.

To assess whether higher-order connectivities form better features than single hop connectivity, we performed an experiment using a simple linear classifier taking $\mathbf{A}, \ldots, \mathbf{A}^L$, as input. In particular, we trained the following network.

$$\tilde{\mathbf{P}} = softmax(\underset{l \in \{1, \cdots, L\}}{\|} \mathbf{A}^l \mathbf{W}) \tag{1}$$

Here $\mathbf{W}$ is a matrix of learnable parameters and $\|$ represents the concatenation operation. We call this model Multi-LINK since LINK is a special case of the model when $L = 1$.

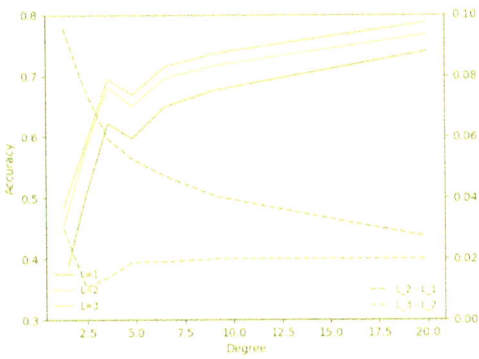

Fig. 1. Multi-LINK accuracy on *Snap-patents* graph with different orders L

Figure 1 shows the accuracy of Multi-LINK on the strongly heterophilic graph *Snap-patents* (details about the graph can be found in Sect. 4) as a function of node degree using different orders L of the adjacency matrix. The functions are mostly increasing and concave, with accuracy increasing rapidly as a function of degree when node degree is low. The figure also shows that using higher-order connectivity significantly improves accuracy for all node degrees and that the improvement from $L = 1$ to 2 is substantially larger than the improvement from $L = 2$ to 3, possibly indicating a saturation in the value of the information provided by increasing the order L.

However, directly using higher-order connectivities in a learning model is inefficient. One issue is that the number of nonzero values in $\mathbf{A}^l$ increases exponentially with l, so it is generally only computationally feasible to use $L = 2$ or 3 for larger graphs in GNNs. Furthermore, the direct use of adjacency information (of any order) as a node feature produces a feature vector of size N, so that any neural network using the vector has input will have $\Omega(N)$ parameters, which means the model is much larger than traditional GNNs and more difficult to train. To deal with these problems, we propose to embed the size $O(N)$ adjacency feature of each node in a lower dimensional space, typically with dimension $D = 512$. We found that a simple and fast embedding method which simply sums every N/D column of A, producing an $N \times D$ matrix, is enough to produce good results. Using this matrix as input to a neural network leads to a network with a constant parameter count as a function of the graph size.

Combining Node Attributes and Topology Features. Another observation underlying our system's design is the fact that, while using topology as features generally outperforms using node attributes on low homophily graphs, neither type of feature strictly dominates the other, even on the same graph. In particular, when looking at different classes in a single graph dataset, we found that a topology based network such as the Multi-LINK algorithm described earlier is substantially better than an attributes based network such as GAT [17] on many classes, but is also markedly worse on certain classes. To study this phenomenon, we used 4 low homophily graphs, *Snap-patents*, *arXiv-year*, *squirrel*, *chameleon* and *Penn94* and compared the accuracy Multi-LINK and GAT on each class in each dataset. The results are shown in Fig. 2, where each point corresponds to a class from one of the graphs, and the x and y coordinates of the point represent its accuracy using Multi-LINK and GAT, respectively. Points on the diagonal would correspond to equal performance using both algorithms. However, the large number of highly off-diagonal points in both the upper left and lower right regions of the plot indicate that the accuracy of the two methods frequently differ substantially, with neither method being strictly better.

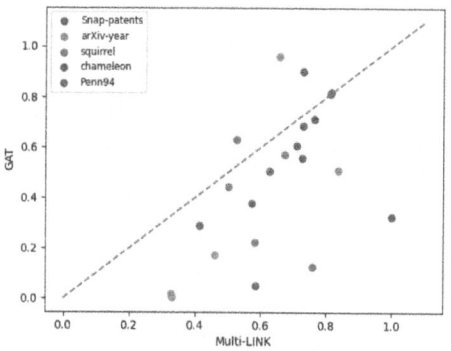

Fig. 2. GAT vs. Multi-LINK class-wise accuracy

To make use of this observation, we propose a method which dynamically combines information from two baseline algorithms, one topology based and the other attributes based. As we show in Sect. 4, this results in a network which is often significantly more accurate than either baseline algorithm alone. Furthermore, our method is flexible and can be used to combine almost any two baseline algorithms. The basic idea for our method is to first estimate the per-class accuracy of the two baseline algorithms on either the test or validation data. We then combine this information with the confidence of the two baselines for each node, where the confidence is the maximum predicted class probability for the node. This combined measure is used to balance between the predictions produced by the two baselines.

3.3   Bin of Paths Embedding

We now formally describe how our algorithm makes use of higher-order connectivity information. Given the highest order L, we first compute the matrices $\mathbf{A}, \cdots, \mathbf{A}^L$. We typically use a value of $L \leq 3$, as we find this gives good results, and this step can be done on the CPU as a preprocessing step fairly efficiently. Consider one of the matrices $\mathbf{A}^l$, for $l \leq L$. For nodes i and j, $\mathbf{A}^l_{i,j}$ is the number of l-hop paths between i and j, and the row vector $\mathbf{A}^l_i$ gives this information for i to all the other nodes. We found that such detailed connectivity information is not necessary to accurately classify i, and is sometimes even somewhat harmful. Instead, we *bin* the connectivities for i into $D \ll N$ bins, by partitioning $\mathbf{A}^l_i$ into D consecutive, disjoint segments, and summing together the $\lceil N/D \rceil$ values in each segment (except possibly the last segment) to form a D dimensional embedding vector for i's l'th order connectivity. To apply this method to the entire matrix $\mathbf{A}^l$, we simply sum together every $\lceil N/D \rceil$ consecutive columns in $\mathbf{A}^l$, to form a new matrix $\mathbf{B}_l \in \mathbb{R}^{N \times D}$. We then concatenate $\mathbf{B}_1, \ldots, \mathbf{B}_L$ to form an overall topology feature matrix.

We found that while the embedding described above typically performs well, it sometimes fails to sufficiently distinguish between different nodes, and thus may harm classification accuracy. For example, consider two nodes i and j, and suppose that one of the bins in i's vector $\mathbf{A}^l_i$ is $[1, 0, 1, 0]$, and the corresponding bin in j's vector $\mathbf{A}^l_j$ is $[0, 1, 0, 1]$. Then after summing the bins, both i and j's bins would be embedded as 2 in $\mathbf{B}_l$, thereby failing to distinguish the two nodes.

To deal with this problem, we first proceed as before and compute matrices $\mathbf{A}^1, \ldots, \mathbf{A}^L$. Next, for each matrix $\mathbf{A}^l$, $1 \leq l \leq L$, we replace any nonzero value in a column j of matrix $\mathbf{A}^l$ with the value j, for all $1 \leq j \leq N$ and $1 \leq l \leq L$. For example, if the bins in the earlier example correspond to columns 1 to 4, then we would replace i's bin by $[1, 0, 3, 0]$ and j's bin by $[0, 2, 0, 4]$. The goal of the replacements is to better distinguish between the bins of different nodes. After the replacement step, we again sum every $\lceil N/D \rceil$ columns of each replaced matrix, producing a matrix $\mathbf{B}'_l$, for each $1 \leq l \leq L$.

The values in higher columns of $\mathbf{B}'_l$ are generally larger than those from lower columns, so directly using rows of $\mathbf{B}'_l$ as features may cause training instability. To deal with this, we normalize $\mathbf{B}'_l$. In particular, we take each column j (containing at least one nonzero value) of $\mathbf{B}'_l$, and divide the column by the sum of all the values in the column. Let the matrix obtained from applying this operation to every column of $\mathbf{B}'_l$ be $\tilde{\mathbf{B}}_l$. We found that using a combination of $\tilde{\mathbf{B}}_l$ and

$\mathbf{B}_l$ better distinguishes between nodes and improves classification accuracy. In particular, we concatenate all the matrices of the form $\mathbf{B}_l + \alpha_l \tilde{\mathbf{B}}_l$ to form

$$\hat{\mathbf{B}} = \underset{l \in \{1, \cdots, L\}}{\|} (\mathbf{B}_l + \alpha_l \tilde{\mathbf{B}}_l) \tag{2}$$

Here $\alpha_l = \frac{E_l}{D}$, where E_l is the number of nonzero values in $\mathbf{B}_l$. The $\alpha_l \tilde{\mathbf{B}}_l$ terms help to distinguish corresponding bins of different nodes. α_l generally increases with l, and we use this heuristic in order to improve distinguishability because we found $\mathbf{B}_l$ often has more indistinguishable bins due to the higher density of the matrix $\mathbf{A}^l$ from which $\mathbf{B}_l$ is formed.

3.4   Confidence and Class-Wise Training Accuracy Weighting

We now formally describe how to combine an attributes based classifier with a topology based one, making use of the observation from Sect. 3.2 that these generally have different accuracies on different classes in a graph. We call the combined model *BoPE-GNN*. For the topology based classifier, we will use *BoPE-Net*, which is an MLP which takes the BoPE embedding described in Sect. 3.3 as input, and produces predicted class probabilities as output. For the attributes based model, which we call the *base model* of BoPE-GNN, we can use any node classifier. We show for example in Sect. 4 that BoPE-GNN can use GAT, LINKX and GloGNN as base models.

We call the part of BoPE-GNN using the base model the *base branch*, and the part using BoPE-Net the *topology branch*. For each node, both branches produce as output predicted class probability vectors. We now describe a way to combine these two predictions, on a per-node basis, with the goal that the combined prediction will be at least as accurate as the better of the two original predictions. Consider a node v_i, and let $\mathbf{p}_i^b$ and $\mathbf{p}_i^t$ be the predicted class vectors for v_i by the base model and BoPE-Net. Let γ_i^b and γ_i^t be the maximum values from vectors $\mathbf{p}_i^b$ and $\mathbf{p}_i^t$. γ_i^b and γ_i^t represent the confidence that the base model and BoPE-Net have in their predictions of v_i's class. In particular, past research [12,19] has shown that higher γ values generally correlate with a higher likelihood that the classifier's prediction is correct. We thus want to incorporate the γ values of the base model and BoPE-Net when combining their predictions.

However, we also need to consider the fact that the base model and BoPE-Net themselves may have different training accuracies, depending on v_i's class. In particular, let κ_i^b and κ_i^t be the predicted classes for v_i by the base model and BoPE-Net. Also, let Λ_i^b and Λ_i^t be the base model and BoPE-Net's training accuracy on classes κ_i^b and κ_i^t, respectively. We describe later how Λ_i^b and Λ_i^t are computed. We compute $w_i = \beta \frac{\gamma_i^b \Lambda_i^b}{\gamma_i^t \Lambda_i^t}$, where β is a positive hyperparameter. The numerator, resp. denominator in w_i are intended to capture the likelihood that the base model, resp. BoPE-Net correctly predicted the class of v_i. In particular, the γ values represent a model's confidence in its prediction, which is generally correlated with its accuracy, while the Λ values represent the model's empirical accuracy on the class which it predicted for v_i. We now use w_i as a weighting parameter for combining the base model and BoPE-Net's predictions for v_i. Formally, we compute the final prediction as:

$$\mathbf{p}_i = softmax(\frac{1}{1+w_i}\mathbf{p}_i^b + \frac{w_i}{1+w_i}\mathbf{p}_i^t) \tag{3}$$

Finally, node v_i's class is predicted as the class with the maximum value in $\mathbf{p}_i$. We call this method of combining the base model and BoPE-Net's predictions Confidence and Class-wise training Accuracy Weighting (CCAW).

The overall BoPE-GNN architecture is shown in Fig. 3. It consists of a base branch containing a base model, which takes as input the node attributes and possibly the graph's adjacency matrix (e.g. algorithms such as GloGNN make explicit use of the adjacency matrix), and a topology branch which runs BoPE-Net. The two branches compute predicted class vectors, which are then combined using CCAW and then trained using cross-entropy loss. Gradients are propagated to both branches, to train both the base model and BoPE-Net concurrently. CCAW makes use of Λ^b and Λ^t values, whose computations are described below. These values are constants and not updated via gradients.

Lastly, we describe how to compute the per class training accuracy values $\Lambda^b(\kappa)$ and $\Lambda^t(\kappa)$, for each class κ. These values are (re)computed during every epoch when we train BoPE-GNN. In particular, before the first epoch we initialize Λ^b and Λ^t to be 0 for every class. Subsequently, after every training epoch, we run the base and BoPE-Net models from that epoch on the training or validation nodes to obtain an estimate of $\Lambda^b(\kappa)$ and $\Lambda^t(\kappa)$ for every κ. While these estimates may not be equal to the values of Λ^b and Λ^t on the test set, we found that they are generally reasonably accurate.

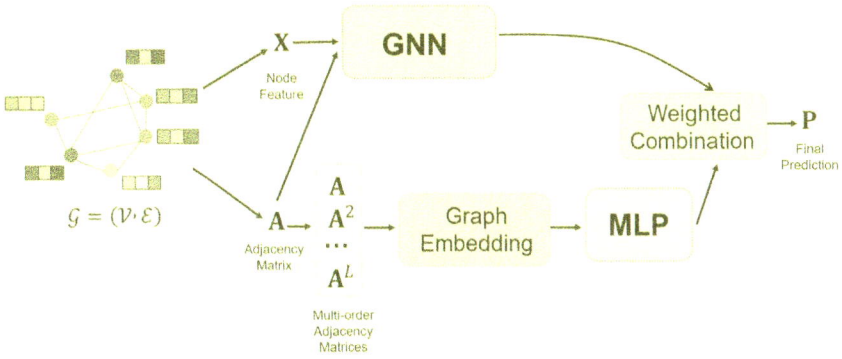

Fig. 3. BoPE-GNN architecture

4   Evaluation

In this section, we evaluate the performance of BoPE-GNN. We first compare its accuracy to existing works. We then compare BoPE-GNN's accuracy to the base models it uses. We conduct ablations on its components and analyze the effect of various hyperparameters. Finally, we evaluate our algorithm's computational efficiency.

Table 1. Datasets statistics

Dataset	# Nodes	# Edges	# Attr.	# Classes	Homophily	Type
Snap-patents [11]	2,923,922	13,975,788	269	5	0.07	Citation network
squirrel [15]	5,201	396,846	2,089	5	0.22	Wikipedia graph
arXiv-Year [11]	169,343	1,166,243	128	5	0.22	Citation network
chameleon [15]	2,277	62,792	2,325	5	0.23	Wikipedia graph
Penn94 [11]	41,554	1,362,229	5	2	0.47	Social network
Twitch-gamer [11]	168,114	6,797,557	7	2	0.55	Social network
Genius [11]	421,961	984,979	12	2	0.62	Social network

4.1 Experimental Settings

We use 7 heterophilic graph datasets with different sizes and homophily levels in our evaluations, including *Genius, Twitch-gamer, Penn94, chameleon, squirrel, arXiv-year* and *Snap-patents*. Partial statistics of the graphs are given in Table 1.

We compare against a number of existing GNN and other ML architectures. These include GCN [9] and GAT [17], which are widely used GNNs originally designed for homophilic graphs, and H_2GCN [25] and GloGNN [10], which are designed for heterophilic graphs and achieve previously state-of-the-art results. Lastly, we compared our method against several existing graph embedding methods, including truncated SVD [8], PRONE [23], and random projection [1], to evaluate the effectiveness of BoPE-GNN's graph embedding.

We search the maximum order of adjacency matrix among $\{1, 2, 3\}$, and the embedding dimension D in $\{64, 128, 256, 1024\}$. We search β in CCAW in $\{0.1, 0.5, 0.7, 0.9\}$. The hidden dimension of GNN is set to 128. The dropout is in the range of $[0.0, 0.9]$ with 0.1 step. For model training, Adam optimizer is adopted with learning rate among $\{0.01, 0.005\}$ and weight decay is among $\{0.001, 0.005\}$. The training epoch is 1500 and early stopping is 200. All experiments are conducted on a Linux machine with an NVIDIA Ampere A100 GPU (80GB RAM), AMD EPYC 7513 CPU (2.6 GHz), and 1TB RAM.

4.2 Node Classification

We first show BoPE-GNN's classification accuracy on different datasets (ROC-AUC for *Genius*) compared to baseline algorithms. We mostly use GAT for the base model of BoPE-GNN, but we also describe some experiments done using LINKX and GloGNN as base. The results of the comparison are shown

Table 2. Classification accuracy of different methods in percent; ± indicates range of one standard deviation. The best result on each dataset is shown in bold. OOM indicates out of memory.

	Snap-patents	squirrel	arXiv-year	chameleon	Penn94	Twitch-gamer	Genius
MLP	31.34 ± 0.05	28.77 ± 1.56	36.70 ± 0.21	46.21 ± 2.99	73.61 ± 0.40	60.92 ± 0.07	86.69 ± 0.09
LINK [24]	60.39 ± 0.07	60.63 ± 1.73	53.97 ± 0.18	68.00 ± 1.95	80.79 ± 0.49	64.85 ± 0.21	73.56 ± 0.14
GCN [9]	45.65 ± 0.04	53.43 ± 2.01	46.02 ± 0.26	64.82 ± 2.24	82.47 ± 0.27	62.18 ± 0.26	87.42 ± 0.37
GAT [17]	45.37 ± 0.44	40.72 ± 1.55	46.05 ± 0.51	60.26 ± 2.50	81.53 ± 0.55	59.89 ± 4.12	55.80 ± 0.87
GCNII [4]	37.88 ± 0.69	38.47 ± 1.58	47.21 ± 0.28	63.86 ± 3.04	82.92 ± 0.59	63.39 ± 0.61	90.24 ± 0.09
H₂GCN [25]	51.52 ± 0.04	36.48 ± 1.86	49.09 ± 0.10	60.11 ± 2.15	81.18 ± 0.58	OOM	90.19 ± 0.17
GPRGNN [5]	40.19 ± 0.03	31.61 ± 1.24	45.07 ± 0.21	46.58 ± 1.71	81.38 ± 0.16	61.89 ± 0.29	90.05 ± 0.31
ACM-GCN [13]	55.15 ± 0.16	54.40 ± 1.88	47.37 ± 0.59	66.93 ± 1.85	82.52 ± 0.96	62.01 ± 1.73	80.33 ± 3.91
LINKX [11]	61.95 ± 0.12	61.81 ± 1.80	56.00 ± 1.34	68.42 ± 1.38	84.71 ± 0.52	66.06 ± 0.19	90.77 ± 0.27
GGCN [22]	OOM	55.17 ± 1.58	OOM	71.14 ± 1.84	OOM	OOM	OOM
GloGNN [10]	62.03 ± 0.21	57.88 ± 1.76	54.68 ± 0.34	71.21 ± 1.84	**85.74 ± 0.42**	66.34 ± 0.29	**90.91 ± 0.13**
BoPE-GAT	**73.64 ± 0.04**	**65.28 ± 0.90**	**58.65 ± 0.38**	**72.63 ± 2.88**	82.86 ± 0.63	63.34 ± 0.49	85.37 ± 2.08
BoPE-LINKX	67.26 ± 0.86	60.75 ± 0.73	55.82 ± 0.44	69.17 ± 0.63	84.92 ± 0.51	66.14 ± 0.18	90.60 ± 0.13
BoPE-GloGNN	64.76 ± 1.37	58.43 ± 1.46	55.21 ± 0.24	71.07 ± 1.64	85.69 ± 0.41	**66.42 ± 0.20**	90.88 ± 0.07

in Table 2, where BoPE-GNN's results are shown in the last three rows, with BoPE-X referring to the use of model X in the base branch of BoPE-GNN.

The results show several things. First, some GNNs designed for homophilic graphs, including GCN and GAT, work poorly on heterophilic graphs, and sometimes even do worse than simpler algorithms such as MLP and LINK. GCNII works relatively well on heterophilic graphs, possibly due to its use of identity mapping and its residual structure.

Second, GNNs designed for heterophilic graphs work better than conventional GNNs. However, some such GNNs incur high computational overhead. For example, H₂GCN and GGCN run out of memory on large-scale graphs with millions of edges, including *Twitch-gamer*, *arXiv-year*, and *snap-patent*.

Third, by using BoPE-GNN on top of base models GAT, LINKX and GloGNN, BoPE-GNN achieves the best performance on almost every dataset. Note that the graphs in Table 2 are ordered from left to right by increasing homophily. The table shows that BoPE-GNN performs especially well on graphs with very low homophily. For example, on *snap-patent*, which has 0.07 homophily, BoPE-GAT achieves 73.64% accuracy, exceeding the previous state-of-the-art GloGNN by 11.61%. Likewise, BoPE-GAT beats existing architectures by 3.47%, 2.65%, and 1.42% on the higher homophily (around 0.22) *squirrel*, *chameleon* and *arXiv-year* graphs. On the remaining graphs *Penn94*, *Twitch-gamer* and *Genius*, which have homophily higher than 0.47, BoPE-GAT performs worse than the state-of-the-art GloGNN. However, by combining BoPE-GNN with GloGNN, we achieve performance that is within 0.1% of GloGNN's accuracy. Part of the reason BoPE-GloGNN does not outperform GloGNN may be that GloGNN itself also makes use of graph topology information, so combining it with BoPE-Net is ineffective. Overall, the results show that BoPE-GNN's approach offers considerable improvement on challenging graphs with low homophily levels.

Table 3. Comparison to different base models

	squirrel	Change	chameleon	Change	arXiv-year	Change	Snap-patent	Change
GCN	53.43 ± 2.01	-	64.82 ± 2.24	-	46.02 ± 0.26	-	45.65 ± 0.04	-
BoPE-GCN	56.93 ± 1.40	3.51	68.03 ± 2.65	3.21	56.43 ± 0.16	10.41	72.25 ± 0.46	26.61
GAT	40.72 ± 1.55	-	60.26 ± 2.50	-	46.05 ± 0.51	-	45.37 ± 0.44	-
BoPE-GAT	**65.28 ± 0.90**	24.56	**72.63 ± 2.88**	12.37	56.61 ± 0.34	10.56	73.64 ± 0.04	28.27
GCNII	38.47 ± 1.58	-	63.86 ± 3.04	-	47.21 ± 0.28	-	37.88 ± 0.69	-
BoPE-GCNII	56.00 ± 0.89	17.53	68.03 ± 1.75	4.17	56.64 ± 0.21	9.43	73.03 ± 0.09	35.15
H₂GCN	36.48 ± 1.86	-	60.11 ± 2.15	-	49.09 ± 0.10	-	51.52 ± 0.04	-
BoPE-H2GCN	58.85 ± 0.59	22.37	67.46 ± 1.02	7.35	**56.74 ± 0.19**	7.65	**74.18 ± 0.05**	22.66

4.3 Improvements on Base GNN Models

In this section, we demonstrate that BoPE-GNN can be combined with a variety of base GNNs and in each case significantly improves on the base's accuracy. In particular, we combined BoPE-GNN with GCN, GAT, GCNII, and H2GCN. The results are shown in Table 3. The large improvement in accuracy across nearly all models and datasets shows that graph topology is a powerful source of information, which BoPE-GNN is typically able to exploit. BoPE-GNN achieved nearly the same accuracy regardless of the base model it was combined with on graphs *arxiv-year* and *snap-patent*, while on the *squirrel* and *chameleon* graphs it worked substantially better when combined with GAT than other GNNs.

4.4 Distribution of CCAW Weights

In order to evaluate the relative importance of the attributes and topology components to BoPE-GNN's accuracy, we studied the distribution of BoPE-GAT's CCAW weights on various graphs, where the weights are computed according to Eq. 3. The cumulative distribution of the weights is shown in Fig. 4. For example, it shows that about 90% of the CCAW weights in BoPE-GAT trained on *arXiv-year* are over 4. A weight over 1 indicates that BoPE-GAT made more use of the topology than attribute features, and this was generally the case across the different datasets. Furthermore, the diversity of the weight distributions shows that BoPE-GAT is able to select weights that are adapted to the different characteristics of the datasets.

4.5 Class-Wise Node Classification Accuracy

A key motivation for BoPE-GNN is to be able to take advantage of the different per-class accuracies provided by attributes versus topology based classifiers. To this end, we compared BoPE-GAT's accuracy on each class with those of the attribute based classifier GAT and the topology based classifier LINK. The results on graphs *arXiv-year* and *snap-patent*, each of which has 5 classes, are shown in Fig. 5. These plots highlight the strong difference in the effectiveness of GAT and LINK on different classes. In addition, they show that BoPE-GAT is

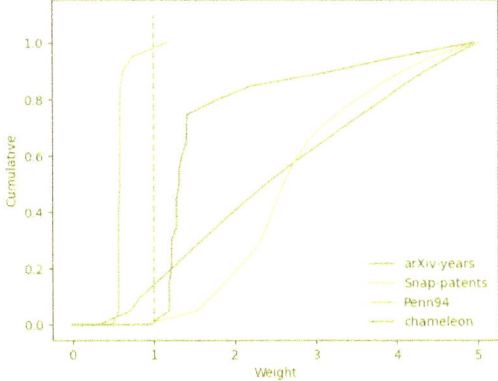

Fig. 4. Cumulative distribution function of CCAW weights

able to achieve accuracy that is close to the higher of GAT and LINK's accuracy on each class, and frequently exceeds these accuracies. This shows that BoPE-GAT can effectively make use of both topology and attribute based information via the CCAW mechanism.

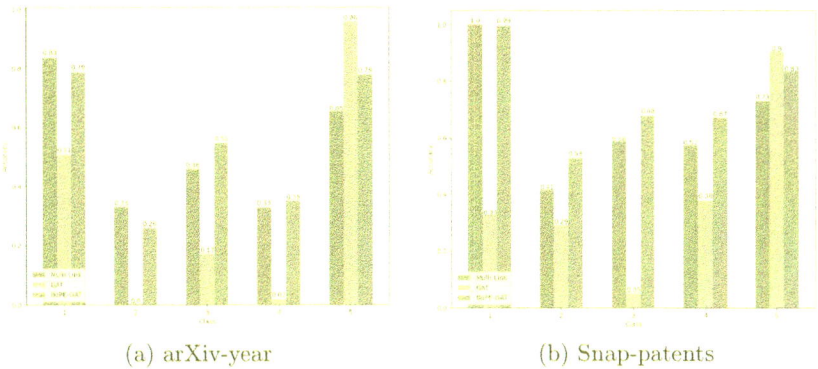

(a) arXiv-year   (b) Snap-patents

Fig. 5. Class-wise accuracy comparison of Multi-LINK, GAT, and BoPE-GAT on *arXiv-year* and *Snap-patents*

4.6   Ablations

We now look at the effectiveness of various parts of the BoPE-GNN architecture using ablations, with GAT as the base model. The results are shown in Table 4. The accuracy of GAT and BoPE-GAT on several datasets is shown in the first and last rows, respectively. BoPE-GAT is generally far more effective than GAT.

The row *BoPE-Net* shows the accuracy using only the topology branch of BoPE-GAT, without running the attributes branch or combining the branches using CCAW. Accuracy is much higher than for GAT, especially on *snap-patent*, where it is almost as accurate as BoPE-GAT. But generally, BoPE-GAT is still significantly better than BoPE-Net.

In the *Adjacency matrix* row, we considered running BoPE-GAT but giving only the adjacency matrix as input to the topology branch, without performing the BoPE embedding. This is significantly less accurate than the full BoPE-GAT, especially on *snap-patent*. We also considered randomly permuting the order of all the nodes in the *Random shuffle* row. This is because the binning procedure in the BoPE embedding produces different outputs for different node orderings. We found that random reordering has a generally modest negative effect on accuracy, depending on the graph. These perturbations can be offset by running BoPE-GAT several times with different node orderings and then using for example, a majority vote to classify nodes. Finally, we evaluated the utility of positional encoding, as given in the $\tilde{\mathbf{B}}$ matrices described in Sect. 3.3. Row *No positional info* shows that positional information can substantially improve accuracy.

The rows *t-SVD*, *ProNE*, *Random projection* and *Deepwalk* consider the effect of using these embeddings instead of BoPE in the topology branch of BoPE-GAT. We found that these can lead to very significant losses in accuracy compared to BoPE. As BoPE is also much faster to compute than the other methods, it offers advantages in both accuracy and speed. Specially, As deepwalk applies random walk to produce the node feature, in Snap-patents there exist many isolated nodes that have no out-degrees, and this makes it hard to produce embeddings by deepwalk.

Table 4. Ablation study for BoPE-GNN

	squirrel	chameleon	arXiv-year	Snap-patents
GAT	40.72 ± 1.55	60.26 ± 2.50	46.05 ± 0.51	45.37 ± 0.44
BoPE-Net	54.72 ± 1.00	65.53 ± 1.41	53.42 ± 0.32	72.15 ± 0.07
Raw adjacency matrix	60.63 ± 1.73	68.00 ± 1.95	53.97 ± 0.18	60.39 ± 0.07
Random shuffling	64.78 ± 1.77	71.05 ± 1.36	55.01 ± 0.42	73.22 ± 0.13
No positional information	60.90 ± 1.12	68.48 ± 1.86	54.06 ± 0.10	73.40 ± 0.09
t-SVD [8]	61.58 ± 1.34	70.09 ± 1.43	45.24 ± 1.14	38.66 ± 1.47
ProNE [23]	65.42 ± 1.43	68.16 ± 2.36	50.48 ± 0.79	28.61 ± 1.32
Random projection [1]	60.54 ± 1.65	67.94 ± 1.93	37.34 ± 0.41	49.75 ± 0.18
DeepWalk [16]	57.68 ± 1.68	69.25 ± 1.17	41.76 ± 3.48	–
BoPE-GAT	65.28 ± 0.90	72.63 ± 2.88	58.65 ± 0.38	73.63 ± 0.04

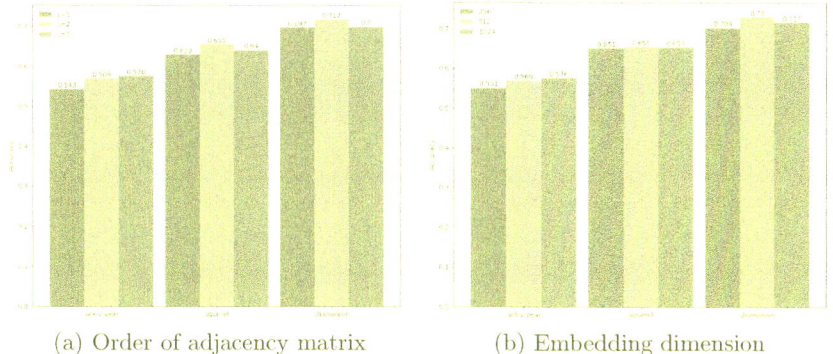

(a) Order of adjacency matrix   (b) Embedding dimension

Fig. 6. Hyper-parameters analysis

4.7   Hyperparameter Analysis

In this section, we analyze the effect of the maximum order of the input adjacency matrices L, as well as the embedding dimension D on BoPE-GNN's accuracy.

As shown in Fig. 6a, the order L has different effects on different datasets. For *arXiv-year*, accuracy improves with order, with a larger increase going from order 1 to 2 than from 2 to 3. Since *arXiv-year* is a relatively large graph and the average degree of each node is relatively small, the adjacency matrix itself may not be providing sufficient information, while higher-order connectivity helps nodes better orient themselves within the graph. On the other graphs *squirrel* and *chameleon*, which are much smaller, we see that while there is an increase in accuracy in going from order 1 to 2, there is a similar magnitude decrease in going from order 2 to 3. This may be due to the fact that the average degree of each node increases exponentially with order, so that for order 3, each node is already connected to a large proportion of all other nodes. Because of this, the order 3 connectivity vectors of different nodes may be quite similar, and thus nodes are less able to distinguish themselves using topological information.

As shown in Fig. 6b, D has different effects on different datasets. For *arXiv-year* and *chameleon*, BoPE-GAT achieves better accuracy as the embedding dimension increases. For *squirrel*, the accuracy first increases with dimension and then decreases. For large graphs like *arXiv-year*, small dimensions may result in over-compression of topology information. On the smaller graphs *chameleon* and *squirrel*, compression is less of an issue. r

4.8   Efficiency Analysis

Here we analyze the efficiency of BoPE, ProNE, and t-SVD. ProNE is a fast matrix factorization method to construct embeddings, and t-SVD is a method that has been used on the feature construction of heterophilic graphs, We compare the feature construction time of the three methods. Firstly, we record the CPU running time of BoPE on Snap-patent, compared with t-SVD in Fig. 7. Our method only takes about 1×10^2 seconds, while the time consumed by ProNE and t-SVD reaches an astonishing 1×10^3 to 1×10^4 seconds. Moreover, with the embedding dimension increase, the increase of our method's time consumption is trivial, while the consumption of ProNE and t-SVD grows fast, making it hard to extend to larger dimensions.

Finally, we evaluate the computational efficiency of BoPE-GNN. We use GAT as our base model, and show the profiling results for the *snap-patent* graph. We also compare BoPE-GAT's efficiency to those of GAT, LINK, and Multi-LINK, as these algorithms are similar to the two branches comprising the BoPE-GAT model. BoPE begins by computing a graph embedding, as described in Sect. 3.3. This is done a single time on the CPU, and takes on the order of tens to the low hundreds of seconds, depending on the size of graph. All the remaining steps are done on the GPU. Table 5 shows that BoPE-GNN is several times slower than GAT and LINK for training and inference, and uses about 1/3 more GPU memory than GAT and about 4× more memory than LINK. Since the models are typically only trained for a few hundred epochs, the total training time for all the models is only a few minutes and hence not a major bottleneck. Considering BoPE-GAT has significantly higher accuracy compared to the baselines, it offers a worthwhile accuracy/speed trade-off.

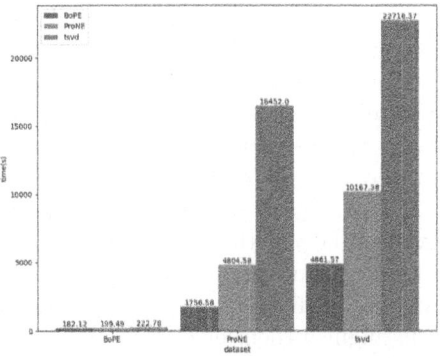

Fig. 7. The embedding time analysis

Table 5. BoPE-GNN efficiency analysis

	Accuracy (%)	Training time (ms/epoch)	Max GPU memory (GB)
GAT	45.37 ± 0.44	56	24.48
LINK	60.39 ± 0.07	97	6.39
Multi-LINK	65.75 ± 0.05	652	35.51
BoPE-Net	72.15 ± 0.07	89	5.71
BoPE-GAT	73.64 ± 0.04	182	32.04

5 Conclusion

We proposes BoPE, an efficient graph embedding method for heterophilic graphs with different sizes and homophily. We then design BoPE-GNN, a graph neural network which achieves state of the art accuracy on heterophilic graphs. BoPE-GNN makes use of higher-order graph connectivity information via BoPE to produce models of constant size. It also introduces a method to combine existing GNNs and our topology based classifier to achieve higher accuracy than either model alone. Extensive experimental evaluations show BoPE-GNN can significantly outperform existing baselines, especially on highly heterophilic graphs. Ablations and hyperparameter analyses were also conducted to demonstrate the effectiveness of BoPE-GNN's different components. In the future, we would like to conduct a theoretical analysis of the benefits of graph topology encoding for heterophilic graph tasks. We also aim to explore BoPE-GNN's applicability to other graph tasks, such as edge prediction on heterophilic graphs.

Disclosure of Interests. The authors have no competing interests to declare that are relevant to the content of this article.

References

1. Bingham, E., Mannila, H.: Random projection in dimensionality reduction: applications to image and text data. In: Proceedings of the Seventh ACM SIGKDD International Conference on Knowledge Discovery and Data Mining, pp. 245–250 (2001)
2. Bo, D., Wang, X., Shi, C., Shen, H.: Beyond low-frequency information in graph convolutional networks. In: Proceedings of the AAAI Conference on Artificial Intelligence, pp. 3950–3957 (2021)
3. Bojchevski, A., Günnemann, S.: Deep gaussian embedding of graphs: unsupervised inductive learning via ranking. arXiv preprint arXiv:1707.03815 (2017)
4. Chen, M., Wei, Z., Huang, Z., Ding, B., Li, Y.: Simple and deep graph convolutional networks. In: International Conference on Machine Learning, pp. 1725–1735. PMLR (2020)
5. Chien, E., Peng, J., Li, P., Milenkovic, O.: Adaptive universal generalized pagerank graph neural network. arXiv preprint arXiv:2006.07988 (2020)

6. Goyal, P., Ferrara, E.: Graph embedding techniques, applications, and performance: a survey. Knowl.-Based Syst. **151**, 78–94 (2018)
7. Hamilton, W., Ying, Z., Leskovec, J.: Inductive representation learning on large graphs. In: Advances in Neural Information Processing Systems, vol. 30 (2017)
8. Kilmer, M.E., Martin, C.D.: Factorization strategies for third-order tensors. Linear Algebra Appl. **435**(3), 641–658 (2011)
9. Kipf, T.N., Welling, M.: Semi-supervised classification with graph convolutional networks. arXiv preprint arXiv:1609.02907 (2016)
10. Li, X., et al.: Finding global homophily in graph neural networks when meeting heterophily. In: International Conference on Machine Learning, pp. 13242–13256. PMLR (2022)
11. Lim, D., et al.: Large scale learning on non-homophilous graphs: new benchmarks and strong simple methods. In: Advances in Neural Information Processing Systems, vol. 34, pp. 20887–20902 (2021)
12. Liu, H., Hu, B., Wang, X., Shi, C., Zhang, Z., Zhou, J.: Confidence may cheat: self-training on graph neural networks under distribution shift. In: Proceedings of the ACM Web Conference 2022, pp. 1248–1258 (2022)
13. Luan, S., et al.: Revisiting heterophily for graph neural networks. In: Advances in Neural Information Processing Systems, vol. 35, pp. 1362–1375 (2022)
14. Mikolov, T., Sutskever, I., Chen, K., Corrado, G.S., Dean, J.: Distributed representations of words and phrases and their compositionality. In: Advances in Neural Information Processing Systems, vol. 26 (2013)
15. Pei, H., Wei, B., Chang, K.C.C., Lei, Y., Yang, B.: Geom-GCN: geometric graph convolutional networks. In: International Conference on Learning Representations (2019)
16. Perozzi, B., Al-Rfou, R., Skiena, S.: Deepwalk: online learning of social representations. In: Proceedings of the 20th ACM SIGKDD International Conference on Knowledge Discovery and Data Mining, pp. 701–710 (2014)
17. Veličković, P., Cucurull, G., Casanova, A., Romero, A., Liò, P., Bengio, Y.: Graph attention networks. In: International Conference on Learning Representations (2017)
18. Wang, M., Hu, G.: A novel method for twitter sentiment analysis based on attentional-graph neural network. Information **11**(2), 92 (2020)
19. Wang, X., Liu, H., Shi, C., Yang, C.: Be confident! towards trustworthy graph neural networks via confidence calibration. In: Advances in Neural Information Processing Systems, vol. 34, pp. 23768–23779 (2021)
20. Wu, F., Souza, A., Zhang, T., Fifty, C., Yu, T., Weinberger, K.: Simplifying graph convolutional networks. In: International Conference on Machine Learning, pp. 6861–6871. PMLR (2019)
21. Wu, J., Chen, H., Cheng, M., Xiong, H.: Curvagn: curvature-based adaptive graph neural networks for predicting protein-ligand binding affinity. BMC Bioinform. **24**(1), 378 (2023)
22. Yan, Y., Hashemi, M., Swersky, K., Yang, Y., Koutra, D.: Two sides of the same coin: heterophily and oversmoothing in graph convolutional neural networks. In: 2022 IEEE International Conference on Data Mining (ICDM), pp. 1287–1292. IEEE (2022)
23. Zhang, J., Dong, Y., Wang, Y., Tang, J., Ding, M.: Prone: fast and scalable network representation learning. In: IJCAI, vol. 19, pp. 4278–4284 (2019)

24. Zheleva, E., Getoor, L.: To join or not to join: the illusion of privacy in social networks with mixed public and private user profiles. In: Proceedings of the 18th International Conference on World Wide Web, pp. 531–540 (2009). https://doi.org/10.1145/1526709.1526781
25. Zhu, J., Yan, Y., Zhao, L., Heimann, M., Akoglu, L., Koutra, D.: Beyond homophily in graph neural networks: current limitations and effective designs. In: Advances in Neural Information Processing Systems, vol. 33, pp. 7793–7804 (2020)

Fast Redescription Mining Using Locality-Sensitive Hashing

Maiju Karjalainen$^{(\boxtimes)}$ ⓘ, Esther Galbrun ⓘ, and Pauli Miettinen ⓘ

University of Eastern Finland, Kuopio, Finland
`maiju.karjalainen@uef.fi`

Abstract. Redescription mining is a data analysis technique that has found applications in diverse fields. The most used redescription mining approaches involve two phases: finding matching pairs among data attributes and extending the pairs. This process is relatively efficient when the number of attributes remains limited and when the attributes are Boolean, but becomes almost intractable when the data consist of many numerical attributes. In this paper, we present new algorithms that perform the matching and extension orders of magnitude faster than the existing approaches. Our algorithms are based on locality-sensitive hashing with a tailored approach to handle the discretisation of numerical attributes as used in redescription mining.

Keywords: Redescription mining · Locality-Sensitive hashing

1 Introduction

A redescription is a pattern that characterises roughly the same entities in two different ways, and redescription mining is the task of automatically extracting redescriptions from the input dataset, given user-defined constraints. Redescription mining has found applications in various fields of science, such as ecometrics. Ecometrics aims to identify and model the functional relationships between traits of organisms and their environments [5,7]. For instance, the teeth of large plant-eating mammals are adapted to the food that is available in their environment, which in turn depends on the climatic conditions, potentially allowing one to reason about the climate in the past based on the fossil record.

To apply redescription mining in this context, the entities in the dataset represent localities, with two sets of attributes recording respectively the distribution of dental traits among species and the climatic conditions at each locality [11,20]. Galbrun et al. [11] mined redescriptions from this dataset using the ReReMi [8] algorithm in about 50 min on a commodity laptop. Replicating the experiment, we obtained a comparable time (bar D of Fig. 1 (left)).

In contrast, the two top bars represent our proposed method, Fier (Fast Initialisation and Extension of Redescriptions), based on locality-sensitive hashing (LSH). Our approach uses the same two-phase procedure as ReReMi: it finds initial pairs in the first phase, and extends them in the second phase (see Sect. 2 for

© The Author(s), under exclusive license to Springer Nature Switzerland AG 2024
A. Bifet et al. (Eds.): ECML PKDD 2024, LNAI 14947, pp. 124–142, 2024.
https://doi.org/10.1007/978-3-031-70368-3_8

details). The top bar, `Fier_full`, uses a LSH-based approach for both phases, while the second bar, `Fier_init`, uses LSH approach only for the first part. The third bar represents a straightforward speed-up of initial pair generation of `ReReMi` that we use as a baseline. It is obtained by replacing on-the-fly discretisation of numerical attributes by discretisation as a pre-processing step.

Figure 1 (left) shows that using `Fier` the whole mining process is completed before the standard `ReReMi` has even finished mining the initial pairs. The process is reduced from 66 min to mere 12 min; mining the initial pairs takes only 25 s.

`Fier`, being significantly faster than the existing methods, allows the use of redescription mining on even larger datasets. Alternatively, the speedup affords more responsive interactive redescription mining [10] and quickly testing parameter and constraint combinations to 'get a feel' for what kind of setup to use with the more exhaustive `ReReMi` algorithm. The price `Fier` has to pay for its speed is its probabilistic nature: unlike `ReReMi`, it does not guarantee to return the best initial pairs nor the (locally) best extensions.

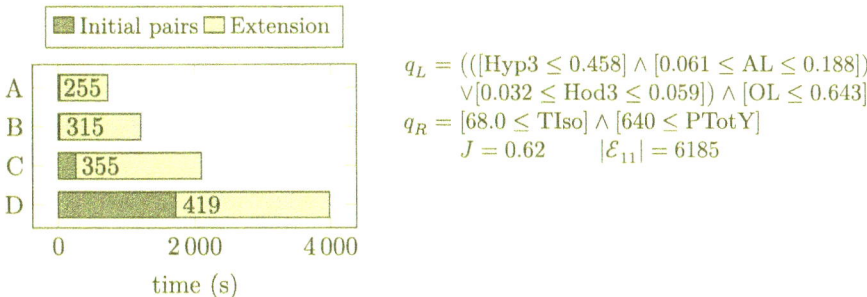

$$q_L = (([\text{Hyp3} \le 0.458] \wedge [0.061 \le \text{AL} \le 0.188])$$
$$\vee [0.032 \le \text{Hod3} \le 0.059]) \wedge [\text{OL} \le 0.643]$$
$$q_R = [68.0 \le \text{TIso}] \wedge [640 \le \text{PTotY}]$$
$$J = 0.62 \qquad |\mathcal{E}_{11}| = 6185$$

Fig. 1. Left: Running times on the DentalW dataset for finding initial pairs (blue) and extending pairs (yellow) using (A) the proposed algorithm (`Fier_full`), (B) `Fier_init` for initial pairs and `ReReMi` for extensions, (C) `ReReMi` with pre-bucketing (`ReReMiBkt`) and (D) standard `ReReMi`. The number within each bar indicates how many initial pairs were found. Right: Example redescription. (Color figure online)

Redescription mining. The input of redescription mining is a pair of *data tables* which we refer to as the left-hand side and right-hand side, denoted respectively $\mathbf{D}_L$ and $\mathbf{D}_R$, over the same *entities*, denoted $\mathcal{E}$. Then, a redescription is a pair of *queries*, denoted q_L and q_R, consisting of *literals* over the attributes of the corresponding table, combined with logical conjunction and disjunction operators, possibly involving negations. The set of entities that satisfy both queries, only the left-hand side query, only the right-hand side query, and neither of them are denoted respectively $\mathcal{E}_{11}$, $\mathcal{E}_{10}$, $\mathcal{E}_{01}$ and $\mathcal{E}_{00}$. The set $\mathcal{E}_{11}$ is also called the support of the redescription, and more in general, we call *support* (supp) of a literal or query the set of entities that satisfy it. The Jaccard index between the sets of entities that satisfy either query is called the *accuracy* of the redescription (we use the terms Jaccard and accuracy interchangeably) and is defined

$$J(q_L, q_R) = \frac{|\mathcal{E}_{11}|}{|\mathcal{E}_{10}| + |\mathcal{E}_{11}| + |\mathcal{E}_{01}|}.$$

Since it is meant to provide two ways to characterise roughly the same entities, accuracy is the main measure of the quality of a redescription. Further constraints that can be applied to redescriptions include a maximum threshold on the number of attributes they involve, since long queries might be difficult to interpret, and minimum thresholds on the number of entities that satisfy both queries and that satisfy neither, since redescriptions that characterise too few or too many of the entities are typically not considered interesting.

Going back to the application in ecometrics, the right-hand side query of the example redescription shown in Fig. 1 (right) characterises the climatic conditions encountered at the locality, in this case requiring high isothermality (TIso) and high annual precipitation (PTotY). The left-hand side query (q_L) is more complex and involves four literals over dental traits, requiring a limited prevalence of hypsodont species (Hyp3) and fairly low fraction of species with acute lophs (AL), or a very low fraction of hypsohorizodont species (Hod3), but allowing up to a rather large fraction of species with obtuse lophs (OL). Hypsodont and hypsohorizodont species refer to species with elongated teeth respectively along the vertical and the horizontal dimension, while acute and obtuse lophs refer respectively to the presence of sharp and blunt edges on the tooth across the chewing direction. In the considered dataset, $|\mathcal{E}_{11}| = 6\,185$ of the 28 886 localities satisfy both queries, representing about $J = 62\%$ of the 10 011 localities that satisfy at least one of the queries.

Related work. In the two decades since redescription mining was introduced [26], various algorithms have been proposed for this task, including based on decision tree induction [22,26,27], itemset mining [12], as well as iterative greedy heuristics [8,12]. The ReReMi algorithm [8] belongs to the latter family.

Other lines of work have focused for instance on selecting a good set of redescriptions [16], designing an interactive tool [10] and providing differentially private methods for redescription mining [17,23]. Meanwhile, redescription mining has been used in applications from domains as diverse as medicine [24], political sciences [9] and palaeontology [11,20].

Also relevant here is the work of Meeng and Knobbe [21] studying the impact of different discretisation strategies on subgroup discovery, a task very similar to redescription mining.

Locality-sensitive hashing was introduced for efficient nearest neighbour search [15]. It has proven to be very useful for various tasks, such as collaborative filtering [4], clustering [2] and privacy preservation [6]. The work most related to the present work is the early application to faster association rule mining [3].

2   The Algorithm

We divide our full algorithm into two parts for the two phases: `Fier_init` computes the initial pairs and `Fier_ext` the extensions. The full algorithm is called `Fier_full`. Before presenting the algorithms, we provide short primers on the `ReReMi` algorithm and on locality-sensitive hashing.

2.1   The `ReReMi` Algorithm

The algorithm on which we build, `ReReMi` [8], mines redescriptions in two main phases. In the first phase, the `InitialPairs` method returns candidate redescriptions with a single literal on either side. In the second phase, up to a chosen number of the best initial pairs are considered. Each is extended in turn through a greedy process that iteratively appends a literal to either of the queries to produce more refined and accurate redescriptions. The `ReReMi` algorithm can handle Boolean, categorical and numerical attributes. In particular, when forming literals for a numerical attribute, it performs on-the-fly discretisation, trying to find the interval that yields the best accuracy for the considered candidate, rather than using pre-determined buckets.

Going a bit further into technical details, the way `ReReMi` finds the initial pairs is simple but time consuming: Iterate over all pairs of attributes, one from either side; For each such pair, consider the possible pairs of literals over the two attributes; Compute the support and accuracy of the corresponding redescription and keep the best matches that satisfy support constraints.

Given a redescription to extend, `ReReMi` considers every available literal and the candidate extensions obtained by appending it (or its negation if allowed) to the current query on the corresponding side, with either the disjunction or the conjunction operator. The support and accuracy of each candidate extension are computed, the best one-step extensions are selected and extended in turn, for each one considering again all the possible ways to extend it with the remaining available literals. Starting with an initial pair, this greedy extension process is applied until no further improvement (in terms of accuracy) can be found or the maximum length of the queries (in terms of the number of involved literals) has been reached. The algorithm then proceeds with the next initial pair.

Computing the support and accuracy for a pair of literals involves computing the cardinality of the intersection and union of the sets of entities satisfying either literal. Similarly for computing the support and accuracy for an extension, although it has been shown [8] that not all entities can change status in the support of an extension. For instance, when extending a redescription by adding a literal to the left-hand side with a conjunction, entities that do not satisfy the current left-hand side query will not satisfy the extended one either. This fact is used to quickly identify the entities that need to be considered when performing the set operations to determine the quality of an extension.

2.2   Primer on LSH

The idea of Locality-Sensitive Hashing (LSH) is to calculate hash values for input items so that similar items get the same hash value with a high probability. The type of hash function depends on the similarity measure used. In this work we use two measures of similarity between binary vectors representing support sets, namely the Jaccard similarity and the Hamming distance.

When using the Jaccard similarity, the corresponding hashing technique is called minhashing [1]. To compute minhash values, we use k random hash functions $h_1, \ldots, h_k$ that map the row indices to values between 0 and the maximum number of rows. To calculate the minhash signature of a vector, we consider the indices of all the rows containing a 1. For every hash function we obtain the hash values for all of these rows and take the smallest value. These concatenated smallest hash values make up the signature of the vector [19].

When using the Hamming distance, we calculate the length-k signature of the vector as the values of the binary vector in k random indices.

To match two binary vectors based on their signatures, the signatures are divided into b smaller sections called bands, each containing r hash values (total $k = r \cdot b$). Two vectors are paired if all r hash values match in at least one band.

The parameters b and r determine the matching probability, i.e. the probability with which two vectors will be paired. If the similarity of the vectors is p, then the matching probability is $1 - (1 - p^r)^b$. The relation between the similarity and the matching probability follows an S-curve; its threshold point can be approximated as $p_T = (1/b)^{1/r}$ [19]. With this approximation, only two parameters need to be set and the third one can be calculated accordingly. To increase the chances of finding pairs with an accuracy above a desired threshold, one can use a slightly lower value, whereas a slightly higher value yields a faster algorithm finding fewer pairs.

2.3   Finding Initial Pairs

For simplicity we first focus on Boolean attributes. The pseudocode for the algorithm is given in Algorithm 1. Starting with the left-hand side dataset $\mathbf{D}_L$, we go through the literals and obtain r_J minhash signatures for each of the corresponding binary support vectors. Literals with the same signature get hashed into the same bin, and we keep track of which side each literal belongs to.

We go through the right-hand side dataset $\mathbf{D}_R$ similarly, except that we discard signatures that point to an empty bin. After calculating signatures for all literals, we go through each signature bin and form candidate pairs between the literals from $\mathbf{D}_L$ and $\mathbf{D}_R$ hashed into that bin.

We repeat this process b_J times, calculating a different minhash signature for the literals for each band. If the same two literals already formed a pair in a previous band, we do not consider it again. After the candidate pairs have been formed, we calculate their actual accuracy and support, and we filter out any candidate that does not satisfy user-defined support and accuracy thresholds.

Algorithm 1. Generating initial pairs from Boolean attributes

Input: Data $\mathcal{D} = (\mathbf{D}_L, \mathbf{D}_R)$, number of bands b_J, width of bands r_J
Output: A set of initial redescriptions $\mathcal{R} = \{(q_L, q_R)\}$
1: **function** Fier_init($\mathcal{D}, b, r$)
2:     **for** $band = 1, \ldots, b$ **do**
3:         $B \leftarrow \emptyset$
4:         generate r hash functions $h_1, \ldots, h_r$ that permute the row indexes
5:         **for** side in $\mathcal{D}$ **do**
6:             **for** attribute a in side **do**
7:                 $sig \leftarrow (\min(h_1(\mathrm{supp}(a))), \ldots, \min(h_r(\mathrm{supp}(a))))$
8:                 $B[sig] \leftarrow B[sig] \cup \{a\}$
9:             **for all** $sig \in B$ **do**
10:                 $\mathcal{R} \leftarrow \mathcal{R} \cup \{\text{pairs of literals } (a_L, a_R) \text{ in } B[sig], a_L \in \mathbf{D}_L, a_R \in \mathbf{D}_R\}$
11:     **return** $\mathcal{R}$

Building initial pairs from categorical attributes is slightly more complicated, as we can create several different literals by considering the different categories, separately or combined. We first consider the literals obtained by considering each category of an attribute a separately and calculate their minhash signatures the same way as for Boolean literals. Next, we create combinations of the categories by joining them with the disjunction operator, e.g. $[a = c_1 \lor a = c_2]$. The number of categories that are combined can be limited with a user-defined parameter. Since we already calculated the signatures for the separate categories, we can obtain the signature for a combination by simply taking the smallest minhash value among the categories included in the combination.

Creating literals out of numerical attributes is similar to categorical attributes, but by specifying intervals such as $[a \leq x]$ or $[x \leq a \leq y]$, instead of categories. We determine the intervals in two steps. First, we discretise the attribute values into n_b small buckets $[buk_l \leq y \leq buk_u]$ (line 5 in Algorithm 2) and calculate signatures for the literals corresponding to the separate buckets. Second, we create extended intervals by combining consecutive buckets, and calculate the signature for an extended interval by taking the smallest minhash values of the buckets included in the interval (line 13). We use support thresholds $\mathrm{supp}_{\mathrm{min}}$ and $\mathrm{supp}_{\mathrm{out}}$ to limit the size of intervals so that they do not cover too few or too many entities. The $\mathrm{supp}_{\mathrm{out}}$ threshold defines the minimum number of entities not covered by the literal.

The discretisation of attribute values into buckets can be done in different ways. We found that the bucketing method did not have a significant impact on the running time nor on the quality of the results. In our experiments we used equal-height binning, where each bucket covers a similar number of entities.

To avoid creating multiple pairs that are very similar to each other, we filter out any subinterval that has the same signature as the interval it is contained in. We do this by keeping track of the signatures we have seen for an attribute (within one band) as we iterate over the intervals.

Algorithm 2. Generating initial pairs from numerical attributes

---

Input: Data $\mathcal{D} = (\mathbf{D}_L, \mathbf{D}_R)$, number of bands b_J, width of bands r_J, number of buckets n_b, min. nb. of entities not in the support supp_{out}, minimum support supp_{min}

Output: A set of initial redescriptions $\mathcal{R} = \{(q_L, q_R)\}$

1: **function** Fier_init($\mathcal{D}, b, r, n_b, \text{supp}_{\text{out}}, \text{supp}_{\text{min}}$)
2:     **for** $band = 1, \ldots, b$ **do**
3:         **for** data table in $\mathcal{D}$ **do**
4:             **for** attribute a in data table **do**
5:                 $[buk_0, buk_2, \ldots, buk_{n_b}] \leftarrow \text{DISCRETISE}(a, n_b)$
6:                 **for** $k = 1, \ldots, n_b$ **do**
7:                     Signatures$[k, :] \leftarrow r_J$ minhash values for $[buk_{k-1} \leq a \leq buk_k]$
8:                 $S \leftarrow \emptyset$
9:                 **for** $l = 0, \ldots, n_b - 1$ **do**
10:                     $u \leftarrow n_b$
11:                     **while** $\text{supp}([buk_l \leq a \leq buk_u]) > |\mathcal{E}| - \text{supp}_{\text{out}}$ **do** $u \leftarrow u - 1$
12:                     **while** $\text{supp}([buk_l \leq a \leq buk_u]) \geq \text{supp}_{\text{min}}$ **do**
13:                         $sig \leftarrow \min_{i=l+1}^{u}(\text{Signatures}[i, :])$
14:                         **if** $u > S[sig]$ **then**
15:                             $B[sig] \leftarrow B[sig] \cup \{[buk_l \leq a \leq buk_u]\}$
16:                             $S[sig] \leftarrow u$
17:                       $u \leftarrow u - 1$
18:         **for all** $sig \in B$ **do**
19:             $\mathcal{R} \leftarrow \mathcal{R} \cup \{$pairs of literals $([buk_l \leq a_L \leq buk_u], [buk'_l \leq a_R \leq buk'_u])$ in $B[sig], a_L \in \mathbf{D}_L, a_R \in \mathbf{D}_R\}$
20:     **return** $\mathcal{R}$

We iterate over the intervals by progressively narrowing them down. Considering the ordered bucket edges $buk_0, buk_1, \ldots, buk_{n_b}$, we start by setting buk_0 as the lower and buk_{n_b} as the upper bound of the interval. We lower the upper bound by removing buckets from the top $buk_{n_b-1}, buk_{n_b-2}, \ldots$ until the support criteria is met. We calculate the signature as the smallest minhash values of the buckets contained in the interval. We lower the upper bound until the support of the interval is below the minimum support criterion. Then, we set buk_1 as the lower bound and repeat the process until we have gone through all lower bounds. We filter out subintervals of already seen intervals during this process, by discarding those intervals whose signature we have already seen, and whose upper bound is smaller than the largest seen upper bound for that signature (line 14).

2.4   Extending Initial Pairs

Computing Signatures for Literals. We start the process by computing the Hamming signatures for all data columns. For this we first randomly select indices over the number of rows in the data, so that we have b_H sets of length r_H random indices to create all signatures from. Note that these are different parameters than b_J and r_J used with the initial pair mining. For each literal to extend with, we store b_H sets of length r_H signatures.

The discretisation of the numerical attributes is done differently from the initial pair search. We start with a small number of buckets to discretise the values into and compute the Hamming signature corresponding to each bucket. Next, we double the number of buckets, perform the discretisation again and compute new signatures. We repeat this process a fixed number of times. We store the signatures for each bucket of each attribute, but we do not store the actual intervals, as they will be re-calculated later. This approach has only a modest effect on the running time, as we only need to compute the signatures for literals once. On the other hand, it sidesteps the issue of choosing the correct bucketing by using many different ones.

The Target Vector. As mentioned earlier, only some entities in the data matter when extending a redescription. Let us consider a redescription (q_L, q_R) which we want to extend on the right side with a conjunction. To improve $J(q_L, q_R)$, we want to find a literal that is satisfied by entities in $\mathcal{E}_{11}$ (so that the numerator does not shrink) and not by entities in $\mathcal{E}_{01}$ (so that the denominator shrinks). Entities in $\mathcal{E}_{10}$ and $\mathcal{E}_{00}$ do not matter, and we call these rows the 'don't-care' rows. In terms of LSH, we can see this as having a target vector that has 1s in rows corresponding to $\mathcal{E}_{11}$ and 0s in rows corresponding to $\mathcal{E}_{01}$. We can easily find data columns that match this target vector using the Hamming distance restricted to these rows (we use Hamming instead of Jaccard to reward matches on 1s and on 0s equally). Similarly, when extending the left side with a disjunction, we want to find literals that have 1s in rows corresponding to $\mathcal{E}_{01}$ and 0s in rows corresponding to $\mathcal{E}_{00}$, with $\mathcal{E}_{11}$ and $\mathcal{E}_{10}$ being the 'don't-care' rows. When extending the right side either by conjunction or disjunction the target vectors are otherwise same as with the left side, except that $\mathcal{E}_{10}$ and $\mathcal{E}_{01}$ switch places. As these subsets of rows are different for each redescription to be extended, we would have to re-compute the signatures for all data columns considering every candidate redescription and both connectives used to extend it on each side. This would clearly cancel the speed benefit. Instead, we replace the 'don't-care' rows in the target vector with 0s when extending with conjunction and with 1 when extending with disjunction. This way we only need to compute the signatures for the data columns once and we can use the same signatures for all extensions.

Extending Redescriptions. The algorithm for extending the redescriptions, denoted `Fier_ext`, is presented in Algorithm 3. We start by storing the initial pairs in a priority queue Q, using accuracy as the key. We always expand the redescription with the current highest accuracy. A redescription can be extended on either of its sides, with a conjunction or a disjunction operator. The corresponding target vectors are computed by COMPUTETARGETVECTOR. COMPUTECANDIDATES uses LSH to find the columns that match the target vectors, recording also the associated side and operator. The same column can match multiple times (in different bands, or different buckets for a numerical attribute), only the first match is stored. The actual extensions are then computed using the same approach as with `ReReMi` (i.e. using the cut-point method [8] for numerical attributes) and the best extension is stored and pushed back to Q if it has not already been extended to maximum length. Compared to `ReReMi`, the speed of `Fier_ext` benefits from not trying every column as a candidate extension, but only those that match with LSH.

2.5   Time Complexity

The time complexity for mining initial pairs from Boolean attributes is $O(N \cdot b \cdot r \cdot |\mathcal{E}| + n_L \cdot n_R)$, where $|\mathcal{E}|$ is the number of entities, N is the total number of attributes, $b \cdot r$ is the total number of hash functions in LSH, n_L is the largest number of left-hand side literals hashed to the same bin, and n_R is the largest number of right-hand side literals hashed to the same bin. The first term is for computing the signatures. The second term is for building the pairs; in the worst case it is quadratic, but in practice much smaller.

Algorithm 3. `Fier_ext`: Extend redescriptions

Input: Initial pairs $\mathcal{P}$, buckets with signatures for each possible extension B, maximum length of a redescription t
Output: A set of extended redescriptions $\mathcal{R}$
1: **function** Fier_ext(P, B, t)
2:     $Q \leftarrow$ heapify the list of initial pairs with accuracy as the key; $R \leftarrow \emptyset$
3:     **while** $Q \neq \emptyset$ **do**
4:         $(q_L, q_R) \leftarrow Q.\text{pop}()$
5:         $V \leftarrow \text{COMPUTETARGETVECTORS}((q_L, q_R))$
6:         $C \leftarrow \text{COMPUTECANDIDATES}(V, B)$
7:         **for** (column, side, operator) $\in C$ **do**
8:             $E \leftarrow \text{COMPUTEEXTENSIONS}((q_L, q_R), \text{column, side, operator})$
9:         **if** $E \neq \emptyset$ **then**
10:             best $\leftarrow$ extension with the highest Jaccard
11:             $\mathcal{R} \leftarrow \mathcal{R} \cup \{\text{best}\}$
12:             **if** len(best) $< t$ **then** $Q.\text{push}(\text{best})$
13:     **return** $\mathcal{R}$

The time complexity for numerical attributes is $O\big(N(|\mathcal{E}| \log |\mathcal{E}| + n_b \cdot b \cdot r \cdot |\mathcal{E}| + n_b^2) + n_L \cdot n_R\big)$; assuming n_b, b, and r are constants, this becomes $O(N|\mathcal{E}| \log |\mathcal{E}| + n_L \cdot n_R)$. Bucketing with equal-height binning takes $O\big(N(|\mathcal{E}| \log |\mathcal{E}|)\big)$. We can think of each bucket as creating a new literal, so creating the signatures takes $O(N \cdot n_b \cdot b \cdot r |\mathcal{E}|)$. Subinterval filtering (Algorithm 2 line 9) takes $O(n_b^2)$ for each of the N attributes. Similarly to Boolean attributes we assume at most n_L and n_R literals hashed to the same bin. The time complexity for the categorical attributes falls between the Boolean and numerical attributes.

The time complexity for mining the extensions is $\tau \cdot n_c \cdot T_R$, where τ is the number of extensions done (at most the number of initial pairs times the maximum length of a redescription), n_c is the maximum number of candidate extensions for a target vector (at most the number of literals in the data but typically much fewer), and T_R is the time ReReMi takes to compute an extension for a given redescription and a literal.

3   Experimental Evaluation

The experimental evaluation asserts that the results obtained by Fier are comparable to ReReMi in quality and that Fier is significantly faster than ReReMi. It is split into three parts: generating the initial pairs (Sect. 3.2), extending initial pairs (Sect. 3.3) and the full system (Sect. 3.4). We also test the sensitivity of Fier to the parameters of locality-sensitive hashing. We did not test the effects of standard redescription mining parameters, such as supp$_{\min}$, that are common to all algorithms of this family and not a property of our proposed approach.

Table 1. Dataset properties.

dataset	$|\mathcal{E}|$, entities	$\mathbf{D}_L$ attributes	$\mathbf{D}_R$ attributes
EuroClim	2 575	12 numerical	12 numerical
NoisyClim	2 575	36 numerical	36 numerical
MammalsW	54 013	48 numerical	4 754 Boolean
Ethno	1 267	23 numerical and categorical	90 numerical
DentalW	28 886	11 numerical	19 numerical
DentalA	6 404	7 numerical	19 numerical
VAA	1 656	9 categorical	107 categorical
CMS$_d$	d	152 numerical	452 numerical

3.1   Experimental Setup

We used twelve different datasets. Their properties are listed in Table 1.

The MammalsW dataset contains information about which mammal species inhabit which areas of the world on one side, and climate information on the other side [14]. That same climate data restricted to Europe, with monthly average temperatures on one side and monthly total precipitations on the other, constitutes the EuroClim dataset. NoisyClim contains three copies of the columns of EuroClim. The first copy is as is, the second and third copies are perturbed by adding noise distributed following $\mathcal{N}(0,1)$ and $\mathcal{N}(0,3)$ respectively.

DentalW is the dataset from the ecometrics study [11] presented in the introductory example. One side contains dental traits aggregated over the resident large herbivorous species whereas the other side contains climate information, at localities across the world. DentalA is a similar dataset, restricted to Southeast Asia and China and considering a slightly different set of dental traits [20].

Ethno is based on Murdock's Ethnographic Atlas [25]. It contains ecological attributes on the left-hand side, and nominal attributes recording cultural features of various tribes on the right hand side. VAA [13] contains the background information and answers to questions regarding political opinions of candidates to the Finnish parliamentary elections of 2011, as collected by an online voting advice application (VAA).

Finally, the CMS data are based on the synthetic data released by the Centers for Medicare and Medicaid Services.[1] We used five versions of the data, all having the same attributes, but increasing number of rows (7199, 14 414, 29 136, 58 204 and 116 395). The differently sized versions of CMS are indicated with the numbers of rows in subscript where relevant.

All algorithms are implemented in Python and the experiments (except when mentioned otherwise) are run with Python 3.6.8 on a machine with 2 AMD EPYC 7702 processors with 64 cores each and 1 TB of main memory. All experiments were run single-threaded. For ReReMi, we used the publicly available version.[2] Our code is publicly available.[3]

Unless otherwise noted, all comparative experiments were run with parameters $b_J = 40$, $r_J = 10$, $\text{supp}_{\min} = 0.1|\mathcal{E}|$, $\text{supp}_{\text{out}} = 0.3|\mathcal{E}|$ and with $n_b = 40$ buckets for numerical attributes. An exception to this are the DentalW and DentalA datasets, for which we used the parameters $b_J = 40$, $r_J = 5$, $n_b = 50$ and $\text{supp}_{\min} = 0.01|\mathcal{E}|$. For DentalW we set supp_{out} to $0.6|\mathcal{E}|$ and for DentalA to $0.3|\mathcal{E}|$. When using Fier, supp_{out} was further multiplied by 1.15 for DentalW and 1.2 for DentalA. Further experimental results are presented in the technical report [18].

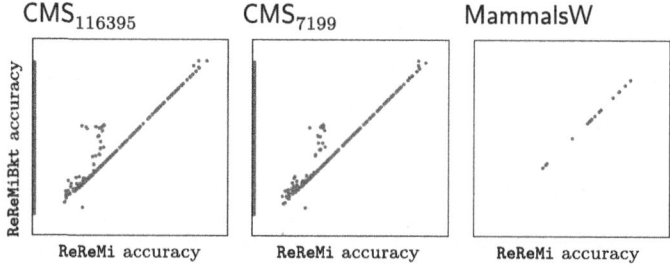

Fig. 2. Comparing the accuracy of pairs found by ReReMiBkt and ReReMi. Each dot represents a pair of columns, and its location indicates the highest-accuracy initial pair ReReMiBkt and ReReMi.

3.2   Finding Initial Pairs

Quality of the pairs. We compare the quality of the pairs found by Fier_init and ReReMiBkt, the ReReMi algorithm with pre-bucketing. ReReMiBkt gives more

---

[1] https://www.cms.gov/Research-Statistics-Data-and-Systems/Downloadable-Public-Use-Files/SynPUFs/DE_Syn_PUF.

[2] https://pypi.org/project/python-clired/.

[3] https://doi.org/10.5281/zenodo.11545892.

comparable running times and the quality of the initial pairs was essentially the same as with ReReMi, as shown in Fig. 2 for the CMS_{116395}, CMS_{7199} and MammalsW datasets. They show that ReReMiBkt and ReReMi get the same results with the MammalsW dataset and there are no dots in the x-axis, which would indicate pairs found only by ReReMi. With the CMS datasets, ReReMi and ReReMiBkt find somewhat different initial pairs. ReReMiBkt finds almost always equally good initial pairs as ReReMi, as well as many initial pairs that ReReMi does not find. This indicates that comparison to ReReMiBkt for the quality of the initial pairs is fair.

Results on the CMS, EuroClim, NoisyClim, Ethno, MammalsW, VAA, DentalW and DentalA datasets are shown in Fig. 3.

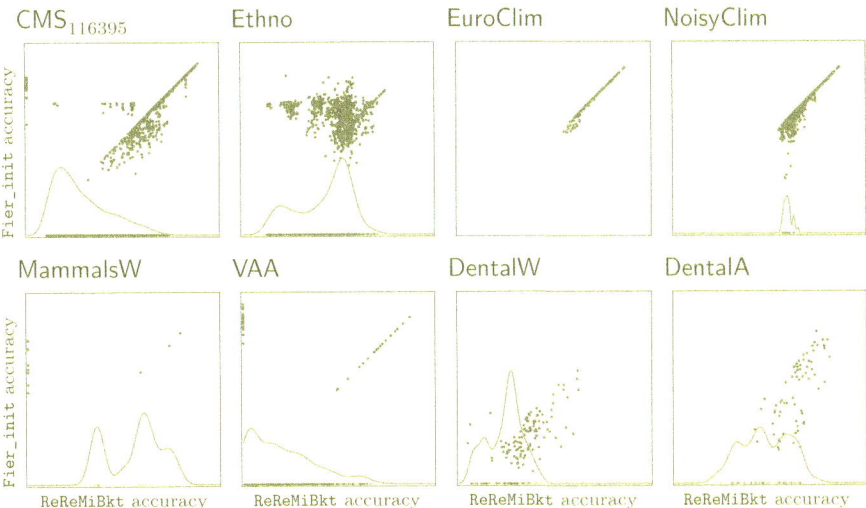

Fig. 3. Comparing the accuracy of pairs found by Fier_init and ReReMiBkt. Each dot represents a pair of columns, and its location indicates the highest-accuracy initial pair Fier_init and ReReMiBkt found. The light blue line at the bottom shows the density of the dots that lie along the x-axis. All axes range across the unit interval. (Color figure online)

The optimal outcome for these experiments is a diagonal line, indicating that Fier_init finds pairs equivalent to those found by ReReMiBkt. This is mostly the case. With CMS, MammalsW and VAA, there are also some dots along the x-axis, meaning that there are pairs of columns that Fier did not find, but most of those dots are at the left end of the x-axis, as indicated by the density plots (light blue). That is, they have low accuracy. DentalW and DentalA show weaker correlation, but still a clear diagonal pattern. Ethno is an outlier, where Fier_init does not find some pairs of literals although they have reasonably high accuracy (indicated by the peak around the middle of the x-axis), but also finds pairs that have significantly higher accuracy than what ReReMiBkt returns.

Speed of execution. The running times on different datasets are presented in Table 2. The running times for DentalW and DentalA are not comparable to other datasets as they were carried out using a different setup (to reflect [11, 20]).

Table 2. Comparing the running times (in seconds) of the algorithms on all datasets. Running times with DentalA and DentalW are not comparable with others, as they were run using a different setup.

Algorithm	EuroClim	NoisyClim	Ethno	VAA	DentalW	DentalA
ReReMi	1 184.89	12 922.03	9 069.96	0.56	1 569.01	1 960.38
ReReMiBkt	341.90	3 163.01	102.70	0.68	320.09	233.02
Fier_init	10.50	68.52	13.65	1.87	20.66	18.24

	MammalsW	CMS_{7199}	CMS_{14414}	CMS_{29136}	CMS_{58204}	CMS_{116395}
ReReMi	138.52	336.61	436.53	626.02	953.50	1 961.12
ReReMiBkt	143.35	308.39	418.40	610.99	943.21	1 899.83
Fier_init	47.66	21.22	40.22	71.02	145.20	290.54

Overall, we see that `Fier_init` is consistently an order of magnitude faster than `ReReMiBkt`, and often two orders of magnitude faster than `ReReMi`. The only exception to this is VAA; the small size of the data and categorical attributes mean that the overhead of the hashing dominates the running time. Another case where the difference is smaller is MammalsW; here, the Boolean attributes mean less benefit for `Fier_init`, and the on-the-fly discretisation of numerical attributes by `ReReMi` is actually faster than pre-bucketing of `ReReMiBkt`.

Looking at the different-sized CMS datasets, we also see that the running time of `Fier_init` grows linearly with the number of rows, matching the asymptotic runtime. Even though `Fier_init` uses more complex data structures for hashing, its memory consumption was only slightly larger than that of `ReReMi` ($\approx 30\%$).

Sensitivity to Parameters. We evaluated the sensitivity of `Fier_init` to the parameters affecting locality-sensitive hashing. Overall, we found it to be very robust.

We tested all combinations of values 20, 40, 60 and 80 for b_J and values 3, 5, 7, 10, 12 and 15 for r_J and evaluated the results both w.r.t. running time and quality of answers. For these experiments we used the EuroClim dataset.

Table 3 shows how the parameters impact the running time and accuracy. We can see that r_J has the largest impact, the running time increasing with smaller values of r_J. This is because r_J determines the length of the minhash signature, and shorter signatures imply a higher chance of matching and forming a candidate pair. On the other hand, the parameters do not have much impact on the average accuracy of the fifth-best result ($\bar{J}@5$).

As the average quality does not change by much, we can conclude that the user can set rows and bands primarily based on how many initial pairs they want to find, but that the settings we have used throughout these experiments ($r_J = 10$, $b_J = 40$) seem to give a good balance.

3.3 Extending Initial Pairs

The next phase of the `Fier` algorithm is the extension of the initial pairs. To test this phase, we use the same pre-mined initial pairs for all algorithms and compare extensions obtained with `Fier_ext` versus with the exhaustive attempts of `ReReMi`. For the extensions, pre-bucketing as done by `ReReMiBkt` is not expected to bring any benefits for `ReReMi`, as its cut-point algorithm is very efficient (cf. Table 2). `Fier_ext` might extend some initial pairs with literals that are somewhat inferior to the ones found by `ReReMi`, or it might fail to find any extensions at all. On the other hand, it should be much faster than `ReReMi`. For these tests we only consider MammalsW, DentalW and two CMS datasets due to space reasons.

Table 3. The effect of LSH parameters to running time (left) and average Jaccard of the fifth best result $J@5$ (right) depending on the number (b) and the width (r) of bands in LSH, for five repetitions and using EuroClim data.

	time (s)						$J@5$					
	r_J						r_J					
b_J	3	5	7	10	12	15	3	5	7	10	12	15
20	33.89	21.34	15.00	6.94	5.68	5.14	0.69	0.70	0.63	0.69	0.50	0.63
40	45.71	33.22	23.09	13.46	11.30	9.54	0.71	0.73	0.65	0.72	0.57	0.67
60	56.91	40.77	29.79	18.59	15.66	13.88	0.73	0.73	0.65	0.71	0.62	0.70
80	68.86	48.69	35.61	23.76	20.35	17.73	0.72	0.71	0.66	0.73	0.63	0.70

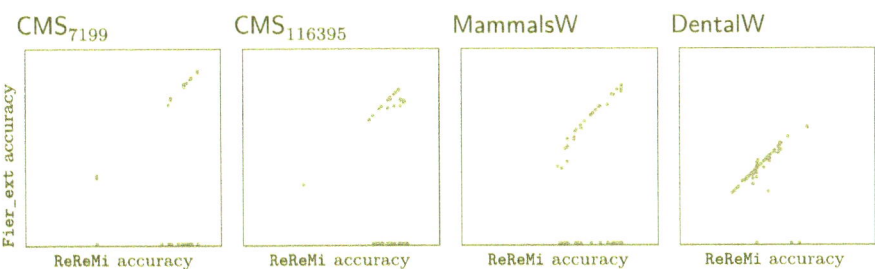

Fig. 4. Comparing the accuracy of once extended initial pairs by `Fier_ext` and `ReReMi`. Each dot represents a pair of columns, and its location indicates the highest-accuracy initial pair `Fier` and `ReReMi` found. All axes range across the unit interval.

Table 4. Accuracy, total number of extension steps, and time in seconds when extending the ReReMi initial pairs multiple times. All results are averages over 5 runs.

			MammalsW			DentalW		
algorithm	r_H	b_H	$J@10$	# ext.	time (s)	$J@10$	# ext.	time (s)
Fier_ext	10	10	0.77	296.4	1471.74	0.60	535.8	643.16
		20	0.79	329.0	1664.60	0.61	547.8	712.50
		40	0.79	354.8	1764.90	0.62	543.8	776.40
	20	10	0.72	112.6	768.05	0.57	500.2	390.47
		20	0.72	130.6	973.91	0.58	523.0	472.29
		40	0.75	237.4	1383.91	0.59	530.6	578.33
	30	10	0.68	60.0	438.41	0.54	402.8	245.93
		20	0.70	65.4	641.48	0.57	462.2	348.41
		40	0.71	101.0	990.65	0.57	512.8	483.04
	40	10	0.63	38.4	285.89	0.51	229.0	125.20
		20	0.68	54.6	473.24	0.53	329.0	221.54
		40	0.70	62.0	727.29	0.54	426.2	395.37
ReReMi			0.83	410.0	3263.28	0.61	517.0	1494.44

			CMS_{7199}			CMS_{116395}		
algorithm	r_H	b_H	$J@10$	# ext.	time (s)	$J@10$	# ext.	time (s)
Fier_ext	10	10	0.82	28.8	90.08	0.77	90.0	1072.10
		20	0.82	40.0	181.25	0.78	106.8	1906.59
		40	0.82	47.0	299.37	0.80	126.4	2646.13
	20	10	0.80	4.0	7.56	0.72	11.4	202.65
		20	0.81	6.4	17.68	0.73	20.4	303.56
		40	0.81	8.6	31.13	0.74	41.8	532.56
	30	10	0.80	0.6	6.82	0.71	5.0	148.69
		20	0.80	1.4	12.58	0.72	5.6	228.74
		40	0.81	3.0	23.88	0.72	8.4	421.66
	40	10	0.79	0.0	7.35	0.71	1.4	117.36
		20	0.79	0.6	13.86	0.71	2.8	223.92
		40	0.80	0.8	25.98	0.71	2.8	396.22
ReReMi			0.83	55.0	695.85	0.81	154.0	7626.95

Extending once. In the first test, we check how similar the extensions are after one round of extensions. The results are presented in Fig. 4. This plot is similar to Fig. 3. As we can see, there are some initial pairs for which Fier_ext does not find any valid extension, but for those that it does find, the results are generally as good as those of ReReMi.

Table 5. Accuracy, number of extension steps, number of resulting redescriptions and time in seconds for full redescription mining. All results are averages over 5 runs.

alg.	r_H	b_H	MammalsW $J@10$	# ext.	# results	time	DentalW $J@10$	# ext.	# results	time
Fier_full	20	20	0.74	94.6	32.4	765	0.57	486.8	216.8	445
		40	0.72	75.0	24.0	901	0.59	505.8	222.6	549
	30	20	0.63	39.0	14.2	446	0.52	423.8	189.0	315
		40	0.69	66.8	20.8	705	0.55	454.6	202.6	447
	40	20	0.63	28.0	10.6	333	0.51	258.8	139.6	199
		40	0.58	37.0	13.2	500	0.53	362.8	178.2	358
ReReMi			0.83	410.0	71.0	3401	0.61	517.0	93.0	3063
ReReMiBkt			0.83	410.0	71.0	3381	0.63	512.0	90.0	1866

alg.	r_H	b_H	CMS_{7199} $J@10$	# ext.	# results	time	CMS_{116395} $J@10$	# ext.	# results	time
Fier_full	20	20	0.75	12.2	18.8	34	0.76	19.6	21.4	463
		40	0.77	23.2	23.0	53	0.76	22.8	23.8	592
	30	20	0.74	3.6	17.0	30	0.75	4.2	17.2	381
		40	0.75	8.8	19.8	41	0.75	6.4	18.4	517
	40	20	0.73	3.0	17.6	31	0.73	0.6	16.0	367
		40	0.72	2.4	17.2	41	0.74	1.4	16.4	510
ReReMi			0.83	55.0	22.0	1032	0.81	154.0	45.0	9588
ReReMiBkt			0.83	55.0	22.0	957	0.83	79.0	29.0	5796

Extending Multiple Times. Genuine redescription mining involves extending the pairs more than just once. At this point it is no longer sensible to compare the accuracies of individual initial pairs, as small differences in each extension step can yield vastly different redescriptions, and the greedy extension done by ReReMi can sometimes yield worse results than Fier_ext. Instead, we consider the average Jaccard of the 10th best result ($\bar{J}@10$). We also measure the number of extension steps done by the algorithms *in total*. That is, if there are 100 initial pairs, and each pair is extended 3 times on both sides, the total number of extension steps is 600. Notice that ReReMi stops extensions when it cannot find any literal to extend with. The results of this experiment are presented in Table 4. Further results are presented in the technical report [18].

As can be seen in Table 4, the average $\bar{J}@10$ for Fier_ext is somewhat less than for ReReMi. With appropriate parameters, however, the difference is small. On the other hand, the running times can differ significantly. With DentalW for instance, ReReMi takes almost 1500 seconds while Fier_ext achieves the same accuracy in half the time (using $r_H = 10, b_H = 20$), or comparable accuracy in less than a quarter of the time (using $r_H = 30, b_H = 20$). We can also see that for the extensions, smaller values of r_H yield higher running times. This is because with smaller r_H, LSH is more random and tends to generate much more potential extensions. On the other hand, higher values of b_H tend to increase the accuracy, as LSH then simply does more repetitions.

3.4 Building Full Redescriptions

The final test is for the full algorithm `Fier_full`, where we use LSH for both initial pairs and extensions. The results are presented in Table 5 with further results in the technical report [18].

For the CMS datasets, `Fier_full` cannot find many extensions. This situation is the same as in Table 4, indicating that it is probably a feature of the extension algorithm rather than of the initial pairs. On the other hand, the average accuracies are still quite high, indicating that the lack of extensions might be a consequence of having high-accuracy initial pairs that cannot be extended with higher accuracy. The running times of the algorithm are very low, showing a significant improvement over `ReReMi` and `ReReMiBkt`.

4 Conclusions

Locality-sensitive hashing in `Fier` significantly speeds up finding the initial pairs and extending them, without sacrificing the quality. Handling intervals from numerical variables requires a more complex approach, especially for the initial pairs, but it pays off with notably improved running times. Based on our experiments, we can recommend using `Fier` for the initialisation of greedy redescription mining without reservation. The extension phase is somewhat more sensitive to hyperparameters and is usually not as significant a bottleneck. For larger data sets or for quick or interactive testing, we can recommend `Fier_full`.

Disclosure of Interests. The authors have no competing interests to declare that are relevant to the content of this article.

References

1. Broder, A.Z., Charikar, M., Frieze, A.M., Mitzenmacher, M.: Min-wise independent permutations. In: STOC'98, pp. 327–336 (1998). https://doi.org/10.1145/276698. 276781
2. Cochez, M., Mou, H.: Twister tries: Approximate hierarchical agglomerative clustering for average distance in linear time. In: SIGMOD'15, pp. 505–517 (2015). https://doi.org/10.1145/2723372.2751521
3. Cohen, E., et al.: Finding interesting associations without support pruning. IEEE Trans. Knowl. Data Eng. **13**(1), 64–78 (2001). https://doi.org/10.1109/69.908981
4. Das, A.S., Datar, M., Garg, A., Rajaram, S.: Google news personalization: scalable online collaborative filtering. In: WWW'07, pp. 271–280 (2007). https://doi.org/10.1145/1242572.1242610

5. Eronen, J.T., et al.: Ecometrics: the traits that bind the past and present together. Integr. Zool. **5**(2), 88–101 (2010). https://doi.org/10.1111/j.1749-4877.2010.00192.x

6. Fernandes, N., Kawamoto, Y., Murakami, T.: Locality sensitive hashing with extended differential privacy. In: ESORICS'21, pp. 563–583 (2021). https://doi.org/10.1007/978-3-030-88428-4_28

7. Fortelius, M., et al.: Fossil mammals resolve regional patterns of Eurasian climate change over 20 million years. Evol. Ecol. Res. **4**(7), 1005–1016 (2002)

8. Galbrun, E., Miettinen, P.: From black and white to full color: extending redescription mining outside the Boolean world. Stat. Anal. Data Min. **5**(4), 284–303 (2012) https://doi.org/10.1002/sam.11145

9. Galbrun, E., Miettinen, P.: Analysing political opinions using redescription mining. In: ICDM'16 Workshops, pp. 422–427 (2016). https://doi.org/10.1109/ICDMW.2016.0066

10. Galbrun, E., Miettinen, P.: Mining Redescriptions with Siren. ACM Trans. Knowl. Discov. Data **12**(1), 6 (2018). https://doi.org/10.1145/3007212

11. Galbrun, E., Tang, H., Fortelius, M., Žliobaitė, I.: Computational biomes: the ecometrics of large mammal teeth. Palaeontol. Electron. (2018) https://doi.org/10.26879/786

12. Gallo, A., Miettinen, P., Mannila, H.: Finding subgroups having several descriptions: algorithms for redescription mining. In: SDM'08, pp. 334–345 (2008) https://doi.org/10.1137/1.9781611972788.30

13. Sanomat, H.: Parliamentary Elections 2011: candidate responses to Helsingin Sanomat candidate selector (2016). http://urn.fi/urn:nbn:fi:fsd:T-FSD2701, version 2.1

14. Hijmans, R.J., Cameron, S.E., Parra, L.J., Jones, P.G., Jarvis, A.: Very high resolution interpolated climate surfaces for global land areas. Int. J. Climatol. **25**, 1965–1978 (2005). www.worldclim.org

15. Indyk, P., Motwani, R.: Approximate nearest neighbors: towards removing the curse of dimensionality. In: STOC'98, pp. 604–613 (1998). https://doi.org/10.1145/276698.276876

16. Kalofolias, J., Galbrun, E., Miettinen, P.: From sets of good redescriptions to good sets of redescriptions. In: ICDM'16, pp. 211–220 (2016). https://doi.org/10.1109/ICDM.2016.0032

17. Karjalainen, M., Galbrun, E., Miettinen, P.: Serenade: an approach for differentially private greedy redescription mining. In: Proceedings of the 20th Anniversary Workshop on KDID at ECML-PKDD'22, pp. 31–46. CEUR Workshop Proceedings (2022)

18. Karjalainen, M., Galbrun, E., Miettinen, P.: Fast redescription mining using locality-sensitive hashing. arXiv 2406.04148 (2024). https://doi.org/10.48550/arXiv.2406.04148

19. Leskovec, J., Rajaraman, A., Ullman, J.: Mining of Massive Data Sets. Cambridge University Press (2020)

20. Liu, L., Galbrun, E., Tang, H., Kaakinen, A., Zhang, Z., Zhang, Z., Žliobaitė, I.: The emergence of modern zoogeographic regions in Asia examined through climate-dental trait association patterns. Nat. Commun. **14**(1) (2023). https://doi.org/10.1038/s41467-023-43807-w

21. Meeng, M., Knobbe, A.: For real: A thorough look at numeric attributes in subgroup discovery. Dat Min. Knowl. Discov. **35**(1), 158–212 (2021). https://doi.org/10.1007/s10618-020-00703-x

22. Mihelčić, M., Džeroski, S., Lavrač, N., Šmuc, T.: A framework for redescription set construction. Expert Syst. Appl. **68**, 196–215 (2017). https://doi.org/10.1016/j.eswa.2016.10.012

23. Mihelčić, M., Miettinen, P.: Differentially private tree-based redescription mining. Data Min. Knowl. Discov. **37**(4), 1548–1590 (2023). https://doi.org/10.1007/S10618-023-00934-8

24. Mihelčić, M., Šimić, G., Babić-Leko, M., Lavrač, N., Džeroski, S., Šmuc, T.: Using redescription mining to relate clinical and biological characteristics of cognitively impaired and Alzheimer's disease patients. PLOS ONE **12**(10) (2017). https://doi.org/10.1371/journal.pone.0187364

25. Murdock, G.P.: Ethnographic atlas: a summary. Ethnology **6**(2), 109–236 (1967)

26. Ramakrishnan, N., Kumar, D., Mishra, B., Potts, M., Helm, R.F.: Turning CARTwheels: an alternating algorithm for mining redescriptions. In: KDD'04, pp. 266–275 (2004). https://doi.org/10.1145/1014052.1014083

27. Zinchenko, T., Galbrun, E., Miettinen, P.: Mining predictive redescriptions with trees. In: ICDM'15 Workshops, pp. 1672–1675 (2015). https://doi.org/10.1109/ICDMW.2015.123

σ-GPTs: A New Approach to Autoregressive Models

Arnaud Pannatier[1,2(✉)], Evann Courdier[1,2], and François Fleuret[3]

[1] Idiap Research Institute, Martigny, Switzerland
arnaud.pannatier@idiap.ch
[2] Ecole Polytechnique Fédérale de Lausanne, Lausanne, Switzerland
[3] Université de Genève, Geneva, Switzerland

Abstract. Autoregressive models, such as the GPT family, use a fixed order, usually left-to-right, to generate sequences. However, this is not a necessity. In this paper, we challenge this assumption and show that by simply adding a positional encoding for the output, this order can be modulated on-the-fly per-sample which offers key advantageous properties. It allows for the sampling of and conditioning on arbitrary subsets of tokens, and it also allows sampling in one shot multiple tokens dynamically according to a rejection strategy, leading to a sub-linear number of model evaluations. We evaluate our method across various domains, including language modeling, path-solving, and aircraft vertical rate prediction, decreasing the number of steps required for generation by an order of magnitude (The code of this work is available at https://github.com/idiap/sigma-gpt).

Keywords: Autoregressive models · Permutations · Transformers · Rejection Sampling

1 Introduction

Transformers demonstrate exceptional autoregressive capabilities across modalities. The traditional take for autoregression is to follow the natural order of the data, for example, left-to-right for text. In the case of vision, the usual scheme is to unfold the images following a raster-scan order and to use transformers to model the obtained sequence. In this work, we make a distinction between the order of the input data and the order of autoregression, highlighting that while they are typically aligned in most applications, they need not be. Our investigation involves training and generating sequences in a randomly shuffled order using transformers. While changing the sequence order is more challenging during training, it also reveals fascinating properties of the models (Fig. 1).

Supplementary Information The online version contains supplementary material available at https://doi.org/10.1007/978-3-031-70368-3_9.

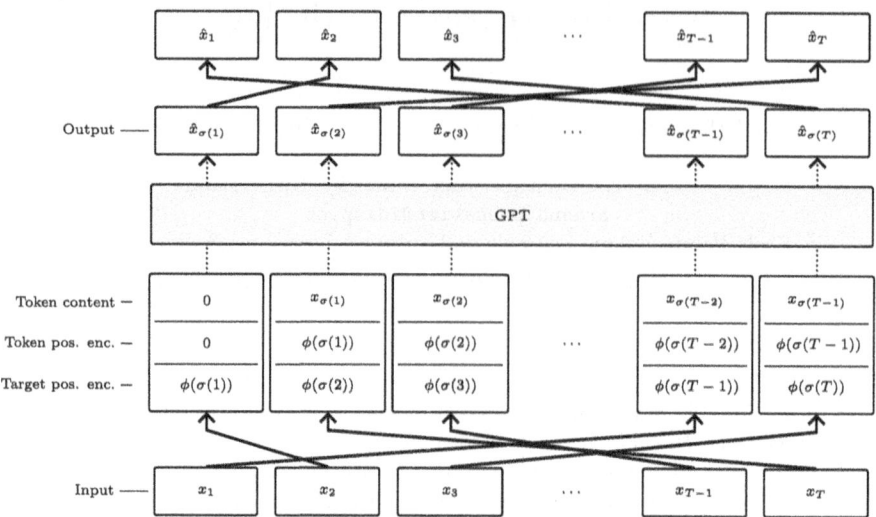

Fig. 1. In our σ-GPT, an arbitrary shuffling order σ can be chosen on-the-fly for every sample. It induces an input order $0, \sigma(1), \sigma(2), \ldots$ and an output order $\sigma(1), \sigma(2), \sigma(3), \ldots$, where the input is first padded with a 0 to ensure a consistent number of tokens. Tokens are shuffled accordingly, and these orders are both encoded separately with two positional encodings concatenated to the input, allowing the model to sample consistently in the autoregressive process. The output is finally shuffled back to the true order.

Table 1. Comparison between our approach, a standard causal transformer encoder (called GPT here), and diffusion models. Our model allows the sampling of a token at any position in the sequence, to model the remaining density according to a partially sampled sequence, naturally supports infilling, and can be used to sample the sequence by burst allowing faster generation. Compared to diffusion models, it can be trained easily using cross-entropy.

	σ-GPT	GPT	Diffusion Models
Sample Anywhere	✔	✗	✗
Conditional Density Estimation	✔	✗	✗
Arbitrary Conditioning	✔	✔	~
Infilling	✔	✗	~
Burst-Sampling	✔	✗	✔
Log-likelihood Training	✔	✔	✗

By breaking away from the standard autoregression order, one can use the model to predict the tokens in any particular order. With this scheme, the model is capable of predicting at any moment of the generation the conditional distribution of the remaining tokens. Having these estimates allows quantifying the possible outcomes of the generation at any given point. More interestingly, they can be leveraged to do rejection sampling, allowing to generate sequences by burst with a dynamical number of steps.

This work is structured as follows, we first introduce σ-GPTs and shuffled autoregression, and show that a model trained with this method combined with a curriculum method can even increase the performance of the underlying model. We then present the additional properties of σ-GPTs, summarized in Table 1, in particular for estimating conditional probabilities and we present our token-based rejection sampling scheme which allows for generating the sequence per burst and its theoretical properties. We evaluate our model and our scheme on three main tasks, which are open text generation, path-solving, and aircraft vertical rate prediction.

Contributions:

– Introduce σ-GPT, a novel architecture, with two positional encodings related respectively to the input and output order, that allows a causal transformer to generate sequences in any order which can be modulated on the fly for any pass through the model.
– Demonstrate that our method can reach similar performance as left-to-right trained autoregressive models when trained with a curriculum scheme.
– Demonstrate that our method can be used to generate samples in any order, allowing for the generation of samples conditioned on any part of the sequence.
– Introduce a novel token-based rejection sampling scheme that leads to the generation of samples per burst.

2 Methodology

2.1 σ-GPTs: Shuffled Autoregression

We propose a novel approach for training autoregressive models, which involves doing next-token prediction on a shuffled input sequence. We present σ-GPT, where σ denotes the permutation used to shuffle the sequence, and by GPT we mean any causal transformer encoder (or causal transformer decoder without cross-attention) such as [16]. To train such a model, each sequence is shuffled randomly during training. The model is then tasked to predict the next token in the shuffled sequence conditioned on all the tokens it has seen before. This training is done as usual with a standard cross-entropy loss. Besides the randomization of the order of the sequence and the addition of a double positional encoding, no other changes are needed to the model or training pipelines. For the rest of the paper, we use 'left-to-right order' to mention the usual order in which models are trained, even in the case of 2D data which are usually mapped to a sequence using a raster-scan order. And we use 'random order' to mean that the input has been shuffled.

2.2   Double Positional Encodings

To be able to model sequences in any order, each token needs to have information about its position and the one of the next token in the shuffled sequence. Specifically, when handling a sequence of tokens alongside a given permutation σ, every token contains three distinct pieces of information: its value $x_{\sigma(t)}$, its current position $\sigma(t)$, and the position $\sigma(t+1)$ of the subsequent token in the shuffled sequence, that are all concatenated. The necessity for double positional encoding arises from the intrinsic characteristics of transformers. Given that each token attends to every previous token in a position-invariant manner, each token needs to contain information about its position in the original sequence, so other tokens can know where they are located. And each token needs to know the position of the next token in the shuffled sequence as it is the target of the prediction. The double positional encoding is the only architectural change needed to train autoregressive models in random order. In this work, we used the standard sinusoidal positional encoding [19] for both the input and output positional encodings.

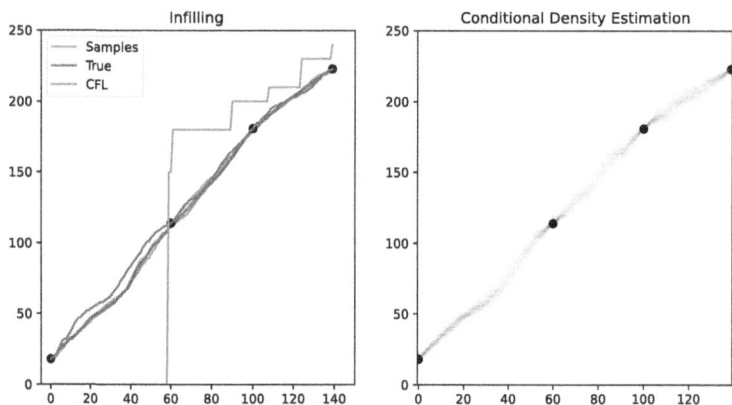

Fig. 2. (Left) We can infill the sequence by conditioning on the known part (black points). (Right) We can also have estimates of the density at any point of the sequence.

2.3   Conditional Probabilities and Infilling

Our method allows making conditional density estimation of the rest of the sequence. It is capable of making predictions all over the task space conditioned on any known subpart of the task. This can be done by prompting the model with the known part of the sequence and then decoding, in parallel and in one pass, the remaining tokens. Such evaluations are not possible with autoregressive models trained in a left-to-right order, as they need to follow the specific order

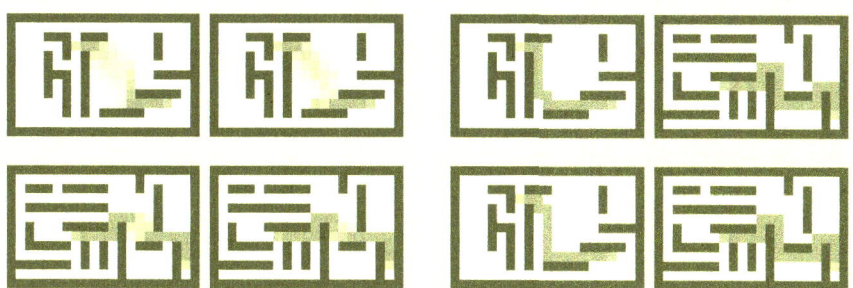

(a) (Left.) The theoretical density of the optimal path in the maze. (Right.) The estimated probability of the class 'path' at every position before starting autoregression. We see that the model has good estimates of the true density.

(b) Two different conditional samplings for each maze. The known part of the path (purple) is prompted first, and the rest of the sequence can be completed coherently.

Fig. 3. Conditional density estimation and infilling on the maze path-solving task.

they've been trained in. Examples, showing that the model usually has good estimates of the unconditioned distribution, can be seen in Figs. 2 and 3a.

Directly related to conditional density estimation, is that our method naturally supports infilling, as it is straightforward to prompt the model with the known part of a signal and to decode auto-regressively or by burst the rest of the signal. Figures 2 and 3a shows example of such samplings.

2.4   Token-Based Rejection Sampling

Autoregressive generation is a slow process as each token has to be generated sequentially. Even with caching strategies, this still scales linearly with the sequence length and it becomes prohibitively expensive for long sequences [20]. As our model allows for the generation of tokens in any order, we can leverage that fact and sample tokens in parallel at every position of the sequence. We can then evaluate the candidate sequence under different orders and accept multiple tokens in one pass. This algorithm runs efficiently on GPU as both the sampling at every position and the evaluation under different orders can be made in parallel, in a forward pass, and using an adapted KV-caching mechanism. We describe this caching mechanism more in detail in Sect. 3 of the supplementary material. When conditioned on partially completed sequences the model outputs distributions that are compatible with different possible outcomes, and when evaluating under different orders for generation, the distribution of tokens is constrained to tokens that are compatible with the previous tokens seen in one given order. As both the sampling and evaluation can be done in parallel, we can compute the acceptance decision efficiently for every token.

Algorithm 1. Token-based rejection sampling, following notation of [4]

Given minimum target length T, y trained σ-GPT, and number of orders N_o

Given a prompt $x_i \in \mathbb{X}$ of length t_0 of initial tokens. ($\mathbb{X}$ can be the empty set)

Set $t = t_0$

while $t < T$ **do**

In parallel, compute distribution conditioned on prompt $p(x_i|\mathbb{X}), \forall i \in t, \ldots, T$

In parallel, sample at every position $\tilde{x}_i \sim p(x_i|\mathbb{X}), \forall i \in t, \ldots, T$

Draw N_o random order σ and in parallel, compute all logits $q(x_i|\mathbb{X}, \tilde{x}_{\sigma_{<i}}), \forall i \in t, \ldots, T$

In parallel sample $T - t$ variables $u_i \sim U[0,1], \forall i \in t, \ldots, T$ from a uniform distribution.

In parallel, compute the acceptance decision $a_i = u_i < \min\left(1, \frac{q\left(\tilde{x}_i|\mathbb{X}, \tilde{x}_{\sigma_{<i}}\right)}{p(\tilde{x}_i|\mathbb{X})}\right)$ for every order.

Select the order that accepts the most tokens before seeing a first rejection.

Keep that order and add the a accepted tokens before the first rejection to the prompt.

Set $t = t + a$

end while

This strategy outputs a decision for each remaining token, but the decisions made by models become sometimes nonsensical when two mutually exclusive tokens are part of the prompt. Once a rejection is seen, all subsequent accepted tokens in the order of evaluation should be discarded. Indeed, the scheme rejects tokens that are incoherent with the ones already seen, and asking a model to make predictions based on incoherent tokens might lead to incoherent decisions. Using multiple orders allows keeping the one that accepts the most tokens in its evaluation. Even if it is dynamic, this algorithm can still easily generate multiple samples at once, by accepting the same amount of tokens for each sequence in the batch. Our rejection sampling algorithm is given in pseudo-code in Algorithm 1.

Other models such as Mask Git [3] or diffusion models [1,8] are doing generation by burst. However, these models usually require fixing the number of steps or a masking schedule beforehand. Our method on the other hand adapts dynamically to the underlying statistics of the data and thus does not require this extra hyper-parameter. We evaluate it on three synthetic cases to showcase this dynamic capability, we present the results in Sect. 3.8.

2.5   Other Orders

Our double positional encoding scheme allows for training and evaluating models in any order. Using a randomized order during training allows conditional density estimation, infilling, and burst-sampling at inference time. However the double positional encoding scheme allows any order to be used, and it can be used to train models in a deterministic order that is not left-to-right. As an example, we use a deterministic 'fractal' order to see how it compares to a random or left-to-right order. This order starts in the middle of the sequence then recursively goes to the first quarter and three-quarters of the sequence, and goes on recursively until all the positions have been visited. Such an order is fully deterministic, yet we make the hypothesis that this order leads to more difficult training for the model as it cannot rely on the locality of the information. We present the results in Sect. 3.5. Note that under perfect models, the order of modeling and decoding should not matter because of the chain rules of probability. We give more details about it in Sect. 1.1 of the supplementary material.

2.6   Denoising Diffusion Models

Denoising diffusion models [8] is a family of generative models that can also be used to generate sequences in a few steps. They are trained to reverse a diffusion process that is applied to the data. Diffusion processes can be both continuous and discrete. In this work, we use as a baseline only the discrete diffusion case, in particular using a uniform diffusion process [1]. To be able to compare the methods fairly, we use the same transformer architecture for both σ-GPT and the diffusion model, changing only the training objective. Compared to σ-GPT, diffusion models are not dynamic and require a fixed number of steps to generate a sequence, independently of the underlying statistics of the data. They also don't natively support conditional density estimation and infilling.

3   Results

3.1   General Performance

We tested our model across three main distinct tasks: language modeling, maze path solving, and aircraft vertical-rate prediction.

- Language Modeling: We used both the GPT-2 (123M) model on the Wikitext-103 dataset [13] and GPT-2 (345M) on OpenWeb Text [5].
- Maze Path Solving: This task involves determining a valid path between a starting and ending point in 13×21 mazes featuring 15 barriers. Presented with an image of an empty maze with start and end points, the model is tasked with producing an image with a legitimate path.
- Aircraft Vertical-Rate Prediction: This task uses real aircraft trajectory data, with its aircraft type. The data represents trajectories conditioned by air traffic control directives. The model's objective is to predict the vertical trajectory from a plane's current altitude to a specified control level.

Additionally to these tasks, we created a synthetic benchmark for evaluating our burst-sampling algorithm.

- Product Dataset: This toy example represents a pure product law case and is made of a sequence of length 100 with two classes (0,1) given by a Bernoulli law with p = 10%.

- Step Dataset: This toy example comprises sequences of two classes (0,1) of length 100 which are 0 everywhere except on a step of length 10 placed randomly in the sequence
- Joint Law dataset: This toy example represents a pure joint law and consists of a sequence of length 100 with 100 different classes, the model should predict a random generation of these different classes.

The general results of our models are presented in Table 2. These results indicate that training in a random order while requiring more compute-time as we describe in Sect. 3.2, reaches similar performances to left-to-right trained models. For the text modeling, to have a fair comparison during training, we monitor the validation perplexity of the sequence evaluated in a left-to-right order. Training in random order for text modeling was plateauing at a higher left-to-right validation perplexity, but using a curriculum scheme allows reaching the same performances, as presented in Sect. 3.3. For the path solving and the vertical rate prediction, the models were able to reach the same left-to-right validation loss during training. In inference, we noticed a one percent drop in accuracy compared to diffusion models and left-to-right trained GPT. For the vertical rate prediction task, the dataset that we used is limited to around 23.000 different sequences, we noticed that the standard left-to-right GPT was sometimes stuck repeating the same altitude, we think this is a modeling issue due to the small data regime. σ-GPT does not seem to suffer as much from this problem and offers a decrease in MSE. We hypothesize that this behavior comes from using a random order in inference which forces the model to fix some tokens over the whole sequence early in the generation. By doing so, the model gains the advantage of having a sketch of the whole sample and then concentrates on completing a coherent sample.

Table 2. General results. We report the validation perplexity for text generation, the test accuracy for the maze solver, and the mean squared error (MSE) for the vertical rate prediction. σ-GPT reaches a similar performance as GPT in text generation and maze solving and it outperforms GPT in the case of the vertical rate prediction. We report the validation perplexity for the text generation. For the path solver, we report the test accuracy on 1000 novel mazes. For the vertical prediction task, we report the mean squared error on the test set. We report the mean and standard deviation for the path-solving and the vertical rate prediction task. We do not report the validation error for the text generation for the discrete diffusion (Dis. Diff) as the training objective is different.

	Text-generation		Path Solving	Vertical Rate
	OWT Val Perp. ($\downarrow$)	Wiki-103 Val Perp. ($\downarrow$)	Accuracy ($\uparrow$)	MSE ($\downarrow$)
GPT	18.14	20.30	99.60 ± 0.70	274.8 ± 70.7
σ-GPT	18.64	16.69	98.30 ± 0.67	141.4 ± 4.1
Dis. Diff.	-	-	99.20 ± 0.67	105.94 ± 1.3

Table 3. Training efficiency. Number of steps/epochs required to reach the same performance and comparison with as GPT trained causally. As learning to predict in any order is a more challenging task, it is expected to need more computing time to reach the same accuracy. We don't report the standard deviation for text generation as we limited the training to one run.

Order	Text-generation	Maze Solver	Climbing Rate
σ-GPT	32500	78.0 ± 6.5	110.7 ± 4.5
GPT	16500	19.3 ± 4.9	25.0 ± 3.6

3.2   Training Efficiency

Modeling sequences in a random order is a more challenging task than modeling in left-to-right order. We think this is due to two main factors, at the beginning of the sequences models cannot rely on adjacent tokens to make educated guesses for the next token. Second some tasks are harder to learn in one direction than another and by modeling the data in any direction, we are always in the harder scenario. We give an example of one task that is harder to learn in one direction in Sect. 1 of the supplementary material.

This implies that we expect and see an increase in the number of steps or epochs required to learn a task. As previously mentioned, we don't see experimentally a drop in the validation performance of our model in the case of the path-finding algorithm or the vertical rate forecasting, but the time to reach the same performance increased. In the case of text modeling, the models plateaued before reaching the same accuracy when trained in random order. We treat that case in the following section. We report in table Table 3, the increase in training steps or epoch to reach the same accuracy. We see that most of the time, the number of epochs or steps needed to reach the same performance drastically increases. We think again that this is due to the increased complexity of modeling the sequence without having to rely on local information.

3.3   Curriculum Learning

For text modeling, we found a gap in validation perplexity in the left-to-right order between models trained purely in a random order and models trained in a left-to-right order. We see in Table 4 that σ-GPT is stuck at larger perplexity in both Open Web Text and WikiText-103 (30.43 vs 18.14 and 39.85 vs 20.30). We found that training for longer and using larger model didn't help in reducing that gap. To solve that problem, we introduced a curriculum learning scheme where the model is shown first more sequences in left-to-right order and progressively learns to model the sequence randomly. Surprisingly using this scheme helped drastically the model which managed to get even better performance than left-to-right trained transformers in the Wikitext-103 case and reduce drastically the gap for models trained on OpenWebText.

Table 4. Curriculum learning. We monitor the Validation Perplexity using a left-to-right order during training to have a comparable evaluation. We see that there is a gap between the model trained purely in a left-to-right fashion (GPT) and others trained in a random order (σ-GPT). Training for longer and larger models didn't help in removing that gap. We introduce a curriculum learning scheme that starts presenting the model with some percentage of the data (written in the corresponding label) in a left-to-right order at the beginning of the training and goes linearly to 100% of sequence in a random order at the end of the training. We see that training with this scheme removes the gap between σ-GPT and regular GPT and it reaches even better than left-to-right performance in the WikiText-103 case.

	Text-generation Val Perp. ($\downarrow$)
	Min. Left-to-Right
Openweb Text - GPT (345 M)	
GPT	18.14
σ-GPT curr. 50%	18.64
σ-GPT no curr.	30.43
WikiText 103 - GPT (128M)	
σ-GPT curr. 50%	16.69
σ-GPT curr. 100%	19.38
σ-GPT curr. 10%	19.45
GPT	20.30
σ-GPT no curr.	39.85

3.4   Open Text Generation: t-SNE of Generated Sequences

To get a qualitative sense of the generated text by the different methods, We generate 3000 sequences of 1024 tokens with each method, embed each sequence using an embedding model, and then project the embeddings to 2D using t-SNE. We present the results in Fig. 4. We used Open-AI `text-embedding-3-small` [14] to embed the generated sequences into a single 1536 vector embedding. We represent as green embeddings of sequences of the validation set, used as reference. We compute the t-SNE using the whole 15'000 embeddings and then plot each method (blue) and the other considered method (small gray dots). We first see that embeddings of GPT, σ-GPT, σ-GPT with burst-sampling, and diffusion are spread over the whole space, showing that the model can generate sequences that are coherent with the validation set.

3.5   Training and Generating in Fractal Order

We describe here the results that we get when training a GPT using a deterministic, but not left-to-right order. We described the order in Sect. 2.5. We train a GPT using this specific order for the different tasks and present the results in Table 5. We found that training in that order was as difficult for the model as

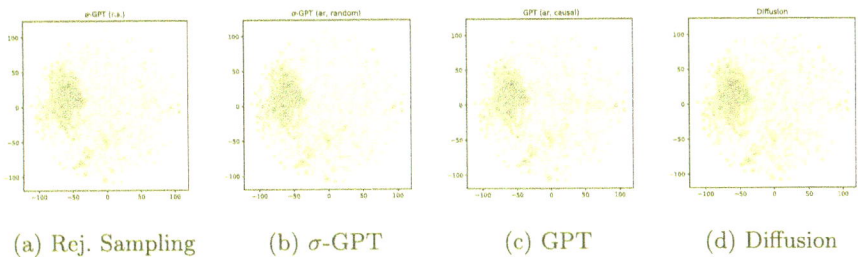

(a) Rej. Sampling   (b) σ-GPT   (c) GPT   (d) Diffusion

Fig. 4. 2D t-SNE of `text-small-3-embeddings` of 3000 sequences generated by each method. We compute the t-SNE of all the embeddings together, and then we display in each graph the embeddings of the validation set (green), the embeddings of the corresponding method (blue), and the embeddings of the other methods (gray). We see that the embeddings of the generated sequences have the same overall distribution compared to validation sets, which seems to indicate that *GPT*, σ-GPT, σ-GPT with burst-sampling, and diffusion models can generate sequences of similar quality. (Color figure online)

training in a random order, and we noticed a small drop in performance compared to σ-GPT. We suspect that this is due to the high discontinuity of the order of the sequence, which is such that two consecutive tokens are seen far away in the sequences. When predicting the first tokens, the model therefore cannot rely on information contained in neighboring tokens to make its prediction. As the training behavior seen in models trained in random and fractal order is similar, we think that the drop in training efficiency comes more from the fact that the model cannot exploit this neighboring information than changing the order at every batch.

Additionally, models trained in a fractal cannot be used as such for infilling and conditional density estimation and therefore cannot be used with our rejection sampling scheme. As the order is fixed for every batch, it might not even need to have a double positional encoding.

3.6 Memorizing

As learning sequences in any direction is harder than modeling them under a predefined order, we also expect that the critical dataset size when the model switches from memorization to generalization will increase. We follow the same hypotheses than [18], namely that the model has two mechanisms, one generalizing and one memorizing the data. As the mechanism of generalizing is more efficient as the dataset grows it will be selected by gradient descent once the size of the dataset gets beyond a critical size. As learning in a random order is a more difficult task, we expect that generalization is more difficult in that setup as well, hence the memorization regime should hold for bigger dataset sizes. We reduce drastically the training dataset size in the case of the path-finding task and we present the results in Fig. 5. Once it gets to 1000 examples both models trained in left-to-right, fractal, and random order are in a memorizing regime,

Table 5. Results for GPT trained in a fractal order compared to a standard left-to-right (GPT) and random (σ-GPT) order. We found that training models in this highly non-continuous order is as hard as training them in a random order, and additionally, models trained in that order cannot be used for conditional density estimation, infilling, or rejection sampling. For text modeling, we report the model perplexity on the validation set, in a random order for σ-GPT and in fractal order for the fractal GPT, as left-to-right validation perplexity is meaningless for fractal GPT which did not see sequences in that order during training.

	Text-generation	Path Solver	Vertical Rate
	Val Perp. ($\downarrow$), Rand. ord.	Test Acc ($\uparrow$)	MSE ($\downarrow$)
σ-GPT	24.46	98.30 ± 0.67	141.4 ± 4.1
Fractal GPT	27.79	98.00 ± 1.94	145.7 ± 2.6

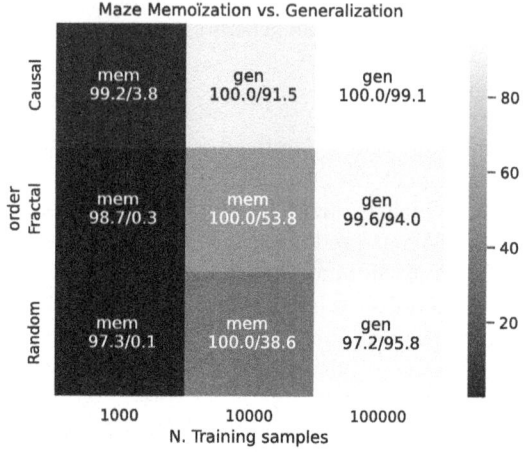

Fig. 5. Number of examples needed to switch from memorization to generalization. The model is trained on a restricted dataset size in the path-finding task. We see that the model trained in a random order needs more examples to switch from memorization to generalization. At 1k samples both models are fully in a memorization regime, at 100k both generalize but in between, at 10k, the model trained in a random order is still in a memorization regime.

getting perfect accuracy on the training data but very low on the validation data. Conversely, once the dataset gets bigger than 100k examples models trained in all the different orders are in a generalization regime. The transition happens in between and we find that it happens faster in the left-to-right order: at around 10k samples, the models trained left-to-right can generalize, while models trained in a random order are still in a memorization regime. We see also that models trained in a fractal order start generalizing faster than models trained in a random order, suggesting that the model can rely on seeing always the same order to generalize more rapidly.

3.7   Infilling and Conditional Density Estimation

We show in Fig. 2 that our model can be used to infill the sequence by conditioning on the known part of the sequence. In this figure, the larger points are part of the prompt and one can see the generated sequence complete sequence that matches the prompt. This figure also shows that our model has good estimates of the density at any point of the sequence. We represent at each point in the sequence the probability of the next sample given the known part of the sequence as shades of gray, the darker the more probable. We see that during generation, the model sees multiple possible outcomes that are coherent with the known part of the sequence. They are then constrained to a single sequence during the sampling. For left-to-right trained models, we can only have estimates of the density for the next tokens and we can't know what the model estimates for the rest of the sequence. In the case of the path-finding task, we show in Fig. 3a that the model conditioned only on an empty maze has good estimates of the true density of the optimal paths, highlighting that the model has already partially solved the problem before starting generation. During sampling, the order of the generation and the sampling procedure influence which path is selected from this joint law. We show as well in Sect. 2.3, that we can constrain the generation of mazes and that the model can generate coherent samples based on some prompted tokens that can be chosen on the fly.

We also give some interactive examples of text generation in the supplementary material.

3.8   Token-Based Rejection Sampling Scheme

We applied our token-based rejection sampling scheme to the problem of text generation, path-finding, and vertical rate forecasting Fig. 6a to 6c. We found that in the three cases, our method was able to generate samples of comparable quality with an order of magnitude fewer steps than the autoregressive method. We compared to discrete diffusion models as well, and we can see that our method always outputs coherent samples, while the samples generated by the diffusion model are sometimes incoherent if the number of steps is not high enough. In the case of path-solving and in the three synthetic tasks, we see that at a comparable number of steps for both methods, our model shows better generation quality.

We tested our token-based rejection sampling scheme on three synthetic cases. We estimate the number of steps required to generate the sequence using a perfect model in Sect. 2 of the supplementary material. We found that our scheme was close to the optimal heuristics in all three cases. In the case of the product dataset, requiring only one step to accept the sequence. In the case of the step dataset, our scheme required four steps to accept the sequence. In the case of the joint law dataset, our scheme required a few more steps, which is expected as the task is more complicated. We see that it manages to generate valid samples with a number of steps close to the optimal heuristics and with a large increase in performance compared to diffusion models at the same step.

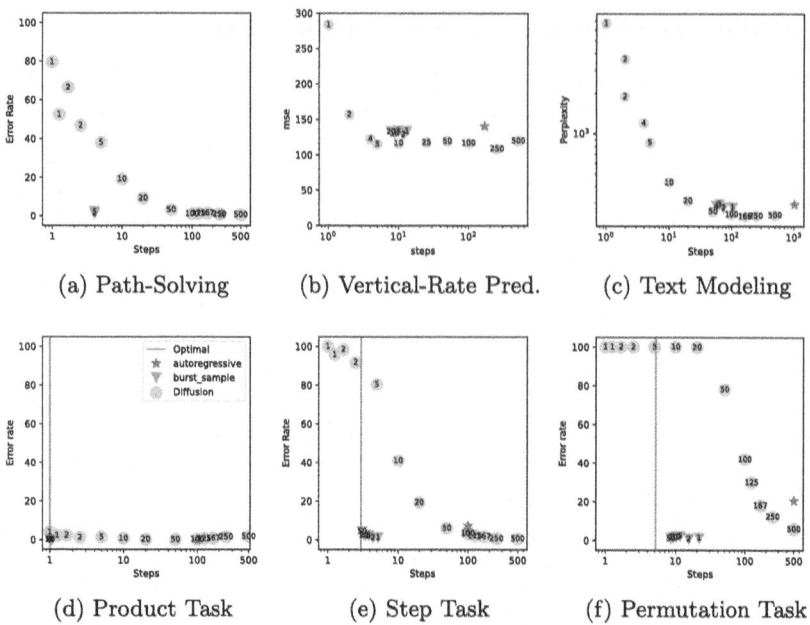

(a) Path-Solving   (b) Vertical-Rate Pred.   (c) Text Modeling

(d) Product Task   (e) Step Task   (f) Permutation Task

Fig. 6. We plot the performance vs steps of our σ-GPT used for autoregression in random order (blue), to σ-GPT with rejection sampling per burst (orange) against diffusion models (gray). We denote in the text the predefined number of steps chosen for generation in the diffusion models. For rejection sampling, we note the number of orders used for the evaluation. We see that increasing the number of orders leads to a decreased number of steps. For the synthetic tasks, we also represent heuristics for the optimal number of steps needed to generate the sequence (gray line), as described in Sect. 2 of the supplementary material, and we see that our scheme is close to this heuristics. (Color figure online)

4   Related Works

Shuffling in Language Models: The objective of σ-GPT is to model shuffled sequences. XLNet [21] uses a similar objective and shuffling of the sequence order as a pretraining task for sentence encoding in the context of natural language understanding. While both approaches are modeling shuffled sequences, they still differ in their implementations: we use a double positional encoding and a regular causal mask instead of masking based on two streams and the modification of the attention matrix. The two models differ as well in their applications, XLNet is used to encode information similarly to BERT while our approach is generative. A recent approach by [6], trains a transformer both on a left-to-right and right-to-left order to solve a classic problem of the transformer model to understand tokens relations in both directions. In a similar setting, it has been shown that text corpora have a preferred left-to-right order of generation [15] and that training in reverse order can lead to a decrease in performance. This is a possible explanation of why training σ-GPT on text generation needed extra care compared to other tasks. In vision, the raster-scan order is mainly used to unfold 2D patches into sequences. However, a recent work by [9] showed that using a set of predefined orders of unfolding the images can improve performance. They pass the order information not by using a double positional encoding, but by a fixed number of beginning-of-sequences tokens.

Burst-Sampling Scheme: Other works are trying to solve the problem of the linear time required by autoregression using burst-sampling. Maskgit [3], for example, uses a BERT-like Masked Language Model (MLM) and a custom decoding scheme, which samples multiple tokens at the same time to generate an output. The number of tokens generated at each pass is fixed by a masking schedule and a confidence-driven decoding scheme is used to choose which tokens to predict next. Another approach, [11] relies on an auxiliary model to guide the generation process. Alternatively, the approach of [10] focuses on generating preliminary drafts of an image, which are then iteratively improved. Current Video generation methods [2,20] are leveraging a MaskGit-like approach [3] for generation because autoregressive generation of video frames would be too costly. Our rejection sampling scheme allows to generate the sequence by burst but in contrast to other schemes, the number of tokens accepted is dynamic and depends on the data being modeled. This allows for faster generation when the underlying data distribution is simple.

Discrete Diffusion Models: Diffusion models are also able to generate sequences in a few steps. We compared our approach with a discrete diffusion baseline [1]. In the original work, discrete diffusion was also used to generate text, however without the possibility of conditional density estimation and infilling. Most of the diffusion approaches do not support conditional density estimation and infilling by default. Consistency Models [17], which be adapted from continuous diffusion models, can be used to infill images but in a continuous case. However, there have been recent discrete text-diffusion models that allow infilling [7,12].

5   Conclusion

Training GPT-like models in different orders offer different desirable properties. It allows for the conditional prediction based on any subset of the tokens of the sequence, it naturally can be used for infilling, and as the model can do partial prediction, we can leverage them to do rejection sampling and accept multiple tokens at the same time during generation. Our findings indicate that conditional prediction learned by the models matches the theoretical partial distribution showing that the model is indeed able to understand and reconstruct the signal in any order. As the training objective of modeling sequences in any order is harder than training in a fixed order, it has an impact on the training efficiency, and in a small dataset size, we show that it leads to more memoïzation. Finally, we showed that our model was able to generate sequences by burst using a novel per-token rejection sampling scheme, reaching optimal heuristics in some cases and decreasing the number of steps needed for generation by an order of magnitude.

Acknowledgement. We thank Youssef Saied for his help and good remarks on the overall project. We thank Romain Fournier for his precious help on some theoretical aspects of the analysis. Arnaud Pannatier was supported by SkySoft ATM for the project "MALAT: Machine Learning for Air Traffic" and the Swiss Innovation Agency Innosuisse under grant number 32432.1 IP-ICT. – Evann Courdier was supported by the "Swiss Center for Drones and Robotics - SCDR" of the Swiss Department of Defence, Civil Protection and Sport via armasuisse S+T under project No 050-38.

References

1. Austin, J., Johnson, D.D., Ho, J., Tarlow, D., Van Den Berg, R.: Structured denoising diffusion models in discrete state-spaces. In: Advances in Neural Information Processing Systems, vol. 34, pp. 17981–17993 (2021)
2. Chang, H., et al.: Muse: text-to-image generation via masked generative transformers (2023)
3. Chang, H., Zhang, H., Jiang, L., Liu, C., Freeman, W.T.: Maskgit: masked generative image transformer. In: Proceedings of the IEEE/CVF Conference on Computer Vision and Pattern Recognition (CVPR), pp. 11315–11325 (2022)
4. Chen, C., Borgeaud, S., Irving, G., Lespiau, J.B., Sifre, L., Jumper, J.: Accelerating large language model decoding with speculative sampling. arXiv preprint arXiv:2302.01318 (2023)
5. Gokaslan, A., Cohen, V.: Openwebtext corpus (2019). http://Skylion007.github.io/OpenWebTextCorpus
6. Golovneva, O., Allen-Zhu, Z., Weston, J., Sukhbaatar, S.: Reverse training to nurse the reversal curse (2024)
7. Gulrajani, I., Hashimoto, T.: Likelihood-based diffusion language models. In: Thirty-Seventh Conference on Neural Information Processing Systems (2023). https://openreview.net/forum?id=e2MCL6hObn
8. Ho, J., Jain, A., Abbeel, P.: Denoising diffusion probabilistic models. In: Advances in Neural Information Processing Systems, vol. 33, pp. 6840–6851 (2020)
9. Kakogeorgiou, I., Gidaris, S., Karantzalos, K., Komodakis, N.: Spot: self-training with patch-order permutation for object-centric learning with autoregressive transformers. In: Proceedings of the IEEE/CVF Conference on Computer Vision and Pattern Recognition (CVPR), pp. 22776–22786 (2024)
10. Lee, D., Kim, C., Kim, S., Cho, M., HAN, W.S.: Draft-and-revise: effective image generation with contextual RQ-transformer. In: Koyejo, S., Mohamed, S., Agarwal, A., Belgrave, D., Cho, K., Oh, A. (eds.) Advances in Neural Information Processing Systems, vol. 35, pp. 30127–30138. Curran Associates, Inc. (2022)
11. Lezama, J., Chang, H., Jiang, L., Essa, I.: Improved masked image generation with token-critic. In: Avidan, S., Brostow, G., Cissé, M., Farinella, G.M., Hassner, T. (eds.) ECCV 2022. LNCS, vol. 13683, pp. 70–86. Springer, Cham (2022). https://doi.org/10.1007/978-3-031-20050-2_5
12. Lou, A., Meng, C., Ermon, S.: Discrete diffusion modeling by estimating the ratios of the data distribution (2024)

13. Merity, S., Xiong, C., Bradbury, J., Socher, R.: Pointer sentinel mixture models. In: International Conference on Learning Representations (2017)
14. OpenAI: New embedding models and API updates (2024). https://openai.com/blog/new-embedding-models-and-api-updates. Accessed 01 Mar 2024
15. Papadopoulos, V., Wenger, J., Hongler, C.: Arrows of time for large language models (2024)
16. Radford, A., Wu, J., Child, R., Luan, D., Amodei, D., Sutskever, I., et al.: Language models are unsupervised multitask learners. OpenAI Blog **1**(8), 9 (2019)
17. Song, Y., Dhariwal, P., Chen, M., Sutskever, I.: Consistency models. In: Proceedings of the 40th International Conference on Machine Learning, ICML 2023. JMLR.org (2023)
18. Varma, V., Shah, R., Kenton, Z., Kramár, J., Kumar, R.: Explaining grokking through circuit efficiency. arXiv preprint arXiv:2309.02390 (2023)
19. Vaswani, A., et al.: Attention is all you need. In: Guyon, I., et al. (eds.) Advances in Neural Information Processing Systems, vol. 30. Curran Associates, Inc. (2017). https://proceedings.neurips.cc/paper_files/paper/2017/file/3f5ee243547dee91fbd053c1c4a845aa-Paper.pdf
20. Villegas, R., et al.: Phenaki: variable length video generation from open domain textual descriptions. In: International Conference on Learning Representations (2023)
21. Yang, Z., Dai, Z., Yang, Y., Carbonell, J., Salakhutdinov, R.R., Le, Q.V.: Xlnet: generalized autoregressive pretraining for language understanding. In: Wallach, H., Larochelle, H., Beygelzimer, A., d'Alché-Buc, F., Fox, E., Garnett, R. (eds.) Advances in Neural Information Processing Systems, vol. 32. Curran Associates, Inc. (2019)

FairFlow: An Automated Approach to Model-Based Counterfactual Data Augmentation for NLP

Ewoenam Kwaku Tokpo[✉] and Toon Calders

University of Antwerp, Antwerp, Belgium
{ewoenamkwaku.tokpo,toon.calders}@uantwerpen.be

Abstract. Despite the evolution of language models, they continue to portray harmful societal biases and stereotypes inadvertently learned from training data. These inherent biases often result in detrimental effects in various applications. Counterfactual Data Augmentation (CDA), which seeks to balance demographic attributes in training data, has been a widely adopted approach to mitigate bias in natural language processing. However, many existing CDA approaches rely on word substitution techniques using manually compiled word-pair dictionaries. These techniques often lead to out-of-context substitutions, resulting in potential quality issues. The advancement of model-based techniques, on the other hand, has been challenged by the need for parallel training data. Works in this area resort to manually generated parallel data that are expensive to collect and are consequently limited in scale. This paper proposes FairFlow, an automated approach to generating parallel data for training counterfactual text generator models that limits the need for human intervention. Furthermore, we show that FairFlow significantly overcomes the limitations of dictionary-based word-substitution approaches whilst maintaining good performance.

Keywords: Natural language processing · Bias mitigation · Counterfactual Data Augmentation

1 Introduction

Despite their growing popularity and unprecedented performance in various application domains, language models (LMs) continue to be plagued with issues of harmful societal biases and stereotypes that have been shown to have detrimental social effects [4]. The biggest contributing factor is the encapsulation of societal biases in everyday language, as is well-documented [1,12,18]. LMs heavily rely on such textual data, now digitalized on various online outlets, as training data, causing them to mirror these biases [24].

In Natural Language Processing (NLP), similar to many machine learning domains, bias mitigation generally occurs at three intervention avenues: the training data, the learning procedure, or the model output [15]. Since model bias

A. Bifet et al. (Eds.): ECML PKDD 2024, LNAI 14947, pp. 160–176, 2024.
https://doi.org/10.1007/978-3-031-70368-3_10

traces its roots to the training data, mitigating bias at the training data level has proven very effective [6,10]. One such approach, Counterfactual Data Augmentation (CDA) [5], seeks to remove spurious correlations between attributes in the training data by evening out the distribution of words that characterize demographic attributes in the context of neutral words that should ideally not be demographically aligned. Specifically, explicit attribute-defining words are replaced with their counterfactual equivalents from complementary demographic groups for every text instance. To illustrate this with an example, an instance of *"She is a nurse"* will be augmented with *"He is a nurse"* in the case of mitigating gender bias. This follows the intuition that in an ideal dataset, the association between gender attributes and target attributes like professions will be even for different gender groups.

Key works, such as [5,26,27], introducing CDA as a bias mitigation technique adopt a word substitution approach based on dictionaries. These word substitution methods are prone to grammatical incoherence because of out-of-context substitutions and omitted word pairs. Because dictionary compilations are often incomplete [8], a direct word-substitution approach will not generalize to omitted words. Take for instance (***Bachelor** and **Masters** degree* v. ***Spinster** and **Mistresses** degree*) and (*she taught **herself*** v. *he taught **herself***) which were common issues we observed with some methods. Additionally, the dictionaries are manually compiled, which not only incurs potential costs but manually compiling counterfactual word pairs for certain demographics may be intrinsically challenging.

Although generative language models like GPT-related models [20] have surged in popularity, their adoption for CDA has been limited due to the relative unavailability of parallel data needed for training. As such, model-based solutions resort to manually compiling parallel training data, a process that is both costly and constrained. This challenge is exacerbated by the fact that training models on limited parallel data can impair performance [28]. Conversational models like ChatGPT generate good counterfactuals in a zero-shot setting. However, these models struggle to generate counterfactuals on a scale large enough for model training. In addition to that, they can be inconsistent in how they generate counterfactuals, as we will discuss in Sect. 5.4.

The primary contribution of this paper is to explore an automated approach to generate parallel training data for a given demographic axis that requires minimal human intervention. Our approach takes from a user a prompt – in the form of a single word-pair – that describes a demographic axis. This pair is subsequently used to model a demographic subspace from which other words that define the demographic attribute can be sampled from a given corpus of text. Using an invertible flow-based model [9], counterfactual words are generated for sampled words. Thereafter, an error correction approach is used in tandem with direct word substitution to generate parallel data to fine-tune a generative language model to generate counterfactual texts. We call our approach and the resultant counterfactual text generation model *FairFlow*. This entire process is simply depicted in a four-step process in Fig. 1. As opposed to existing works, which will be discussed in Sect. 2, FairFlow does not rely on human-generated parallel data for training and eliminates the need for manually compiled word-pair dictionaries.

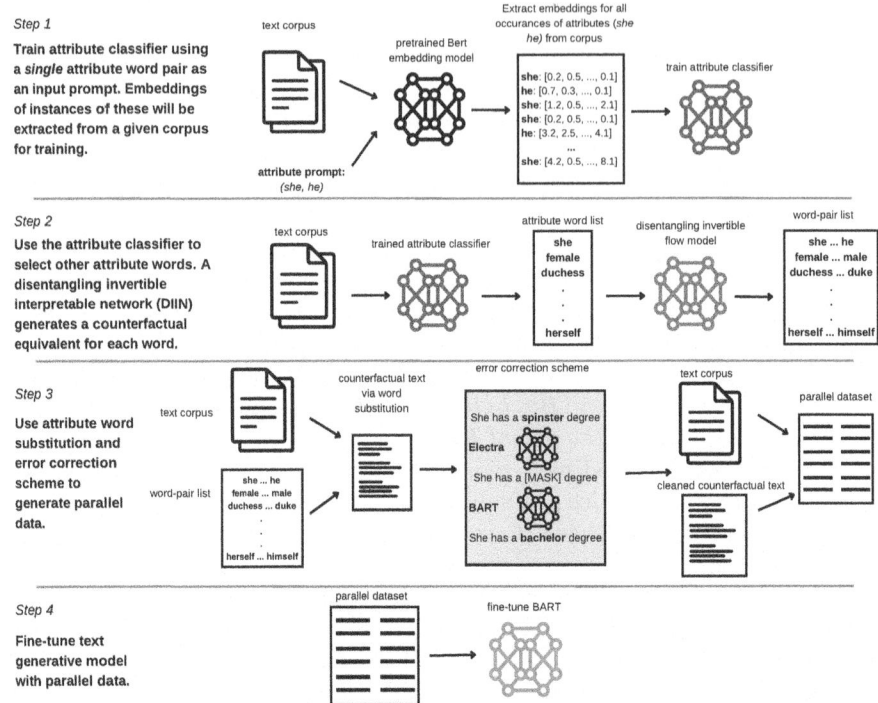

Fig. 1. An end-to-end description of Fairflow, described in four steps: 1) train a classifier to identify attribute words from a corpus; 2) generate counterfactual equivalents for attribute words using an invertible generative flow model; 3) use a word substitution scheme and our proposed error-correction scheme to make the parallel text more fluent and realistic; 4) fine-tune a generative model with the generated parallel data.

In summary, this paper explores and proposes techniques to develop a robust model-based counterfactual generator in the absence of parallel training data. Key contributions include:

1. An automated approach to compiling dictionaries of word pairs that only requires a user to input a word-pair prompt that describes a demographic axis.
2. We proposed an error correction approach to generate parallel data from dictionary word substitutions.
3. We train a counterfactual model using our generated parallel data and show that the error correction approach not only improves the grammatical composition of the model but also improves the generalization of the model.

We make our implementation code and materials for FairFlow available[1].

2   Background and Related Literature

Early works on CDA used simple rule-based word-substitution approaches for counterfactual data augmentation. Specifically, they created dictionaries of attribute word pairs and used matching rules to swap words [6]. Later works began to incorporate grammatical information like part-of-speech tags to swap attribute words [26]. In the absence of interventions for named entities, Lu et al. [5] do not augment sentences or text instances containing proper nouns, and named entities as generating counterfactuals without proper name interventions could result in semantically incorrect sentences. Zhao et al. [26] circumvented this by anonymizing named entities by replacing them with special tokens. Lamenting on the aforementioned lack of parallel corpus for training neural models, Zmigrod et al. [27] used a series of unsupervised techniques such as dependency trees, lemmata, part-of-speech tags, and morpho-syntactic tags for counterfactual generation. Hall-Maudsley et al. [16] improve on Zmigrod et al. by incorporating a names intervention method to resolve the challenges of generating counterfactuals for named entities. They achieve this using a bipartite graph to match first names.

Because the aforementioned techniques rely on dictionary word replacement techniques and ignore the context of the text, they are prone to generating ungrammatical texts. Additionally, the inability of these techniques to resolve out-of-dictionary words not only preserves certain attribute correlations but also introduces errors. We illustrate two instances of such limitations using the word substitution approach by Hall-Maudsley et al. on the Bias-in-bios dataset [6]; 1) *"Memory received her **Bachelor** and **Masters** of Accountancy..."* produces *"Memory received his **Spinster** and **Mistresses** of Accountancy..."* due to the polysemous nature of *bachelor* and *master*; 2) *"Laura discovered her passion for programming after teaching **herself** some Python..."*, is transformed into *"Anthony discovered his passion for programming after teaching **herself** some Python..."* as the gender pronouns *herself* and *himself* are excluded from the dictionary compiled by Hall-Maudsley et al.

More recently, sequence-to-sequence model-based approaches to counterfactual generation have been proposed [19,25]. Wu et al. [25] propose Polyjuice, a generative counterfactual model for diverse use cases like counterfactual explanations. They generate parallel data by pairing naturally occurring sentences in a corpus based on edit distances. Although effective for explanations, such an approach is not applicable for bias mitigation as attribute words, in the case of the latter, have to be specifically defined and replaced. Specifically for bias mitigation, Qian et al. [19] introduce the *perturber*, which is a Bart [14] model fine-tuned on a human-generated parallel text. However, their approach only generates counterfactuals for specific user-defined entities in a text. e.g. *original: "Torii chose to remain behind, pledging that he and his **men** would fight..."*,

---

[1] https://github.com/EwoeT/FairFlow.

*rewrite: "Tara chose to remain behind, pledging that she and her **men** would fight ...".* As earlier stated, such manually compiled datasets are expensive and are only available on small scales, which can degrade performance [28]. Additionally, similar manual efforts must be solicited for every language domain for which counterfactuals have to be generated. As opposed to existing works, the main advantage of our work is the non-reliance on human-generated parallel data and word lists.

3   Approach

Our entire approach can be summarized in four steps as illustrated in Fig. 1. The process commences with training a classifier to detect attribute words in a corpus, after which counterfactuals for these attribute words are generated using an invertible flow model. Parallel data is thereafter created by using a combination of word substitution and an error-correction scheme. Finally, a generative model is fine-tuned using the generated parallel data. We expound on these steps in the following subsections.

3.1   Attribute Classifier Training

To select a list of words that characterize a given demographic axis, e.g. gender, we first train an attribute classifier that approximates the attribute subspace. To do this, the user first inputs a prompt in the form of a single pair of words that describes a given demographic axis, e.g., (she, he) in the case of gender. Using a pretrained contextualized word embedding model, contextualized word representations are generated for each appearance of the input words within a given text corpus - we take *BERT-base-uncased* [7] as our choice of representation model. These embeddings are used to train a classifier to approximate the demographic subspace. Formally, consider the word-pair (x_a, x_b) that define a demographic axis, we obtain two sets $Z_a = \{z_{a_1}, z_{a_2}, ..., z_{a_n}\}$ and $Z_b = \{z_{b_1}, z_{b_2}, ..., z_{b_n}\}$ where $z_{a_i} \in R^d$ and $z_{b_i} \in R^d$ are context-specific vector representations of instances of x_a, x_b respectively, generated from a text corpus V by a pretrained embedding model E; so that $E(x_i, c_i) = z_i$ if x_i is an instance of a word x and c_i is its context. We estimate the demographic subspace by training a classifier H to maximizing the objective $\sum_{z_i \in \{Z_a \cup Z_b\}} log(P(y|z_i))$, where $y = \{a, b\}$ is the class label of z_i. H is parameterized as a feed-forward neural network with one hidden layer and Gelu non-linear activation.

3.2   Generating Word-Pair List

Selecting Attribute Words. Given a demographic subspace, we select all words that lie within the attribute-defining regions of the subspace. This process is formally described as follows. Given our initial corpus V, we select words $x_i \in V$ based on the criterion $P(y|E(x_i, c_i); \Theta_H) > \phi$ where Θ_H represents the parameters that define H and ϕ is a predefined threshold. Z_a is thus expanded to include all words that have at least an instance satisfying $P(y = a|E(x_i, c_i); \Theta_H) > \phi$ and Z_b to include all words with at least an instance satisfying $P(y = b|E(x_i, c_i); \Theta_H) > \phi$. Although some neutral words may be included in these sets, they do not produce any counterfactual equivalent in the next stage, hence making no difference.

Generating Counterfactual Word-Pairs with DIIN. The first step in generating counterfactual equivalents for the set of words Z_a and Z_b is to define a transformation T from the original embedding space into an "interpretable" space where an embedding is factorizable into independent components. We train T to constrain attribute information *only* to the first k dimensions (we will collectively refer to these dimensions as K) of a word in the interpretable space. By so doing, K can be swapped to alter the attribute (e.g. gender) of the word. We implement T using a flow-based generative model [9,13,17]; specifically, we use the disentangling invertible interpretation network (DIIN) architecture by Esser et al. [11].

Formally, given the contextualized representation z of a word x, the goal is to learn a transformation T that maps the original representation $z \in R^d$ to an interpretable representation $\tilde{z} \in R^d$ s.t. $T(z) = \tilde{z}$. The interpretable representation $\tilde{z}$ is sampled from a base distribution $\tilde{z} \sim p_{\tilde{z}}(\tilde{z})$ – a standard Gaussian distribution in this case. Using the change of variable theorem, T is learned by maximizing the log-likelihood

$$\log(p_{\mathcal{Z}}(z)) = \log(p_{\tilde{\mathcal{Z}}}(T(z))) + \log(|\det(\frac{\partial T(z)}{\partial z})|) \tag{1}$$

To constrain attribute information only to K, we pair embeddings of words that have the same attribute F and train T to generate similar values for both embeddings in their first k dimensions in the interpretable space. Mathematically, Given a pair of embeddings (z_{a_1}, z_{a_2}) that belong to the same demographic group such that $F_{z_{a_1}} = F_{z_{a_2}}$, the objective is achieved by minimizing the loss function:

$$\begin{aligned}\mathcal{L}(z_{a_1}, z_{a_1}|F) = &||T(z_{a_1})_D||^2 - \log(\det(T(z_{a_1}))) \\ &+ ||T(z_{a_2})_{(D\setminus K)}||^2 - \log(\det(T(z_{a_2}))) \\ &+ \frac{||T(z_{a_2})_K - \sigma T(z_{a_1})_K||^2}{1 - \sigma^2}\end{aligned} \tag{2}$$

where D is a term to collectively refer to all d components of the embedding. $\sigma \in (0,1)$ is a positive correlation factor that determines the strength of the correlation between $z_{a_{2K}}$ and $z_{a_{1K}}$. We also use the dimensionality estimation approach of Esser et al. to estimate the dimensionality of K.

Once our invertible flow model has been trained to constrain F to the first k dimensions of $\tilde{z}$ (in the interpretable space), we replace $z_{a_{iK}}$ which is the first k dimensions of $\tilde{z}_{a_i}$ with K'_b; such that $z_{a_{iK}} \to K'_b$, where $K'_b = \frac{1}{N}\sum_{i=0}^{N} z_{b_{iK}}$ is the average of the first k dimensions of the complementary demographic group. This process is depicted in Fig. 2. We use a majority voting scheme to then

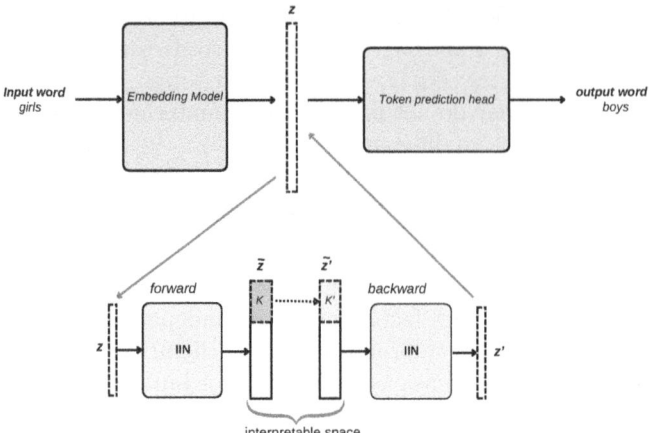

Fig. 2. Counterfactual word generation using an invertible interpretation flow network IIN.

select the most frequent equivalent generated for each word. An output example of this process obtained using a { *"she"*, *"he"*} prompt is shown in Fig. 3. We then extend this list using the names intervention approach of Hall-Maudsley et al. to generate counterfactuals for names.

3.3   Error Correction

With the word pairs generated from the previous phase, we use the word substitution approach of Hall-Maudsley et al. to build a base corpus. To transform this base corpus into fluent and realistic text labels for our parallel training data, we proposed an error correction scheme which we describe below in two steps.

Erratic Token Detection. The idea here is to detect and mask tokens that have a low probability of appearing in the context of a given text; following $t_i = t_{<mask>}$ if $P(t_i|T \setminus t_i) < \theta$, where T is the sequence of tokens, t_i is the ith token in T, and θ is a predefined threshold value. We define the resulting masked text as T_Π. This is achieved using a pretrained Electra model [3]. Electra is an LM pretrained using a text corruption scheme – text instances are corrupted by randomly replacing a number of tokens with plausible alternatives from BERT. Electra is then trained to predict which tokens are real and fictitious.

Since the use of wordpiece tokenization causes issues (as a word can be broken down into multiple subtokens) if a subtoken is selected for masking, we replace the entire sequence of associated subtokens with a <mask> token. For instance, *"The men are duchesses"*, in a wordpiece tokenization could be decomposed to *["The", "men", "are", "duchess", "##es"]*, Consequently, when *"duchess"* is identified as an erratic token, the masking scheme replaces the entire subsequence *["duchess", "##es"]*, thereby, generating *"The men are <mask> "*.

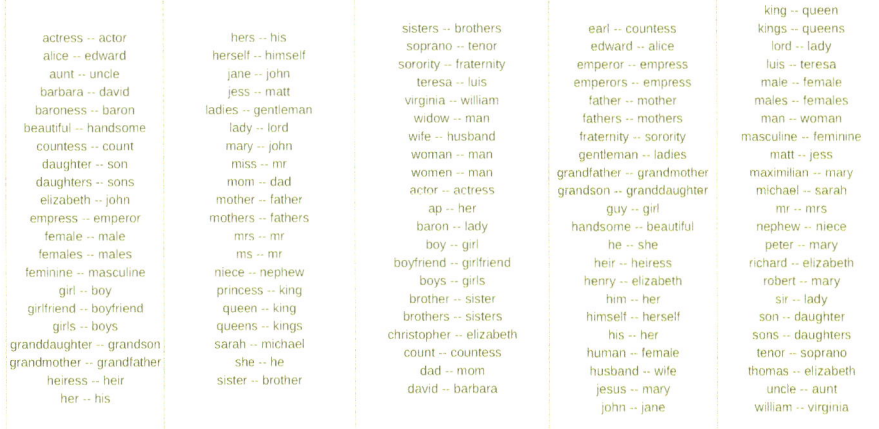

Fig. 3. An automatically compiled dictionary using the input prompt { *"she"*, *"he"*}. Words are discovered using the attribute classifier, and the counterfactuals are generated using the disentangling invertible interpretation network.

Text Insertion with BART. Having obtained our masked intermediary texts, we generate plausible token replacements for each masked token. Since a `<mask>` token could correspond to multiple subword tokens, the replacement generator should be capable of generating multiple tokens for a single `<mask>` instance, making it suitable to use a generative model – pretrained BART [14] – to predict these replacement tokens. Because Masked Language Modeling is one of BART's pretraining objectives, we can utilize it in its pretrained form without the need for finetuning. Given T_{II} from the previous step, the BART model tries to predict the correct infilling x using the context of T_{II}.

3.4   Training the Generative Model

The final stage of the approach is to fine-tune a BART model using the parallel data obtained from the previous steps. The BART generator takes the original text as input and is trained to autoregressively generate the counterfactual of the source text using the corresponding parallel counterfactual texts as labels in a teacher-forcing manner [23]. We formulate this as:

$$\mathcal{L}_{generator} = -\sum_{t=1}^{k} log P(y_t | Y_{<t}, X) \qquad (3)$$

where X and Y are the source and target texts, respectively, $y_t \in Y$ is the tth token in the target text, and $Y_{<t}$ refers to all tokens in Y preceding y_t.

4   Experimental Set-Up

This section describes key implementation details of our work and the evaluation framework. We specifically evaluate gender bias in the binary sense within the English language domain.

4.1   Training Set-Up

The main corpus for training the attribute classifier and the disentangling invertible flow model comprises Wikipedia articles via Wikimedia dumps[2].

4.2   Evaluation Datasets

For the appraisal of our model, we used the datasets discussed below. These datasets, upon which various CDA interventions were applied, were used to train a classification model on a downstream task. These datasets were only used for evaluation purposes and were not included in training Fairflow.

1. **Bias-in-bios**: This dataset provided by De-Arteaga et al. [6] contains Wikipedia profiles of professionals. The dataset originally contained labels corresponding to 28 distinct professions alongside the gender labels of the profiled individuals. We reclassified the professions into binary labels, aligning them with male-dominated and female-dominated occupations according to gender distribution. This categorization was done for two reasons. The first was to simplify the classification task from multiclass to binary. Secondly, this enabled us to easily induce bias by creating an imbalance between gender and class labels.
2. **ECHR**: The ECHR dataset by Chalkidis et al.[3] [2] contains case facts from the European Court of Human Rights (ECHR) on human rights breaches by European states. It further contains information on the gender of the applicant, human rights articles that were violated, and the defendant state (Central-Eastern European states *v.* all other states). The primary classification task here was to predict the defendant's state based on the case facts.
3. **Jigsaw**: This dataset[4] contains public comments from the now defunct online platform Civil Comments. The primary classification task for this dataset was toxicity detection.

For all the evaluation datasets, we maintained a balanced gender and class label distribution in the test sets as shown in Table 1. The training sets for the Bias-in-bios and the Jigsaw datasets were sampled with an imbalance to induce bias following the observations of Dixon et al. [10]. The training set for ECHR was left relatively balanced with the additional purpose of providing a baseline.

---

[2] https://dumps.wikimedia.org.
[3] https://huggingface.co/datasets/coastalcph/fairlex.
[4] https://www.kaggle.com/c/jigsaw-unintended-bias-in-toxicity-classification.

Table 1. Evaluation dataset statistics: The test sets are balanced with regard to gender and labels.

Dataset	Task	Train			Test		
		Number (K)	Positive class %	Females in Pos. %	Number (K)	Positive class %	Females in Pos. %
Bias-in-bios	Career	18	50	12	4	50	50
ECHR	State	7	18	41	1	50	50
Jigsaw	Toxicity	5	47	77	1	50	50

4.3   Comparative Techniques

We implemented two variants of FairFLow: *FairFLowV1* and *FairFLowV2*, and compared them to three CDA setups. 1) *original* is the unaugmented original text; 2) *Hall-M* uses the direct word-substituion approach proposed by Hall-Maudsley et al. [16]; 3) *Hall-M + BART* is a BART model fine-tuned with counterfactuals generated by Hall-Maudsley et al.; 4) *FairFlowv1* is a BART model fine-tuned with our error correction scheme applied to counterfactuals from Hall-Maudsley et al.; it follows the same approach of FairLow in Fig. 1 but with a manually compiled dictionary. 5) *FairFlowv2* is a BART-model fine-tuned with our full approach in Fig. 1. We take *Hall-M* and *Hall-M + BART* as our baseline approaches. We excluded *perturber* by Qian et al. [19] from our evaluation since the objective of their approach significantly differs from ours; as elaborated in Sect. 2. Also, conversational models like ChatGPT[5] [in their current form] struggled to generate large amounts of text needed for the evaluation; hence, we do not include quantitative results using them.

5   Evaluation and Results

We quantitatively evaluated our approach using three main criteria: *utility*, *extrinsic bias mitigation*, and *task performance*.

5.1   Utility

By utility, we refer to how realistic and effective the generated counterfactuals are by computing their fluency (perplexity) and gender transfer accuracy.

---

[5] https://chat.openai.com/.

dem. axis	original	Hall-M	Hall-M + Bart	ChatGPT	Meta-llama	FairFlowV2
gender (she_he)	In 2011, she won two prestigious competitions: Miss Ukraine-Earth and Miss Earth. In addition, Christina entered the Top-10 most beautifulul girls of the world. Kristina is from Zaporizhzhya.	In 2011, he won two prestigious competitions: Miss Ukraine-Earth and Miss Earth. In addition, Joe entered the Top-10 most beautiful boys of the world. Gilbert is from Zaporizhzhya.	In 2011, he won two prestigious competitions: Miss Ukraine-Earth and Miss Earth. In addition, Joe entered the Top-10 most beautiful boys of the world. Alberto is from Zaporizhzhya.	In 2011, he won two prestigious competitions: Mr. Ukraine-Earth and Mr. Earth. In addition, Christian entered the Top-10 most handsome guys of the world. Christian is from Zaporizhzhya.	In 2011, he won two prestigious competitions: Mister Ukraine-Hero and Mister Hero. In addition, Christopher entered the Top-10 most handsome men of the world. Christopher is from Zaporizhzhya.	In 2011, he won two prestigious competitions: Mr Ukraine-Earth and Mr Earth. In addition, he entered the top-10 most handsome boys of the world. Irving is from Zaporizhzhya.
religion (catholic _muslim)	In 1579 he converted from Orthodoxy to Roman Catholicism	unavailable	unavailable	In 1579 he converted from Orthodoxy to Islam.	In 1579 he converted from Islam to Islam.	In 1579 he converted from Orthodoxy to Sunni Islam.

Fig. 4. Text samples from Bias-in-bios and Wikipedia demonstrate that *FairFlow* and *ChatGPT-4* generate more robust counterfactual texts. Compared to *ChatGPT-4*, *Meta-llama-3-8B-Instruct* generates more inaccurate counterfactuals.

Grammatical Correctness and Fluency. We used a referenceless fluency metric due to the relative unavailability of parallel data. As we noted earlier, the parallel data used by Qian et al. only contains counterfactuals for only specific user-defined entities and is thus not suitable for evaluating our work. Similar to Wu et al. [25], we score fluency by computing the perplexity of the generated text using pretrained GPT-2 [21]. A low perplexity implies that a given text conforms well to the probabilistic distribution of natural text as learned by the pretrained language model.

Based on our earlier assertion about how out-of-context substitutions impair fluency, our error correction approach should expectedly increase fluency (reduce perplexity). We confirm this in Table 2 as we see that fluency is consistently improved in both *FairFlowV1* and *FairFlowV2*.

Transfer Accuracy. Here, similar to Tokpo et al. [22], we computed the percentage of texts that were converted from the source attribute to the target attribute, i.e., female to male or vice versa. We fine-tuned a BERT model to predict the gender of the text. We quantified gender transfer accuracy as $1 - probability\_of\_original\_attribute$. We expect the original text to have a very low transfer accuracy, as its attributes would remain the same. As shown in Table 2, FairFlowV2 especially shows strong fluency scores whilst maintaining a

Table 2. PPL (*left*) of generated text using various CDA techniques. Lower scores indicate better fluency. Gender transfer accuracy (*right*) of the various CDA interventions. This indicates the percentage of counterfactual instances that were correctly resolved to new gender styles. *The original samples have very low accuracies because original gender is preserved.*

Approach	PPL ↓			Transfer Accuracy ↑		
	Bios	Jigsaw	ECHR	Bios	Jigsaw	ECHR
*Original**	41.023	69.67	32.88	0.04	15.96	36.14
Hall-M	43.51	76.37	33.70	98.60	**79.00**	75.10
Hall-M + BART	47.59	83.76	39.93	98.70	78.50	71.10
FairFlowV1	42.77	65.80	33.70	**98.91**	77.99	74.69
FairFlowV2	**39.86**	**63.99**	**33.33**	98.51	70.736	**76.51**

good transfer accuracy. This shows that automating the dictionary generation process does not materially impair transfer accuracy.

5.2   Extrinsic Bias Mitigation

We trained a BERT classifier using the downstream classification tasks corresponding to the respective datasets and computed the *True Positive rate difference (TPRD)* and *False Positive rate difference (FPRD)* between two gender groups as in the case of De-Arteaga et al. [6]. $TPRD = P(\hat{y} = 1|y = 1, A = a) - P(\hat{y} = 1|y = 1, A = a')$ and $FPRD = P(\hat{y} = 1|y = 0, A = a) - P(\hat{y} = 1|y = 0, A = a')$. Where y is the true label, $\hat{y}$ is the predicted label, and A is the gender group variable.

We show in Table 3 consistently high TPRD scores for FairFlow1; this further buttresses the evidence that our approach to error correction works effectively and enhances bias mitigation whilst improving fluency. Similar to our findings for transfer accuracy, we find that automating dictionary compilation does not compromise bias mitigation much, as FaiFlowV2 maintains a good mitigating effect.

5.3   Task Performance

We carried out the task performance test to observe the extent to which bias mitigation impacts the task model's performance. Because we maintain a balanced distribution for our test sets, we expect the fairer models to have better performance. Specifically, we computed the accuracy and F1 scores for the default classification task of the respective datasets. In Table 4, FairFlow1 shows the most improved performance in general, particularly in accuracy. We again show from the strong performance of FairFlowV2, how effective an automatically generated dictionary could be.

Table 3. Extrinsic fairness: TPRD – True positive rate difference between male and female text instances. FPRD – False positive rate difference between male and female text instances.

Approach	TPRD ↓			FPRD ↓		
	Bios	Jigsaw	ECHR	Bios	Jigsaw	ECHR
*Original**	0.133	0.120	0.000	0.151	0.160	0.0
Hall-M	0.055	0.010	0.030	0.071	0.070	0.0
Hall-M + BART	0.051	0.025	0.010	0.074	**0.060**	0.0
FairFlowV1	**0.044**	**0.005**	**0.000**	**0.065**	0.065	0.0
FairFlowV2	0.057	0.040	0.010	0.070	0.080	0.0

Table 4. Task performance: Accuracy and F1 scores of classification tasks. FairFLow1 shows better performance scores in general. FairFlow2 maintains a significant bias mitigating effect despite an automated dictionary approach.

Approach	ACC ↑			F1 ↑		
	Bios	Jigsaw	ECHR	Bios	Jigsaw	ECHR
*Original**	91.20	88.50	97.60	47.92	48.95	52.86
Hall-M	92.53	90.25	97.83	48.38	48.62	52.98
Hall-M + BART	92.64	90.62	97.36	**48.48**	49.49	52.73
FairFlowV1	**92.97**	**90.75**	98.08	48.32	**49.69**	53.11
FairFlowV2	92.81	90.00	**98.32**	48.36	49.21	**53.24**

5.4   Qualitative Analysis and Key Observations

By analyzing samples from FairFlow, ChatGPT, and the comparative models, we find that FairFLow and ChatGPT have the most grammatically coherent counterfactuals. Additionally, we find that:

1. ***Automating the dictionary compilation process does not materially impair counterfactual generation.*** As shown in Fig. 4, even with a dictionary that was automatically compiled, FirFlowV2 generates fluent and plausible counterfactuals. This is aided by the combination of the error correction scheme, which makes it more robust to grammatical errors and helps it generalize better.
2. ***A model fine-tuned on erroneous data mimics those errors.*** We observe that the error correction approach incorporated in FairFlow makes the model more robust, fluent, and grammatically coherent. The direct word replacement technique (*Hall-M*) is unable to replace out-of-dictionary words. The output of *Hall-M + BART* mirrors the same errors as *Hall-M*, showing that a generative model fine-tuned on erroneous data will mimic those errors.
3. ***ChatGPT generates good counterfactuals but has practical limitations.*** We observe that, in general, ChatGPT generates good counterfactuals

in zero-shot settings but struggles to generate counterfactuals on a scale suitable for training a model. This limited the extent of quantitative evaluation we could perform on it. Secondly, ChatGPT is inconsistent in generating counterfactuals for names, as it tends to skip some names for which counterfactuals could have been generated. This is more so if the names refer to public figures, which occasionally leads to grammatical incoherent outputs. However, the advantage of such occurrences is that they preserve factuality better, which may be a more desirable attribute in certain contexts. We also observed some irregular counterfactuals from *Meta-llama-3-8B-Instruct*, as shown in Fig. 4. Some of the counterfactuals it generated impacted the original context of the text, which should have been retained.

6   Conclusion

In this paper, we highlight some issues that pertain to dictionary-based word-substitution counterfactual data augmentation techniques. We discuss how these techniques, relying on manually compiled dictionaries, are prone to grammatical incoherence and lack generalization outside dictionary terms. We discuss how a model-based approach is primarily inhibited by the relative unavailability of parallel corpora for training. In light of this: 1) we propose an automated dictionary generation approach that can automatically extract and generate word-pairs from a corpus with little human intervention; 2) we propose an error correction approach that can be used to generate fluent and grammatically coherent parallel text to train a generative model for CDA; 3) we combine these approaches to fine-tune a BART model for the purpose of generating counterfactual texts (we call the resulting model *FiarFLow*); 4) we show that our error correction approach significantly improves the fine-tuned model's fluency and bias-mitigating effect; 5) we also show that automating the dictionary compilation process comes at little cost to the performance of the CDA model and is a viable solution in settings where human intervention is challenging.

Limitations
The primary limitation of our work is the lack of exploration into more diverse demographic and language domains. The work mostly focuses on (binary) gender bias in English, which is a significant limitation, considering how nuanced gender can be in other languages. Due to the relative unavailability of CDA test resources in other demographic domains, such as race, the scope of evaluation in these areas is limited. Our future work will be directed towards addressing these research directions.

Another limitation of this work is its reliance on the tokenization scheme used by the embedding model, which means that words expressed in multiple subtokens are not included in the automatic compilation of the dictionary.

Acknowledgements. Ewoenam Kwaku Tokpo received funding from the Flemish Government under the "Vlaams AI-Onderzoeksprogramma" (Flanders AI Research Program). We also thank Marco Favier for sharing his insights and engaging in valuable discussions.

Ethics Statement. From an ethical perspective, the primary point to keep in mind regarding the use of counterfactual models is their impact on factuality. Since CDA approaches are designed to be *counterfactual*, they should be used cautiously in sensitive domains where factuality is essential. Secondly, CDA bias mitigation techniques like FairFlow do not automatically guarantee fairness; hence, they must be used with that understanding.

References

1. Beukeboom, C.J., Burgers, C.: Linguistic bias. In: Oxford Encyclopedia of Communication, pp. 1–19. Oxford University Press (2017)
2. Chalkidis, I., Passini, T., Zhang, S., Tomada, L., Schwemer, S.F., Søgaard, A.: FairLex: a multilingual benchmark for evaluating fairness in legal text processing. In: Proceedings of the 60th Annual Meeting of the Association for Computational Linguistics, Dublin, Ireland (2022)
3. Clark, K., Luong, M.T., Le, Q.V., Manning, C.D.: ELECTRA: pre-training text encoders as discriminators rather than generators. arXiv preprint arXiv:2003.10555 (2020)
4. Dastin, J.: Amazon scraps secret AI recruiting tool that showed bias against women. In: Ethics of Data and Analytics, pp. 296–299. Auerbach Publications (2022)
5. Lu, K., Mardziel, P., Wu, F., Amancharla, P., Datta, A.: Gender bias in neural natural language processing. In: Nigam, V., et al. (eds.) Logic, Language, and Security. LNCS, vol. 12300, pp. 189–202. Springer, Cham (2020). https://doi.org/10.1007/978-3-030-62077-6_14
6. De-Arteaga, M., et al.: Bias in bios: a case study of semantic representation bias in a high-stakes setting. In: Proceedings of the Conference on Fairness, Accountability, and Transparency, pp. 120–128 (2019)
7. Devlin, J., Chang, M.W., Lee, K., Toutanova, K.: BERT: pre-training of deep bidirectional transformers for language understanding. arXiv preprint arXiv:1810.04805 (2018)
8. Dinan, E., Fan, A., Williams, A., Urbanek, J., Kiela, D., Weston, J.: Queens are powerful too: mitigating gender bias in dialogue generation. In: Proceedings of the 2020 Conference on Empirical Methods in Natural Language Processing (EMNLP), pp. 8173–8188 (2020)
9. Dinh, L., Krueger, D., Bengio, Y.: NICE: non-linear independent components estimation. arXiv preprint arXiv:1410.8516 (2014)
10. Dixon, L., Li, J., Sorensen, J., Thain, N., Vasserman, L.: Measuring and mitigating unintended bias in text classification. In: Proceedings of the 2018 AAAI/ACM Conference on AI, Ethics, and Society, pp. 67–73 (2018)
11. Esser, P., Rombach, R., Ommer, B.: A disentangling invertible interpretation network for explaining latent representations. In: Proceedings of the IEEE/CVF Conference on Computer Vision and Pattern Recognition, pp. 9223–9232 (2020)
12. Fiedler, K.: Social Communication. Psychology Press (2011)
13. Kobyzev, I., Prince, S.J., Brubaker, M.A.: Normalizing flows: an introduction and review of current methods. IEEE Trans. Pattern Anal. Mach. Intell. **43**(11), 3964–3979 (2020)

14. Lewis, M., et al.: BART: denoising sequence-to-sequence pre-training for natural language generation, translation, and comprehension. In: Proceedings of the 58th Annual Meeting of the Association for Computational Linguistics, pp. 7871–7880. Association for Computational Linguistics, July 2020. https://doi.org/10.18653/v1/2020.acl-main.703. https://aclanthology.org/2020.acl-main.703

15. Lohia, P.K., Ramamurthy, K.N., Bhide, M., Saha, D., Varshney, K.R., Puri, R.: Bias mitigation post-processing for individual and group fairness. In: ICASSP 2019-2019 IEEE International Conference on Acoustics, Speech and Signal Processing (ICASSP), pp. 2847–2851. IEEE (2019)

16. Maudslay, R.H., Gonen, H., Cotterell, R., Teufel, S.: It's all in the name: mitigating gender bias with name-based counterfactual data substitution. In: Proceedings of the 2019 Conference on Empirical Methods in Natural Language Processing and the 9th International Joint Conference on Natural Language Processing (EMNLP-IJCNLP), pp. 5267–5275. Association for Computational Linguistics, Hong Kong, China, November 2019. https://doi.org/10.18653/v1/D19-1530. https://aclanthology.org/D19-1530

17. Papamakarios, G., Nalisnick, E., Rezende, D.J., Mohamed, S., Lakshminarayanan, B.: Normalizing flows for probabilistic modeling and inference. J. Mach. Learn. Res. **22**(57), 1–64 (2021)

18. Porter, S.C., Rheinschmidt-Same, M., Richeson, J.A.: Inferring identity from language: linguistic intergroup bias informs social categorization. Psychol. Sci. **27**(1), 94–102 (2016)

19. Qian, R., Ross, C., Fernandes, J., Smith, E.M., Kiela, D., Williams, A.: Perturbation augmentation for fairer NLP. In: Proceedings of the 2022 Conference on Empirical Methods in Natural Language Processing, pp. 9496–9521 (2022)

20. Radford, A., Narasimhan, K., Salimans, T., Sutskever, I., et al.: Improving language understanding by generative pre-training (2018)

21. Radford, A., et al.: Language models are unsupervised multitask learners (2018)

22. Tokpo, E.K., Calders, T.: Text style transfer for bias mitigation using masked language modeling. In: Ippolito, D., Li, L.H., Pacheco, M.L., Chen, D., Xue, N. (eds.) Proceedings of the 2022 Conference of the North American Chapter of the Association for Computational Linguistics: Human Language Technologies: Student Research Workshop, pp. 163–171. Association for Computational Linguistics, Hybrid: Seattle, Washington, July 2022. https://doi.org/10.18653/v1/2022.naacl-srw.21. https://aclanthology.org/2022.naacl-srw.21

23. Williams, R.J., Zipser, D.: A learning algorithm for continually running fully recurrent neural networks. Neural Comput. **1**(2), 270–280 (1989). https://doi.org/10.1162/neco.1989.1.2.270

24. Wolf, M.J., Miller, K., Grodzinsky, F.S.: Why we should have seen that coming: comments on Microsoft's Tay "experiment," and wider implications. ACM SIGCAS Comput. Soc. **47**(3), 54–64 (2017)

25. Wu, T., Ribeiro, M.T., Heer, J., Weld, D.S.: Polyjuice: generating counterfactuals for explaining, evaluating, and improving models. In: Proceedings of the 59th Annual Meeting of the Association for Computational Linguistics and the 11th International Joint Conference on Natural Language Processing (Volume 1: Long Papers), pp. 6707–6723 (2021)

26. Zhao, J., Wang, T., Yatskar, M., Ordonez, V., Chang, K.W.: Gender bias in coreference resolution: Evaluation and debiasing methods. In: Proceedings of the 2018 Conference of the North American Chapter of the Association for Computational Linguistics: Human Language Technologies, Volume 2 (Short Papers), pp. 15–20. Association for Computational Linguistics, New Orleans, Louisiana, June 2018. https://doi.org/10.18653/v1/N18-2003. https://aclanthology.org/N18-2003
27. Zmigrod, R., Mielke, S.J., Wallach, H., Cotterell, R.: Counterfactual data augmentation for mitigating gender stereotypes in languages with rich morphology. In: Proceedings of the 57th Annual Meeting of the Association for Computational Linguistics, pp. 1651–1661. Association for Computational Linguistics, Florence, Italy, July 2019. https://doi.org/10.18653/v1/P19-1161. https://aclanthology.org/P19-1161
28. Zoph, B., Yuret, D., May, J., Knight, K.: Transfer learning for low-resource neural machine translation. In: Proceedings of the 2016 Conference on Empirical Methods in Natural Language Processing, pp. 1568–1575 (2016)

GrINd: Grid Interpolation Network for Scattered Observations

Andrzej Dulny[✉], Paul Heinisch, Andreas Hotho, and Anna Krause

CAIDAS, University Würzburg, Würzburg, Germany
{andrzej.dulny,anna.krause}@uni-wuerzburg.de,
paul.heinisch@stud-mail.uni-wuerzburg.de,
hotho@informatik.uni-wuerzburg.de

Abstract. Predicting the evolution of spatiotemporal physical systems from sparse and scattered observational data poses a significant challenge in various scientific domains. Traditional methods rely on dense grid-structured data, limiting their applicability in scenarios with sparse observations. To address this challenge, we introduce GrINd (Grid Interpolation Network for Scattered Observations), a novel network architecture that leverages the high-performance of grid-based models by mapping scattered observations onto a high-resolution grid using a Fourier Interpolation Layer. In the high-resolution space, a NeuralPDE-class model predicts the system's state at future timepoints using differentiable ODE solvers and fully convolutional neural networks parametrizing the system's dynamics. We empirically evaluate GrINd on the DynaBench benchmark dataset, comprising six different physical systems observed at scattered locations, demonstrating its state-of-the-art performance compared to existing models. GrINd offers a promising approach for forecasting physical systems from sparse, scattered observational data, extending the applicability of deep learning methods to real-world scenarios with limited data availability.

Keywords: Physics · Dynamical Systems · Fourier · NeuralPDE · DynaBench

1 Introduction

Understanding and accurately predicting the evolution of spatiotemporal physical systems is an important challenge in various scientific domains, encompassing fields such as climate modeling, weather forecasting, computational fluid simulation, biology, and many more [7,10,15,23]. These systems are typically guided by physical laws summarized as partial differential equations, which describe the rate of change of the systems' quantities as a function of their partial derivatives in space [13]. Traditionally, numerical simulations have served as the cornerstone of predictive modeling for these systems, with various methods such as the method of lines, spectral methods, finite volume methods, and the finite element method designed to tackle specific challenges and specific classes of equations [27,45].

A. Bifet et al. (Eds.): ECML PKDD 2024, LNAI 14947, pp. 177–193, 2024.
https://doi.org/10.1007/978-3-031-70368-3_11

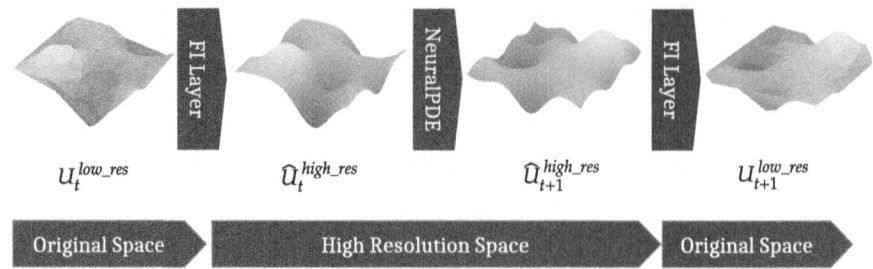

$$u_t^{low\_res} \qquad \hat{u}_t^{high\_res} \qquad \hat{u}_{t+1}^{high\_res} \qquad u_{t+1}^{low\_res}$$

Original Space ▸   High Resolution Space ▸   Original Space ▸

Fig. 1. Summary of our approach. The low-resolution observations are first mapped onto a high-resolution grid using a Fourier Interpolation Layer. In this high resolution space a predictive model (NeuralPDE) forecasts the evolution of the system which is then mapped back to the original observation space.

In recent years, a notable paradigm shift has emerged, with a range of new deep learning-based methods being proposed to replace or augment the classical numerical methods. Models like Physics Informed Neural Networks [41] can be used to solve differential equations by employing neural networks as solution approximators. Other approaches like SINDy [11] or PDE-Net [35] solve the task of reconstructing the equation that describes the evolution of the system by using symbolic regression and sparsity enforcement. Yet another range of methods focus solely on forecasting the evolution of the system. Models like PanguWeather [9], FourCastNet [25] or GraphCast [26] leverage large-scale neural architectures to learn the behavior of the system directly from data without any physical information.

Crucially, the performance of these types of models depends on the availability of large amounts of high-quality data that in most cases is assumed to be distributed across a spatial grid. In fact most methods proposed for forecasting physical systems work only on grid structured data [17] and a persistent challenge arises in scenarios where observational data is inherently sparse and spatially scattered, a common occurrence in many real-world applications. As Dulny et al. [17] pointed out, the task of predicting the evolution of a physical system based only on low-resolution scattered observations remains a challenging and unsolved task.

To tackle this challenge, we introduce GrINd (**Grid Interpolation Network** for Scattered Observations), an abstract network architecture which leverages the high performance of grid-based models, by mapping the scattered observations onto a high-resolution grid using a Fourier Interpolation Layer. In the high-resolution space a NeuralPDE-class model [16] predicts the state of the system at future timepoints using a combination of differentiable ODE solvers [14] and the method of lines [45] with fully convolutional neural networks parametrizing the dynamics of the system. Our approach is summarized in Fig. 1.

We empirically evaluate the performance of our model on the DynaBench benchmark dataset [17] encompassing six different physical systems observed on scattered locations and show its state-of-the-art performance compared to other non-grid based models.

Overall, our contributions can be summarized as follows:

1. We introduce GrINd - a novel network architecture for forecasting physical systems from sparse, scattered observational data.
2. We combine a Fourier Interpolation Layer and a NeuralPDE model for efficient predictions in high-resolution space
3. We empirically evaluate our model on the DynaBench dataset, demonstrating state-of-the-art performance compared to existing models.

We make the source code containing our model as well as all experiments in this paper publicly available.[1]

2   Related Work

Learning the behavior of real physical systems from empirical data using machine learning techniques has become of increasing interest in recent years. The work in this area can generally be divided into two overlapping areas: data-driven solution and data-driven discovery of partial differential equations. Examples of the former include data-driven models for predicting the evolution of various physical systems such as weather with transformer models [9,25,26], fluid simulation with graph neural networks [44] or simply solving specific PDEs using Physics Informed Neural Networks [41]. Examples of the latter include SINDy [11] which uses sparse regression to find the underlying governing equations of a system and PDE Net [35] which uses a set of learnable filters and symbolic regression to reconstruct the equation. The relevance of the area also emerges from the increasing number of benchmark datasets available for physical systems [17,36,42].

Most works use high-resolution grid data which offers many advantages, allowing the use of well-developed models and architectures such as a modified vision transformer [1,9,25,26,29,32]. Another approach by Ayed et al. [4] uses a hidden-state neural-based model to forecast dynamical systems using a ResNet. The Finite Volume Neural Network (FINN) [39] predicts the evolution of diffusion-type systems by explicitly modeling the flow between grid points utilizing the Finite Volume Method. NeuralPDE [16] combines CNNs and the method of lines solver to find a solution for the underlying PDE.

Only a limited number of approaches have been proposed to tackle the challenging task of forecasting a physical system from scattered data. Iakovlev et al. [20] introduce continuous-time representation and learning of the dynamics of PDE-driven systems using a GNN. The MGKN (multipole graph kernel network) [33] proposed by Li et al. unifies GNNs with multiresolution matrix factorization to capture long-range correlations. PhyGNNet [21] divides the computational domain into meshes and uses the message-passing mechanism, as well as the discrete difference method, based on Taylor expansion and least squares

---

[1] https://professor-x.de/grind.

regression, to predict future states. Graph networks, however, can become computationally expensive when reaching a certain size, common in real-world tasks.

To the best of our knowledge, our approach is the first model to take advantage of Fourier Analysis to interpolate observations onto a higher resolution grid. Several deep learning approaches have been proposed to perform training entirely in the Fourier domain such as Fourier Convolutional Neural Networks [40] or the Fourier Neural Operator [34]. However, all works taking advantage of the Fast Fourier Transform (FFT) still require the input data to be grid-structured.

3   Method

In this section, we describe the technical details of our proposed approach. The overall architecture of our approach consisting of the Fourier interpolation layer and the predictive NeuralPDE model can be seen schematically in Fig. 1.

3.1   Fourier Interpolation Layer

One of the core components of our approach is the Fourier interpolation layer, which takes as input the values of a function u at several non-uniform locations $X = \{\mathbf{x}_1, \ldots, \mathbf{x}_H\}$ and outputs the interpolated values of the function $\tilde{u}$ at grid locations X_{grid}. To calculate the interpolation in a differentiable way, we leverage the approximation properties of the Fourier series summarized in Theorem 1.

While in principle any functional basis of the L^2 function space can be used for approximation, we opt for the Fourier basis as it has been previously studied in related work in the context of partial differential equations [2,24,34] and several algorithms could potentially be used for faster computations [5,6,48].

Theorem 1. *Given a periodic complex function $u\colon \Omega \to \mathbb{C}$ on the interval box $\Omega = [0,1]^M$, such that $\int_\Omega ||u||^2 d\mathbf{x} < \infty$, the following series:*

$$\hat{u}_N(\mathbf{x}) = \sum_{\mathbf{k} \in K} c_\mathbf{k}(u) e^{2i\pi \mathbf{k} \cdot \mathbf{x}} \tag{1}$$

with

$$c_\mathbf{k}(u) = \int_\Omega u(\mathbf{x}) e^{2i\pi \mathbf{k} \cdot \mathbf{x}} d\mathbf{x}$$

$$N = (N_1, \ldots, N_M)$$

$$K = K_{N_1, \ldots, N_M} := K_{N_1} \times \ldots \times K_{N_M}$$

$$K_{N_i} = \begin{cases} \{-\frac{N_i}{2}, \ldots, \frac{N_i}{2} - 1\} & \text{for } N_i \text{ even} \\ \{-\frac{N_i-1}{2}, \ldots, \frac{N_i-1}{2}\} & \text{for } N_i \text{ odd} \end{cases}$$

converges pointwise to u as $N_i \to \infty$, i.e.

$$\text{for each } \mathbf{x} \in [0,1]^M : \lim_{\min N_i \to \infty} \hat{u}_N(\mathbf{x}) = u(\mathbf{x})$$

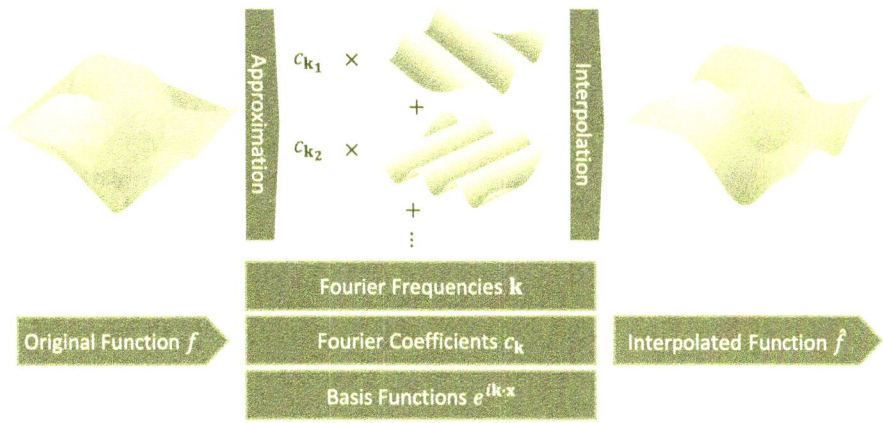

Fig. 2. Fourier Interpolation Layer. We use a Fourier series approximation of the original function u by fitting the Fourier coefficients $c_{\mathbf{k}}$ using LLS. The Fourier coefficients can be used to evaluate the approximation at any arbitrary collection of points $\mathbf{x}$ thus making it possible to interpolate the function onto a high-resolution grid.

Proof. The proof of this theorem can be found in [46] or [38].

This theorem underpins various numerical methods and techniques, including the Fast Fourier Transform (FFT), which plays an important role in signal processing [8], data compression [43], and solving differential equations efficiently [24]. However, applying FFTs directly to non-uniformly sampled data presents challenges, as the traditional approach assumes data points are uniformly distributed along the domain [5,48]. To address this limitation, we employ an estimation technique to compute the Fourier coefficients $c_{\mathbf{k}}$ for non-uniformly sampled points as summarized in Fig. 2.

Specifically, given the Fourier series expressed as Eq. (1) and the function u measured at several non-uniform locations $X = \{\mathbf{x}_1, \ldots, \mathbf{x}_H\}$ we are trying to estimate the coefficients $c_{\mathbf{k}}, \mathbf{k} \in K$ such that the approximation error

$$\sum_{\mathbf{x} \in X} \|u(\mathbf{x}) - \hat{u}_N(\mathbf{x})\|^2 \tag{2}$$

is minimized. $N = (N_1, \ldots, N_M)$ represents the number of frequencies in each dimension. We note that while the number of frequencies can be chosen separately for each dimension $1, \ldots, M$ in practice we keep this parameter constant as $N_1 = \ldots = N_M =: N_{freq}$.

We formulate the estimation process as a linear least squares (LLS) problem where the coefficients $c_{\mathbf{k}}$ are the unknowns. The LLS problem can be expressed as follows:

$$\min_{\mathbf{c} \in \mathbb{R}^{N_1 \cdots N_M}} \|\mathbf{A}\mathbf{c} - \mathbf{b}\|^2 \tag{3}$$

where, $\mathbf{A}$ is a matrix representing the basis functions $e^{2i\pi \mathbf{k}\cdot\mathbf{x}}$ evaluated at the non-uniform sample points $\mathbf{x}$, i.e. $A = [a_{jl}] = [e^{2i\pi \mathbf{k}_l \cdot \mathbf{x}_j}]_{\mathbf{k}_l \in K, \mathbf{x}_j \in X}$, $\mathbf{c}$ is a vector containing the unknown Fourier coefficients: $\mathbf{c} = [c_{\mathbf{k}_l}]_{\mathbf{k}_l \in K}$ and $\mathbf{b}$ is a vector containing the values of the function u evaluated at the non-uniform sample points $\mathbf{x}$: $\mathbf{b} = [u(\mathbf{x}_j)]_{\mathbf{x}_j \in X}$.

Solving this linear least squares system allows us to estimate the Fourier coefficients efficiently, enabling the interpolation $\tilde{u}$ of the function u onto a uniform grid of high resolution in Fourier space. We achieve this by evaluating the Fourier series approximation given by Eq. (1) at new high-resolution grid location points $\mathbf{x} \in X_{grid}$ as follows:

Assuming $\mathbf{c}^*$ is the solution to the LLS problem given in Eq. (3), we calculate the interpolated values at $\mathbf{x} \in X_{grid}$ as

$$\tilde{u}(\mathbf{x}) = \Re\left(\sum_{\mathbf{k} \in K} c_{\mathbf{k}}^* e^{2i\pi \mathbf{k}\cdot\mathbf{x}}\right) \tag{4}$$

where $\Re(z) = a$ is the real part of a complex number $z = a + bi \in \mathbb{C}$.

This procedure originally only applies to the interpolation of periodic functions. In Appendix A we show how our approach can be extended to non-periodic functions.

Overall we call this interpolation procedure the FI (Fourier Interpolation) Layer which in summary takes as input a set of arbitrary locations $X \subset \mathbb{R}^M$ together with the corresponding values $u(X)$ of a function u and a set of target locations $X_{\text{target}} \subset \mathbb{R}^M$, transforms them into the Fourier coefficients $[c_{\mathbf{k}}]_{\mathbf{k} \in K}$. These are then used to evaluate the Fourier approximation at the set of target locations X_{target}. Overall this can be summarized as:

$$\text{FI}(X, u(X), X_{\text{target}}) \xrightarrow{\min_{\mathbf{c} \in \mathbb{R}^{|K|}} ||\mathbf{Ac}-\mathbf{b}||^2} \mathbf{c}^* \xrightarrow{\Re(\sum_{\mathbf{k} \in K} c_{\mathbf{k}}^* e^{2i\pi \mathbf{k}\cdot\mathbf{x}})} \tilde{u}(X_{\text{target}}) \tag{5}$$

3.2   NeuralPDE

NeuralPDE were originally proposed by Dulny et al. [16] for modeling dynamical systems, particularly those governed by partial differential equations (PDEs). The core idea lies in representing the underlying PDE using the Method of Lines, which discretizes the spatial dimensions and transforms the PDE into a system of ordinary differential equations (ODEs). This discretization allows spatial derivatives to be represented as convolutional operations, making CNNs a natural choice for parametrization, which are then trained using a differentiable ODE solver [14].

Specifically, given measurements at grid locations $\mathbf{U}(t) = [U_{ij}(t)]$ of an underlying physical system u which is assumed to be governed by some partial differential equation:

$$\frac{\partial u}{\partial t} = F(u, \nabla u, \nabla^2 u, \dots) \tag{6}$$

the task is to learn the function F from data and use it to forecast the state of the system at future timepoints $t + 1, \ldots, t + T$.

The method of lines treats the system $\mathbf{U}$ as a collection of ordinary differential equations (ODEs), with the value of the system at each of the grid locations as one variable in the system. The governing Eq. (6) then becomes an ODE

$$\frac{\mathrm{d}\mathbf{U}}{\mathrm{d}t} = \tilde{F}(\mathbf{U}) \tag{7}$$

with all partial derivatives $\nabla u, \nabla^2 u, \ldots$ replaced by finite difference approximations, at grid locations, e.g. $\frac{\partial U_{ij}}{\partial x} \sim U_{ij+1} - U_{ij-1}$.

The function $\tilde{F}$ can now be parametrized using a convolutional network θ, as every finite difference approximation can be learned using a convolution operator [16], e.g. $\frac{\partial U_{ij}}{\partial x} \sim U_{ij+1} - U_{ij-1}$ corresponds to the convolution $\begin{bmatrix} 0 & 0 & 0 \\ -1 & 0 & 1 \\ 0 & 0 & 0 \end{bmatrix}$.

The NeuralPDE model N_θ is trained using a differentiable ODE solver [14] to forecast the physical system u.

3.3 GrINd

The overall architecture of our model is displayed in Fig. 1. We combine two Fourier Interpolation Layers with a spatiotemporal prediction model on gridded data to form the **Grid Interpolation Network for Scattered Observations** (GrINd). We select NeuralPDE [16] as the spatiotemporal prediction model, as it showed best performance on the DynaBench Benchmark [17].

We first use a Fourier Interpolation Layer FI_1 to transform the scattered, low-resolution observations U_t^{low} at locations X_{low} at time t into a high-resolution grid interpolation $\hat{U}_t^{\mathrm{high}}$ at grid locations X_{high}. In the high-resolution space we use the NeuralPDE model N_θ to forecast the state of the system at time $t + 1$ as $\hat{U}_{t+1}^{\mathrm{high}}$. Finally we transform the high resolution predictions back into the original space using another Fourier Interpolation Layer FI_1 to obtain the final predictions U_{t+1}^{low}. Overall this can be written as:

$$\mathrm{GrINd}(U_t^{\mathrm{low}}) = \mathrm{FI}_2(X_{\mathrm{high}}, \mathrm{N}_\theta(\mathrm{FI}_1(X_{\mathrm{low}}, U_t^{\mathrm{low}}, X_{\mathrm{high}})), X_{\mathrm{low}}) \tag{8}$$

4 Experiments

In this section we describe the experimental setup used for evaluating the performance of our proposed model.

4.1 Data

For our experiments we use the DynaBench [17] dataset proposed by Dulny et al., which serves as a standardized benchmark for evaluating machine learning models on dynamical systems with sparse observations. The dataset consists of six physical systems described by partial differential equations: advection, Burgers', Gas Dynamics, Kuramoto-Sivashinsky, Reaction-Diffusion and the Wave equation. The equations have been simulated on a high-resolution grid with periodic boundary conditions using the finite differences method [30] and recorded at spatially scattered locations. This mimics a real-world setting where measurement stations are typically also sparse and not located on a grid.

For our experiment we use the same version of the dataset on which the original benchmark experiments were performed [17]. It contains 7000 simulations of each physical system with varying initial conditions (splitted into 5000 train, 1000 validation and 1000 test simulations), each recorded over 201 time steps at 900 measurement locations.

For evaluating the performance of our proposed Fourier Interpolation layer, we use the raw high-resolution version of the dataset as ground truth for comparison with the output of the interpolation layer. Similar to Dulny et al., we also compare our results to models trained and evaluated on a version of the dataset containing observations on a 30×30 grid.

4.2   Baseline Models

We compare the performance of our proposed approach to the original models evaluated in the DynaBench benchmark [17]. These include four graph- and point-cloud based models: Point Transformer [50], Point GNN [47], Graph Kernel Network [3] and Graph PDE [20]. In our results we omit three models included in the original benchmark as they are strictly outperformed by all other models.

The baselines also include two models designed to work on grid data: NeuralPDE [16] and a Residual Network [19]. We omit the CNN baseline included in the original benchmark as it showed subpar performance.

Additionally we include the *persistence* baseline from the original benchmark which corresponds to predicting no change of the system at all.

4.3   Model Configuration

We use $N_{freq} = 18$ Fourier Frequencies in each Fourier Interpolation Layer, as this values shows the lowest interpolation error on most physical systems from the DynaBench dataset. The target interpolation points are distributed along a 64×64 grid, which is the original resolution of the simulations in the DynaBench dataset.

The NeuralPDE model architecture is taken from [17].

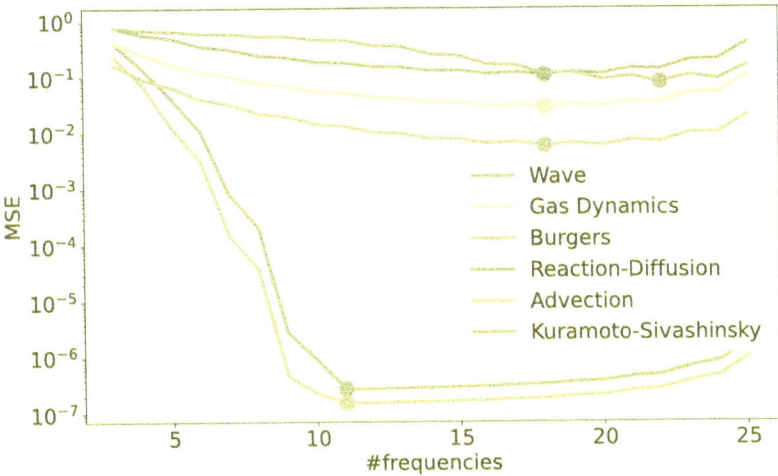

Fig. 3. Results of the interpolation experiment. The plot shows the Mean Squared Error between the interpolation given by our Fourier Interpolation Layer and the full resolution dataset as a function of the number of fourier frequencies used. The lowest interpolation error for each dataset has been marked with a dot.

4.4 Training

We replicate the original training setting from the DynaBench benchmark as closely as possible. The models are trained to predict the next simulation step and evaluated in a closed-loop rollout on 16 simulation steps. We train the models with the Mean Squared Error loss using the Adam Optimizer [22] with a learning rate of 0.0001.

All models are trained for a maximum of 100 000 optimization steps while monitoring the 16 step rollout MSE on a separate validation dataset every 1000 steps with early stopping if the metric did not improve after 5 consecutive checks. Afterwards the best checkpoint with respect to the validation metric is used for testing.

We implemented our models using PyTorch [37] and Pytorch Lightning [18]. The source code of our models and the performed experiments is available under https://professor-x.de/grind.

5 Results and Discussion

In this section we present the experiments we performed and discuss the results.

5.1 Interpolation Accuracy

Firstly, to showcase the feasibility of our proposed approach, we evaluate the performance of our proposed Fourier Interpolation Layer on the task of directly interpolating a set of scattered observations onto a high resolution grid. To this end, we use the DynaBench dataset with 900 measurement locations as input to a single Fourier Interpolation Layer with interpolation target points distributed on a 64×64 grid over the simulation domain. We compare the output of the Fourier Interpolation Layer to the *full* resolution version of the DynaBench dataset which was used to sample the scattered observations. We quantify the interpolation error using the *Mean Squared Error* (MSE) metric. The experiment is repeated for different selection of frequencies N_{freq}, varying between 3 and 25 (cf. Sect. 3.1).

Figure 3 illustrates the outcomes of the interpolation experiment conducted across six distinct physical systems within the DynaBench dataset. Across all systems, a consistent trend emerges, characterized by a distinctive "U" shape in the performance curves. This trend is intuitively comprehensible: when a limited number of Fourier frequencies are chosen, the Fourier approximation fails to capture the intricate nature of the function adequately. Conversely, employing a high number of Fourier frequencies leads to overfitting of the Fourier Interpolation Layer, resulting in elevated reconstruction errors.

Furthermore, it becomes apparent that the optimal number of frequencies varies depending on the complexity of the system, suggesting that individual selection for each dataset is ideal. However, the experiments also show a relative flat minimum ranging from 10 to 20 frequencies per dimension, suggesting that our model is relatively robust with respect to this hyperparameter. Our results for the Dynabench datasets with a lower number of observations show similar results (cf. Appendix B). Therefore, we suggest setting the total number of Fourier frequencies in the range between $\frac{1}{3}$ and $\frac{2}{3}$ of the number of observations.

For our experiments on the DynaBench datasets we use 18 frequencies per dimension which achieves the empirical minimal reconstruction error.

5.2 DynaBench

In this section, we present the results of our proposed GrINd model on the Dyna-Bench dataset, comparing its performance against the baseline models described in Sect. 4.2.

Figure 4 shows the rollout results of our model over 16 steps compared to the baselines from the DynaBench dataset. Overall, with the exception of the *Reaction-Diffusion* and *Kuramoto-Sivashinsky* equations, our model outperforms all models that work on scattered data (PointGNN, Point Transformer, KernelNN and GraphPDE) by a large margin. This is especially true over longer prediction horizons.

In case of the *Advection*, *Gas Dynamics* and *Wave* equation, our model performs on par with grid models, which get access to a grid structured array of observation locations. While the grid models still get access to the same amount of measurement location (30×30 grid vs 900 scattered observations), this additional structure has been shown to be beneficial for predicting the evolution of a physical system [17].

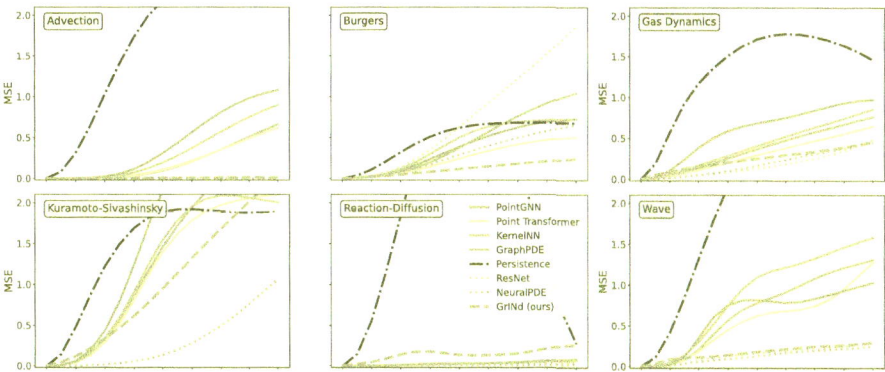

Fig. 4. Rollout results on the DynaBench dataset for 16 steps. The graph and point cloud baselines are plotted with solid lines, the grid baselines with dotted lines, the persistence baseline with a dash-dotted line and our model with a dashed line. Results of our model have been averaged over 10 runs. Baseline results taken from [17]. For better readability MSE scores over 2.0 are not displayed.

Finally, in case of the *Burgers'* equation our model even outperforms the best grid model from the DynaBench benchmark. This is possible because our model interpolates the system onto a higher-resolution grid than was used to train the baseline models. We hypothesize that in this case the performance gain from using a higher grid outweighs the performance loss from the interpolation error (cf. Sect. 5.1).

Table 1. MSE after 1 prediction step. The best performing model for each equation has been underlined. Additionally, the best non-grid model has been underwaved. A = Advection, B = Burgers', GD = Gas Dynamics, KS = Kuramoto-Sivashinsky, RD = Reaction-Diffusion, W = Wave. Results of our model have been averaged over 10 runs. Baseline results taken from [17].

model	A	B	GD	KS	RD	W
GrINd(ours)	$1.92 \cdot 10^{-4}$	$6.56 \cdot 10^{-3}$	$4.93 \cdot 10^{-2}$	$4.54 \cdot 10^{-2}$	$2.33 \cdot 10^{-2}$	$3.61 \cdot 10^{-2}$
GraphPDE	$1.37 \cdot 10^{-4}$	$1.07 \cdot 10^{-2}$	$1.95 \cdot 10^{-2}$	$7.20 \cdot 10^{-3}$	$1.42 \cdot 10^{-4}$	$2.07 \cdot 10^{-3}$
KernelNN	$6.31 \cdot 10^{-5}$	$1.06 \cdot 10^{-2}$	$1.34 \cdot 10^{-2}$	$6.69 \cdot 10^{-3}$	$1.87 \cdot 10^{-4}$	$5.43 \cdot 10^{-3}$
Point TF	$4.42 \cdot 10^{-5}$	$1.03 \cdot 10^{-2}$	$7.25 \cdot 10^{-3}$	$4.90 \cdot 10^{-3}$	$1.41 \cdot 10^{-4}$	$2.38 \cdot 10^{-3}$
PointGNN	$2.82 \cdot 10^{-5}$	$8.83 \cdot 10^{-3}$	$9.02 \cdot 10^{-3}$	$6.73 \cdot 10^{-3}$	$1.36 \cdot 10^{-4}$	$1.39 \cdot 10^{-3}$
NeuralPDE	$8.24 \cdot 10^{-7}$	$1.12 \cdot 10^{-2}$	$3.73 \cdot 10^{-3}$	$5.37 \cdot 10^{-4}$	$3.03 \cdot 10^{-4}$	$1.70 \cdot 10^{-3}$
ResNet	$2.16 \cdot 10^{-6}$	$1.48 \cdot 10^{-2}$	$3.21 \cdot 10^{-3}$	$4.90 \cdot 10^{-4}$	$1.57 \cdot 10^{-4}$	$1.46 \cdot 10^{-3}$
Persistence	$8.12 \cdot 10^{-2}$	$3.68 \cdot 10^{-2}$	$1.87 \cdot 10^{-1}$	$1.42 \cdot 10^{-1}$	$1.47 \cdot 10^{-1}$	$1.14 \cdot 10^{-1}$

Table 2. MSE after 16 prediction steps. The best perfoming model for each equation has been underlined. Additionally, the best non-grid model has been underwaved. A = Advection, B = Burgers', GD = Gas Dynamics, KS = Kuramoto-Sivashinsky, RD = Reaction-Diffusion, W = Wave. Results of our model have been averaged over 10 runs. Baseline results taken from [17].

model	A	B	GD	KS	RD	W
GrIND (ours)	$1.23 \cdot 10^{-1}$	$2.38 \cdot 10^{-1}$	$4.51 \cdot 10^{-1}$	$2.33 \cdot 10^{0}$	$2.53 \cdot 10^{-1}$	$2.90 \cdot 10^{-1}$
GraphPDE	$1.08 \cdot 10^{0}$	$7.30 \cdot 10^{-1}$	$9.69 \cdot 10^{-1}$	$2.10 \cdot 10^{0}$	$8.00 \cdot 10^{-2}$	$1.03 \cdot 10^{0}$
KernelNN	$8.97 \cdot 10^{-1}$	$7.27 \cdot 10^{-1}$	$8.54 \cdot 10^{-1}$	$2.00 \cdot 10^{0}$	$6.35 \cdot 10^{-2}$	$1.58 \cdot 10^{0}$
Point TF	$6.17 \cdot 10^{-1}$	$5.04 \cdot 10^{-1}$	$6.43 \cdot 10^{-1}$	$2.10 \cdot 10^{0}$	$5.64 \cdot 10^{-2}$	$1.27 \cdot 10^{0}$
PointGNN	$6.61 \cdot 10^{-1}$	$1.04 \cdot 10^{0}$	$7.59 \cdot 10^{-1}$	$2.82 \cdot 10^{0}$	$5.82 \cdot 10^{-2}$	$1.31 \cdot 10^{0}$
NeuralPDE	$2.70 \cdot 10^{-4}$	$6.60 \cdot 10^{-1}$	$4.43 \cdot 10^{-1}$	$1.06 \cdot 10^{0}$	$2.24 \cdot 10^{-2}$	$2.48 \cdot 10^{-1}$
ResNet	$8.65 \cdot 10^{-5}$	$1.86 \cdot 10^{0}$	$4.80 \cdot 10^{-1}$	$1.07 \cdot 10^{0}$	$7.05 \cdot 10^{-3}$	$2.99 \cdot 10^{-1}$
Persistence	$2.39 \cdot 10^{0}$	$6.79 \cdot 10^{-1}$	$1.46 \cdot 10^{0}$	$1.90 \cdot 10^{0}$	$2.76 \cdot 10^{-1}$	$2.61 \cdot 10^{0}$

Single Prediction Steps. We additionally numerically evaluate the performance of our model both on single and multi-step predictions. Table 1 displays the Mean Squared Error (MSE) after a single prediction step for each physical system in the DynaBench dataset. While in this case our GrINd model still achieves competitive performance, it is however outperformed by other grid and non-grid based models with the exception of the *Burgers'* equation. We observe that in this case the numerical interpolation error, as examined in Sect. 5.1 is larger than the numerical forecasting error for the baseline models. We hypothesize that the gain from mapping the observations onto a high-resolution grid is not sufficient to outweigh the interpolation error. This is also supported by the conclusion from the original DynaBench benchmark, which suggests that for shorter prediction horizons grid models perform on par with non-grid models.

Multiple Prediction Steps. We further assess the predictive capabilities of GrINd by evaluating its performance after 16 prediction steps. As shown in Table 2, for longer prediction horizons GrINd consistently outperforms non-grid based models in terms of MSE. For the *Burgers'* equation, our model even outperforms the best grid baseline. As opposed to single step predictions, longer prediction horizons require more numerical stability which is not guaranteed by models working on unstructured data. In this case the benefit of using a grid representation outweighs the error associated with using the Fourier Interpolation Layer. The results for 16 prediction steps highlight our model's robustness in capturing the dynamics of physical systems over longer forecasting horizons.

Overall, the results demonstrate that GrINd is a promising approach for forecasting physical systems from sparse, scattered observational data. This is especially the case for longer prediction horizons, where our model shows more numerical stability than other models able to work with scattered data, similar to grid models like NeuralPDE [16] or ResNet [19].

5.3   Limitations

One of the limitations of our method is its computational cost. For a given Fourier interpolation layer with K^2 total Fourier coefficients, N_1 input points and N_2 target points, the complexity of the interpolation algorithm consists of $O(K^4 N_1)$ operations to calculate the Fourier coefficients using a least squares algorithm [31] and $O(N_2 K^2)$ operations to calculate the interpolated values in the target space. Considering the optimal number of frequencies is linearly dependent on the number of input points (cf. Sect. 5.1) the algorithm has a cubic complexity in terms of number of input points.

However, our model is designed for applications in domains with sparse observations, where the number of input points is low. In our experiments we found that in practice the Fourier Interpolation layer does not take up considerable compute time compared to the actual prediction model.

6   Conclusion and Future Work

In this paper, we introduced GrINd (Grid Interpolation Network for Scattered Observations), a novel approach designed to address the challenge of forecasting physical systems from sparse, scattered observational data. GrINd leverages the high performance of grid-based deep learning approaches for the task of forecasting physical systems from sparse and scattered data. It combines a novel Fourier Interpolation Layer with a NeuralPDE model to predict the evolution of spatiotemporal physical systems.

The empirical evaluations on the DynaBench benchmark dataset reveal that GrINd outperforms existing models, particularly for longer prediction horizons, showcasing the potential of using interpolation-based methods for modeling physical systems, as grid-based models show more numerical stability. Our work also opens up several avenues for future research and development that we summarize in the following.

In future work we would like to investigate the performance of our model on real-world data. One example application includes the weather prediction task, where predominantly the ERA5 reanalysis dataset [12] is used for training deep learning models. While the dataset combines observations with grid-based numerical simulations, our model allows for incorporating live and historical observations directly, by interpolating the scattered measuring stations onto a grid.

While we use the NeuralPDE model in our architecture, as it showed best perfomance in the DynaBench benchmark [17], in principle our approach allows for any grid-based model to be used in its placer. Future investigations could explore the integration of other such models to further improve predictive performance.

Another promising direction for future work is using a permutation-invariant neural network such as Deep Sets [49] or Set Transformers [28] to predict the Fourier Coefficients. This could improve the interpolation layer by enabling pre-training on a wide range of data. Additionally, learning the interpolation layer parameters alongside the NeuralPDE model could lead to better alignment between the interpolation and prediction processes, enhancing overall forecasting accuracy.

Instead of interpolating the data onto a high-resolution grid before learning the dynamics, an alternative approach could be to directly learn the dynamics in Fourier space using neuralODEs [14]. By operating directly in the frequency domain, the model may capture underlying patterns and dependencies more effectively, potentially improving forecasting accuracy.

In summary, GrINd represents a significant contribution in forecasting physical systems from sparse observational data, by leveraging the already high performance of grid models. However, there are still ample opportunities for further research and development to enhance the approach and address additional challenges in predictive modeling for spatiotemporal physical systems.

Disclosure of Interests. The project underlying this publication was funded by the German Federal Ministry of Education and Research under the grant number 16DKWN0099B (MAGNET4Cardiac7T). The responsibility for the content of this publication lies with the authors.

References

1. Alkin, B., Furst, A., Schmid, S., Gruber, L., Holzleitner, M., Brandstetter, J.: Universal physics transformers (2024)
2. Anandkumar, A.: Neural operators for solving pdes and inverse design. In: Proceedings of the 2023 International Symposium on Physical Design, ISPD 2023, p. 195. Association for Computing Machinery, New York, NY, USA, March 2023. https://doi.org/10.1145/3569052.3578911
3. Anandkumar, A., et al.: Neural operator: graph kernel network for partial differential equations. In: ICLR 2020 Workshop on Integration of Deep Neural Models and Differential Equations, June 2019
4. Ayed, I., de Bézenac, E., Pajot, A., Brajard, J., Gallinari, P.: Learning dynamical systems from partial observations. CoRR abs/1902.11136 (2019). https://doi.org/10.48550/arXiv.1902.11136
5. Barnett, A.H.: Aliasing error of the exp (β1-z2) kernel in the nonuniform fast Fourier transform. Appl. Comput. Harmon. Anal. **51**, 1–16 (2021). https://doi.org/10.1016/j.acha.2020.10.002
6. Barnett, A.H., Magland, J., af Klinteberg, L.: A parallel nonuniform fast Fourier transform library based on an "exponential of semicircle" kernel. SIAM J. Sci. Comput. **41**(5), C479–C504 (2019). https://doi.org/10.1137/18m120885x
7. Bauer, P., Thorpe, A., Brunet, G.: The quiet revolution of numerical weather prediction. Nature **525**(7567), 47–55 (2015). https://doi.org/10.1038/nature14956
8. Bi, G., Zeng, Y.: Transforms and fast algorithms for signal analysis and representations. Birkhäuser Boston (2004). https://doi.org/10.1007/978-0-8176-8220-0
9. Bi, K., Xie, L., Zhang, H., Chen, X., Gu, X., Tian, Q.: Accurate medium-range global weather forecasting with 3D neural networks. Nature **619**(7970), 533–538 (2023). https://doi.org/10.1038/s41586-023-06185-3
10. Boussard, J., et al.: Towards causal representations of climate model data (2023)

11. Brunton, S.L., Proctor, J.L., Kutz, J.N.: Discovering governing equations from data by sparse identification of nonlinear dynamical systems. Proc. Natl. Acad. Sci. **113**(15), 3932–3937 (2016). https://doi.org/10.1073/pnas.1517384113

12. C3S: Era5 hourly data on single levels from 1940 to present (2018). https://doi.org/10.24381/CDS.ADBB2D47. https://cds.climate.copernicus.eu/doi/10.24381/cds.adbb2d47

13. Cannon, R.H.: Dynamics of Physical Systems. McGraw-Hill (1967). Google-Books-ID: Rix6s2VIlOkC

14. Chen, R.T.Q., Rubanova, Y., Bettencourt, J., Duvenaud, D.K.: Neural ordinary differential equations. In: Advances in Neural Information Processing Systems, vol. 31. Curran Associates, Inc. (2018)

15. Cullen, M.J., Davies, T., Mawson, M.H., James, J.A., Coulter, S.C., Malcolm, A.: An overview of numerical methods for the next generation U.K. NWP and climate model. Atmosphere-Ocean **35**(sup1), 425–444 (1997). https://doi.org/10.1080/07055900.1997.9687359

16. Dulny, A., Hotho, A., Krause, A.: NeuralPDE: modelling dynamical systems from data. In: Bergmann, R., Malburg, L., Rodermund, S.C., Timm, I.J. (eds.) KI 2022. LNCS, pp. 75–89. Springer, Cham (2022). https://doi.org/10.1007/978-3-031-15791-2_8

17. Dulny, A., Hotho, A., Krause, A.: DynaBench: a benchmark dataset for learning dynamical systems from low-resolution data. In: Koutra, D., Plant, C., Gomez Rodriguez, M., Baralis, E., Bonchi, F. (eds.) ECML PKDD 2023. LNCS, vol. 14169, pp. 438–455. Springer, Cham (2023). https://doi.org/10.1007/978-3-031-43412-9_26

18. Falcon, W.: The PyTorch Lightning team: PyTorch Lightning (2019). https://doi.org/10.5281/zenodo.3828935

19. He, K., Zhang, X., Ren, S., Sun, J.: Deep residual learning for image recognition. In: 2016 IEEE Conference on Computer Vision and Pattern Recognition (CVPR), pp. 770–778, June 2016. https://doi.org/10.1109/CVPR.2016.90. ISSN: 1063-6919

20. Iakovlev, V., Heinonen, M., Lähdesmäki, H.: Learning continuous-time PDEs from sparse data with graph neural networks. In: International Conference on Learning Representations (2021)

21. Jiang, L., Wang, L., Chu, X., Xiao, Y., Zhang, H.: PhyGNNet: solving spatiotemporal PDEs with physics-informed graph neural network. In: Proceedings of the 2023 2nd Asia Conference on Algorithms, Computing and Machine Learning, CACML 2023, pp. 143–147. Association for Computing Machinery, New York, NY, USA, May 2023. https://doi.org/10.1145/3590003.3590029

22. Kingma, D., Ba, J.: Adam: a method for stochastic optimization. In: International Conference on Learning Representations (ICLR), San Diega, CA, USA (2015)

23. Kleinstreuer, C.: Modern Fluid Dynamics. FMIA, vol. 87. Springer, Dordrecht (2010). https://doi.org/10.1007/978-1-4020-8670-0

24. Kopriva, D.A.: Implementing Spectral Methods for Partial Differential Equations. Springer, Dordrecht (2009). https://doi.org/10.1007/978-90-481-2261-5

25. Kurth, T., et al.: FourCastNet: accelerating global high-resolution weather forecasting using adaptive Fourier neural operators (2022)

26. Lam, R., et al.: Learning skillful medium-range global weather forecasting. Science (2023). https://doi.org/10.1126/science.adi2336

27. Larsson, S., Thomée, V.: Partial Differential Equations with Numerical Methods. Springer, Heidelberg (2003). https://doi.org/10.1007/978-3-540-88706-5

28. Lee, J., Lee, Y., Kim, J., Kosiorek, A., Choi, S., Teh, Y.W.: Set transformer: a framework for attention-based permutation-invariant neural networks. In: Chaudhuri, K., Salakhutdinov, R. (eds.) Proceedings of the 36th International Conference on Machine Learning. Proceedings of Machine Learning Research, vol. 97, pp. 3744–3753. PMLR, 09–15 June 2019

29. Lessig, C., Luise, I., Schultz, M.: AtmoRep: large scale representation learning for atmospheric data. In: EGU General Assembly Conference Abstracts. EGU General Assembly Conference Abstracts, p. EGU–3117, May 2023. https://doi.org/10.5194/egusphere-egu23-3117

30. LeVeque, R.J.: Finite difference methods for ordinary and partial differential equations. Soc. Ind. Appl. Math. (2007). https://doi.org/10.1137/1.9780898717839

31. Li, L.: A new complexity bound for the least-squares problem. Comput. Math. Appl. **31**(12), 15–16 (1996). https://doi.org/10.1016/0898-1221(96)00072-7. https://www.sciencedirect.com/science/article/pii/0898122196000727

32. Li, Z., Shu, D., Farimani, A.: Scalable transformer for PDE surrogate modeling (2023)

33. Li, Z., et al.: Multipole graph neural operator for parametric partial differential equations. In: Advances in Neural Information Processing Systems, vol. 33, pp. 6755–6766. Curran Associates, Inc. (2020)

34. Li, Z., et al.: Fourier neural operator for parametric partial differential equations. In: International Conference on Learning Representations (2021)

35. Long, Z., Lu, Y., Dong, B.: PDE-Net 2.0: learning PDEs from data with a numeric-symbolic hybrid deep network. J. Comput. Phys. **399**(C) (2019). https://doi.org/10.1016/j.jcp.2019.108925

36. Nathaniel, J., et al.: ChaosBench: a multi-channel, physics-based benchmark for subseasonal-to-seasonal climate prediction (2024)

37. Paszke, A., et al.: Pytorch: an imperative style, high-performance deep learning library. In: Advances in Neural Information Processing Systems, vol. 32, pp. 8024–8035. Curran Associates, Inc. (2019)

38. Potts, D., Schmischke, M.: Approximation of high-dimensional periodic functions with Fourier-based methods. SIAM J. Numer. Anal. **59**(5), 2393–2429 (2021). https://doi.org/10.1137/20M1354921. arXiv:1907.11412 [cs, math]

39. Praditia, T., Karlbauer, M., Otte, S., Oladyshkin, S., Butz, M.V., Nowak, W.: Finite volume neural network: modeling subsurface contaminant transport. CoRR abs/2104.06010 (2021). https://doi.org/10.48550/arXiv.2104.06010

40. Pratt, H., Williams, B., Coenen, F., Zheng, Y.: FCNN: Fourier convolutional neural networks. In: Ceci, M., Hollmén, J., Todorovski, L., Vens, C., Džeroski, S. (eds.) ECML PKDD 2017. LNCS (LNAI), vol. 10534, pp. 786–798. Springer, Cham (2017). https://doi.org/10.1007/978-3-319-71249-9_47

41. Raissi, M., Perdikaris, P., Karniadakis, G.E.: Physics informed deep learning (part i): data-driven solutions of nonlinear partial differential equations (2017). https://doi.org/10.48550/arXiv.1711.10561

42. Rasp, S., et al.: WeatherBench 2: a benchmark for the next generation of data-driven global weather models (2023)

43. Redinbo, G., Manomohan, R.: Fault-tolerant FFT data compression. In: Proceedings. 2000 Pacific Rim International Symposium on Dependable Computing, pp. 110–119 (2000). https://doi.org/10.1109/PRDC.2000.897293

44. Sanchez-Gonzalez, A., Godwin, J., Pfaff, T., Ying, R., Leskovec, J., Battaglia, P.W.: Learning to simulate complex physics with graph networks. In: Proceedings of the 37th International Conference on Machine Learning, ICML 2020, JMLR.org (2020)

45. Schiesser, W.E.: The Numerical Method of Lines: Integration of Partial Differential Equations. Academic Press, San Diego (1991)
46. Serov, V.: Fourier Series, Fourier Transform and Their Applications to Mathematical Physics. Springer, Cham (2017). https://doi.org/10.1007/978-3-319-65262-7
47. Shi, W., Rajkumar, R.: Point-GNN: graph neural network for 3D object detection in a point cloud. In: Proceedings of the IEEE/CVF International Conference on Computer Vision (ICCV), pp. 1711–1719 (2020)
48. Shih, Y.H., Wright, G., Andén, J., Blaschke, J., Barnett, A.H.: cuFINUFFT: a load-balanced GPU library for general-purpose nonuniform FFTs. In: 2021 IEEE International Parallel and Distributed Processing Symposium Workshops (IPDPSW), pp. 688–697 (2021). https://doi.org/10.1109/IPDPSW52791.2021.00105
49. Zaheer, M., Kottur, S., Ravanbakhsh, S., Poczos, B., Salakhutdinov, R.R., Smola, A.J.: Deep sets. In: Advances in Neural Information Processing Systems, vol. 30. Curran Associates, Inc. (2017)
50. Zhao, H., Jiang, L., Jia, J., Torr, P.H., Koltun, V.: Point transformer. In: Proceedings of the IEEE/CVF International Conference on Computer Vision (ICCV), pp. 16259–16268, October 2021

MEGA: Multi-encoder GNN Architecture for Stronger Task Collaboration and Generalization

Faraz Khoshbakhtian[1,2]($\boxtimes$), Gaurav Oberoi[2], Dionne Aleman[1], and Siddhartha Asthana[2]

[1] University of Toronto, Toronto, Canada
`faraz.khoshbakhtian@mail.utoronto.ca, dionne.aleman@utoronto.ca`
[2] Mastercard AI Garage, Gurugram, India
`{faraz.khoshbakhtian,gaurav.oberoi,siddhartha.asthana}@mastercard.com`

Abstract. Self-supervised learning in graphs has emerged as a promising avenue for harnessing unlabeled graph data, leveraging pretext tasks to generate informative node representations. However, the reliance on a single pretext task often constrains generalization across various downstream tasks and datasets. Recent advancements in multi-task learning on graphs aim to tackle this limitation by integrating multiple pretext tasks, framing the problem as a multi-objective optimization to train a shared set of parameters. However, these approaches frequently encounter task interference, where competing tasks degrade overall performance by conflicting with each other due to the limited expressivity of the model. In this work, we introduce MEGA, a novel multi-encoder graph neural network architecture designed to alleviate task interference by providing distinct parameter spaces for the decoupled training of each task. This architecture allows for independent learning from multiple pretext tasks, followed by a simple self-supervised dimensionality reduction technique to combine the insights gleaned. Through extensive experiments, we demonstrate the superiority of our approach, showcasing an average performance improvement of 3.8% across three commonly used downstream tasks (i.e., link prediction, node classification, and partition prediction) and nine benchmark datasets.

Keywords: Network Representation Learning · Self-supervised Learning · Graph Neural Networks · Task Interference

1 Introduction

Graphs play a ubiquitous role in real-world applications, ranging across diverse domains such as social networks, biological networks, and transportation systems [42]. The omnipresence of networks underscores the importance of effectively learning from graph structures to extract meaningful insights. As a result,

---

Supplementary Information The online version contains supplementary material available at https://doi.org/10.1007/978-3-031-70368-3_12.

A. Bifet et al. (Eds.): ECML PKDD 2024, LNAI 14947, pp. 194–208, 2024.
https://doi.org/10.1007/978-3-031-70368-3_12

significant attention has been directed towards the field of graph representation learning, with the objective of uncovering valuable information embedded within graph structures for subsequent analysis and modelling tasks [4,11,44].

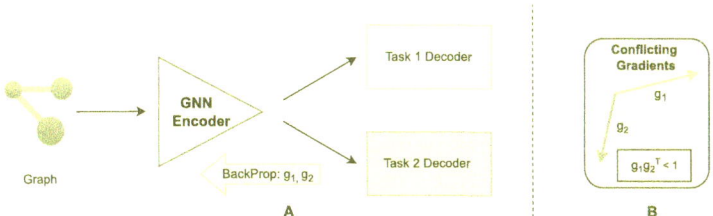

Fig. 1. A) In a typical multi-task self-supervised learning graph framework aimed at learning downstream task generalized representations, gradients from both the losses are backpropagated to the shared GNN encoder. B) These gradients often exhibit conflicting directions, leading to interference in the optimal learning of encoder parameters.

In line with developments in the broader machine learning (ML) literature, self-supervised learning (SSL) has emerged as an active area of study in graph representation learning. The scarcity and cost of calculating high-quality labels in many real-world scenarios have motivated the use of self-supervised pretext tasks to train graph neural network (GNN) models on larger volumes of unlabeled data [23]. The learned representations from these pretext tasks are then utilized for downstream modeling tasks, enhancing performance without the need for large labeled datasets [7,15,19,44,47].

Recent advancements in graph representation learning have highlighted the limitations of single SSL pretext tasks in generalizing across diverse downstream tasks. This has sparked interest in multi-task self-supervised learning (MT-SSL) approaches [15,16]. In MT-SSL, models are trained on multiple pretext tasks, aiming to enhance robustness and generalizability by leveraging insights from diverse tasks [5,41]. Multi-task learning (MTL) has been explored from various design perspectives, including reducing inference computation costs [41] and improving task generalization in self-supervised setups [16]. Self-supervised downstream task generalization is an open and active area of the network representation learning and MT-SSL provides a viable pathway by integrating learning from multiple pretext tasks, thereby enhancing the adaptability and robustness of the generated representations.

Despite the potential benefits of MTL, a critical issue that needs to be addressed is *task interference*, where tasks compete for model expressivity, leading to degraded performance in some or all tasks. Task interference is caused by the conflict in gradients arising when multiple tasks are optimized together via a shared parameter space [38]. Recent methods propose different approaches to parameter sharing and conflict resolution to mitigate conflicting gradients. Some works use hard-parameter sharing schemes, where all tasks share the same encoder parameters [15,16,24] and regulate competition using an optimization algorithm, while others opt-in for soft parameter sharing methods, allocating separate or overlapping parameter spaces for different tasks [6,39,45]. Present works regulate task competition using either multi-objective optimization (MOO) techniques [16] for finding optimal pretext loss weights, or end-to-end training [41].

In the graph learning literature, recent works have adopted the hard-parameter sharing and MOO philosophy for multi-task SSL frameworks [15,16]. This approach entails that the pretext tasks compete for the expressivity of the same parameter space and the competition is regulated via an optimization algorithm. Moreover, while MOO-based MTL approaches add significant computational overhead, they often do not lead to improved performance [20,48]. Recent works suggest that unitary scalarization (i.e., giving the same weight to all losses) of losses can achieve similar or improved performance compared to MOO methods without the added complexity [20,34]. Figure 1 demonstrates a typical hard-parameter sharing setup in graph SSL with two pretext tasks sharing a GNN encoder, where task interference arises from conflicting gradients in the encoder. Another challenge in MT-SSL for graphs is the limited representative capacity of GNNs, which suffer from issues like oversmoothing, over-squashing, and theoretical expressivity bounds [35,40,49].

In this work, we present **M**ulti-**E**ncoder **G**NN **A**rchitecture (MEGA), a novel MT-SSL graph architecture aimed at overcoming the limitations of existing approaches and enhancing task generalization. MEGA features a simple, yet efficient multi-encoder setup, where each task receives a dedicated parameter space to fully leverage the expressivity of the corresponding GNN. Additionally, we employ a self-supervised dimensionality reduction pipeline to project the task-specific embeddings into a compact representation space. Our modular approach prevents conflicting gradients and eliminates the need for MOO by using separate encoders for individual tasks, fostering task collaboration. MEGA is end-to-end trainable, parallelizable, and scalable, making it suitable for real-world applications. Our contributions include:

- We introduce MEGA, a novel architecture for graph representation learning that leverages a multi-encoder setup. This design enables each pretext task to operate within its own dedicated parameter space, effectively eliminating task interference from its source.
- Our architecture includes a simple dimensionality reduction technique that effectively captures the nuances between task embeddings, fostering collaboration rather than competition among tasks.
- We demonstrate the scalability and parallelizability of our model, showcasing its applicability to large-scale graph representation learning tasks.
- Through extensive experiments, we validate the effectiveness of our approach in improving performance on various graph-based downstream tasks, compared to existing MTL-SSL methods.[1]

---

[1] Our code is available on GitHub at https://github.com/faraz2023/MEGA.

2   Related Works

Graph Neural Networks: GNNs have emerged as powerful tools for learning from graph-structured data, enabling the compression and extraction of meaningful information [4,23,42]. GNNs exploit the graph topology to learn node embeddings that capture the structural and positional information of nodes [11]. Modern GNNs employ the message-passing paradigm, updating each node's embedding by aggregating representations of its neighbors and itself [10,49]. Notable architectures include the graph convolutional network (GCN) [18], GraphSAGE for inductive settings [10], and the graph attention network (GAT) with learnable attention weights [43].

Self-supervised Learning on Graphs: The rise of SSL on graphs has been fueled by the abundance of unlabeled graph data ranging from molecular to interaction networks [23,32,46,47]. SSL enables the extraction of valuable information through pre-trained models on pretext tasks, generating embeddings for various applications such as node classification or link prediction [14,51]. Various pretext tasks like mutual information (MI) maximization [44], feature reconstruction [13], and whitening decorrelation [54] have been developed for self-supervised GNNs.

Each pretext task has its advantages and limitations in SSL for graph tasks. For instance, MI maximization in DGI [44] excels in graph classification but may not do well for node classification tasks [28,30,44]. Similarly, feature reconstruction [13] may struggle when input features are sparse. Therefore, relying solely on a single pretext task may not suffice for downstream task generalization.

Multi-task Learning: In their seminal work, [5] introduced the first MTL architecture with a shared backbone encoder, a technique known as hard parameter sharing. However, this approach often leads to negative transfer, where tasks compete for the limited expressivity of the shared model, resulting in suboptimal performance [21,25].

To address negative transfer, one strategy employs a Pareto optimality framework to balance task trade-offs using MOO techniques in a hard-parameter sharing setup [22,37]. However, recent studies have highlighted the limitations of MOO-based MTL approaches, noting their failure to effectively mitigate task interference despite significant computational overhead [20,48]. Another strategy, soft-parameter sharing, uses separate parameter spaces for each task or groups of tasks, allowing greater flexibility in learning task-specific features and mitigating negative transfer [6,26,39].

Multi-task Self-supervised Learning on Graphs: Integration of MTL and SSL is a well-studied area of research in domains such as natural language processing [33,36] and computer vision [8,27]. These approaches train models over multiple upstream pretext tasks to generate rich representations that can promote strong downstream task generalization [16]. MT-SSL on graphs is an emerging area of research and some recent works have shown the promise of training GNNs on multiple pretext tasks in enhancing downstream task generalization [15,16]. Notably, AutoSSL [15] and ParetoGNN [16] optimize a joint loss function in a hard-parameter sharing setting (i.e., a single shared backbone encoder). The joint loss function consists of a linear combination of task-specific losses and the weights for each loss are calculated by MOO techniques such as multiple gradient descent; this technique is also know as loss scalarization. In this work, we focus on ParetoGNN as our main baseline as AutoSSL is created based on the assumption of graph homophily which does not stand for all networks and degrades generalizability for downstream tasks.

ParetoGNN suffers from two primary limitations. First, MOO loss scalarization approaches, despite their computational overhead, do not consistently outperform simpler scalarization methods in joint loss calculation, and they struggle with task interference due to conflicting gradients [20,38]. We empirically validate these findings by training a single-encoder multi-task GNN using unitary scalarization (i.e., equal weight for all pretext tasks), denoted as `SingleEnc`, and compare its performance with ParetoGNN. Our results align with existing research indicating that MOO struggles to resolve conflicting gradients and consequent task interference [38,39]. Second, the limited expressivity of GNNs makes a single-encoder approach suboptimal for diverse pretext tasks, as problems like oversmoothing [35] and oversquashing [40] can distort messages in deep or wide encoders. To address these issues, MEGA employs a multi-encoder setup, where each pretext task is assigned to a dedicated GNN encoder, minimizing negative transfer and task interference.

3   Methods

3.1   Preliminaries

We first define basic notation and concepts used throughout this work.

Graph Notation: A graph can be represented as $G = (V, E)$, where V is the set of nodes and $E \subseteq V \times V$ is the set of edges. Each node $v \in V$ is associated with a feature vector $\mathbf{x}_v \in \mathbb{R}^D$, where $D = d_0$ is the dimensionality of the node features. The graph structure is represented by an adjacency matrix $\mathbf{A} \in \{0,1\}^{|V| \times |V|}$, where $\mathbf{A}ij = 1$ if there is an edge between nodes i and j, and $\mathbf{A}_{ij} = 0$ otherwise. The input feature matrix for the entire graph is denoted by $\mathbf{X} \in \mathbb{R}^{|V| \times d_0}$, where each row corresponds to the feature vector of a node in the graph.

Message-Passing GNN: We use the classical GCN layer [18] with the following propagation mechanism:

$$h_i^{\ell+1} = \sigma \left(\sum_{j \in N_i} \frac{1}{\sqrt{|N_i| |N_j|}} h_j^\ell W^\ell \right) \tag{1}$$

where $W^{(\ell)} \in \mathbb{R}^{d_\ell \times d_{\ell+1}}$ is a trainable weight matrix and σ is an element-wise non-linear activation function (e.g., ReLU [2]). $h_j^\ell \in \mathbb{R}^{d_\ell}$ denotes the feature of node j input to the $(\ell + 1)$th layer and $h_i^{\ell+1} \in \mathbb{R}^{d_{\ell+1}}$ is the calculated feature vector for node i in layer $(\ell + 1)$; $h_j^0 = \mathbf{x}_j$ and $d_\ell, d_{\ell+1}$ are feature sizes for layers

ℓ, $\ell + 1$, respectively. N_i denotes the collection of neighbors node i including self-connection. The output feature is normalized by $1/\sqrt{|N_i||N_j|}$. A typical GCN layer only aggregates the information from 1-hop neighbors.

3.2 Task Interference Problem in MT-SSL

The objective of MT-SSL is to concurrently learn multiple pretext tasks $\tau_i\big|_{i=1}^k$ such that the advancement in learning one task contributes positively to the learning of others. This is typically achieved through a framework where a shared encoder, characterized by parameters θ_e, is trained alongside task-specific decoders denoted by parameters $\theta_1, \theta_2, ..., \theta_K$. Each pretext task τ_i is associated with a distinct loss function $\mathcal{L}_i(\theta_e, \theta_i)$. The overarching objective of MT-SSL is to determine the optimal parameter set $\theta = \theta_e, \theta_1, \theta_2, ..., \theta_K$ that maximizes performance across all pretext tasks. This optimization problem is reformulated as a MOO task, aimed at simultaneously optimizing the performance of multiple objectives, namely the loss functions associated with each pretext task:

$$\theta^* = \arg\min_{\theta} \sum_{i=1}^{K} w_i \mathcal{L}_i(\theta_e, \theta_i) \tag{2}$$

where w_i denotes parameters either learned by the MOO algorithm or specified manually. In practical scenarios, global optima may not be achievable, and gradients of tasks conflict with each other, leading to tasks competing for model capacity.

Conflicting Gradients. Let g_i and g_j represent the gradients of tasks τ_i and τ_j, respectively, with respect to the shared encoder parameters θ_e. These gradients are then defined as

$$g_i = \nabla_{\theta_e} \mathcal{L}_i(\theta_e, \theta_i) = \frac{\partial \mathcal{L}_i(\theta_e, \theta_i)}{\partial \theta_e} \tag{3}$$

$$g_j = \nabla_{\theta_e} \mathcal{L}_j(\theta_e, \theta_j) = \frac{\partial \mathcal{L}_j(\theta_e, \theta_j)}{\partial \theta_e} \tag{4}$$

During backpropagation, these gradients are utilized to update the parameters of the encoder θ_e in the direction of the negative gradient, as $\theta_e \leftarrow \theta_e - \lambda g$, where λ is the learning rate. The gradients g_i and g_j are considered conflicting when they point in opposite directions, thereby optimizing the parameters of the encoder in contradictory directions. Mathematically, g_i and g_j are conflicting if their dot product is negative, i.e., $g_i g_j^\top < 0$.

Task Interference. When gradients of two tasks conflict, they engender interference and competition for parameter space. Let gradient g_i drive updates to the encoder parameters as $\theta_e' = \theta_e - \lambda g_i$. Consequently, the loss for task τ_j is influenced by gradient g_i, resulting in an updated loss function $\mathcal{L}_j(\theta_e', \theta_j)$. The corresponding change in loss, denoted as $\Delta \mathcal{L}_j$, can be expressed as

$$\Delta \mathcal{L}_j = \mathcal{L}_j(\theta_e', \theta_j) - \mathcal{L}_j(\theta_e, \theta_j) \tag{5}$$

$$= \mathcal{L}_j(\theta_e - \lambda g_i, \theta_j) - \mathcal{L}_j(\theta_e, \theta_j) \tag{6}$$

Using Taylor Series expansion, $\mathcal{L}'_j(\theta'_e, \theta_j)$ can be expanded as

$$\mathcal{L}_j(\theta_e - \lambda g_i, \theta_j) = \mathcal{L}_j(\theta_e, \theta_j) + \frac{\partial \mathcal{L}_j(\theta_e, \theta_j)}{\partial \theta_e} \cdot (-\lambda g_i)^\top + O(\lambda^2) \qquad (7)$$

$$= \mathcal{L}_j(\theta_e, \theta_j) - \lambda g_i g_j^\top + O(\lambda^2) \qquad (8)$$

Therefore, the interference in task τ_j due to gradient g_i is quantified by $\Delta \mathcal{L}_j = -\lambda g_i g_j^\top + O(\lambda^2)$. In the case of conflicting gradients g_i, g_j and a sufficiently small learning rate λ, both losses $\mathcal{L}_i$ and $\mathcal{L}_j$ experience an increase.

Eliminating Task Interference. Recent studies have introduced methods to mitigate task interference in MTL by grouping only correlated tasks together [39] or designating highly conflicting layers as task-specific layers [38]. We propose that the most effective approach to eliminate task interference is to address conflicts in gradients at their origin. This resolution can be achieved by incorporating task-specific encoders, mirroring the structure of task-specific decoders. In our multi-encoder architecture, task interference during the pre-training phase is eliminated entirely, enabling multiple pretext tasks to learn independently and subsequently combine their knowledge in synergy with each other as described in Sect. 3.3.

3.3 MEGA Architecture

Our architecture, MEGA, is designed to enable collaboration among multiple self-supervised tasks while maintaining task-specific expressivity. The MEGA pipeline (Fig. 2) is structured around parallelizable pretext task modules, each comprising three main components: a task-specific graph augmentation module, a dedicated GNN encoder, and a task projection head (decoder).

Each task module is responsible for optimizing its corresponding pretext task. Unlike previous work such as [16] that made pretext tasks compete for model expressivity, in our architecture, each pretext task exploits its own dedicated GNN encoder. Each task module can support the integration of any GNN encoder; in our work, we employ a two-layer GCN [18] for its simplicity and effectiveness. Our architecture is designed to be parallelizable and is end-to-end trainable. Our approach enables parallel processing of the task-specific modules while significantly reducing the computational overhead typically associated with MOO methods [16,48].

After training the multi-encoder architecture on the pretext tasks, we extract the embeddings from each task module and aggregate them using concatenation. This aggregated embedding is then subjected to a self-supervised dimensionality reduction step to obtain the final node representations. We evaluate the performance of various dimensionality reduction techniques, including autoencoders [53], variational autoencoders [29], and principal component analysis (PCA) [1]. Our empirical results demonstrate that PCA consistently outperforms the other methods in preserving the essential structural features captured by task-specific embeddings (see Appx. C.) for ablations on the dimensionality reduction component of MEGA).

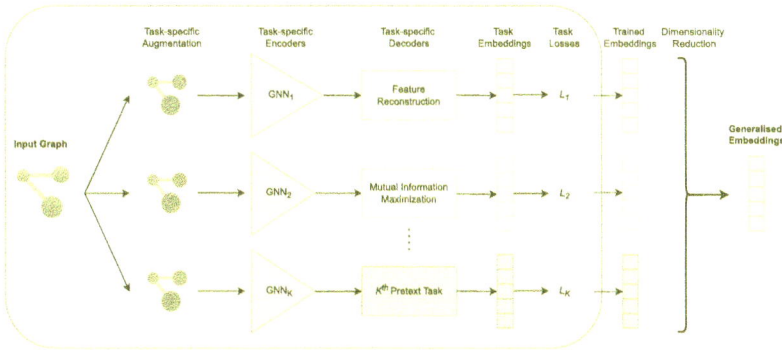

Fig. 2. MEGA Architecture: Multiple pretext task modules, each comprising three main components: a task-specific graph augmentation module, a dedicated GNN encoder, and a task projection head (decoder). This architecture allows each pretext task to exploit its own dedicated parameter space, eliminating task interference.

3.4   Pretext Tasks

MEGA is a flexible framework for MT-SSL on graphs that can integrate various SSL pretext tasks. MEGA employs decoupled training for each task in the MTL setup, allowing for distinct graph sampling strategies, custom encoder modules, and unique task augmentation/output projection heads. We train MEGA on five pretext tasks, categorized into generative reconstruction, whitening decorrelation, and mutual information maximization. These tasks are trained with sub-graph sampling from the original graph as a form of data augmentation and for memory efficiency. Formal definitions for these pretext tasks is available in Appx. E.[2]

Generative Reconstruction: This category includes tasks aimed at reconstructing node features and topological information, which are critical for encoding rich graph information into node representations. We employ two tasks in this category:

- *Feature Reconstruction (FeatRec):* This task involves masking a subset of node features and reconstructing them based on their local sub-graph context. The objective is to ensure that the learned representations capture the essential attributes of nodes, facilitating the recovery of masked features [13].
- *Topology Reconstruction (TopoRec):* In addition to node features, preserving topological information is vital. This task focuses on reconstructing the links between connected nodes, thereby encoding pairwise topological relationships into the representations [55].

---

[2] We thank the authors of [16] for sharing their implementation and codebase publicly.

Mutual Information Maximization: This category includes tasks that maximize the mutual information between different views of the graph or its sub-components, capturing intrinsic patterns within the graph structure:

- *Node-Graph Mutual Information (MI-NG):* This task seeks to minimize the distance between the graph-level representation of an intact sub-graph and its node representations while maximizing the distance between the graph-level representation and corrupted node representations [3,44].
- *Node-Subgraph Mutual Information (MI-NSG):* Here, the goal is to maximize the similarity between representations of two views of a sub-graph associated with the same anchor nodes while minimizing the similarity between representations of sub-graphs associated with different anchor nodes [12].

Whitening Decorrelation (RepDecor): This task independently augments the same sub-graph into two views, and minimizes the distance between corresponding nodes in the two views while enforcing the feature-wise covariance of all nodes to be equal to the identity matrix [16]. This task is closely related and inspired by the success of whitening decorrelation in learning representative embeddings without requiring negative pairs or offline encoders [9,52].

4   Experiments

4.1   Experiment Setting

Datasets. We evaluate MEGA on nine real-world benchmark datasets with diverse network characteristics: PUBMED, AMAZON-PHOTO, WIKI-CS, COAUTHOR-CS, CHAMELEON, ACTOR, AMAZON-COMPUTER, OGBN-ARXIV, and SQUIRREL. These datasets vary in size, ranging from a few thousand to over a hundred thousand nodes, and feature dimensions from tens to thousands of attributes. A detailed description of these datasets, including their statistical properties and sources, is provided in Appx. A.

Baselines. We benchmark MEGA against seven baselines: ParetoGNN [16], a single-encoder GNN trained on five pretext tasks with unitary loss scalarization (SingleEnc), and five models each focused on a single pretext task (see Sect. 3.4, and Appx. E.). This comparison spans MTL and single-task SSL approaches, assessing MEGA's task generalization capability.

SSL Training Hyper-parameters. All models utilize a common GNN encoder architecture comprising two GCN layers [18]. To ensure a fair comparison with ParetoGNN [25], we adopt the same set of hyperparameters as recommended by the authors across all models. The detailed configuration of these hyperparameters is provided in Appx. D.

Downstream Tasks. We assess the performance of all models on three commonly used downstream tasks in graph-based learning: node classification, link prediction, and partition prediction. The evaluation metrics employed for these tasks are accuracy for node and partition prediction, and area under the receiver

Table 1. Performance metrics for label prediction, link prediction, and partition prediction across multiple datasets. MEGA surpasses baseline architectures, achieving the highest average rank across datasets and tasks. Highlight: Best performance. Average and standard deviation computed over 5 seeds.

Method	PUBMED	AM.PHOTO	WIKI.CS	CO.CS	CHAM.	ACTOR	AM.COMP	ARXIV	SQUIRREL	RANK
				AVERAGE PERFORMANCE						
RepDecor	55.41	87.43	65.15	45.99	78.71	34.73	73.18	47.79	70.11	7.4
TopoRec	75.69	86.04	75.35	84.53	77.35	44.35	81.99	69.29	64.00	6.9
MI-NSG	60.99	89.11	79.22	79.99	78.94	46.77	84.08	76.05	68.52	6.1
MI-NG	78.52	91.66	79.50	86.48	81.14	41.04	87.05	80.62	69.42	4.1
FeatRec	77.27	90.21	80.93	85.38	81.68	40.20	86.12	81.38	75.86	3.7
ParetoGNN	82.69	89.87	80.39	85.90	81.55	47.08	87.36	80.57	75.16	3.3
SingleEnc	82.59	90.46	81.04	87.48	80.96	48.24	87.96	80.62	75.09	2.7
MEGA	85.08	94.16	85.48	85.37	86.06	46.80	91.55	84.76	78.50	1.6
				NODE CLASSIFICATION (Accuracy)						
RepDecor	$62.18_{\pm0.3}$	$88.31_{\pm0.7}$	$69.00_{\pm0.7}$	$38.35_{\pm1.3}$	$67.76_{\pm2.3}$	$23.41_{\pm1.6}$	$76.63_{\pm0.4}$	$31.06_{\pm8.3}$	$52.83_{\pm1.4}$	7.6
TopoRec	$83.10_{\pm0.2}$	$89.16_{\pm1.5}$	$74.42_{\pm1.8}$	$85.70_{\pm0.4}$	$62.89_{\pm1.7}$	$27.53_{\pm0.4}$	$83.79_{\pm1.3}$	$64.17_{\pm1.8}$	$45.23_{\pm1.7}$	6.3
MI-NSG	$62.64_{\pm0.8}$	$91.16_{\pm1.7}$	$78.48_{\pm0.8}$	$85.77_{\pm0.8}$	$70.39_{\pm2.0}$	$26.93_{\pm0.8}$	$85.61_{\pm1.6}$	$68.57_{\pm1.8}$	$53.87_{\pm1.7}$	5.5
MI-NG	$84.95_{\pm0.3}$	$94.22_{\pm0.9}$	$82.32_{\pm1.5}$	$91.24_{\pm0.4}$	$68.16_{\pm1.9}$	$25.95_{\pm0.8}$	$87.89_{\pm0.8}$	$72.12_{\pm0.6}$	$50.53_{\pm1.8}$	4.4
FeatRec	$81.16_{\pm0.5}$	$93.69_{\pm1.0}$	$83.34_{\pm0.5}$	$92.32_{\pm0.4}$	$70.48_{\pm2.0}$	$24.92_{\pm0.8}$	$87.88_{\pm1.2}$	$72.46_{\pm0.5}$	$59.31_{\pm1.7}$	4.0
ParetoGNN	$85.61_{\pm0.3}$	$94.27_{\pm0.7}$	$82.25_{\pm0.8}$	$92.47_{\pm0.1}$	$71.67_{\pm1.7}$	$25.43_{\pm1.0}$	$87.98_{\pm1.4}$	$72.36_{\pm0.3}$	$58.66_{\pm1.9}$	2.9
SingleEnc	$85.81_{\pm0.2}$	$94.18_{\pm0.5}$	$81.25_{\pm1.3}$	$92.26_{\pm0.2}$	$70.53_{\pm2.8}$	$25.83_{\pm1.1}$	$89.41_{\pm0.3}$	$72.16_{\pm0.3}$	$58.96_{\pm1.5}$	3.3
MEGA	$85.12_{\pm0.1}$	$94.09_{\pm0.8}$	$84.44_{\pm0.3}$	$92.47_{\pm0.3}$	$72.85_{\pm0.9}$	$27.50_{\pm1.4}$	$91.00_{\pm0.9}$	$73.24_{\pm0.1}$	$62.25_{\pm1.7}$	1.7
				LINK PREDICTION (AUC-ROC)						
RepDecor	$59.04_{\pm1.1}$	$89.03_{\pm0.1}$	$66.45_{\pm1.8}$	$58.35_{\pm0.3}$	$82.35_{\pm1.6}$	$64.65_{\pm2.0}$	$66.52_{\pm4.4}$	$66.96_{\pm2.8}$	$89.03_{\pm0.1}$	6.7
TopoRec	$81.50_{\pm3.8}$	$80.03_{\pm2.6}$	$84.41_{\pm3.0}$	$85.11_{\pm2.3}$	$86.22_{\pm2.5}$	$65.46_{\pm1.8}$	$79.67_{\pm3.6}$	$70.63_{\pm1.7}$	$89.35_{\pm0.8}$	4.5
MI-NSG	$74.01_{\pm1.7}$	$86.50_{\pm4.7}$	$86.43_{\pm0.9}$	$73.10_{\pm2.5}$	$80.23_{\pm3.0}$	$67.40_{\pm1.2}$	$80.89_{\pm0.6}$	$80.21_{\pm2.5}$	$88.82_{\pm0.4}$	5.4
MI-NG	$76.45_{\pm0.5}$	$88.43_{\pm3.0}$	$82.56_{\pm2.9}$	$79.70_{\pm4.4}$	$86.49_{\pm1.3}$	$60.96_{\pm0.5}$	$86.92_{\pm3.1}$	$87.26_{\pm2.6}$	$91.46_{\pm1.2}$	3.8
FeatRec	$73.99_{\pm2.5}$	$84.68_{\pm2.9}$	$83.84_{\pm1.4}$	$75.48_{\pm1.8}$	$83.34_{\pm2.2}$	$59.73_{\pm1.9}$	$82.81_{\pm3.8}$	$88.93_{\pm3.7}$	$90.41_{\pm0.1}$	5.1
ParetoGNN	$78.89_{\pm1.8}$	$83.16_{\pm4.1}$	$84.94_{\pm1.2}$	$76.36_{\pm2.2}$	$83.82_{\pm1.8}$	$63.49_{\pm0.7}$	$84.96_{\pm3.4}$	$87.25_{\pm5.8}$	$89.87_{\pm0.7}$	4.6
SingleEnc	$79.57_{\pm2.4}$	$84.98_{\pm1.9}$	$87.51_{\pm3.7}$	$81.38_{\pm2.2}$	$83.18_{\pm2.1}$	$65.46_{\pm2.0}$	$85.75_{\pm2.3}$	$87.70_{\pm3.8}$	$89.32_{\pm1.1}$	3.6
MEGA	$88.74_{\pm0.3}$	$95.17_{\pm0.1}$	$95.15_{\pm0.1}$	$73.47_{\pm3.2}$	$94.68_{\pm0.1}$	$58.17_{\pm1.7}$	$93.99_{\pm0.1}$	$96.20_{\pm0.1}$	$94.11_{\pm0.1}$	2.3
				PARTITION PREDICTION (Accuracy)						
RepDecor	$45.01_{\pm0.3}$	$84.93_{\pm0.9}$	$60.01_{\pm0.8}$	$41.28_{\pm0.6}$	$86.01_{\pm1.0}$	$16.12_{\pm0.8}$	$76.39_{\pm0.9}$	$45.35_{\pm3.8}$	$68.47_{\pm1.6}$	7.6
TopoRec	$62.46_{\pm0.4}$	$88.92_{\pm1.2}$	$67.22_{\pm1.5}$	$82.79_{\pm2.1}$	$82.94_{\pm1.5}$	$40.05_{\pm1.1}$	$82.50_{\pm0.8}$	$73.06_{\pm0.5}$	$57.43_{\pm1.0}$	6.8
MI-NSG	$46.31_{\pm0.9}$	$89.66_{\pm0.6}$	$72.74_{\pm1.1}$	$81.12_{\pm1.1}$	$86.18_{\pm2.2}$	$45.99_{\pm0.7}$	$85.75_{\pm1.0}$	$79.37_{\pm0.5}$	$62.86_{\pm1.3}$	6.1
MI-NG	$74.14_{\pm0.4}$	$92.31_{\pm0.8}$	$73.62_{\pm0.4}$	$88.51_{\pm0.6}$	$88.77_{\pm0.9}$	$36.21_{\pm0.6}$	$86.34_{\pm1.5}$	$82.48_{\pm0.2}$	$66.28_{\pm2.0}$	4.5
FeatRec	$76.66_{\pm0.5}$	$92.26_{\pm0.7}$	$75.60_{\pm1.0}$	$88.36_{\pm0.4}$	$91.23_{\pm1.3}$	$35.93_{\pm0.8}$	$87.66_{\pm0.7}$	$82.75_{\pm0.3}$	$77.87_{\pm1.0}$	3.3
ParetoGNN	$83.55_{\pm0.4}$	$92.18_{\pm0.5}$	$73.97_{\pm1.1}$	$88.88_{\pm0.8}$	$89.17_{\pm1.3}$	$52.30_{\pm0.9}$	$89.15_{\pm0.6}$	$82.09_{\pm0.4}$	$76.96_{\pm0.8}$	3.1
SingleEnc	$82.40_{\pm0.3}$	$92.21_{\pm0.6}$	$74.35_{\pm1.3}$	$88.81_{\pm0.5}$	$89.17_{\pm1.2}$	$53.45_{\pm1.8}$	$88.72_{\pm0.6}$	$82.01_{\pm0.4}$	$76.98_{\pm1.4}$	3.1
MEGA	$81.39_{\pm0.5}$	$93.23_{\pm0.5}$	$76.86_{\pm0.8}$	$90.18_{\pm0.6}$	$90.66_{\pm1.2}$	$54.72_{\pm1.9}$	$89.65_{\pm0.9}$	$84.83_{\pm0.2}$	$79.15_{\pm1.6}$	1.3

operating characteristic curve (AUC-ROC) for link prediction. These tasks are widely adopted in the literature on SSL for graphs [16,18,44,55].

Downstream Evaluation. For all models and downstream tasks, we adhere to the standard procedure of freezing the encoder weights after SSL training for inference [44]. We employ early stopping [50] during the SSL phase to preserve the optimal weights for each model. For the evaluation of each model on each downstream task, we train a simple fully-connected multi-layer perceptron (MLP) [31] with a single hidden layer and *ReLU* activation function. The MLP architecture is chosen to exploit nonlinear relationships in the latent space. For each dataset, we use a 20%/80% split for testing and training, respectively. We omit the validation set to ensure a fair comparison, as all downstream models follow the same simple MLP architecture. For partition prediction, we employ the METIS partitioning algorithm [17] with 10 partitions. For label prediction, we train separate models with testing edges removed to prevent label leakage. We conduct 10 experiments with different seeds for each downstream task, SSL method, and dataset. Overview of the hardware and software used for these experiments is available in Appx. B.

Table 2. Performance of MEGA and `ParetoGNN` approaches on the Chameleon dataset for node classification. Increasing the model capacity of `ParetoGNN` by increasing the number of layers and hidden dimension doesn't yield performance improvements. Highlight: Best performance. Average and standard deviation computed over 5 seeds.

Approach	Layers	Hidden Dim	Parameters	Accuracy
MEGA	2	[512, 256]	7,742,496	$72.85_{\pm 0.9}$
ParetoGNN	2	[512, 256]	2,983,243	$71.67_{\pm 1.7}$
ParetoGNN	5	[512, 512, 512, 512, 256]	3,774,286	$67.50_{\pm 2.1}$
ParetoGNN	6	[512, 512, 512, 512, 512, 256]	4,037,967	$67.10_{\pm 1.5}$
ParetoGNN	7	[512, 512, 512, 512, 512, 512, 256]	4,301,648	$65.80_{\pm 1.7}$
ParetoGNN	2	[1024, 256]	4,306,251	$71.25_{\pm 0.3}$
ParetoGNN	3	[1024, 1024, 256]	5,357,900	$70.48_{\pm 2.2}$
ParetoGNN	4	[1024, 1024, 1024, 256]	6,409,549	$68.16_{\pm 1.6}$
ParetoGNN	5	[1024, 1024, 1024, 1024, 256]	7,461,198	$66.40_{\pm 1.2}$
ParetoGNN	2	[2024, 256]	6,890,251	$71.21_{\pm 0.5}$
ParetoGNN	3	[2024, 2024, 256]	10,992,900	$69.82_{\pm 1.6}$

4.2 Results

Table 1 presents the comparative performance of MEGA and seven baseline models across three downstream tasks and nine datasets.

Our analysis indicates that, on average across datasets, none of the individual pretext tasks outperformed the MTL methods. Specifically, ParetoGNN achieved an average rank of 3.3, SingleEnc obtained a rank of 2.7, and MEGA exhibited the highest performance with an average rank of 1.6. These results underscore MEGA potential for strong downstream task generalization through MT-SSL. The improvements stem from MEGA's ability to eliminate conflicting gradients from their source.

Within individual tasks, `FeatRec` demonstrates the highest performance with an average rank of 3.7. However, it is noteworthy that no single pretext task consistently outperforms all downstream tasks. For example, `MI-NG` excels in link prediction but shows less effectiveness in partition prediction.

Consistent with the findings of [20], we observe that the unitary loss scalarization approach utilized by `SingleEnc` outperforms ParetoGNN [16] within MT-SSL methods. These results underscore the challenges arising from conflicting gradients in MOO and hard-parameter sharing MT-SSL frameworks.

To validate that MEGA's enhanced performance is not solely due to increased model parameters, we train various versions of the ParetoGNN model with comparable sizes to MEGA. Subsequently, we evaluate their performance on node classification using the Chameleon dataset, as presented in Table 2. The results demonstrate that MEGA outperforms all overparameterized versions of ParetoGNN, providing further validation of its robust task generalization capabilities.

5   Conclusion

This work introduces MEGA, a novel multi-encoder GNN architecture designed to address the limitations of existing MT-SSL approaches on graphs. MEGA's architecture enables independent learning of multiple pretext tasks by providing distinct parameter spaces for each task, effectively eliminating task interference and enhancing task generalization. By providing dedicated encoders for each pretext task, MEGA effectively captures diverse aspects of graph structure and node features, leading to improved performance in node classification, link prediction, and partition prediction tasks. Notably, MEGA outperforms both single-task trained encoders and hard-parameter sharing MT-SSL approaches, showcasing its strong task generalization capabilities. Our extensive experiments demonstrate MEGA improves performance across a variety of downstream tasks and datasets, with an average improvement of 3.8% compared to existing methods. The flexibility of MEGA's architecture allows for the integration of any GNN encoder and pretext task, making it a versatile framework for SSL on graphs. Moreover, our approach is scalable and parallelizable, enabling efficient training and deployment in real-world scenarios. Future work could explore the integration of a wider range of pretext tasks into the MEGA framework. Additionally, we aim to enhance the synergy within pretext task learning via end-to-end training to further boost its generalization capabilities.

References

1. Abdi, H., Williams, L.J.: Principal component analysis. Wiley Interdisc. Rev. Comput. Stat. **2**(4), 433–459 (2010)
2. Agarap, A.F.: Deep learning using rectified linear units (ReLU). arXiv preprint arXiv:1803.08375 (2018)
3. Bachman, P., Hjelm, R.D., Buchwalter, W.: Learning representations by maximizing mutual information across views. In: Advances in Neural Information Processing Systems, vol. 32 (2019)
4. Bronstein, M.M., Bruna, J., LeCun, Y., Szlam, A., Vandergheynst, P.: Geometric deep learning: going beyond Euclidean data. IEEE Signal Process. Mag. **34**(4), 18–42 (2017)
5. Caruana, R.: Multitask learning. Mach. Learn. **28**, 41–75 (1997)
6. Chen, T., et al.: AdaMV-MoE: adaptive multi-task vision mixture-of-experts. In: Proceedings of the IEEE/CVF International Conference on Computer Vision, pp. 17346–17357 (2023)
7. Chen, T., Kornblith, S., Norouzi, M., Hinton, G.: A simple framework for contrastive learning of visual representations. In: International Conference on Machine Learning, pp. 1597–1607. PMLR (2020)
8. Doersch, C., Zisserman, A.: Multi-task self-supervised visual learning. In: Proceedings of the IEEE International Conference on Computer Vision, pp. 2051–2060 (2017)
9. Ermolov, A., Siarohin, A., Sangineto, E., Sebe, N., et al.: Whitening for self-supervised representation learning. Proc. Mach. Learn. Res. **139** (2021)
10. Hamilton, W., Ying, Z., Leskovec, J.: Inductive representation learning on large graphs. In: Advances in Neural Information Processing Systems, vol. 30 (2017)
11. Hamilton, W.L., Ying, R., Leskovec, J.: Representation learning on graphs: methods and applications. IEEE (2017)
12. Hassani, K., Khasahmadi, A.H.: Contrastive multi-view representation learning on graphs. In: International Conference on Machine Learning, pp. 4116–4126. PMLR (2020)
13. Hou, Z., et al.: GraphMAE: self-supervised masked graph autoencoders. In: Proceedings of the 28th ACM SIGKDD Conference on Knowledge Discovery and Data Mining, pp. 594–604 (2022)
14. Hu, W., et al.: Strategies for pre-training graph neural networks. In: International Conference on Learning Representations (2019)
15. Jin, W., Liu, X., Zhao, X., Ma, Y., Shah, N., Tang, J.: Automated self-supervised learning for graphs. In: International Conference on Learning Representations (2021)
16. Ju, M., et al.: Multi-task self-supervised graph neural networks enable stronger task generalization. In: The Eleventh International Conference on Learning Representations (2022)
17. Karypis, G., Kumar, V.: METIS: a software package for partitioning unstructured graphs, partitioning meshes, and computing fill-reducing orderings of sparse matrices. Computer Science and Engineering Technical Reports (1997)
18. Kipf, T.N., Welling, M.: Semi-supervised classification with graph convolutional networks. In: International Conference on Learning Representations (2016)
19. Kolesnikov, A., Zhai, X., Beyer, L.: Revisiting self-supervised visual representation learning. In: Proceedings of the IEEE/CVF Conference on Computer Vision and Pattern Recognition, pp. 1920–1929 (2019)

20. Kurin, V., De Palma, A., Kostrikov, I., Whiteson, S., Mudigonda, P.K.: In defense of the unitary scalarization for deep multi-task learning. In: Advances in Neural Information Processing Systems, vol. 35, 12169–12183 (2022)

21. Lee, G., Yang, E., Hwang, S.: Asymmetric multi-task learning based on task relatedness and loss. In: International Conference on Machine Learning, pp. 230–238. PMLR (2016)

22. Lin, X., Zhen, H.L., Li, Z., Zhang, Q.F., Kwong, S.: Pareto multi-task learning. In: Advances in Neural Information Processing Systems, vol. 32 (2019)

23. Liu, Y., et al.: Graph self-supervised learning: a survey. IEEE Trans. Knowl. Data Eng. **35**(6), 5879–5900 (2022)

24. Lopes, I., Vu, T.H., de Charette, R.: Cross-task attention mechanism for dense multi-task learning. In: Proceedings of the IEEE/CVF Winter Conference on Applications of Computer Vision, pp. 2329–2338 (2023)

25. Meng, Z., Yao, X., Sun, L.: Multi-task distillation: towards mitigating the negative transfer in multi-task learning. In: 2021 IEEE International Conference on Image Processing (ICIP), pp. 389–393. IEEE (2021)

26. Misra, I., Shrivastava, A., Gupta, A., Hebert, M.: Cross-stitch networks for multi-task learning. In: Proceedings of the IEEE Conference on Computer Vision and Pattern Recognition, pp. 3994–4003 (2016)

27. Ni, M., et al.: M3P: learning universal representations via multitask multilingual multimodal pre-training. In: Proceedings of the IEEE/CVF Conference on Computer Vision and Pattern Recognition, pp. 3977–3986 (2021)

28. Peng, Z., et al.: Graph representation learning via graphical mutual information maximization. In: Proceedings of The Web Conference 2020, pp. 259–270 (2020)

29. Pinheiro Cinelli, L., Araújo Marins, M., Barros da Silva, E.A., Lima Netto, S.: Variational autoencoder. In: Variational Methods for Machine Learning with Applications to Deep Networks, pp. 111–149. Springer, Cham (2021). https://doi.org/10.1007/978-3-030-70679-1_5

30. Poduval, P., Oberoi, G., Verma, S., Agarwal, A., Singh, K., Asthana, S.: Bip-NRL: mutual information maximization on bipartite graphs for node representation learning. In: Koutra, D., Plant, C., Gomez Rodriguez, M., Baralis, E., Bonchi, F. (eds.) ECML PKDD 2023. LNCS, vol. 14172, pp. 728–743. Springer, Cham (2023). https://doi.org/10.1007/978-3-031-43421-1_43

31. Popescu, M.C., Balas, V.E., Perescu-Popescu, L., Mastorakis, N.: Multilayer perceptron and neural networks. WSEAS Trans. Circuits Syst. **8**(7), 579–588 (2009)

32. Prasad, U., Kumari, N., Ganguly, N., Mukherjee, A.: Analysis of the co-purchase network of products to predict amazon sales-rank. In: Reddy, P.K., Sureka, A., Chakravarthy, S., Bhalla, S. (eds.) BDA 2017. LNCS, vol. 10721, pp. 197–214. Springer, Cham (2017). https://doi.org/10.1007/978-3-319-72413-3_13

33. Radford, A., et al.: Language models are unsupervised multitask learners. OpenAI blog **1**(8), 9 (2019)

34. Royer, A., Blankevoort, T., Ehteshami Bejnordi, B.: Scalarization for multi-task and multi-domain learning at scale. In: Advances in Neural Information Processing Systems, vol. 36 (2024)

35. Rusch, T.K., Bronstein, M.M., Mishra, S.: A survey on oversmoothing in graph neural networks. arXiv preprint arXiv:2303.10993 (2023)

36. Sanh, V., et al.: Multitask prompted training enables zero-shot task generalization. In: International Conference on Learning Representations (2021)

37. Sener, O., Koltun, V.: Multi-task learning as multi-objective optimization. In: Advances in Neural Information Processing Systems, vol. 31 (2018)

38. Shi, G., Li, Q., Zhang, W., Chen, J., Wu, X.M.: Recon: reducing conflicting gradients from the root for multi-task learning. arXiv preprint arXiv:2302.11289 (2023)
39. Standley, T., Zamir, A., Chen, D., Guibas, L., Malik, J., Savarese, S.: Which tasks should be learned together in multi-task learning? In: International Conference on Machine Learning, pp. 9120–9132. PMLR (2020)
40. Topping, J., Di Giovanni, F., Chamberlain, B.P., Dong, X., Bronstein, M.M.: Understanding over-squashing and bottlenecks on graphs via curvature. In: International Conference on Learning Representations (2021)
41. Vandenhende, S., Georgoulis, S., Van Gansbeke, W., Proesmans, M., Dai, D., Van Gool, L.: Multi-task learning for dense prediction tasks: a survey. IEEE Trans. Pattern Anal. Mach. Intell. 44(7), 3614–3633 (2021)
42. Veličković, P.: Everything is connected: graph neural networks. Curr. Opin. Struct. Biol. 79, 102538 (2023)
43. Veličković, P., Cucurull, G., Casanova, A., Romero, A., Liò, P., Bengio, Y.: Graph attention networks. In: International Conference on Learning Representations (2018)
44. Veličković, P., Fedus, W., Hamilton, W.L., Liò, P., Bengio, Y., Hjelm, R.D.: Deep graph infomax. In: International Conference on Learning Representations (2018)
45. Wang, H., et al.: Graph mixture of experts: learning on large-scale graphs with explicit diversity modeling. In: Advances in Neural Information Processing Systems, vol. 36 (2024)
46. Wu, Z., et al.: MoleculeNet: a benchmark for molecular machine learning. Chem. Sci. 9(2), 513–530 (2018)
47. Xie, Y., Xu, Z., Zhang, J., Wang, Z., Ji, S.: Self-supervised learning of graph neural networks: a unified review. IEEE Trans. Pattern Anal. Mach. Intell. 45(2), 2412–2429 (2022)
48. Xin, D., Ghorbani, B., Gilmer, J., Garg, A., Firat, O.: Do current multi-task optimization methods in deep learning even help? In: Advances in Neural Information Processing Systems, vol. 35, pp. 13597–13609 (2022)
49. Xu, K., Hu, W., Leskovec, J., Jegelka, S.: How powerful are graph neural networks? In: International Conference on Learning Representations (2018)
50. Yao, Y., Rosasco, L., Caponnetto, A.: On early stopping in gradient descent learning. Constr. Approx. 26(2), 289–315 (2007)
51. You, J., Leskovec, J., He, K., Xie, S.: Graph structure of neural networks. In: International Conference on Machine Learning, pp. 10881–10891. PMLR (2020)
52. Zbontar, J., Jing, L., Misra, I., LeCun, Y., Deny, S.: Barlow Twins: self-supervised learning via redundancy reduction. In: International Conference on Machine Learning, pp. 12310–12320. PMLR (2021)
53. Zhai, J., Zhang, S., Chen, J., He, Q.: Autoencoder and its various variants. In: 2018 IEEE International Conference on Systems, Man, and Cybernetics (SMC), pp. 415–419. IEEE (2018)
54. Zhang, H., Wu, Q., Yan, J., Wipf, D., Yu, P.S.: From canonical correlation analysis to self-supervised graph neural networks. In: Advances in Neural Information Processing Systems, vol. 34, pp. 76–89 (2021)
55. Zhang, M., Chen, Y.: Link prediction based on graph neural networks. In: Proceedings of the 32nd International Conference on Neural Information Processing Systems, pp. 5171–5181 (2018)

MetaQuRe: Meta-learning from Model Quality and Resource Consumption

Raphael Fischer[1]([✉]) [iD], Marcel Wever[2] [iD], Sebastian Buschjäger[1] [iD], and Thomas Liebig[1] [iD]

[1] Lamarr Institute for Machine Learning and Artificial Intelligence,
TU Dortmund University, 44227 Dortmund, Germany
{raphael.fischer,sebastian.buschjaeger,thomas.liebig}@tu-dortmund.de
[2] Munich Center for Machine Learning, LMU Munich, 80539 Munich, Germany
marcel.wever@ifi.lmu.de

Abstract. Automated machine learning (AutoML) allows for selecting, parametrizing, and composing learning algorithms for a given data set. While resources play a pivotal role in neural architecture search, it is less pronounced by classical AutoML approaches. In fact, they generally focus on only maximizing predictive quality and disregard the importance of finding resource-efficient solutions. To push resource awareness further, our work explicitly explores how measures such as running time or energy consumption can be better considered in AutoML. Firstly, we propose a novel method for algorithm selection that balances multiple performance aspects (including resource demand) as prioritized by the user with the help of compositional meta-learning. Secondly, to foster research on green meta-learning and AutoML, we release the `MetaQuRe` data set, which contains information on predictive (Qu)ality and (Re)source consumption of models evaluated across hundreds of data sets and four execution environments. We use this data to put our methodology into practice and conduct an in-depth analysis of how our approach and data set can help in making AutoML more resource-aware, which represents our third contribution. Lastly, we publish `MetaQuRe` alongside an extensive code base, allowing for reproducing all results, expanding our data with results from custom environments, and exploring `MetaQuRe` interactively. In short, our work demonstrates both the importance as well as benefits of rethinking AutoML and meta-learning in a resource-aware way, thus paving the path for making future ML solutions more sustainable.

Keywords: Meta-learning · AutoML · Resource-aware ML · Green AI · Sustainable AI · Algorithm Selection

1 Introduction

Automated machine learning (AutoML) tackles the problem of automatically selecting and parametrizing machine learning (ML) algorithms that maximize

A. Bifet et al. (Eds.): ECML PKDD 2024, LNAI 14947, pp. 209–226, 2024.
https://doi.org/10.1007/978-3-031-70368-3_13

performance, usually in terms of predictive quality, on a user-given data set [19]. Classical AutoML systems propose solutions in the form of ML pipelines [11,25,29,34,42], consisting of different feature pre-processing steps and either a single learning algorithm or an ensemble of learners [8,11]. Internally, such systems often deploy meta-learning [19,36] to guide the parameter and algorithm selection, such that costly evaluations (e.g., testing a candidate via cross-validation) can be replaced with predictions of the meta-model. The subfield of neural architecture search (NAS), on the other hand, has specialized in searching suitable architectures of deep neural networks (DNNs) [7,38], often combined with tuning further hyperparameters [41].

While today's AutoML systems are very effective, they are well-known to consume large amounts of compute resources [35]. This stands in stark contrast to the demand for more resource-aware [12] and sustainable [40] ML advances. While in NAS literature, adapting the architectures to specific hardware constraints is a central research direction [2], resource awareness is less of a topic in classical AutoML systems. Here, several works aim to make AutoML methods themselves more efficient [9,10,24], however, the computational cost and resource efficiency of ML pipeline candidates is usually not subject of the decision process. In fact, information on compute utilization generally is dramatically underreported in scientific papers [13,32] and public ML experiment databases [37], which are important resources for meta-learning [19].

Our work addresses the necessity for "green" and resource-aware AutoML [35] via four central contributions, which can be summarized as follows:

- We propose a methodological framework for resource-aware meta-learning, which selects ML algorithms under consideration of multiple performance aspects, as prioritized by the user.
- We release a novel data set named `MetaQuRe`, which comprises extensive information on (Qu)ality and (Re)source demand of popular ML algorithms. Being assembled from evaluations across several execution environments, it is the first-ever data set allowing for resource-aware meta-learning.
- We conduct an experimental study that underpins the necessity of our work, showcases practical insights from our data set, and demonstrates the feasibility of our meta-learning framework.
- We offer an extensive code base that allows for exploring `MetaQuRe` interactively, reproducing all results, and extending our data set with performance evaluations on custom environments.

To already provide some more depth, our methodology proposes *compositional meta-learning* as a means for not only recommending suitable ML algorithms for given tasks, but also estimating candidate performance along multiple dimensions, which relate to predictive quality or resource consumption. This enables our method to also adapt to use case priorities, which is important considering that algorithm selection might be subject to manifold objectives (e.g., searching for the most accurate, fastest, or resource-efficient model). To unify performance measures and select algorithms under consideration of the targeted execution environment we utilize the concept of index scaling. The completely novel `MetaQuRe` data set enabled us to put our methodology into practice and was assembled by evaluating 10 different ML algorithms on 200 data sets across four different execution environments. Our experiments demonstrate the feasibility of our method and showcase how it produces highly accurate and efficient ML solutions. At the same time, it is more resource-friendly than competing AutoML approaches such as AutoGluon [8], Naive AutoML [24], and TabPFN [17]. Along with our complete code base and an interactive exploration tool, `MetaQuRe` can be found at github.com/raphischer/metaqure. We believe that our work and data set are considered a valuable contribution to the state-of-the-art and foster more work that makes AutoML more resource-aware.

2   Related Work

Before presenting our methodology and practical findings in depth, we first provide an overview of related literature. Automating ML is a costly endeavor, and related literature is well aware of the importance of resource awareness [32] and sustainability [40], both in terms of running AutoML systems as well as for judging output model performance [35]. To increase efficiency of AutoML systems, research has evolved along three main dimensions – improving sampling-efficiency [11], evaluation efficiency [5], and the combination thereof [1,9]. Naive AutoML (NAM) [24] breaks up the overall AutoML problem into sub-problems and tackles each of them independently, yielding competitive results to more sophisticated AutoML systems at a reduced cost for execution. Another approach called AutoGluon (AGL) [8] focuses on the configuration of a single ensemble and greedily adds more base learners as long as they improve predictive performance. While yielding state-of-the-art performance in terms of predictive accuracy [15], the configured ML model is becoming more costly with every base learner added to the ensemble.

Another line of research aims to incorporate domain expertise [22,33] or leverage meta-learning to not start from scratch but be informed by past AutoML runs on other data sets [6,10,11,36] to alleviate the costs for AutoML. Other meta-learning approaches aim to increase the efficient use of resources by estimating the costs of ML algorithms in terms of running time [26] or whether the algorithms are likely to fail for a particular data set [28]. The most recent TabPFN [17] proposes a totally different take on AutoML by introducing a prior-fitted transformer model that, after priming on the train data, can be used to produce predictions on the test data with a single forward pass. In comparison to NAM, AGL, or other AutoML systems, this approach arguably requires much fewer resources as it does not require any online learning on the data at hand. However, this conclusion conceals the costs of training TabPFN in the first place, and deploying a transformer architecture in practice might still demand substantial resources. Moreover, TabPFN puts hard limits on the dimensions of applicable data sets, supporting only up to 100 numerical features, 10 classes, and 1000 data points.

All of the mentioned approaches have in common that they only aim at reducing the overall costs of running AutoML, i.e., finding a well-performing ML pipeline efficiently. The efficiency of individual candidates however is not explicitly considered. Moreover, these approaches are agnostic of the available compute resources on which the final model shall be deployed. In NAS, another subbranch of AutoML, the importance of resource awareness is not only acknowledged but one of the central themes [2]. Since nowadays deep learning models in several cases are also deployed on the edge [14], NAS was more and more extended for constraining architecture size. While various techniques have been proposed in this literature, they are specifically tailored to this particular model class and do not easily translate to general ML pipelines. In short, we found established AutoML systems for building these pipelines to be, to the best of our knowledge, neither considerate of the specific deployment environment, nor of the environmental cost that ML pipeline candidates might cause.

3   Methodology

In the following, we introduce our methodology for performing meta-learning and AutoML in a more resource-aware fashion. We start by formalizing the problem of classic automated algorithm selection and afterward discuss how it can be extended to also be resource-aware. We then introduce the concept of index scaling, which is a key enabler for our solution to algorithm selection. In short, it leverages compositional meta-learning, which we explain in detail in Sect. 3.4. Our methodology concludes with some additional considerations regarding the transparency of our method.

3.1   Automated Algorithm Selection

For solving a learning problem on a specific data set $d \in \mathbb{D}$, a broad arsenal of ML algorithms $a \in \mathbb{A}$ can be utilized. The optimal choice a^* for given data, however, is not only initially unknown, but also subject to the execution environment $e \in \mathbb{E}$ [12]. It corresponds to a specific hardware platform and software stack, possibly strongly impacting algorithms' practical performance. To illustrate this, consider, for example, that one would want to perform classification on `credit-g` data (d), for which many different ML algorithms a can be utilized (e.g., linear regressors or random forests). However, their practical performance is not only subject to the specific software implementation (e.g., `scikit-learn` [30]), but also underlying hardware (for example due to differences in processor logic). Being faced with a specific learning configuration $c = (d, e)$, selecting the optimal algorithm a_c^* can be understood as an optimization problem that is subject to a quality measure μ (e.g., accuracy):

$$a_c^* = \arg \max_{a \in \mathbb{A}} \mu(a, c), \text{ with } \mu : \mathbb{A} \times (\mathbb{D} \times \mathbb{E}) \to \mathbb{R} \qquad (1)$$

Naively, a_c^* can be identified by performing an exhaustive search over $\mathbb{A}$. Note that we understand this algorithm pool as a finite set of algorithms with fixed hyperparameters. Due to the presence of continuous hyperparameters, simultaneously finding good hyperparameters would ask for a different approach, which is why we leave this part of the AutoML problem for future work. Practically deploying an algorithm produces a trained ML model, however, we chose to here denote both via a and use the terms interchangeably, depending on the context. We want to also highlight that while not explicitly denoted, certain combinations of d, e, and a might be inadmissible – as examples, consider forecasting time series (d) with a classification algorithm (a), or learning a highly complex DNN (a) on ultra-low power devices (e), which is hardly feasible in practice.

3.2 Incorporating Resource Awareness

Optimal algorithm performance is often assessed from quality measures $\mathbb{M}_Q = \{\mu \colon \mathbb{A} \times (\mathbb{D} \times \mathbb{E}) \to \mathbb{R}\}$ such as accuracy or f_1 score. However, this neglects the importance of considering algorithmic resource efficiency. To explicitly consider such trade-offs, algorithm selection should ideally also reflect on resource measures[1] $\mathbb{M}_R = \{\mu \colon \mathbb{A} \times (\mathbb{D} \times \mathbb{E}) \to \mathbb{R}\}$, like running time, energy draw, or model size. To acknowledge both quality- and resource-related measures in the model search, we extend Eq. (1) as follows:

$$a_c^* \in \arg \max_{(a \in \mathbb{A})} S(a, c), \text{ where } S(a, c) = \sum_i w_i \cdot \mu_i(a, c), \text{ with } \mu_i \in \mathbb{M} \quad (2)$$

Thus, algorithmic performance is understood in a multi-dimensional fashion, and evaluations are assessed along m different measures $\boldsymbol{\mu} = (\mu_1, \mu_2, \dots, \mu_m)$, with $\mu_i \in \mathbb{M}$, $\mathbb{M} = \mathbb{M}_R \cup \mathbb{M}_Q$, and $\mathbb{M}_R \cap \mathbb{M}_Q = \emptyset$. The overall performance score $S(a, c)$ is the convex combination of all measures with corresponding priorities w_i, which are subject to the use case at hand. This allows for identifying a_c^* based on a specific objective $\boldsymbol{\omega} = (w_1, w_2, \dots, w_m)$, with $\sum_i w_i = 1$ and $\forall w_i \colon 0 \le w_i \le 1$. For example, when being interested in the fastest or most accurate solution, Eq. (2) can be re-transformed into Eq. (1) by setting exactly one $w_i = 1$ and all others to zero. On the other hand, it also allows for searching algorithms that balance all measures well. Note that with this extension, there very well might not be a single optimal algorithm, but rather a set of Pareto-optimal compromises $a_c^* \in \arg\max_{(a \in \mathbb{A})} S(a, c)$.

3.3 Relative Index Scaling

A central problem arising from Eq. (2) lies in the diversity of measures, which can occur on vastly different scales and magnitudes. Without further processing, one could for example assess the number of parameters in millions, percent for accuracy, or energy draw in Wattseconds. Obviously, aggregating such measures via a (weighted) sum does not make sense. This problem becomes even more

---

[1] Note that our wording here diverges from the mathematical definition of measures – we use it to describe properties measured from practically deploying algorithms.

evident considering that values for resource consumption are impacted dramatically when switching between different environments (e.g., a powerful workstation versus a lightweight edge device [14]). In addition, Eq. (2) assumes that higher measures represent a better performance, whereas many measurements naturally work the opposite way, i.e., smaller is better (e.g., running time). We propose to solve these issues by projecting performance measures onto a relative index scale [12,14]:

$$\tilde{\mu}_i(a, c) = \left(\frac{\mu_i^*(c)}{\mu_i(a, c)}\right)^{\sigma_i} = \left(\frac{\max_{a' \in \mathbb{A}_{\mathbb{H}}}(\sigma^i \cdot \mu_i(a', c))}{\sigma^i \cdot \mu_i(a, c)}\right)^{\sigma_i} \tag{3}$$

Concisely, this maps the real-valued measures μ onto values in relation to the best historically recorded measure $\mu_i^*(c)$, which receives $\tilde{\mu}_i = 1$. Note how this best empiric result is subject to the environment (as part of c), which makes algorithm performance comparable across diverse environments. With $\mathbb{A}_{\mathbb{H}} \subseteq \mathbb{A}$, we denote algorithms for which historical results on c are available. To unify the understanding of measure improvement, Eq. (3) introduces an indicator constant, with $\sigma_i = 1$ for measures that want to be minimized and $\sigma_i = -1$ for all others. Index scaling projects all measurements onto the unit scale for further processing, i.e., $\tilde{\mu}_i : \mathbb{A} \times \mathbb{D} \times \mathbb{E} \to [0, 1]$, where higher values always indicate measure improvement[2]. Naturally, Eq. (3) can also be used inversely for calculating real-valued measures μ_i from index-scaled results $\tilde{\mu}_i$. We argue that both scales are important – the former is more helpful for users who have an intuitive understanding of the units behind measures, whereas index scaling allows for additional comparisons and aggregation.

3.4   Compositional Meta-learning

Index scaling makes solving Eq. (2) feasible by replacing the μ_i with $\tilde{\mu}_i$, however, the naive approach (i.e., evaluating all a) remains very costly. As a more efficient solution, we propose a novel method that utilizes *compositional meta-learning* to alleviate the effort of assessing measures by costly ML evaluations. To be concise, it solves

$$\widehat{a_c^*} \in \arg\max_{a \in \mathbb{A}_{\mathbb{H}}} \hat{S}(a, X_c) = \arg\max_{a \in \mathbb{A}_{\mathbb{H}}} \sum_i w_i \cdot \hat{\tilde{\mu}}_i(a, X_c), \tag{4}$$

with $\hat{\tilde{\mu}} = (\hat{\tilde{\mu}}_1, \hat{\tilde{\mu}}_2, \ldots, \hat{\tilde{\mu}}_m)$ and $\hat{\tilde{\mu}} : \mathbb{A}_{\mathbb{H}} \times \mathbb{X}_{\mathbb{D}} \times \mathbb{X}_{\mathbb{E}} \to [0, 1]^m$ being the index-scaled surrogates of μ. Where the latter can only obtained from measuring the practical performance of an algorithm a on c, the surrogates $\hat{\tilde{\mu}}$ are instead estimated from corresponding meta-features $X_c \in \mathbb{X}_{\mathbb{D}} \times \mathbb{X}_{\mathbb{E}}$. They could for example encode information on specific characteristics of data sets [4,20,39] and hardware or software components of the environment [14].

---

[2] To prevent zero division, we set $\tilde{\mu}_i = 1$ if $\mu_i = \mu_i^* = 0$ (e.g., an error of zero, the best result), and $\tilde{\mu}_i = 0$ if $\mu_i = 0$ and $\sigma_i = 1$ (e.g., 0% accuracy, the worst result).

For training meta-learners, an evaluation history $\mathbb{H}$ needs to be assembled. It comprises performance measures obtained from evaluating algorithms on configurations along respective meta-features, i.e., $\mathbb{H} = \{(\boldsymbol{\mu}(a,c), \tilde{\boldsymbol{\mu}}(a,c), X_c)\}$, with $(a,c) \in \mathbb{A}_{\mathbb{H}} \times \mathbb{D}_{\mathbb{H}} \times \mathbb{E}_{\mathbb{H}} \subseteq \mathbb{A} \times \mathbb{D} \times \mathbb{E}$. Finding good meta-learners for estimating $\tilde{\boldsymbol{\mu}}$ poses a problem of algorithm selection of its own. An exhaustive search is, however, less problematic here – firstly, these models only need to be trained once and can then be utilized for efficiently selecting algorithms for any X_c and $\boldsymbol{\omega}$. In addition, the data dimensionality of $\mathbb{H}$ is usually rather manageable, as even large data sets in $\mathbb{D}_{\mathbb{H}}$ will be represented by small sets of meta-features (our contributed data set for example consists of 8000 instances with less than 100 features). Lastly, considering the demand for trustworthy ML [3], it is very reasonable to constrain the pool of candidate meta-learners to simple and interpretable algorithms [31], which also allows for fast evaluation. For optimal results, the meta-learners $\hat{\tilde{\boldsymbol{\mu}}}$ should minimize the empirical estimation error on the real-valued measures:

$$\min_{\hat{\tilde{\boldsymbol{\mu}}}} \sum_{X_a, X_c, \mu(a,c) \in \mathbb{H}} |\boldsymbol{\mu}(a,c) - \hat{\boldsymbol{\mu}}(a, X_c)| \tag{5}$$

The real-valued estimates $\hat{\boldsymbol{\mu}}$ can be obtained by reversing Eq. (3) and inputting the index-scaled surrogate values $\hat{\tilde{\boldsymbol{\mu}}}(a, X_c)$, which are originally meta-learned. With surrogate models trained on $\mathbb{H}$, our method can now automatically and efficiently solve Eq. (4) for a specific objective $\boldsymbol{\omega}$ and learning configuration c by (1) estimating the index-scaled measures $\hat{\tilde{\boldsymbol{\mu}}}(a, X_c)$ for all candidates a, (2) assessing each candidate's overall score $\hat{S}(a, X_c)$, (3) sorting the results, and (4) evaluating the most promising candidate to finally obtain $\boldsymbol{\mu}(\hat{a}_c^*, c)$.

3.5   Additional Remarks

First of all, note how our approach does not make any assumptions as to how the meta-features X_c are calculated. We discuss experiments with different approaches [4,20,39] in Sect. 5.2, but purposely decided to formulate our method most universally in this aspect. In our practical implementation of compositional meta-learning, we moreover decided to train individual meta-learners for the different algorithms. Instead, one could utilize universal models that estimate algorithm performance based on additional meta-information X_a. This is plausible when the algorithm pool can be well-described by meta-features, however, in MetaQuRe, we specifically chose very diverse algorithms and therefore decided to train individual regressors. Besides allowing for automated and resource-aware algorithm selection under consideration of user priorities encoded via $\boldsymbol{\omega}$, we additionally argue our method to be transparent and interpretable, as it provides several by-product explanations for every decision. This is a convenient co-effect of our compositional approach, which accompanies every recommendation with estimates for all performance measures, explaining as to *why* the selected algorithm is suitable for the task. As we propose to only employ interpretable meta-learners [31], users of our framework can even understand the reasoning behind every single measure estimate (e.g., via feature importance for the input meta-features). In addition, the means for interfering with the decision process by adjusting the objective enables users to better understand the decisions and measure the trades of candidates. The estimated performance could even be communicated to less knowledgeable users, for example, in the form of labels [13,27], which is readily offered in our implemented tools. Considering the demand for trustworthiness [3] and explainability [18] of ML solutions, we think this to be an additional noteworthy advantage of our method.

4   Data on Model Quality and Resource Consumption

When drafting our method, we immediately faced the problem that hardly any data for resource-aware meta-learning exists [13], as information on compute utilization is extremely under-reported in databases like OpenML [37]. To address this flaw, we contribute MetaQuRe – a new data set that allows for meta-learning from evaluations assessed in terms of algorithmic prediction quality and resource consumption. For assembling MetaQuRe, we evaluated a range of popular ML algorithms on 200 classification data sets from various domains, as offered by scikit-learn [30], OpenML [37], and the UCI repository[3]. We specifically chose $|\mathbb{A}| = 10$ algorithms that are of both universal and diverse nature – to be specific, we deployed k-nearest neighbor (kNN), support vector machine (SMV), random forest (RF), extra random forest (XRF), adaptive boosting (AB), Gaussian naive Bayes (GNB), ridge regression (RR), logistic regression (LR), linear stochastic gradient descent (SGD), and multiplayer perceptron (MLP) classifiers, as offered by scikit-learn 1.4.0 [30].

To reflect the implications of ML deployment across different execution environments, each algorithm × data set combination was evaluated on four platforms: a 2023 workstation (Intel i9-13900K, 64 GB RAM), a 2015 desktop (Intel i7-6700, 32 GB RAM), a lightweight 2021 laptop (Intel i7-10610U, 32 GB RAM), and an ML-specialized 2023 NVIDIA Jetson AGX Orin (ARMv8 rev 1 v81, 64 GB RAM). Performance was assessed along $m = 10$ different measures, namely accuracy (ACC), f_1 score (F1), recall (REC), precision (PRC), number of parameters (PAR), model file size on disc (FS), as well as energy draw (EN) and running time (RT), both measured for the complete training (ENT, RTT) and for performing a single inference (ENI, RTI). Suitable hyperparameters for all algorithms were identified via basic grid search on the train split, whereas the quality measures were computed on the test split. All reported measures are averaged over three runs with different seeds to account for randomness stemming from algorithm logic and data splitting. We utilized CodeCarbon[4] and jetson-stats[5] for profiling the energy draw. The data set and complete code for reproducing or extending our results are available at

---

[3] https://archive.ics.uci.edu.
[4] https://github.com/mlco2/codecarbon.
[5] https://github.com/rbonghi/jetson_stats.

github.com/raphischer/metaqure. We want to highlight that the repository additionally includes a browser-based tool to explore `MetaQuRe` in more depth than this paper allows. Following the call for reporting any paper's environmental impact [32], we estimate the total carbon emissions of this work to a maximum of $7 \cdot 24 \cdot 1 \cdot 0.354 \approx 59$ CO$_2$e – one full week ($7 \cdot 24$ h) of running all environments in parallel (1 kW), with a carbon efficiency of (0.354 kg/kWh for Germany) [21]. Note that the final assemblage of our data set was completed in four days and our method (including the cross-validated algorithm search) can be trained on `MetaQuRe` in mere minutes, however, we spent much time on development and decided to rather over- than under-estimate our computational expenses.

5   Experimental Results

We start our experimental investigation by discussing exemplary results from `MetaQuRe`. In the second half, we discuss the results of performing compositional meta-learning on this data.

5.1   Insights from `MetaQuRe`

To start off, let us explore how the different algorithms perform on a single, exemplary data set (`parkinsons`). Figure 1 shows the training energy draw versus accuracy trades of each algorithm across two environments. As expected, model accuracy is hardly affected by the hardware choice, except for SGD, which interestingly becomes more accurate when deployed on the `ARMv8` (this could for example be caused by differing CPU logic). We also clearly see the advantage of index scaling, as introduced in Sect. 3.3 – compared to the real-valued energy draw (left), the relative values (right) are placed on the more easily comparable unit scale. The best measures per environment (maximum accuracy of kNN, minimum energy draw of GNB) receive $\tilde{\mu}_i = 1$, and everything else is placed in relation to these values. As an example, SVM on the `Intel i9` draws about 0.168 Ws, so following Eq. (3) and in comparison with the best empiric result (GNB at 0.129 Ws), it is placed at $x = \frac{0.129}{0.168} \approx 0.76$. With this scaling, it becomes evident that several algorithms on the `Intel i9` are relatively close to GNB when it comes to energy draw. The differences are much more drastic on the `ARMv8`, where all algorithms require at least 300 times the energy of training GNB (0.011 Ws). The node colors indicate the overall performance score $S(a, c)$ under consideration of all 10 performance measures, of which two are displayed as axes – GNB and RR (on the i9) seem to perform reasonably well across all measures.

To explore the complete measure trade-offs, we display selected algorithm performances via star plots in Fig. 2. First of all, this unveils the strengths and weaknesses of any algorithm. GNB on `parkinsons` (left) performs relatively well across all measures, however predictive quality and file size seem to be the drawbacks. Note how index-scaling eases the comparison across environments – all values are bounded and higher values always indicate improvement. SGD

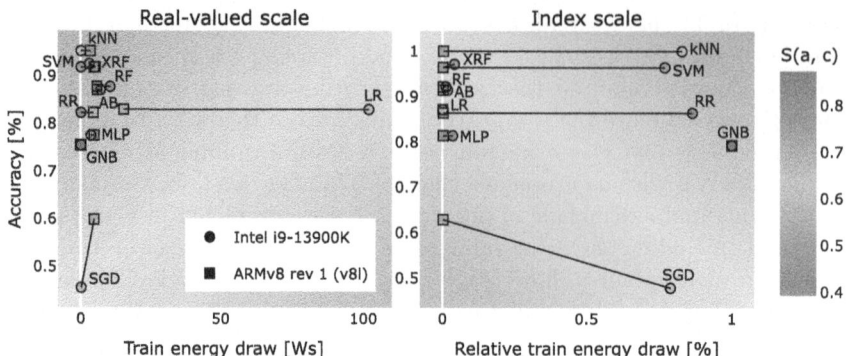

Fig. 1. Comparison of model performance in terms of training energy draw and accuracy on **parkinsons** data, both for real measurements (left) and relative index-scaled values (right). Node colors indicate the overall performance score.

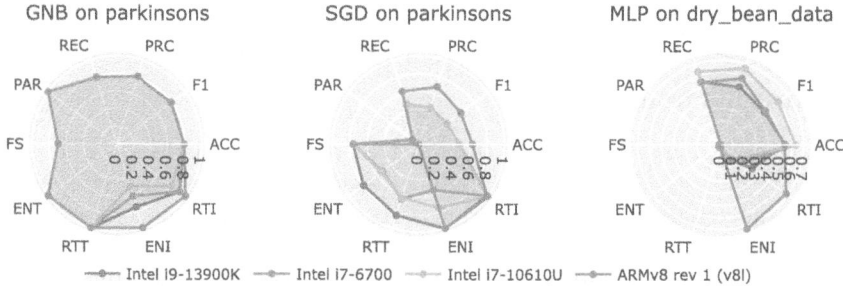

Fig. 2. Comparison of index-scaled algorithm performance across the environments along all ten measures. Deployment on the ARMv8 interestingly not only impacts the (relative) resource consumption but also the predictive quality.

(middle) for example might have better predictive quality on the ARMv8 (cf. Fig. 1), however, its relative resource consumption during training is much worse. A similar trade is happening with MLP on dry_bean_data, as the inference resource demand improves on the ARMv8 at cost of predictive quality. While we here are limited to discussing these single exemplary results, we invite readers to explore MetaQuRe in full depth with our interactive tool, which generates similar visualizations to the ones in this Section automatically.

Lastly, let us explore how each measure is affected by switching environments across all data sets, displayed in Fig. 3. Index scaling unifies the values of all measures and thus allows for comparing the standard deviation of measures for each algorithm × data configuration, across all environments. It shows that even the relative values can change a lot – as expected, the measures related to model size (PAR, FS) are very consistent, quality measures (ACC, F1, PRC, REC) are affected in several outlier cases, and performance in terms of resource consumption (ENT, RTT, ENI, RTI) varies vastly.

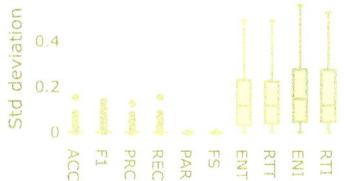

Fig. 3. Impact of environment choice on the standard deviation of index-scaled measures.

Table 1. Search objectives with weights

Objective	Weights	
O1: Most accurate	$w_i =$	$\begin{cases} 1 & \text{for ACC} \\ 0 & \text{otherwise} \end{cases}$
O2: Lowest energy	$w_i =$	$\begin{cases} 1 & \text{for ENT} \\ 0 & \text{otherwise} \end{cases}$
O3: Best balanced	$w_i =$	$\begin{cases} \frac{1}{8} & \text{for ACC, F1, REC, PRC} \\ \frac{1}{12} & \text{otherwise} \end{cases}$

5.2 Learning from `MetaQuRe`

By utilizing our data set as evaluation history $\mathbb{H}$, we can put compositional meta-learning for automated algorithm selection, as introduced in Sect. 3.4, into practice. For that, we defined five cross-validation splits (using the original data sets as group information) and identified optimal meta-learner algorithms via Eq. (5) based on a single train split. Our pool of meta-learners comprises 18 simple and interpretable models, namely variants of linear regressors, support vector regressors, and decision trees (depth limited to five), each equipped with different types and strengths of regularization. With meta-learners trained on different train splits of $\mathbb{H}$, we can make estimates for all recorded measures by predicting $\hat{\boldsymbol{\mu}}$ for each individual test split. Having access to all estimates and being faced with a specific search objective $\boldsymbol{\omega}$, we can now aggregate the estimated overall scores $\hat{S}(a, X_c)$ and identify the most promising candidate via Eq. (4). In our evaluations, we explore three objectives O1-O3 with corresponding weights w_i, as given in Table 1. For O3, measures are weighted such that both types of measures (four quality and six resource measures) in sum equally impact the overall score. Note that these are just examples – in practice, such objectives need to be elicited in collaboration with the respective user, e.g., via preference learning approaches [16].

Based on our introduced scoring, we can compare the suitability of any algorithm for the three objectives across all tasks. Figure 4 displays the resulting distributions of true best a_c^* (blue) and estimated best $\hat{a}_c^*$ (fuchsia) algorithms. It demonstrates how depending on the search criteria, different algorithms have a better chance of performing well – generally, SVM and RF are most accurate, GNB has the lowest train energy draw, and GNB and RR balance all measures well. Whereas most AutoML systems only focus on maximizing predictive capabilities, our method's recommendations change drastically with the use case objective. The differences between the bars indicate that our meta-learning does not always select the optimal algorithm, however it clearly models the important patterns of algorithm performance. We can also see that the optimal algorithm choice for resource-aware objectives (second and third row) is indeed impacted by the environment choice, as are our method's recommendations.

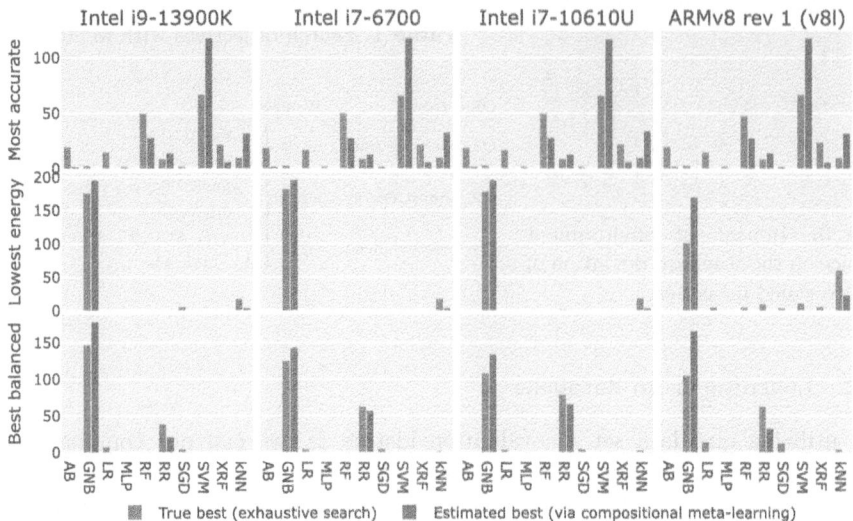

Fig. 4. Distributions of optimal model choice per data set, along with the recommendations of our model, across three different objectives (rows). Selecting the algorithm with the lowest train energy draw is clearly impacted by the choice of environment (columns). As expected, algorithm suitability across data sets changes dramatically with the objective, demonstrating the necessity for explicitly considering them in AutoML.

The quality of compositional meta-learning can be further quantified by investigating the absolute estimation errors across $\mathbb{H}$. We display them in Fig. 5, showcasing how difficult each individual measure is to predict. We here also demonstrate the importance of index-scaling the performance measure values for meta-learning – the errors of recalculated values $\hat{\mu}$ (blue) are much lower compared to directly meta-learning the real-valued measures μ (fuchsia).

We next explore the impact of choosing meta-features for data sets, as our method is not relying on a fixed extraction method (recall Sect. 3.5). For that, we tested (1) manual engineering using various statistics over the data ($|X| = 24$, similar to [39]), (2) principal component reduction (PCA, $|X| = 15$), (3) learned features via Dataset2Vec ($|X| = 32$) [20], as well as joining the features of (1) and (3) ($|X| = 56$). The resulting dimensionality-reduced meta-feature spaces are displayed in Fig. 6, obtained via uniform manifold approximation and projection (UMAP) [23], along with some general information on the MetaQuRe data sets (encoded via scatter color and size). To build the complete X_c, we joined each data set feature set with one-hot encoded information on the environment. In the lower row, we present the resulting mean absolute errors of predicting the overall performance scores $|\hat{S}(a, X_c) - S(a, c)|$ for all algorithms and data sets, based on our three objectives. We see that the meta-feature engineering only marginally impacts the prediction and that the objectives are not equally hard – the best-balanced model performance (O3) is the easiest to predict, and the most energy-saving training (O2) is the hardest. In all other experiments, we

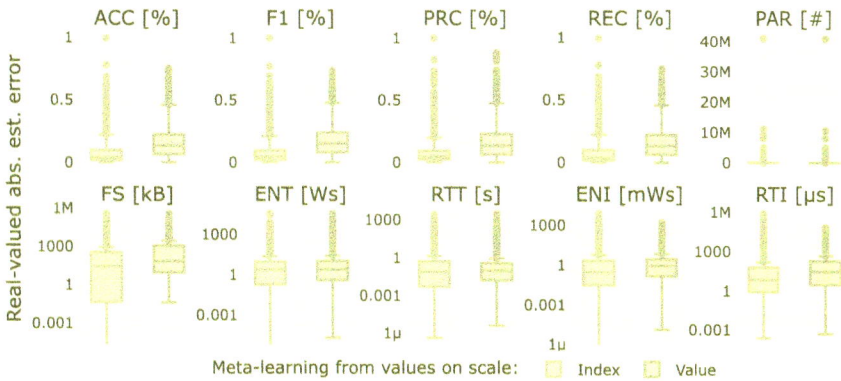

Fig. 5. Estimation errors when predicting algorithm performance along the individual measures. Meta-learning from the index-scaled values (blue) results clearly in lower errors than directly learning the real measurements (fuchsia). (Color figure online)

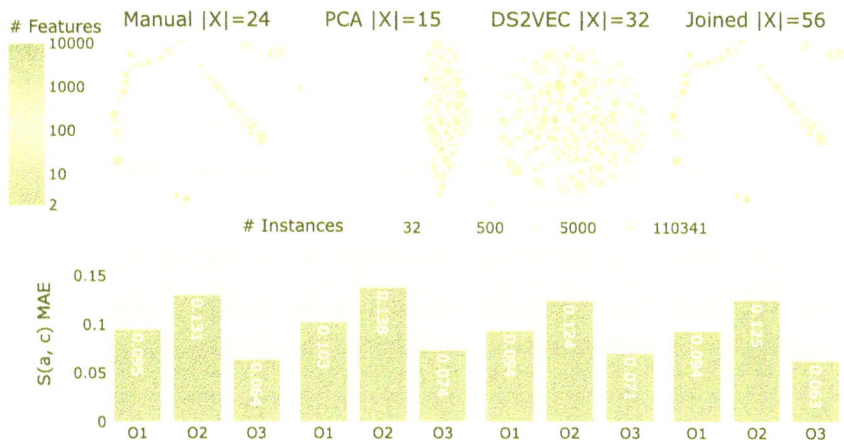

Fig. 6. UMAP embedding [23] of different meta-features for the data sets evaluated for assembling `MetaQuRe`, along with resulting prediction errors for our three search objectives. Combining all feature sets results in the lowest errors.

utilized the joined set of meta-features, as it resulted in the lowest errors across all objectives.

Lastly, we compare our method against renowned AutoML systems as introduced in Sect. 2, namely AutoGluon (AGL) [8], Naive AutoML (NAM) [24], and Tab(PFN) [17], as well as an (EXH)austive search baseline. As all competitors are optimized for obtaining the most accurate results, we compare them with outputs from the respective objective (O1). Recall that the AutoML competitors work very differently from our method – AGL and NAM do not search within our fixed algorithm pool, but instead construct completely novel learning pipelines. As budget, we offered them the time required to perform an exhaus-

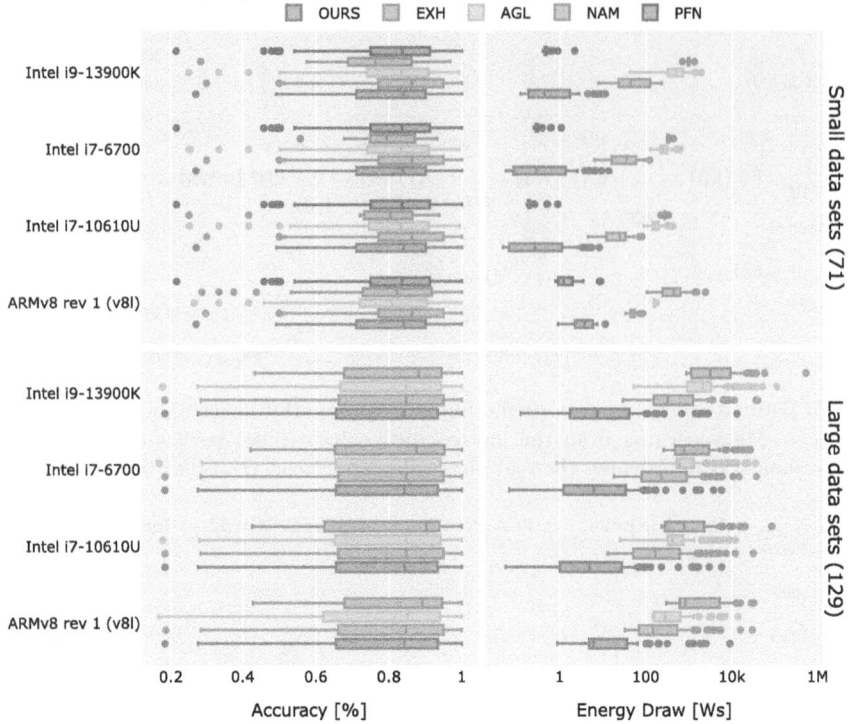

Fig. 7. Comparison of energy consumption and accuracy achieved by our method against AutoML baselines. While obtaining slightly less accurate models (due to our very basic algorithm pool), our method consumes much less energy. PFN has a comparable energy draw, however could only be deployed to 72 of our data sets and also required immense undisclosed resources for training the prior-data transformer.

tive search (i.e., training all algorithms from our pool, per data set), however for most data sets, this was even lower than the minimum budget requested by the libraries. PFN, on the other hand, does not even train a model but instead uses its prior-data-fitted transformer to predict classes in a single forward pass. Due to its tight constraints on input data, it could only be deployed on 71 of our 200 data sets – the resulting comparison is displayed in the top half of Fig. 7. On the left, we see that our model is on par with AGL and PFN in terms of accuracy, with NAM being significantly worse and EXH retrieving even more accurate algorithms. However, all other approaches (except for PFN) consume much higher amounts of resources across all environments (right side). Interestingly, exhaustively testing our algorithm pool achieves optimal accuracy at relatively low energy demand, indicating that the more costly AutoML frameworks are overpowered for these rather simple data sets. PFN seems to be most efficient on most evaluations, however note, that the immense effort for training the transformer network in the first place was not reported and thus cannot be considered here – and yet, our method is close to their energy draw. In the second half, we showcase the results of the remaining 129 data sets, which are too big for PFN. Note that PFN explicitly restricts the number of classes and features in supported data sets, so subsampling (to make them feasible for PFN) is hardly an option. For these rather complex data sets, NAM achieves most accurate results but also requires the most energy, closely followed by AGL. Our selection method is by far the most energy-saving approach while recommending algorithms with reasonably high accuracy.

6   Conclusion

In summary, our paper responds to the pressing call for resource awareness in AutoML with a series of pivotal contributions. Firstly, we introduced a methodological framework for resource-considerate meta-learning, which empowers users to find ML algorithms in alignment with their specific performance demands. Internally, it leverages index scaling and compositional meta-learning to unify and balance the various estimated measures of candidate algorithms, thus alleviating the effort of deploying them practically. Secondly, we introduce the `MetaQuRe` data set, a rich resource that for the first time allows to not only meta-learn from predictive quality but also resource requirements of diverse ML algorithms across several execution environments. Thirdly, our empirical insights into `MetaQuRe` showcased how algorithms exhibit benefits and weaknesses, depending on the objective and data at hand. This investigation evidences the imperative nature of resource-aware AutoML, which we address with our novel meta-learning approach. Notably, we were also able to demonstrate its practical efficacy – our method performs competitively to other AutoML approaches while requiring much fewer resources. Lastly, we contribute an extensive code base which fosters the reproducibility of findings and enables fellow researchers to adapt our ideas to their specific problems. We also see a lot of potential for future work, like devising dedicated methods for refining the algorithm pool, including further environments and data sets in `MetaQuRe`, exploring additional objectives (e.g., via preference learning), expanding the meta-feature space (e.g., via landmarking), or testing our data benefits for other AutoML tasks such as algorithm ranking or hyperparameter optimization. In conclusion, our endeavors mark a significant stride towards a more sustainable and efficient future of ML, and we believe that they can empower fellow researchers to be more resource-aware when navigating the complexities of AutoML.

References

1. Awad, N.H., Mallik, N., Hutter, F.: DEHB: evolutionary hyberband for scalable, robust and efficient hyperparameter optimization. In: Proceedings of the 30th International Joint Conference on Artificial Intelligence (IJCAI), pp. 2147–2153 (2021). https://doi.org/10.24963/ijcai.2021/296

2. Benmeziane, H., Maghraoui, K.E., Ouarnoughi, H., Niar, S., Wistuba, M., Wang, N.: A comprehensive survey on hardware-aware neural architecture search (2021). https://arxiv.org/abs/2101.09336

3. Chatila, R., et al.: Trustworthy AI. In: Braunschweig, B., Ghallab, M. (eds.) Reflections on Artificial Intelligence for Humanity. LNCS (LNAI), vol. 12600, pp. 13–39. Springer, Cham (2021). https://doi.org/10.1007/978-3-030-69128-8_2

4. Demšar, J.: Statistical comparisons of classifiers over multiple data sets. J. Mach. Learn. Res. (JMLR) **7**(1), 1–30 (2006). http://jmlr.org/papers/v7/demsar06a.html

5. das Dores, S.C.N., Soares, C., Ruiz, D.D.: Bandit-based automated machine learning. In: 7th Brazilian Conference on Intelligent Systems (BRACIS), pp. 121–126 (2018). https://doi.org/10.1109/BRACIS.2018.00029

6. Drori, I., et al.: AutoML using metadata language embeddings (2019). http://arxiv.org/abs/1910.03698

7. Elsken, T., Metzen, J.H., Hutter, F.: Neural architecture search: a survey. J. Mach. Learn. Res. (JMLR) **20**(55), 1–21 (2019). http://jmlr.org/papers/v20/18-598.html

8. Erickson, N., Mueller, J., Shirkov, A., Zhang, H., Larroy, P., et al.: AutoGluon-tabular: robust and accurate AutoML for structured data (2020). https://arxiv.org/abs/2003.06505

9. Falkner, S., Klein, A., Hutter, F.: BOHB: robust and efficient hyperparameter optimization at scale. In: Proceedings of the 35th International Conference on Machine Learning (ICML) (2018). http://proceedings.mlr.press/v80/falkner18a.html

10. Feurer, M., Eggensperger, K., Falkner, S., Lindauer, M., Hutter, F.: Auto-sklearn 2.0: hands-free AutoML via meta-learning. J. Mach. Learn. Res. (JMLR) **23**, 261:1–261:61 (2022). http://jmlr.org/papers/v23/21-0992.html

11. Feurer, M., Klein, A., Eggensperger, K., Springenberg, J.T., Blum, M., Hutter, F.: Efficient and robust automated machine learning. In: Advances in Neural Information Processing Systems 28 (NIPS Proceedings), pp. 2962–2970 (2015). https://proceedings.neurips.cc/paper/2015/hash/11d0e6287202fced83f79975ec59a3a6-Abstract.html

12. Fischer, R., Jakobs, M., Mücke, S., Morik, K.: A unified framework for assessing energy efficiency of machine learning. In: Workshop Proceedings of the European Conference on Machine Learning and Data Mining (ECML PKDD) (2022). https://doi.org/10.1007/978-3-031-23618-1_3

13. Fischer, R., Liebig, T., Morik, K.: Towards more sustainable and trustworthy reporting in machine learning. Data Mining Knowl. Discovery (2024). https://doi.org/10.1007/s10618-024-01020-3

14. Fischer, R., van der Staay, A., Buschjäger, S.: Stress-testing USB accelerators for efficient edge inference (2024). https://doi.org/10.21203/rs.3.rs-3793927/v1

15. Gijsbers, P., Bueno, M.L.P., Coors, S., LeDell, E., Poirier, S., et al.: AMLB: an AutoML benchmark. J. Mach. Learn. Res. (JMLR) **25**(101), 1–65 (2024). http://jmlr.org/papers/v25/22-0493.html

16. Giovanelli, J., Tornede, A., Tornede, T., Lindauer, M.: Interactive hyperparameter optimization in multi-objective problems via preference learning. Proc. AAAI Conf. Artif. Intell. **38**(11), 12172–12180 (2024). https://ojs.aaai.org/index.php/AAAI/article/view/29106

17. Hollmann, N., Müller, S., Eggensperger, K., Hutter, F.: TabPFN: a transformer that solves small tabular classification problems in a second (2023). https://arxiv.org/abs/2207.01848

18. Holzinger, A., Saranti, A., Molnar, C., Biecek, P., Samek, W.: Explainable AI methods - a brief overview, pp. 13–38 (2022). https://doi.org/10.1007/978-3-031-04083-2_2

19. Hutter, F., Kotthoff, L., Vanschoren, J.: Automated Machine Learning - Methods, Systems, Challenges. Springer, Cham (2019). https://doi.org/10.1007/978-3-030-05318-5

20. Jomaa, H.S., Schmidt-Thieme, L., Grabocka, J.: Dataset2Vec: learning dataset meta-features. Data Mining Knowl. Discovery **35**(3), 964–985 (2021). https://doi.org/10.1007/s10618-021-00737-9

21. Lacoste, A., Luccioni, A., Schmidt, V., Dandres, T.: Quantifying the carbon emissions of machine learning (2019). http://arxiv.org/abs/1910.09700

22. Mallik, N., et al.: PriorBand: practical hyperparameter optimization in the age of deep learning. In: Advances in Neural Information Processing Systems 36 (NeurIPS Proceedings), vol. 36, pp. 7377–7391 (2023). https://proceedings.neurips.cc/paper_files/paper/2023/file/1704fe7aaff33a54802b83a016050ab8-Paper-Conference.pdf

23. McInnes, L., Healy, J., Melville, J.: UMAP: uniform manifold approximation and projection for dimension reduction (2020). https://arxiv.org/abs/1802.03426

24. Mohr, F., Wever, M.: Naive automated machine learning. Mach. Learn. **112**(4), 1131–1170 (2023). https://doi.org/10.1007/s10994-022-06200-0

25. Mohr, F., Wever, M., Hüllermeier, E.: ML-Plan: automated machine learning via hierarchical planning. Mach. Learn. **107**(8-10), 1495–1515 (2018). https://doi.org/10.1007/s10994-018-5735-z

26. Mohr, F., Wever, M., Tornede, A., Hüllermeier, E.: Predicting machine learning pipeline runtimes in the context of automated machine learning. IEEE Trans. Pattern Anal. Mach. Intell. **43**(9), 3055–3066 (2021). https://doi.org/10.1109/TPAMI.2021.3056950

27. Morik, K., Kotthaus, H., Fischer, R., Mücke, S., Jakobs, M., et al.: Yes we care! - Certification for machine learning methods through the care label framework. Front. Artif. Intell. **5** (2022). https://doi.org/10.3389/frai.2022.975029

28. Nguyen, T.D., Maszczyk, T., Musial, K., Zöller, M.A., Gabrys, B.: Avatar - machine learning pipeline evaluation using surrogate model. In: Advances in Intelligent Data Analysis XVIII, pp. 352–365 (2020). https://doi.org/10.1007/978-3-030-44584-3_28

29. Olson, R.S., Moore, J.H.: TPOT: a tree-based pipeline optimization tool for automating machine learning. In: Hutter, F., Kotthoff, L., Vanschoren, J. (eds.) Automated Machine Learning. TSSCML, pp. 151–160. Springer, Cham (2019). https://doi.org/10.1007/978-3-030-05318-5_8

30. Pedregosa, F., Varoquaux, G., Gramfort, A., Michel, V., Thirion, B., et al.: Scikit-learn: machine learning in Python. J. Mach. Learn. Res. (JMLR) **12**(85), 2825–2830 (2011). http://jmlr.org/papers/v12/pedregosa11a.html

31. Rudin, C.: Stop explaining black box machine learning models for high stakes decisions and use interpretable models instead. Nat. Mach. Intell. **1**(5), 206–215 (2019). https://doi.org/10.1038/s42256-019-0048-x

32. Schwartz, R., Dodge, J., Smith, N.A., Etzioni, O.: Green AI. Commun. ACM **63**(12), 54–63 (2020). https://doi.org/10.1145/3381831
33. Souza, A., Nardi, L., Oliveira, L.B., Olukotun, K., Lindauer, M., Hutter, F.: Bayesian optimization with a prior for the optimum. In: Oliver, N., Pérez-Cruz, F., Kramer, S., Read, J., Lozano, J.A. (eds.) ECML PKDD 2021. LNCS (LNAI), vol. 12977, pp. 265–296. Springer, Cham (2021). https://doi.org/10.1007/978-3-030-86523-8_17
34. Thornton, C., Hutter, F., Hoos, H.H., Leyton-Brown, K.: Auto-WEKA: combined selection and hyperparameter optimization of classification algorithms. In: Proceedings of the 19th International Conference on Knowledge Discovery and Data Mining (KDD), pp. 847–855 (2013). https://doi.org/10.1145/2487575.2487629
35. Tornede, T., Tornede, A., Hanselle, J., Mohr, F., Wever, M., Hüllermeier, E.: Towards green automated machine learning: status quo and future directions. J. Artif. Intell. Res. (JAIR) **77**, 427–457 (2023). https://doi.org/10.1613/jair.1.14340
36. Vanschoren, J.: Meta-learning: a survey (2018). http://arxiv.org/abs/1810.03548
37. Vanschoren, J., van Rijn, J.N., Bischl, B., Torgo, L.: OpenML: networked science in machine learning. ACM SIGKDD Explor. Newsl. **15**(2), 49-60 (2014). https://doi.org/10.1145/2641190.2641198
38. Wistuba, M., Rawat, A., Pedapati, T.: A survey on neural architecture search (2019). http://arxiv.org/abs/1905.01392
39. Wistuba, M., Schilling, N., Schmidt-Thieme, L.: Two-stage transfer surrogate model for automatic hyperparameter optimization. In: Proceedings of the European Conference on Machine Learning and Data Mining (ECML PKDD), pp. 199–214 (2016). https://doi.org/10.1007/978-3-319-46128-1_13
40. van Wynsberghe, A.: Sustainable AI: AI for sustainability and the sustainability of AI. AI Ethics (2021). https://doi.org/10.1007/s43681-021-00043-6
41. Zela, A., Klein, A., Falkner, S., Hutter, F.: Towards automated deep learning: efficient joint neural architecture and hyperparameter search (2018). http://arxiv.org/abs/1807.06906
42. Zöller, M., Huber, M.F.: Benchmark and survey of automated machine learning frameworks. J. Artif. Intell. Res. (JAIR) (2021). https://doi.org/10.1613/jair.1.11854

Propagation Structure-Semantic Transfer Learning for Robust Fake News Detection

Mengyang Chen[1,2], Lingwei Wei[1(✉)], Han Cao[1,2], Wei Zhou[1], Zhou Yan[3], and Songlin Hu[1,2]

[1] Institute of Information Engineering, Chinese Academy of Sciences, Haidian District, Beijing 100085, China
{chenmengyang,weilingwei,caohan,zhouwei,husonglin}@iie.ac.cn,
{chenmengyang22,caohan22}@mails.ucas.ac.cn
[2] School of Cyber Security, University of Chinese Academy of Sciences, Huairou District, Beijing 101408, China
[3] State Key Laboratory of Communication Content Cognition, People's Daily Online, Chaoyang District, Beijing 100020, China
yanzhou@people.cn

Abstract. Fake news generally refers to false information that is spread deliberately to deceive people, which has detrimental social effects. Existing fake news detection methods primarily learn the semantic features from news content or integrate structural features from propagation. However, in practical scenarios, due to the semantic ambiguity of informal language and unreliable user interactive behaviors on social media, there are inherent semantic and structural noises in news content and propagation. Although some recent works consider the effect of irrelevant user interactions in a hybrid-modeling way, they still suffer from the mutual interference between structural noise and semantic noise, leading to limited performance for robust detection. To alleviate this issue, this paper proposes a novel Propagation Structure-Semantic Transfer Learning framework (PSS-TL) for robust fake news detection under a teacher-student architecture. Specifically, we design dual teacher models to learn semantics knowledge and structure knowledge from noisy news content and propagation structure independently. Besides, we design a Multi-channel Knowledge Distillation (MKD) loss to enable the student model to acquire specialized knowledge from the teacher models, thereby avoiding mutual interference. Extensive experiments on two real-world datasets validate the effectiveness and robustness of our method.

Keywords: Fake news detection · propagation structure learning · transfer learning · social networks

1 Introduction

Fake news generally refers to false information that is spread deliberately to deceive people [27]. The rapid advancement of online media has sparked a rise in

A. Bifet et al. (Eds.): ECML PKDD 2024, LNAI 14947, pp. 227–244, 2024.
https://doi.org/10.1007/978-3-031-70368-3_14

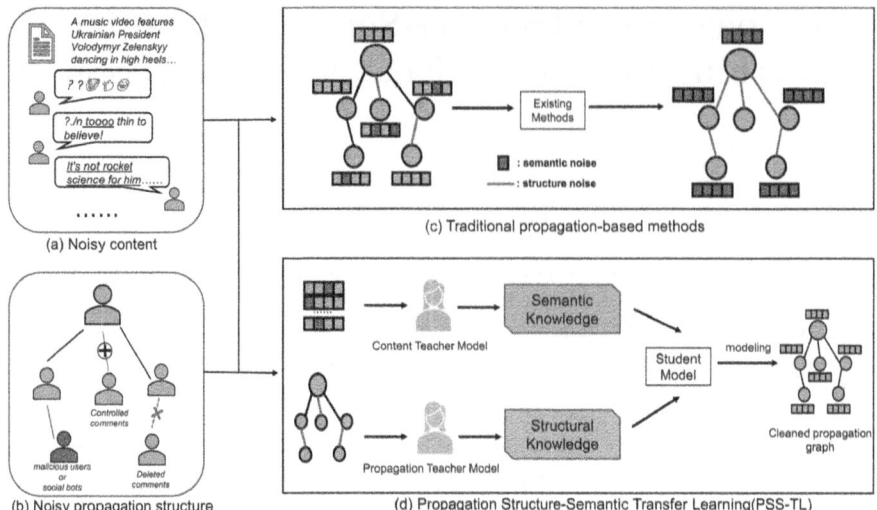

Fig. 1. The motivation of this paper. (a): Noisy content, including garbled characters, spelling errors, and idioms, contributes to semantic noise. (b) Unreliable interactions among users lead to structural noise in news propagation trees. (c) Previous works generally learn high-level features in a hybrid way. They would suffer from the mutual inference between the learning of noisy contents and incomplete propagation trees, leading to unreliable node representations. (d) Our PSS-TL adopts a dual teacher-student distillation framework, where two teachers independently transfer reliable semantic and structural features to the student model for robust detection.

fake news, which inflicts significant harm on society [5,6,31], making fake news detection into a research hotspot.

Existing methods of fake news detection mainly focus on textual content such as news text and contexts [2,20,37] and propagation information such as interactions between users [1,10,16,21,28,33]. Recent works make attempts to achieve robust detection by considering the effect of noise in practical scenes. There are inherent noises in news and its corresponding propagation trees. Due to the semantic ambiguity inherent in informal language, spelling errors, and garbled characters, there is semantic noise in the news and its comments (see Fig. 1(a)). Moreover, unreliable interactive behaviors among potential malicious users and social bots, deleted or controlled comments and limited collection of propagation data may also bring structural noise in propagation trees [22,36] (see Fig. 1(b))[1]. These studies mainly focus on the impact of noises in news propagation graphs and have made efforts to solve this problem through refining propagation structures [33–35] or training a robust detection model through contrastive learning or adversarial learning [9,18,29].

---

[1] Here, we divide the propagation graph of news into two parts: the content being spread (the content of news and comments in the propagation graph) and the propagation structure (the interactive relationships between news and comments).

However, in extracting high-level semantic and structural features through hybrid modeling, these models confront a significant challenge: the **mutual interference** arising between semantic and structural noise. Semantic noise introduces inaccuracies in feature representation, thereby skewing the computation of structural weights. In turn, structural noise, perpetuated by flawed adjacency relationships, infiltrates the feature-learning process, compounding the disruptive effects.

Therefore, the representations learned by these models are prone to containing unreliable knowledge. As shown in Fig. 1(c), during the modeling of propagation graphs, the noise in both nodes and edges, representing textual content and user behaviors respectively, will mutually impact each other and cause an error accumulation after the message passing. It will further make the model learn unreliable representations, i.e., most node embeddings may be poisoned, which will result in the limited performance of detection.

In this paper, we propose a novel Propagation Structure-Semantic Transfer Learning framework (PSS-TL) to alleviate the mutual interference between structural noise and semantic noise for robust fake news detection under the teacher-student architecture. Different from the above methods, we separately learn semantics and propagation through dual-channel transfer learning (see Fig. 1(d)). The separate modeling by two teachers would successfully prevent interaction between noisy semantic features and structural features, therefore alleviating the mutual interference between the extraction of semantic and structural features. The student model could receive reliable knowledge provided by authoritative teachers for better fake news detection.

Specifically, we design dual-teacher models to separately learn semantic and structure knowledge from the news content and the corresponding propagation trees. The content teacher captures effective semantic knowledge from news content using a multi-layer perceptron. Meanwhile, the propagation teacher extracts propagation knowledge by traversing a global propagation graph of the news, which is based on common users between the news and the positional encoding vectors of the news, utilizing graph convolutional networks. For the student model, we design a local-global propagation interaction module to fully utilize both local and global propagation information of news. It first integrates features by modeling local interactions in each propagation graph. Subsequently, it spreads this representation across the global graph to obtain a combined representation that integrates local and global propagation information.

To better transfer reliable semantic knowledge and propagation structure knowledge to the student model for detection, we design the Multi-channel Knowledge distillation (MKD) loss, which consists of two components: teacher supervision loss and targeted guidance loss. The teacher supervision loss consists of distillation losses between the logits generated by the teacher model and the student model, which allows teacher models to provide specialized knowledge to the student model. Targeted guidance loss includes alignment losses between the latent representations of two teacher models and the student model. It aims to maximize the consistency of representations of the same news between each teacher model and student model, making knowledge transfer from teacher models to student models more efficient. In this way, the student model can learn structure and semantic knowledge separately, thereby alleviating the mutual interference between structural noise and semantic noise.

We conduct experiments on two real-world public fake news datasets. The experimental results show that our PSS-TL significantly outperforms comparison baselines and achieves better detection performance, indicating the effectiveness and superiority of our method. Extensive experiments further indicate better generalization ability across domains and promising robust detection performance under various noisy scenarios.

The contributions of this work can be summarized as follows: 1) We propose a novel Propagation Structure-Semantic Transfer Learning framework (PSS-TL) for improving robust fake news detection. It alleviates the mutual interference between structural noise and semantic noise in propagation modeling via a dual teacher transfer learning network. 2) We design a new multi-channel knowledge distillation loss for better training the student model to pertinently transfer reliable structure and semantic knowledge to the student network. 3) Experimental results demonstrate the effectiveness and superiority of PSS-TL for fake news detection. Our PSS-TL obtains better robust detection performance under various noise scenarios[2].

2   Related Work

The goal of detecting fake news is to identify and assess the authenticity of a piece of information. Existing methods for detecting fake news mainly focus on two aspects: textual content and news propagation.

Text-based fake news detection methods extract semantic patterns from news textual content for detection through the employment of feature engineering [2,20,24] and a wide array of deep learning architectures, including neural networks [13,25] and pre-trained language models [11,12]. Some works also integrate tasks such as stance detection and sentiment analysis with fake news detection, enabling multi-task learning [7,17].

Propagation-based fake news detection methods capture the propagation patterns of news by modeling the interactions between news and comments into time series [16,19] or topological structures such as propagation trees [10,21] and propagation graphs [1,33,34]. Some studies further explore multi-relational interactions between the users and news in the propagation graph [4,39].

Recently, to alleviate the incomplete propagation issue, some works enhance the robustness of Graph Neural Network (GNN) models and augment the original propagation graph. Studies such as [9,18,29] propose different augmentations to the propagation structure for graph contrastive learning to enhance the robustness of GNN models. Others, like [32–35], optimize the original propagation graph through specific objective functions. However, these methods fail

---

[2] The code is available at github.com/IMCMY99/PSS-TL.

to consider the mutual interference between structural noise and semantic noise when modeling real-world propagation, as semantic noise spreads through noisy propagation structure and further intensifies the issue of noise in propagation, resulting in the limited performance of detection.

3   Propagation Structure-Semantic Transfer Learning Framework

In this section, we propose a novel Propagation Structure-Semantic Transfer Learning framework (PSS-TL), to independently address noise in both semantics and propagation structures, thereby avoiding mutual interference between them.

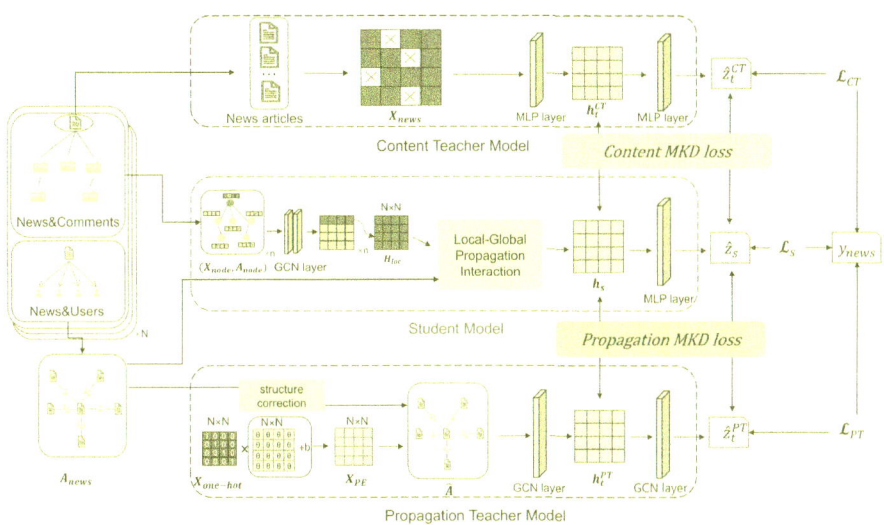

Fig. 2. The overall architecture of PSS-TL.

Problem Statement. Fake news detection is to verify the authenticity of a given news article, we take fake news detection as a binary classification problem, where each sample consists of news, comments, users, propagation structures, and user engagements, with each sample annotated with a ground truth label indicating its authenticity. Formally, Dataset $\mathcal{D}$ consists of N samples and each sample is represented by $\mathcal{G} = (\mathcal{V}, \mathcal{E}, \mathcal{U}_{\mathcal{G}}, \mathcal{S}_{\mathcal{G}})$, where $\mathcal{V} = \{r, c_1, ..., c_N\}$ represents the features of the news r and its comments $(c_1, ..., c_N)$, $\mathcal{U}_{\mathcal{G}}$ represents the users engaged in the news, $\mathcal{E}$ represents a set of explicit interactive behaviors, e.g., retweet, and $\mathcal{S}_{\mathcal{G}} = \{s_u, u \in \mathcal{U}_{\mathcal{G}}\}$ represents the engagements of users in news, in which $s_u \in \mathcal{S}_{\mathcal{G}}$ is the number of times user u interact with news. The task objective of fake news detection is to learn a classifier f to classify samples and determine whether the news is true (labeled as 0) or false (labeled as 1), i.e.,

$$f : \mathcal{G} \longrightarrow y, \quad y \in \{0, 1\}.$$

3.1   Overview

As shown in Fig. 2, PSS-TL has three components: a content teacher model, a propagation teacher model with structure correction and a local-global propagation interaction module. The content teacher captures semantic knowledge through a multi-layer perceptron. The propagation teacher extracts structural knowledge from a global perspective through the common participating users among news, with the structure correction mechanism which is based on the frequency of common users. The Local-Global Propagation Interaction (LGPI) module is designed to fully integrate local propagation information and global propagation information. We design the Multi-channel Knowledge distillation (MKD) loss for model training. Through MKD loss, the professional knowledge in the teacher models is fully independently extracted to the student model, allowing the student model to learn semantic and structural knowledge separately, avoiding mutual interference between semantics and propagation structures.

3.2   Dual Teacher Models

As there will be mutual interference of semantic noise and structural noise when modeling content and propagation jointly, we adopt dual-teacher models to learn semantic and structural knowledge from content and propagation containing noise, respectively.

Content Teacher Model. We utilize multi-layer perceptron as the backbone of the content teacher model for automatic semantic feature learning from noisy content. Given the input content X_{news}, the hidden representation h_t^{CT} and output $\hat{z}_t^{CT}$ can be computed as,

$$\begin{aligned} h_t^{CT} &= \mathrm{MLP}_1(X_{\mathrm{news}}) \\ \hat{z}_t^{CT} &= \mathrm{ReLU}(\mathrm{MLP}_2(h_t^{CT})). \end{aligned} \tag{1}$$

Propagation Teacher Model. Inspired by Yuan et al. [38], we construct a global propagation graph $A_{\mathrm{news}} \in \mathbb{R}^{N \times N}$ based on the shared participation of users among news items and refine it to learn structural knowledge, which can effectively reduce computational resource consumption.

Firstly, we construct a user-engagements graph $E \in \mathbb{R}^{N \times |\mathcal{U}|}$, where $\mathcal{U} = \cup_{\mathcal{G}} \mathcal{U}_{\mathcal{G}}$ contains all users in the dataset $\mathcal{D}$ and $E_{i,j}$ represents the number of times users u_j participate in news $\mathcal{G}_i$. And then we get the global propagation graph A_{news} by:

$$A_{\text{news}} = EE'. \tag{2}$$

Then we design a positional encoding vector to avoid the interference from semantic noise on structure learning. Formally, for each sample, we define a learnable propagation vector using a one-hot positional encoding vector $X_{\text{one-hot}} \in \mathbb{R}^{n \times n}$ as follows:

$$X_{PE} = W X_{\text{one-hot}} + b, \tag{3}$$

where W is a learnable weight matrix and b is a learnable bias vector.

Some studies have found that the common users between news are related to the veracity of the news pairs. inspired by Karrer et al. [14] and Newman [23], we employ a multi-layer perceptron to get the refined global propagation structure $\hat{A}_{\text{news}}$ by increasing the weight of the edges between news items of the same veracity and decreasing those of different veracity in the original global propagation structure A_{news}. The formal computation is as follows:

$$\hat{A}_{\text{news}} = A_{\text{news}} M + I, \tag{4}$$

where I is the identity matrix and M is the edge retention matrix, calculated based on the degrees of the pairwise propagating nodes:

$$M_{ij} = \begin{cases} \text{Softmax}\big(\text{MLP}(A_{\text{news},ij}, d_i, d_j)\big) & \text{if } A_{\text{news},ij} \neq 0 \\ 0 & \text{else} \end{cases}. \tag{5}$$

Given the propagation vector X_{PE} and $\hat{A}_{\text{news}}$, we utilize a two-layer graph convolutional network (GCN) to compute the hidden representation h_{str} of all news and the output $\hat{z}_{str}$:

$$\begin{aligned} h_t^{PT} &= \text{GCN}_1(\hat{A}_{\text{news}}, X_{PE}) \\ \hat{z}_t^{PT} &= \text{ReLU}(\text{GCN}_2(h_t^{PT})). \end{aligned} \tag{6}$$

Training of Dual Teacher Models. We train the above teacher models by using the Cross-Entropy classification loss of the predicted results outputted by the teacher models. The objective can be defined as follows:

$$\mathcal{L}_{tea} = \text{CE}(\hat{z}_t^{tea}, y_{\text{news}}), \quad tea \in \{CT, PT\}, \tag{7}$$

where $\text{CE}(\cdot)$ denotes the cross-entropy loss, and $y_{\text{news}} \in \{0, 1\}$ represents the ground-truth label of the news.

3.3 Local-Global Propagation Interaction Enhanced Student Model

To better utilize the local and global propagation information of news, we design the Local-Global Propagation Interaction module.

As the news propagation graphs are fed into the student model and local propagation information of all news is output, the learned local propagation information is further spread on the global propagation structure constructed by the propagation teacher through the LGPI module, as shown in Eq. (9).

Taking GCN as an example, the node features X_{node} of each news and its propagation graph A_{node} are fed into GCN to obtain local representations H_{loc}. Then, the local representations $H_{loc \times n}$ of all news items and the global propagation graph A_{news} are fed into the LGPI module to incorporate inter-news relationships, resulting in a representation h_s that captures the contextual information among news. This representation is then fed into a linear layer to produce the predicted output $\hat{z}_s$. The specific steps are as follows:

$$H_{loc} = \mathrm{ReLU}(\mathrm{GCN}_{loc}(X_{\mathrm{node}}, A_{\mathrm{node}})) \tag{8}$$

$$h_s = W_1(A_{\mathrm{news}}H_{loc \times N}) + b_1 \tag{9}$$

$$\hat{z}_s = \mathrm{ReLU}(W_2 h_s + b_2), \tag{10}$$

where W_1 and W_2 are parameter matrices, and b_1 and b_2 are bias terms.

3.4   Multi-channel Knowledge Distillation Training Objective

To better transfer the reliable knowledge obtained by each teacher to the student model, we develop a novel Multi-channel Knowledge Distillation (MKD) loss to perform effective and targeted supervision in the training stage of the student model. It mainly involves teacher supervision loss and targeted guidance loss.

Teacher Supervision Loss. $\mathcal{L}_{\mathrm{sup}}$ measures the distillation loss that arises from the teacher model's supervision to the student model's output. It quantifies the degree of alignment between the two models' predictions, i.e.,

$$\mathcal{L}_{\mathrm{sup}} = D_{\mathrm{KL}}\left(\mathrm{Softmax}\left(\frac{\hat{z}_s}{\rho}\right), \mathrm{Softmax}\left(\frac{\hat{z}_t}{\rho}\right)\right), \tag{11}$$

where $D_{\mathrm{KL}}(\cdot)$ represents the Kullback-Leibler divergence between two distributions. $\hat{z}_s$ and $\hat{z}_t$ refer to the predicted output generated by the student model and a teacher model. ρ is the temperature parameter to adjust the smoothness of the output generated by the teacher models.

Targeted guidance loss $\mathcal{L}_{\mathrm{tar}}$ serves as additional supplementation of knowledge from the teacher model to the student model. It achieves consistency between the hidden representations of the same news between the teacher model and the student model while minimizing consistency between the hidden representations of different news, i.e.,

$$\mathcal{L}_{\mathrm{tar}} = \frac{1}{|V|} \sum_{v_i \in V} \log \frac{e^{\langle h_{s,i}, h_{t,i} \rangle}}{e^{\langle h_{s,i}, h_{t,i} \rangle} + \sum_{j \neq i} e^{\langle h_{s,i}, h_{t,j} \rangle}}, \tag{12}$$

where h_s and h_t refer to the hidden representations in the modeling of the student model and teacher model, respectively.

Thus, the total objective of MKD loss can be defined as,

$$\mathcal{L}_{\mathrm{MKD}} = \mathcal{L}_{\mathrm{sup}} + \mathcal{L}_{\mathrm{tar}}. \tag{13}$$

To sum up, the total objective of training the student model can be,

$$\mathcal{L}_s = \mathcal{L}_{\mathrm{CLS}} + \underbrace{\lambda\mathcal{L}_{\mathrm{sup}}^{PT} + \beta\mathcal{L}_{\mathrm{tar}}^{PT}}_{\mathcal{L}_{\mathrm{MKD}}^{PT}} + \underbrace{(1-\lambda)\mathcal{L}_{\mathrm{sup}}^{CT} + (1-\beta)\mathcal{L}_{\mathrm{tar}}^{CT}}_{\mathcal{L}_{\mathrm{MKD}}^{CT}}, \tag{14}$$

where $\mathcal{L}_{\mathrm{CLS}}$ refers to the cross-entropy loss based on the predicted and ground-truth labels for fake news detection. λ and $\beta \in [0,1]$ are two hyperparameters to control the emphasis of supervision loss and targeted guidance loss, respectively. $\mathcal{L}_{\mathrm{MKD}}^{CT}$ and $\mathcal{L}_{\mathrm{MKD}}^{PT}$ are the reliable knowledge distillation losses from content and propagation teacher models to the student model, respectively.

Table 1. The statistics of datasets.

Datasets	PolitiFact	GossipCop	COAID
Number of News	314	5,464	186
Number of True News	157	2,732	166
Number of False News	157	2,732	18
Number of Posts	40,740	308,798	15,996
Number of Users	30,812	75,914	11,467

4 Experiment

4.1 Experimental Setups

Datasets. We evaluate our method on two real-world datasets, PolitiFact and GossipCop [26]. PolitiFact and GossipCop include news articles from fact-checking websites PolitiFact and GossipCop. To leverage user engagement, we utilize news data collected by Dou et al. [4]. Besides, to achieve cross-domain detection, we utilize COAID [3] as the target domain for generalization evaluation, which consists of 185 claims about Covid-19 healthcare and 15,996 posts. We divided the datasets into training, validation, and testing sets in a ratio of 7:1:2. The statistics of these datasets are shown in Table 1.

Comparison Methods. We compare with the following representative fake news detection methods.
GCN [15] learns the representation of each news article by performing graph convolutional operations on the news propagation graph. **GAT** [30] employs a

Graph Attention Network to encode the propagation structure of news. **Graph-SAGE** [8] models the news propagation graph using a Graph Sampling Aggregation Network, learning the representation of each article by randomly sampling and aggregating the features of neighboring nodes. **Bi-GCN** [1] constructs a bidirectional propagation graph based on the news propagation graph to enhance the effectiveness of representation learning. **EBGCN** [33] is a Bayesian graph convolutional network-based rumor detection methodology that models uncertainty within rumor propagation. **UPSR** [34] constructs an uncertainty-aware propagation structure reconstruction graph based on the original news propagation graph, enhancing the effectiveness of representation learning through Gaussian estimation. **UPFD** [4] combines various signals such as news text features and user preference features by jointly modeling news content and a propagation tree using a connection. **DECOR** [35] constructs a news social graph based on the engagement of news users and optimizes the social graph using a stochastic blockmodel to enhance the effectiveness of graph-based representation learning for fake news detection.

Implementation Details. All experiments were conducted on a single Tesla V100, using PyTorch version 1.12.1 and Geometric version 2.3.1. We employ a pre-trained BERT (*bert-base-uncased*) to encode the textual features of each news article and comment. The dimension of the hidden vectors is set to 64, and the content teacher model utilizes a two-layer MLP, while the propagation teacher model employs a two-layer GCN. The learning rate for the propagation teacher model and student model is set to $5e-4$ and that for the content teacher is set to $5e-5$. We search the optimal parameters λ and β from {0.1, 0.2, 0.3, 0.4, 0.5, 0.6, 0.7, 0.8, 0.9}, and the temperature parameter ρ in {1, 2, 5, 7, 10}. We implement comparison methods under the same environment according to the parameter setting reported in their original papers. We run five times and report the average results.

We use accuracy and macro-average F1 score as the evaluation metrics for each model. In subsequent sub-experiments, all methods employ GCN as the backbone GNN.

4.2 Main Results

The overall experimental results are shown in Table 2. From the tables, it can be observed that under different backbones of student models, PSS-TL achieves state-of-the-art detection performance, showing the superiority of our method.

From the results, we have the following observations:

1) Some methods such as DECOR, EBGCN, and UPSR that account for noisy user behaviors in the modeling of propagation, exhibit significantly better performance than those that ignore the effect of propagation noise, such as BiGCN and basic GNNs.
2) Our method outperforms DECOR, EBGCN, and UPSR on both datasets. This can be attributed to the dual-teacher knowledge distillation architecture

Table 2. Results (%) for fake news detection on PolitiFact and GossipCop. The best result is in boldface. For each method, we run five times and report the average results.

Models	PolitiFact		GossipCop	
	Accuracy	Macro-F1	Accuracy	Macro-F1
GCN	81.63 ± 1.82	81.46 ± 1.89	94.88 ± 0.11	94.84 ± 0.11
GAT	82.72 ± 0.79	82.61 ± 0.81	95.35 ± 0.17	95.27 ± 0.17
GraphSAGE	82.81 ± 1.54	82.68 ± 1.57	95.21 ± 1.48	95.16 ± 1.50
BiGCN	84.62 ± 0.46	82.98 ± 0.46	96.06 ± 0.98	95.94 ± 0.99
EBGCN	89.87 ± 3.00	89.29 ± 3.16	96.70 ± 0.71	96.75 ± 0.69
UPSR	91.46 ± 2.43	90.86 ± 2.43	96.86 ± 0.61	96.72 ± 0.63
UPFD				
w/GCN	82.99 ± 2.78	82.69 ± 2.92	95.35 ± 0.38	95.30 ± 0.38
w/GAT	82.99 ± 0.84	82.90 ± 0.86	96.01 ± 0.41	95.97 ± 0.41
w/GraphSAGE	85.81 ± 3.13	85.60 ± 3.05	95.78 ± 0.36	95.75 ± 0.36
DECOR				
w/GCN	88.71 ± 0.00	88.47 ± 0.00	97.22 ± 0.12	97.21 ± 0.12
w/GAT	91.29 ± 0.82	91.03 ± 0.79	96.70 ± 0.06	96.70 ± 0.06
w/GraphSAGE	88.39 ± 0.64	88.12 ± 0.69	96.60 ± 0.07	96.60 ± 0.07
PSS-TL(Ours)				
w/GCN	$\mathbf{93.23} \pm 0.18$	92.47 ± 0.19	$\mathbf{97.71} \pm 0.19$	97.62 ± 0.20
w/GAT	93.05 ± 0.12	$\mathbf{92.94} \pm 0.17$	$\mathbf{97.71} \pm 0.16$	$\mathbf{97.71} \pm 0.16$
w/ GraphSAGE	92.83 ± 0.14	92.86 ± 0.17	97.23 ± 0.027	97.23 ± 0.17

effectively preventing interference between the semantic and structural noises. The dual teacher models can transfer reliable knowledge to the student model for better detection.

3) When the student model adopts various GNN backbones, our PSS-TL gains promising performance consistently on both datasets. It shows the insensitivity of our framework to different GNN architectures.

4.3 Ablation Study

We conducted ablation experiments to evaluate the key components of our method. **w/o Content Teacher** refers to removing the content teacher component. **w/o Propagation Teacher** indicates removing the propagation teacher component. **w/o $\mathcal{L}_{tar}$** and **w/o $\mathcal{L}_{sup}$** refer to removing the targeted guidance loss and teacher supervision loss from dual teacher models to the student model in our MKD loss, respectively. **w/o LGPI** means removing the Local-Global Propagation Interaction module in the student model.

The results are shown in Table 3. Our full PSS-TL achieves better performance on both datasets. It can be observed that: 1) On PolitiFact and GossipCop, removing the propagation teacher or content teacher leads to a more

Table 3. Results (%) of ablation study. The best result is in boldface.

Methods	PolitiFact		GossipCop	
	Accuracy	Macro-F1	Accuracy	Macro-F1
PSS-TL	**93.23**	**92.47**	**97.71**	**97.62**
w/o Content Teacher	90.32	89.29	97.44	97.32
w/o Propagation Teacher	91.94	90.91	97.25	97.13
w/o L_{tar}	91.94	90.91	97.25	97.18
w/o L_{sup}	88.71	86.27	97.16	97.03
w/o LGPI	90.32	88.89	97.20	97.19

pronounced decrease in performance, which indicates that the two teachers are capable of effectively capturing semantics and propagation characteristics within noisy data respectively. 2) When removing teacher supervision loss, our method achieves worse performance on both evaluation metrics, demonstrating the usefulness of teachers' supervision to the student. 3) Meanwhile, the performance of w/o L-tar is significantly lower than PSS-TL, validating our hypothesis that targeted guidance can provide additional knowledge supplementation for the student. 4) As the student model relies solely on local propagation information for detection, w/o LGPI is inferior to PSS-TL, confirming the effectiveness of the LGPI module in our method.

4.4 Generalization Evaluation

We conduct cross-domain fake news detection to evaluate the generalization of our method. The results are shown in Table 4 where GCN is adopted as the GNN backbone for PSS-TL and DECOR.

The PSS-TL method exhibits remarkable performance, outperforming other techniques. Specifically, compared to DECOR, PSS-TL achieves a 6.25% enhancement in accuracy when transferring from GossipCop to COAID, and a 2.95% improvement when transferring from PolitiFact to COAID. The results indicate that PSS-TL has a kind of generalization ability, which can be attributed to its effective transfer learning mechanisms. Moreover, our method simultaneously models the structural information of user interactions and the semantic information of content, thereby obtaining more general and more discriminative detection features for cross-domain detection.

4.5 Robustness Evaluation

To further validate the effectiveness and robustness of our method under three noisy scenarios, single semantic noises, single structural noises and mixed noises. we introduce random masking to the node features and edges in the news propagation graph and conduct tests on datasets with different ratios of noise.

Table 4. Performance comparison of PSS-TL and DECOR in cross-domain detection. *PolitiFact* ⟶ COAID refers to training on the source domain (i.e., PolitiFact) and testing on the target domain (i.e., COAID).

Methods	PolitiFact ⟶ COAID		GossipCop ⟶ COAID		COAID ⟶ COAID	
	Accuracy	Macro-F1	Accuracy	Macro-F1	Accuracy	Macro-F1
DECOR	89.47	93.94	84.21	90.91	97.37	93.70
UPSR	84.21	91.43	73.68	68.42	94.74	97.06
PSS-TL	**92.11**	**95.65**	**89.47**	**94.12**	**100.00**	**100.00**

Figures 3, 4 and 5 show the results of different methods under semantic, structural and mixed noises, respectively. Our method consistently outperforms other methods and exhibits more stable performance under different noisy scenarios and noise rates. In the semantic noise scenario, PSS-TL performs relatively better on PolitiFact and even maintains its performance on GossipCop, whereas some propagation-based methods such as Bi-GCN suffer significantly on PolitiFact. In the structural noise scenario, some methods such as UPSR may benefit from the structural noise but still perform less than PSS-TL. As shown in Fig. 5 and Fig. 6, for mixed noise scenarios, some methods for addressing propagation noise have faced challenges, for instance, when the noise ratio is 0.5 on PolitiFact, EBGCN, UPSR, and DECOR have been subjected to more severe challenges than the other two types of noise scenarios. This discrepancy might be attributed to the mutual interference between semantic noises and structural noises. Nevertheless, our method still suffers less in such a scenario, which validates the effectiveness and robustness of our method under various scenarios.

4.6 Parameter Analysis

We explore the performance of PSS-TL against three vital parameters, *i.e.*, the two parameters to balance the influence of different teachers in teacher supervision (λ) and targeted guidance (β), and the temperature parameter in teacher

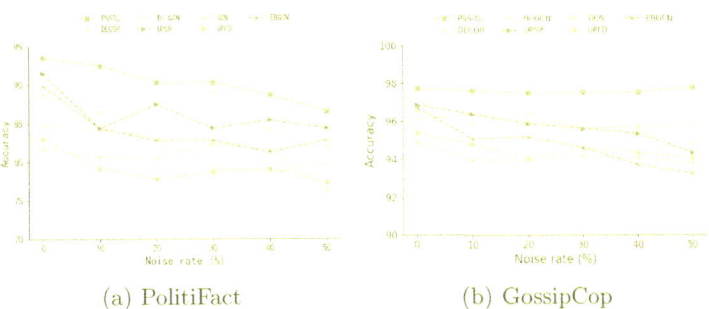

(a) PolitiFact   (b) GossipCop

Fig. 3. Robust detection results (%) against different ratios of semantic noises.

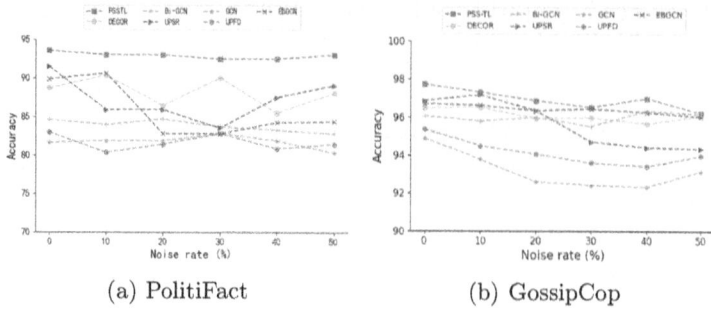

Fig. 4. Robust detection results (%) against different ratios of structural noises.

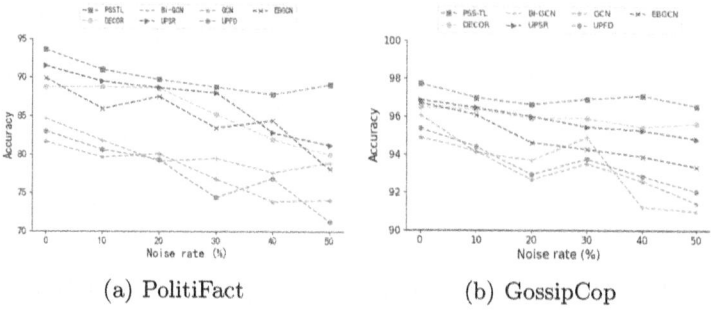

Fig. 5. Robust detection results (%) against different ratios of mixed noises (i.e., semantic and structural noises).

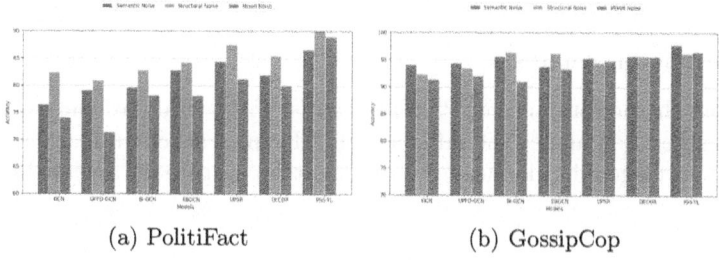

Fig. 6. Robust detection results (%) against different models with different noise (The ratio of noise is 0.5).

supervision (ρ). Figure 7 and Fig. 8 show the results of PSS-TL with different values of λ and β.

Effect of λ. The larger λ is, the more the student model tends to be supervised by the propagation teacher. The best settings are 0.1 and 0.2 on GossipCop and PolitiFact, respectively. This is due to the greater impact of noise with more nodes in the graph, necessitating more guidance from the propagation teacher.

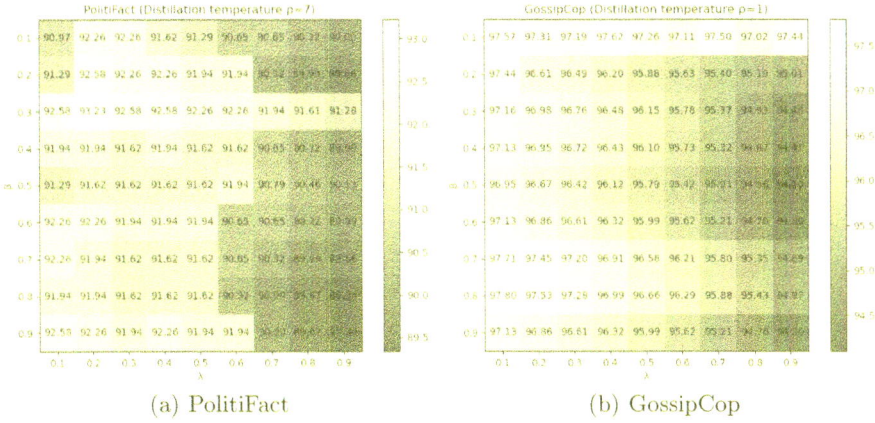

Fig. 7. Parameter analysis of λ and β.

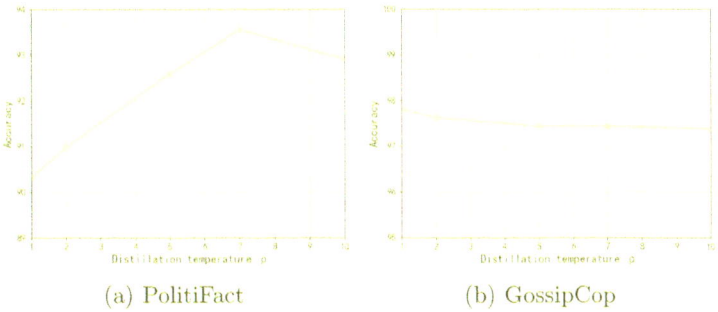

Fig. 8. Parameter analysis of ρ.

Effect of β. β represents the extent to which the propagation teacher supplements the student model's knowledge in targeted guidance. The optimal setting is 0.8 and 0.3 on GossipCop and PolitiFact, respectively. The differences between β and λ highlight the necessity of weighing guidance for different teachers in targeted instruction.

Effect of ρ. The temperature parameter ρ adjusts the smoothness of the output generated by teacher models. The larger the ρ, the student model will receive more knowledge from the teacher model. we fix the best settings for λ and β. The optimal setting is 1 and 7 on GossipCop and PolitiFact, respectively. It suggests that on the PolitiFact dataset, student models need to acquire more knowledge because the interaction between content noise and propagation noise has a greater impact on this dataset.

5   Conclusion

This paper investigates the issue of mutual interference between feature noise and propagation structural noise in fake news detection. We propose a multi-channel teacher-student knowledge transfer method, which enables the student model to learn content knowledge and propagation knowledge of news separately based on the guidance of the teacher model, without being interfered with by the noise of both. Experiments on two real-world datasets demonstrate the effectiveness of the proposed method.

Acknowledgements. This work was supported by the National Key Research and Development Program of China (No. 2022YFC3302102), and the Postdoctoral Fellowship Program of CPSF (No. GZC20232969).

References

1. Bian, T., Xiao, X., Xu, T., et al.: Rumor detection on social media with bi-directional graph convolutional networks. In: AAAI, vol. 34, pp. 549–556 (2020)
2. Castillo, C., Mendoza, M., Poblete, B.: Information credibility on twitter. In: WWW, pp. 675–684 (2011)
3. Cui, L., Lee, D.: CoAID: COVID-19 healthcare misinformation dataset. arXiv preprint arXiv:2006.00885 (2020)
4. Dou, Y., Shu, K., Xia, C., et al.: User preference-aware fake news detection. In: SIGIR, pp. 2051–2055 (2021)
5. Faris, R., Roberts, H., Etling, B., Bourassa, N., Zuckerman, E., Benkler, Y.: Partisanship, propaganda, and disinformation: online media and the 2016 US presidential election. Berkman Klein Center Research Publication **6** (2017)
6. Fisher, M., Cox, J.W., Hermann, P.: Pizzagate: from rumor, to hashtag, to gunfire in DC. Washington Post **6**, 8410–8415 (2016)
7. Hamed, S.K., Ab Aziz, M.J., Yaakub, M.R.: Fake news detection model on social media by leveraging sentiment analysis of news content and emotion analysis of users' comments. Sensors **23**(4), 1748 (2023)
8. Hamilton, W., Ying, Z., Leskovec, J.: Inductive representation learning on large graphs. NIPS **30** (2017)
9. He, Z., Li, C., Zhou, F., et al.: Rumor detection on social media with event augmentations. In: SIGIR, pp. 2020–2024 (2021)
10. Hu, D., Wei, L., Zhou, W., et al.: A rumor detection approach based on multi-relational propagation tree. J. Comput. Res. Dev. **58**(7), 1395–1411 (2021)
11. Jwa, H., Oh, D., Park, K., Kang, J.M., Lim, H.: exBAKE: automatic fake news detection model based on bidirectional encoder representations from transformers (BERT). Appl. Sci. **9**(19), 4062 (2019)
12. Kaliyar, R.K., Goswami, A., Narang, P.: FakeBERT: fake news detection in social media with a BERT-based deep learning approach. Multimedia Tools Appl. **80**(8), 11765–11788 (2021)
13. Karimi, H., Tang, J.: Learning hierarchical discourse-level structure for fake news detection. In: NAACL, pp. 3432–3442 (2019)
14. Karrer, B., Newman, M.E.: Stochastic blockmodels and community structure in networks. Phys. Rev. E **83**(1), 016107 (2011)

15. Kipf, T.N., Welling, M.: Semi-supervised classification with graph convolutional networks. In: ICLR (2016)
16. Liu, Y., fang Brook Wu, Y.: Early detection of fake news on social media through propagation path classification with recurrent and convolutional networks. In: AAAI (2018)
17. Luvembe, A.M., Li, W., Li, S., et al.: Dual emotion based fake news detection: a deep attention-weight update approach. IPM **60**(4), 103354 (2023)
18. Ma, G., Hu, C., Ge, L., et al.: Towards robust false information detection on social networks with contrastive learning. In: CIKM, pp. 1441–1450 (2022)
19. Ma, J., Gao, W., Mitra, P., et al.: Detecting rumors from microblogs with recurrent neural networks (2016)
20. Ma, J., Gao, W., Wei, Z., et al.: Detect rumors using time series of social context information on microblogging websites. In: CIKM, pp. 1751–1754 (2015)
21. Ma, J., Gao, W., Wong, K.F.: Rumor detection on twitter with tree-structured recursive neural networks. ACL (2018)
22. Ma, J., Gao, W., Wong, K.F.: Detect rumors on twitter by promoting information campaigns with generative adversarial learning. In: WWW, pp. 3049–3055 (2019)
23. Newman, M.E.: Network structure from rich but noisy data. Nat. Phys. **14**(6), 542–545 (2018)
24. Popat, K.: Assessing the credibility of claims on the web. In: WWW, pp. 735–739 (2017)
25. Ruchansky, N., Seo, S., Liu, Y.: CSI: a hybrid deep model for fake news detection. In: CIKM, pp. 797–806 (2017)
26. Shu, K., Mahudeswaran, D., Wang, S., et al.: FakeNewsNet: a data repository with news content, social context, and spatiotemporal information for studying fake news on social media. Big Data **8**(3), 171–188 (2020)
27. Shu, K., Sliva, A., Wang, S., et al.: Fake news detection on social media: a data mining perspective. ACM SIGKDD Explor. Newsl. **19**(1), 22–36 (2017)
28. Song, C., Shu, K., Wu, B.: Temporally evolving graph neural network for fake news detection. IPM **58**(6), 102712 (2021)
29. Sun, T., Qian, Z., Dong, S., et al.: Rumor detection on social media with graph adversarial contrastive learning. In: WWW, pp. 2789–2797 (2022)
30. Veličković, P., Cucurull, G., Casanova, A., Romero, A., Liò, P., Bengio, Y.: Graph attention networks. In: ICLR (2018)
31. Vosoughi, S., Roy, D., Aral, S.: The spread of true and false news online. Science **359**(6380), 1146–1151 (2018)
32. Wei, L., Hu, D., Zhou, W., Wang, X., Hu, S.: Modeling the uncertainty of information propagation for rumor detection: a neuro-fuzzy approach. IEEE Trans. Neural Netw. Learn. Syst. **35**(2), 2522–2533 (2024). https://doi.org/10.1109/TNNLS.2022.3190348
33. Wei, L., Hu, D., Zhou, W., et al.: Towards propagation uncertainty: Edge-enhanced Bayesian graph convolutional networks for rumor detection. In: ACL, pp. 3845–3854, August 2021
34. Wei, L., Hu, D., Zhou, W., et al.: Uncertainty-aware propagation structure reconstruction for fake news detection. In: COLING, pp. 2759–2768 (2022)
35. Wu, J., Hooi, B.: DECOR: degree-corrected social graph refinement for fake news detection. In: KDD, pp. 2582–2593 (2023)
36. Yang, X., Lyu, Y., Tian, T., et al.: Rumor detection on social media with graph structured adversarial learning. In: IJCAI, pp. 1417–1423 (2021)
37. Yu, F., Liu, Q., Wu, S., et al.: A convolutional approach for misinformation identification. In: IJCAI, pp. 3901–3907 (2017)

38. Yuan, C., Ma, Q., Zhou, W., et al.: Jointly embedding the local and global relations of heterogeneous graph for rumor detection. In: ICDM. IEEE (2019)
39. Yuan, C., Ma, Q., Zhou, W., et al.: Early detection of fake news by utilizing the credibility of news, publishers, and users based on weakly supervised learning. In: COLING, pp. 5444–5454 (2020)

Exploring Contrastive Learning for Long-Tailed Multi-label Text Classification

Alexandre Audibert[(✉)], Aurélien Gauffre, and Massih-Reza Amini

Université Grenoble Alpes, CNRS/LIG, 150 place du Torrent,
38401 Saint-Martin d'Hères, France
{alexandre.audibert,aurelien.gauffre,
massih-reza.amini}@univ-grenoble-alpes.fr

Abstract. Learning an effective representation in multi-label text classification (MLTC) is a significant challenge in natural language processing. This challenge arises from the inherent complexity of the task, which is shaped by two key factors: the intricate connections between labels and the widespread long-tailed distribution of the data. To overcome this issue, one potential approach involves integrating supervised contrastive learning with classical supervised loss functions. Although contrastive learning has shown remarkable performance in multi-class classification, its impact in the multi-label framework has not been thoroughly investigated. In this paper, we conduct an in-depth study of supervised contrastive learning and its influence on representation in MLTC context. We emphasize the importance of considering long-tailed data distributions to build a robust representation space, and we identify two critical challenges associated with contrastive learning: the "lack of positives" and the "attraction-repulsion imbalance". Building on these insights, we introduce a novel contrastive loss function for MLTC. It attains Micro-F1 scores that either match or surpass those obtained with other frequently employed loss functions, and demonstrates a significant improvement in Macro-F1 scores across four multi-label datasets.

Keywords: Supervised Contrastive Learning · Multi-Label Text Classification

1 Introduction

In recent years, multi-label text classification has gained significant popularity in the field of natural language processing (NLP). Defined as the process of assigning one or more labels to a document, MLTC plays a crucial role in numerous real-world applications such as document classification, sentiment analysis, and news article categorization.

Despite its similarity to multi-class mono-label text classification, MLTC presents two fundamental challenges: handling multiple labels per document

A. Bifet et al. (Eds.): ECML PKDD 2024, LNAI 14947, pp. 245–261, 2024.
https://doi.org/10.1007/978-3-031-70368-3_15

and addressing datasets that tend to be long-tailed. These challenges highlight the inherent imbalance in real-world applications, where some labels are more present than others, making it hard to learn a robust semantic representation of documents.

Numerous approaches have emerged to address this issue, such as incorporating label interactions in model construction and devising tailored loss functions. Some studies advocate expanding the representation space by incorporating statistical correlations through graph neural networks in the projection head [25,28]. Meanwhile, other approaches recommend either modifying the conventional Binary Cross-Entropy (BCE) by assigning higher weights to certain samples and labels or introducing an auxiliary loss function for regularization [32]. Concurrently, recent approaches based on supervised contrastive learning employed as an auxiliary loss managed to enhance semantic representation in multi-class classification [5,10].

While contrastive learning represents an interesting tool, its application in MLTC remains challenging due to several critical factors. Firstly, defining a positive pair of documents is difficult due to the interaction between labels. Indeed, documents can share some but not all labels, and it can be hard to clearly evaluate the degree of similarity required for a pair of documents to be considered positive. Secondly, the selection of effective data augmentation techniques necessary in contrastive learning proves to be a non-trivial task. Unlike images, where various geometric transformations are readily applicable, the discrete nature of text limits the creation of relevant augmentations. Finally, the data distribution in MLTC often shows an unbalanced or long-tailed pattern, with certain labels being noticeably more common than others. This might degrade the quality of the representation [9,33]. Previous research in MLTC has utilized a hybrid loss [2], combining supervised contrastive learning with classical BCE [15], without exploring the effects and properties of using only contrastive learning on the representation space. Derived from the motivation to solely employ contrastive learning techniques, two challenges emerge: the "lack of positive" and the "attraction-repulsion imbalance". The "lack of positive" issue arises when instances lack positive pairs in contrastive learning. This issue is intrinsically linked to supervised contrastive learning, which necessitates multiple positive examples to be effective [13]. This effect is exacerbated in the MLTC framework by factors such as a long-tailed distribution, small batches due to model complexity, and the absence of augmentation, which can result in a complete absence of positive pairs for certain examples. The "attraction-repulsion imbalance" is characterized by the dominance of head labels in repulsion term [33]. We emphasize that in multi-label scenarios, the notion of attraction is also unbalanced. This imbalance arises because if an instance encompasses both long-tail and head-tail labels, a significant portion of the positive term will be associated with the head-tail label, thereby degrading the representation.

In this paper, we address these challenges directly by incorporating label prototypes, a memory queue to mitigate the "lack of positives" issue, and employing a re-weighting strategy to address the "attraction-repulsion imbalance". Our research offers an extensive investigation into the exclusive utilization of contrastive learning for MLTC, and introduces a novel supervised contrastive approach as a loss function, introducing the following key contributions:

- We propose and study the effect of implementing a fully supervised contrastive loss function in the context of Multi-Label Text Classification (MLTC).
- We conduct a comprehensive examination of the influence of contrastive learning on the representation space, specifically in the absence of BCE and data augmentation.
- We put forth a substantial ablation study, illustrating the crucial role of considering the long-tailed distribution of data in resolving challenges such as the "attraction-repulsion imbalance" and "lack of positive instances".
- We introduce a novel contrastive loss function for MLTC that attains Micro-F1 scores on par with or superior to existing loss functions, along with a marked enhancement in Macro-F1 scores.
- Finally, we examine the quality of the representation space and the transferability of the features learned through supervised contrastive learning.

The structure of the paper is as follows: in Sect. 2, we provide an overview of related work. Section 3 introduces the notations used throughout the paper and outlines our approach. In Sect. 4, we present our experimental setup, while Sect. 5 provides results obtained from four datasets. Finally, Sect. 6 presents our conclusions.

2   Related Work

In this section, we delve into an exploration of related work on supervised contrastive learning, multi-label text classification, and the application of supervised contrastive learning to MLTC.

2.1   Supervised Contrastive Learning

The idea of supervised contrastive learning has emerged in the domain of vision with the work of [13] called *SupCon*. This study demonstrates how the application of a supervised contrastive loss may yield results in multi-class classification that are comparable, and in some cases even better, to the traditional approaches. The fundamental principle of contrastive learning involves enhancing the representation space by bringing an anchor and a positive sample closer in the embedding space, while simultaneously pushing negative samples away from the anchor. In supervised contrastive learning, a positive sample is characterized as an instance that shares identical class with the anchor. In [9], a comparison was made between the classical cross-entropy loss function and the *SupCon* loss. From this study, it appeared that both loss functions converge to the same representation under balanced settings and mild assumptions on the encoder. However, it was observed that the optimization behavior of *SupCon* enables better generalization compared to the cross-entropy loss.

In situations where there is a long-tailed distribution, it has been found that the representation learned via the contrastive loss might not be effective. One way to improve the representation space is by using class prototypes [5,9,33]. Although these methods have shown promising results, they primarily tackle challenges in multi-class classification problems.

2.2 Multi-label Classification

Learning MLTC using the binary cross-entropy loss function, while straightforward, continues to be a prevalent approach in the literature. A widely adopted and simple improvement to reduce imbalance in this setting is the use of focal loss [16]. This approach prioritizes difficult examples by modifying the loss contribution of each sample, diminishing the loss for well-classified examples, and accentuating the importance of misclassified or hard-to-classify instances. An alternative strategy involved employing the asymmetric loss function [20], which tackles the imbalance between the positive and negative examples during training. This is achieved by assigning different penalty levels to false positive and false negative predictions. This approach enhances the model's sensitivity to the class of interest, leading to improved performance, especially in datasets with imbalanced distributions.

Other works combine an auxiliary loss function with BCE, as in multi-task learning, where an additional loss function serves as regularization. For instance, [32] suggest incorporating an auxiliary loss function that specifically addresses whether two labels co-occur in the same document. Similarly, [1] propose a label-correlation-aware loss function designed to maximize the separation between positive and negative labels inside an instance.

Rather than manipulating the loss function, alternative studies suggest adjusting the model architecture. A usual approach involves integrating statistical correlations between labels using Graph Neural Network [19,25,28]. Additionally, a promising avenue of research looks into adding label parameters to the model, which would enable the learning of a unique representation for every label as opposed to a single global representation [1,12,27].

2.3 Supervised Contrastive Learning for Multi-label Classification

The use of supervised contrastive learning in multi-label classification has recently gained interest within the research community. All the existing studies investigate the effects of supervised contrastive learning by making some kind of prior assumption about label interactions in the learned representation space. Some studies have suggested employing contrastive learning to process features associated with different labels [6,29]. This contrastive loss, in conjunction with the BCE loss function, acts as a form of regularization. Typically, such losses are employed for regularization purposes and fail to reach the optimal representation without BCE.

[15] propose five different supervised contrastive loss functions that are used jointly with BCE to improve semantic representation of classes. In addition, [23] suggest using a KNN algorithm during inference in order to improve performance. Some studies use supervised contrastive learning with a predefined hierarchy of labels [24,26,31].

While contrastive loss functions in mono-label multi-class scenarios push apart representations of instances from different classes, directly applying this approach to the multi-label case may yield suboptimal representations, particularly for examples associated with multiple labels. This can lead to a deterioration in results, particularly in long-tail scenarios.

In contrast to other methods, our approach does not rely on any prior assumptions about label interactions. We address the long-tail distribution challenge in MLTC by proposing several key changes in the supervised contrastive learning loss.

3   Method

We begin by introducing the notations and then present our approach. In the following, B is defined as the set of indices of examples in a batch, and L represents the number of labels. The representation of the ith document in a batch is denoted as z_i. The associated label vector for example i is $y_i \in \{0,1\}^L$, with y_i^j representing its jth element. Furthermore, we denote by $I_B = \{z_i \mid i \in B\}$ the set of document embeddings in the batch B.

3.1   Contrastive Baseline $\mathcal{L}_{Base}$

Before introducing our approach, we provide a description of our baseline for comparison, denoted as $\mathcal{L}_{Base}$, and defined as follows:

$$\mathcal{L}_{Base} = -\frac{1}{|B|} \sum_{z_i \in I_B} \frac{1}{N(i)} \sum_{z_j \in I_B \setminus z_i} \frac{|y_i \cap y_j|}{|y_i \cup y_j|} \log \frac{\exp(z_i \cdot z_j / \tau)}{\sum_{z_k \in I_B \setminus z_i} \exp(z_i \cdot z_k / \tau)} \quad (1)$$

This loss is a simple extension of the *SupCon* loss [13] with an additional term introduced to model the interaction between labels, corresponding to the Jaccard Similarity. τ represents the temperature, '$\cdot$' represents the cosine similarity, and $N(i)$ is the normalization term defined as:

$$N(i) = \sum_{j \in B \setminus i} \frac{|y_i \cap y_j|}{|y_i \cup y_j|}$$

It is to be noted that $\mathcal{L}_{Base}$, does not consider the inherent long-tailed distribution of multi-label dataset.

3.2   Motivation

Our work can be dissected into two improvements compared to the conventional contrastive loss proposed for MLTC.

Each of these improvements aims to tackle the long-tailed distribution inherent in the data and alleviate concerns related to the absence of positive instances and the imbalance in the attraction-repulsion dynamics. These improvements are outlined as follows.

Lack of Positive Instances: We use a memory system by maintaining a queue $Q = \{\tilde{z}_j\}_{j \in \{1,...,K\}}$, which stores the learned representations of the K preceding instances from the previous batches obtained from a momentum encoder. This is in line with other approaches [4,11] that propose to increase the number of positive and negative pairs used in a contrastive loss. Additionally, we propose to incorporate a set of L trainable label prototypes $C = \{c_i \mid i \in \{1,...,L\}\}$. This strategy guarantees that each example in the batch has at least as many positive instances as the number of labels it possesses. These two techniques are particularly advantageous for the labels in the tail of the distribution, as they guarantee the presence of at least some positive examples in every batch.

Attraction-Repulsion Imbalance: Previous work highlights the significance of assigning appropriate weights to the repulsion term within the contrastive loss [33]. In the context of multi-label scenarios, our proposal involves incorporating a weighting scheme into the repulsion term (denominator terms in the contrastive loss function), to decrease the impact of head labels. For an anchor example i with respect to any other instances $k \neq i$ in the batch and in the memory queue, we define the weighting of the repulsion term as:

$$g_i(z_k, \beta) = \begin{cases} 1 & \text{if } z_k \in C, \\ \beta & \text{otherwise.} \end{cases} \tag{2}$$

with $0 < \beta < 1$. This function assigns equal weights to all prototypes, allocating less weight to all other examples present in both the batch and the queue.

In contrastive learning for mono-label multi-class classification, the attraction term is consistently balanced, as each instance is associated with only one class. While, in MLTC, a document can have multiple labels, some in the head and others in the tail of the class distribution. Our approach not only weights positive pairs based on label interactions but also considers the rarity of labels within the set of positive pairs. Instead of iterating through each instance, we iterate through each positive label of an anchor defining a positive pair, as an instance associated with this label.

3.3   Multi-label Supervised Contrastive Loss

To introduce properly our loss function, we use the following notation: $H = I \cup Q$ represents the set of embeddings in the batch and in the queue; $\Delta(z_i) = \{k \in$

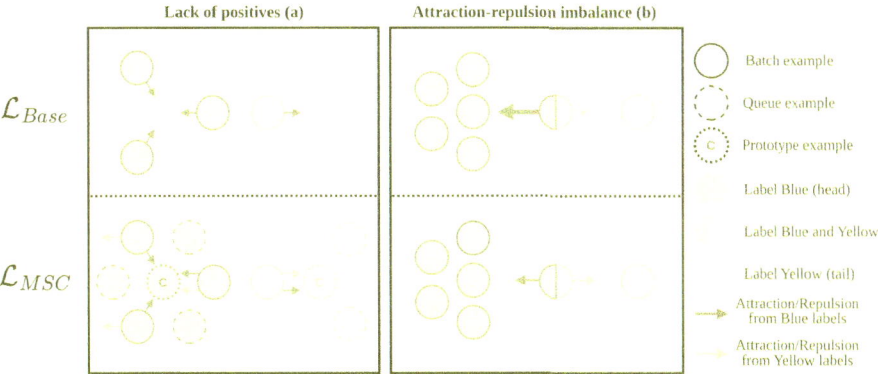

Fig. 1. Illustration of how "lack of positives" and "attraction-repulsion imbalance" problem are addressed by $\mathcal{L}_{Base}$ (classical contrastive loss for MLTC) and $\mathcal{L}_{MSC}$. (a) Adding prototypes and a queue in $\mathcal{L}_{MSC}$ ensures a consistent positive pairing and expands positive and negative samples diversity. (b) Reweighting negative pairs addresses the imbalance between head and tail labels. For clarity, only the attraction/repulsion on the sample in the middle is depicted, without queue and prototypes. Color blue (respectively yellow) corresponds to a label in the head (respectively tail) of the distribution. (Color figure online)

$[1, L]|y_i^k = 1\}$ represents the set of labels for example i; and $P(j,i) = \{z_l \in H|y_l^j = 1\}\backslash z_i$ represents the set of representations for examples belonging to label j, excluding the representation of example i. Our balanced multi-label contrastive loss can then be defined as follows :

$$\mathcal{L}_{MSC} = \frac{1}{|B|} \sum_{i \in B} \ell(z_i) \tag{3}$$

where $\ell(z_i)$ is the individual loss for example i defined as :

$$\ell(z_i) = -\frac{1}{|y_i|} \sum_{j \in \Delta(z_i)} \frac{1}{N(i,j)} \sum_{z_l \in P(j,i) \cup c_j}$$
$$f(z_i, z_j) \log \frac{\exp(z_i \cdot z_l/\tau)}{\sum_{z_k \in H \cup C \backslash z_i} g_i(z_k, \beta) \exp(z_i \cdot z_k/\tau)} \tag{4}$$

$g_i(z_k, \beta)$ are our tailored weights for repulsion terms defined previously. f represents the weights between instances and $N(i,j)$ is a normalization term, both are defined as:

$$f(z_i, z_j) = \begin{cases} 1 & \text{if } z_j \in C \\ \frac{1}{|y_i \cup y_j|} & \text{otherwise.} \end{cases} \qquad N(i,j) = \sum_{z_l \in P(j,i) \cup c_j} f(z_i, z_l) \tag{5}$$

This f defined in Eq. 5 is build so that the equation coincides with the Jaccard similarity in scenarios where labels are balanced.

Figure 1 illustrates the influence of addressing the lack of positives and attraction-repulsion imbalance with our proposed multi-label contrastive loss, denoted as $\mathcal{L}_{MSC}$, compared to the original supervised contrastive loss, $\mathcal{L}_{Base}$ on the exact same training examples in two different situations.

It is to be noted that until now, the learning of a representation space for documents through a pure contrastive loss has remained uncharted. Despite numerous studies delving into multi-label contrastive learning, none have exclusively employed contrastive loss without the traditional BCE loss.

4 Experimental Setup

This section begins with an introduction to the datasets employed in our experiments. Subsequently, we will provide a description of the baseline approaches against which we will compare our proposed balanced multi-label contrastive loss, along with the designated metrics.

4.1 Datasets

We consider the following four multi-label datasets.

1. **RCV1-v2** [14]: RCV1-v2 comprises categorized newswire stories provided by Reuters Ltd. Each newswire story may be assigned multiple topics, with an initial total of 103 topics. We have retained the original training/test split, albeit modifying the number of labels.
2. **AAPD** [30]: The Arxiv Academic Paper Dataset (AAPD) includes abstracts and associated subjects from 55,840 academic papers, where each paper may have multiple subjects.
3. **UK-LEX** [3]: United Kingdom (UK) legislation is readily accessible to the public through the United Kingdom's National Archives website[1]. The majority of these legal statutes have been systematically organized into distinct thematic categories such as health-care, finance, education, transportation, and planning.
4. **BGC**: The Blurb Genre Collection[2] denotes a dataset comprising promotional descriptions of literary works.

Table 1 presents an overview of the main characteristics of these datasets, ordered based on the decreasing number of labels per example.

4.2 Comparison Baselines

To facilitate comparison, our objective is to assess our approach against the current state-of-the-art from two angles. We first examine methods that focus on the learning of a robust representation, and then we assess approaches that are centered around BCE and its extensions.

---

[1] https://www.legislation.gov.uk.
[2] https://www.inf.uni-hamburg.de/en/inst/ab/lt/resources/data/blurb-genre-collection.html.

Table 1. Datasets statistics. The table shows the number of examples (in thousands) within the training, validation, and test sets, as well as the number of class labels L, the average number of labels per example $\overline{L}$, and the average word count per document $\overline{W}$.

Dataset	|Train|	|Val|	|Test|	L	$\overline{L}$	$\overline{W}$
RCV1	19.7k	3.5k	781k	91	3.2	241
AAPD	42.5k	4.8k	8.5k	54	2.4	163
UK-LEX	20.0k	8.0k	8.5k	69	1.7	1154
BGC	58.7k	14.7k	18.3k	146	3.0	158

Baseline: Learning a Good Representation Space. We assess our balanced multi-label contrastive learning by comparing it with the following loss functions that were introduced for learning improved representation spaces.

- $\mathcal{L}_{MLM}$, represents the classical masked language model loss associated with the pre-training task of transformer-based models [17].
- $\mathcal{L}_{Base}$, serves as our baseline for contrastive learning, as presented in the previous section.
- $\mathcal{L}_{BQueue}$, corresponds to $\mathcal{L}_{Base}$ with additional positive instances using a queue.
- $\mathcal{L}_{BQProto}$, represents the strategy that involves integrating prototypes into the previous $\mathcal{L}_{BQueue}$ loss function.

Standard Loss Function for Multi-label. The second type of losses that we consider in our comparisons are based on BCE.

- $\mathcal{L}_{BCE}$, denotes the BCE loss, computed as follows :

$$\mathcal{L}_{BCE} = -\frac{1}{N} \sum_{i=1}^{N} \frac{1}{L} \sum_{j=1}^{L} y_i^j \log(\hat{y}_i^j) + (1 - y_i^j) \log(1 - \hat{y}_i^j)$$

where, $\{\hat{y}_i^1, ..., \hat{y}_i^L\}$ represent the model's output probabilities for the ith instance in the batch.

- $\mathcal{L}_{FCL}$, denotes the focal loss, as introduced by [16], which is an extension of $\mathcal{L}_{BCE}$. It incorporates an additional hyperparameter $\gamma \geqslant 0$, to regulate the ability of the loss function to emphasize over difficult examples.

$$\mathcal{L}_{FCL} = -\frac{1}{N} \sum_{i=1}^{N} \frac{1}{L} \sum_{j=1}^{L} y_i^j (1 - \hat{y}_i^j)^\gamma \log(\hat{y}_i^j) + (1 - y_i^j)(\hat{y}_i^j)^\gamma \log(1 - \hat{y}_i^j)$$

- $\mathcal{L}_{ASY}$, represents the asymmetric loss function [20] proposed to reduce the impact of easily predicted negative samples during the training process through dynamic adjustments, such as'down-weights' and'hard-thresholds'. The computation of the asymmetric loss function is as follows:

$$\mathcal{L}_{ASY} = -\frac{1}{N}\sum_{i=1}^{N}\frac{1}{L}\sum_{j=1}^{L} y_i^j (1 - s_i^j)^{\gamma^+} \log(s_i^j)(1 - y_i^j) + (s_i^j)^{\gamma^-} \log(1 - s_i^j)$$

with $s_i^j = \max(\hat{y}_i^j - m, 0)$. The parameter m corresponds to the hard-threshold, whereas γ^+ and γ^- are the down-weights.
- $\mathcal{L}_{ZLPR}$, in contrast to the prior loss function aimed at enhancing BCE, another form of loss leveraging softmax and pairwise interaction was introduced in [22]. It can be computed as follows:

$$\mathcal{L}_{ZLPR} = -\frac{1}{N}\sum_{i=1}^{N} \log(1 + \langle y_i, \exp(-s_i)\rangle) + \log(1 + \langle 1 - y_i, \exp(s_i)\rangle) \quad (6)$$

4.3   Implementation Details

Our implementation is Pytorch-based[3], involving the truncation of documents to 300 tokens as input for a pre-trained model. As in previous works [22,24], we employed various encoder-only models as backbones. For AAPD, RCV1 datasets, we utilized the Roberta-base [17] as the backbone, implementing it through Hugging Face's resources[4]. For the UK-LEX dataset, we employed Legal-BERT, also provided by Hugging Face[5] and for BGC we employed bert-base-uncased [8][6]. The source code for this implementation is publicly available on GitHub for research purpose[7].

In all experiments, the pre-trained model maintains a dropout rate of 0.1, and weight decay is not applied to bias and LayerNorm parameters. The learning rate for parameters other than the backbone remains fixed at $5e^{-5}$. Gradient Clipping with a parameter value of 1 is implemented, and no data augmentation is utilized.

In the baseline, we employed the standard linear scheduler, and the number of epochs was selected from $\{10, 40, 80\}$. As is commonly practiced, we employed a linear scheduler. During training, the model with the best F1-micro score is kept for testing, while the model achieving the best average results (averaged over seeds) on validation is retained for testing part, we also tested the standard

---

[3] https://pytorch.org.
[4] https://huggingface.co/roberta-base.
[5] https://huggingface.co/nlpaueb/legal-bert-base-uncased.
[6] https://huggingface.co/google-bert/bert-base-uncased.
[7] https://github.com/audibert-alexandre-fra/ECML-Contrastive-Learning-Multi-Label.

parameters. For the focal loss we set $\gamma = 2$ and for the Asymmetric loss we set $\gamma^+ = 0, \gamma^- = 3, m = 0.3$.

Contrastive Learning tends to converge to a better representation with more iterations, which is why we consistently set the number of epochs to 80 in all experiments. We assessed the representation space of three checkpoints and retained the best one for testing. The available checkpoints include the last checkpoint, the one with the lowest loss in training, and the one with the lowest loss in validation. The checkpoints with the best micro-F1 is kept. As a common practice for contrastive learning, a cosine scheduler is used. As in SupCon [13], we use a projection head composed of two fully connected layers with ReLU as activation function: $W_2 \cdot \text{ReLU}(W_1 \cdot x)$ where $W_1 \in \mathbb{R}^{h \times h}$ and $W_2 \in \mathbb{R}^{d \times h}$ where h is the dimension of the hidden space and d is set to 256 in our experiments. As in SupCon the projection head is discarded to evaluate the representation space. For the hyperparameter, we set the size of the MoCo queue equal to 512 and the momentum encoder is update with a momentum equal to 0.999 as in [11]. Finally, in our experiments, we set β to 0.1; this parameter was not subject to search.

As common practice, we designated the [CLS] token as the final representation for the text, utilizing a fully connected layer as a decoder on this representation. Our approach involves a batch size of 32, and the learning rate for the backbone is chosen from the set $\{5e^{-5}, 2e^{-5}\}$. Throughout all experiments, we use AdamW optimizer [18], setting the weight decay set to 0.01 and implementing a warm-up stage that comprises 5% of the total training. For evaluating the representation space, we trained logistic regressions with AdamW separately for each individual label. To expedite training and conserve memory, we employed 16-bit automatic mixed precision.

Once the optimal model checkpoint achieved through supervised contrastive learning is identified, we remove the projection head and proceed to train a linear layer using BCE. The configuration remains consistent with the "Common process for all experiments," and we explore learning rates within the range of $5e^{-5}, 2e^{-5}$ and epoch numbers within 5, 10.

The evaluation of results is conducted on the test set using traditional metrics in MLTC, namely the hamming loss, Micro-F1 score and Macro-F1 score [32].

5   Experimental Results

We start our evaluation by conducting an ablation study, comparing various loss functions proposed for representation learning, as outlined in Sect. 4.2.

Table 2 summarizes these results obtained across various temperatures and seeds. The score achieved with $\mathcal{L}_{MLM}$ is merely 10 points lower in the Micro-F1 score compared to the best results, highlighting the effectiveness of the representation space found during the pre-training phase. Our approach primarily focuses on the Macro-F1 score, targeting the prevalent long-tailed distribution in MLTC data. As the table shows, each additional component we have introduced contributes around one point to the Macro-F1 score. Maintaining a balance between

Table 2. Evaluation of progressive complexity in contrastive loss functions. Micro-F1 (μ-F$_1$), Macro-F1 (M-F$_1$), and Hamming Loss (multiplied by 10^3) metrics are averaged over nine values (three seeds and three temperatures 0.07, .1, .2) - except for $\mathcal{L}_{MLM}$ averaged on three seeds.

Loss	AAPD			RCV1		
	μ-F$_1$	M-F$_1$	Ham	μ-F$_1$	M-F$_1$	Ham
$\mathcal{L}_{MLM}$	63.86	45.62	28.48	80.06	58.42	13.5
$\mathcal{L}_{Base}$	72.25	56.42	24.4	87.89	73.7	8.51
$\mathcal{L}_{BQueue}$	72.73	57.92	24.15	87.56	72.9	8.72
$\mathcal{L}_{BQProto}$	73.3	59.126	**23.69**	88.00	74.82	8.44
$\mathcal{L}_{MSC}$ (ours)	**73.59**	**60.00**	23.74	**88.40**	**76.82**	**8.21**

attraction and repulsion terms proves crucial, particularly for RCV1-v2, where it resulted in a 2-point improvement in the Macro-F1 score.

Our proposed loss function, $\mathcal{L}_{MSC}$, exhibited superior performance over the baseline $\mathcal{L}_{Base}$ for all metrics, emphasizing the importance of addressing both the "Lack of Positive" issue and the "Attraction-Repulsion Imbalance" for an optimal representation space. Throughout our experiments, setting the temperature to 0.1 consistently yielded the best results across all baselines. Consequently, we adopted this setting for all subsequent experiments.

5.1 Comparison with Standard MLTC Losses

Table 3 presents a comparison of performance between the standard BCE-based loss functions outlined in Sect. 4.2 and our approach.

$\mathcal{L}_{MSC}$ outperforms all baselines based on BCE in Macro-F1 score. The asymmetric loss function achieves comparable results only for the AAPD and BGC dataset, albeit with the worst score in other metrics. Regarding Micro-F1, the performance of the $\mathcal{L}_{MSC}$ is equivalent for the AAPD dataset and slightly better for RCV1-v2, UK-LEX and BGC compared to the best score of the three standard losses. In standard approaches, $\mathcal{L}_{ZLPR}$ stands out as the strongest baseline because this loss considers label interactions more comprehensively. Even though $\mathcal{L}_{ZLPR}$ is quite efficient, $\mathcal{L}_{MSC}$ exhibits better performance on three datasets in terms of F1-macro and on two datasets in terms of F1-micro. This indicates that supervised contrastive learning without the addition of another loss function can be at least equivalent to the state-of-the-art loss, highlighting the importance of considering label interactions.

5.2 Fine-Tuning After Supervised Contrastive Learning

To evaluate the quality of the representation space given by the contrastive learning phase, we explored the transferability of features through a fine-tuning stage. This study introduces two novel baselines: $\mathcal{L}_{Base-FT}$ and $\mathcal{L}_{MSC-FT}$, which are

Table 3. Comparative Analysis of multi-label loss functions. Metrics used are Micro-F1 (μ-F$_1$), Macro-F1 (M-F$_1$), and Hamming Loss (multiplied by 10^3). *FT* stands for the fine-tuning version of the loss mentioned above.

Loss	AAPD			RCV1			UK-LEX			BGC		
	μ-F$_1$	M-F$_1$	Ham	μ-F$_1$	M-F$_1$	Ham	μ-F$_1$	M-F$_1$	Ham	μ-F$_1$	M-F$_1$	Ham
Supervised Loss												
$\mathcal{L}_{ASY}$	72.92	60.63	25.3	86.63	75.02	10.02	70.53	60.58	14.43	79.92	66.09	86.18
$\mathcal{L}_{FCL}$	73.85	59.91	22.61	88.36	76.69	8.19	73.23	61.17	**11.54**	80.61	65.48	77.78
$\mathcal{L}_{BCE}$	73.89	59.98	22.53	88.17	76.06	8.17	72.61	60.97	11.95	80.04	64.85	80.31
$\mathcal{L}_{ZLPR}$	**74.16**	60.27	**22.43**	88.61	76.78	7.98	72.72	61.71	12.14	80.88	**66.5**	77.64
Contrastive Loss												
$\mathcal{L}_{Base}$	72.51	56.67	24.13	87.86	73.79	8.48	72.3	59.66	12.31	80.23	61.64	79.63
$\mathcal{L}_{-FT}$	73.09	58.55	23.61	88.41	76.08	8.18	72.45	60.66	12.23	80.49	64.3	79.21
Ours												
$\mathcal{L}_{MSC}$	73.84	**60.75**	23.72	88.54	77.05	8.12	**73.5**	**62.06**	11.83	**81.21**	66.00	**76.32**
$\mathcal{L}_{-FT}$	74.00	60.41	23.01	**88.65**	**77.18**	7.99	72.97	61.33	12.04	80.99	65.47	77.13

obtained by fine-tuning the representation learn with contrastive learning instead of doing a simple linear evaluation.

From Table 3, it comes that in all cases, $\mathcal{L}_{MSC-FT}$ achieved superior results in both micro-F1 and macro-F1 scores compared to $\mathcal{L}_{Base-FT}$. These results show that the features learned with $\mathcal{L}_{MSC}$ are robust and offer an enhanced starting point for fine-tuning, in contrast to the traditional $\mathcal{L}_{MLM}$. Conversely, the performance of $\mathcal{L}_{Base-FT}$ was either worse or comparable to that of BCE, which underlies the benefits of our new loss function.

5.3   Representation Analysis

To quantify the quality of the latent space learned by our approach, we evaluate how well the embeddings are separated in the latent space according to their labels using two established metrics : Silhouette score [21] and Davies-Bouldin index [7]. These metrics collectively assess the separation between clusters and cohesion within clusters of the embeddings.

We treat each unique label combination in the dataset as a separate class to apply these metrics to the multi-label framework. Such expansion can potentially dilute the effectiveness of traditional clustering metrics by creating too many classes. To mitigate this, our analysis focuses on subsets of the most prevalent label combinations, retaining only half of the most represented label combination.

Table 4 presents our findings. A direct comparison between the baseline contrastive method $\mathcal{L}_{Base}$, and our proposed $\mathcal{L}_{MSC}$ method (prior to fine-tuning) reveals a significant enhancement in both metrics score. The integration of fine-

Table 4. Clustering Metrics for different loss functions on 10^4 embeddings from RCV1-v2 test set. Only 50% of most represented label combinations are kept.

Metric	$\mathcal{L}_{MLM}$	$\mathcal{L}_{BCE}$	$\mathcal{L}_{Base}$	$\mathcal{L}_{Base-FT}$	$\mathcal{L}_{MSC}$	$\mathcal{L}_{MSC-FT}$
Sil ↑	−0.14	0.15	0.07	0.13	0.10	**0.16**
DBI ↓	2.83	2.02	2.23	2.00	2.07	**1.98**

tuning using BCE significantly enhances $\mathcal{L}_{Base}$ and $\mathcal{L}_{MSC}$ for both metrics, which demonstrates the effectiveness of the hybrid approach. Using our loss with fine-tuning is the only method able to surpass BCE in both metrics. This underscores its efficacy in creating well-differentiated and cohesive clusters in the latent space.

To apply clustering evaluation metrics such as the Silhouette score or the Davies-Bouldin index to multi-label embeddings, it is necessary to create one class for each unique multi-label combination, resulting in up to 2^L classes. Although 50% of these were retained in Table 4, we now explore a more general scenario by varying this proportion as reported in Fig. 2.

Our approach, $\mathcal{L}_{MSC}$, consistently outperforms $\mathcal{L}_{Base}$, except for a single proportion value of 20%, for Silhouette score. This could be attributed to the fact that our approach attempts to address the tail labels, which are typically discarded when keeping smaller proportions of top label combination.

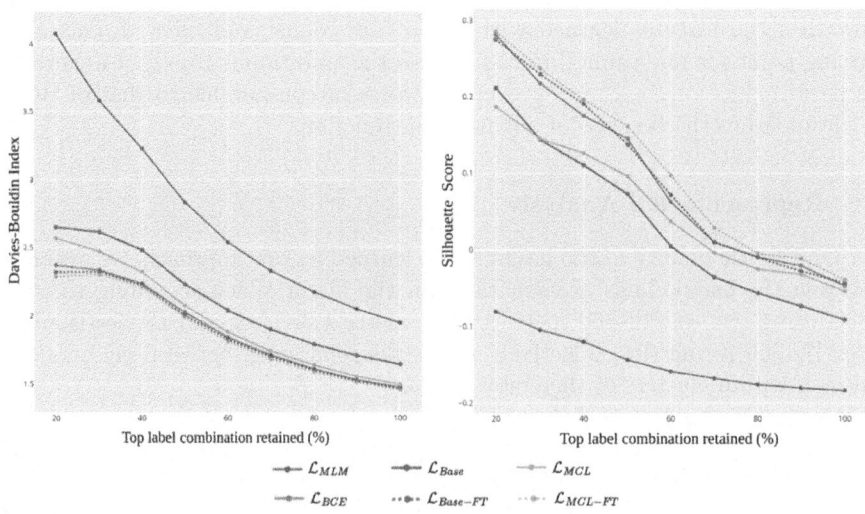

Fig. 2. Clustering quality metrics of different approaches across top classes retained.

6   Conclusion

In this paper, we have introduced a supervised contrastive learning loss for MLTC which outperforms standard BCE-based loss functions. Our method highlights the importance of considering the long-tailed distribution of data, addressing issues such as the "lack of positives" and the "attraction-repulsion imbalance". We have designed a loss that takes these issue into consideration, outperforming existing standard and contrastive losses in both micro-F1 and macro-F1 across four standard multi-label datasets. Moreover, we also verify that these considerations are also essential for creating an effective representation space. Additionally, our findings demonstrate that initializing the model's learning with supervised contrastive pretraining yields better results than existing contrastive pre-training methods.

Acknowledgement. This work was supported by the ANR (Agence Nationale de Recherche) grants Lawbot (ANR-20-CE38-0013).

References

1. Alhuzali, H., Ananiadou, S.: SpanEmo: casting multi-label emotion classification as span-prediction. In: Proceedings of the 16th Conference of the European Chapter of the Association for Computational Linguistics: Main Volume. Association for Computational Linguistics, April 2021
2. Bai, J., Kong, S., Gomes, C.P.: Gaussian mixture variational autoencoder with contrastive learning for multi-label classification. In: International Conference on Machine Learning. PMLR (2022)
3. Chalkidis, I., Søgaard, A.: Improved multi-label classification under temporal concept drift: rethinking group-robust algorithms in a label-wise setting. arXiv preprint arXiv:2203.07856 (2022)
4. Chen, X., Fan, H., Girshick, R., He, K.: Improved baselines with momentum contrastive learning. arXiv preprint arXiv:2003.04297 (2020)
5. Cui, J., Zhong, Z., Liu, S., Yu, B., Jia, J.: Parametric contrastive learning. In: Proceedings of the IEEE/CVF International Conference on Computer Vision (2021)
6. Dao, S.D., Ethan, Z., Dinh, P., Jianfei, C.: Contrast learning visual attention for multi label classification. arXiv preprint arXiv:2107.11626 (2021)
7. Davies, D.L., Bouldin, D.W.: A cluster separation measure. IEEE Trans. Pattern Anal. Mach. Intell. **PAMI-1**(2), 224–227 (1979)
8. Devlin, J., Chang, M., Lee, K., Toutanova, K.: BERT: pre-training of deep bidirectional transformers for language understanding. CoRR **abs/1810.04805** (2018)
9. Graf, F., Hofer, C., Niethammer, M., Kwitt, R.: Dissecting supervised contrastive learning. In: International Conference on Machine Learning. PMLR (2021)
10. Gunel, B., Du, J., Conneau, A., Stoyanov, V.: Supervised contrastive learning for pre-trained language model fine-tuning. arXiv preprint arXiv:2011.01403 (2020)
11. He, K., Fan, H., Wu, Y., Xie, S., Girshick, R.: Momentum contrast for unsupervised visual representation learning. In: Proceedings of the IEEE/CVF Conference on Computer Vision and Pattern Recognition, pp. 9729–9738 (2020)

12. Kementchedjhieva, Y., Chalkidis, I.: An exploration of encoder-decoder approaches to multi-label classification for legal and biomedical text. In: Findings of the Association for Computational Linguistics: ACL 2023. Association for Computational Linguistics, Toronto, July 2023
13. Khosla, P., et al.: Supervised contrastive learning. Advances in Neural Information Processing Systems (2020)
14. Lewis, D.D.: RCV1-v2/LYRL2004: the LYRL2004 distribution of the RCV1-v2 text categorization test collection (2004)
15. Lin, N., Qin, G., Wang, G., Zhou, D., Yang, A.: An effective deployment of contrastive learning in multi-label text classification. In: Findings of the Association for Computational Linguistics: ACL 2023. Association for Computational Linguistics, Toronto, July 2023
16. Lin, T.Y., Goyal, P., Girshick, R., He, K., Dollár, P.: Focal loss for dense object detection. In: Proceedings of the IEEE International Conference on Computer Vision (2017)
17. Liu, Y., et al.: RoBERTa: a robustly optimized bert pretraining approach. arXiv preprint arXiv:1907.11692 (2019)
18. Loshchilov, I., Hutter, F.: Decoupled weight decay regularization. arXiv preprint arXiv:1711.05101 (2017)
19. Ma, Q., Yuan, C., Zhou, W., Hu, S.: Label-specific dual graph neural network for multi-label text classification. In: Proceedings of the 59th Annual Meeting of the Association for Computational Linguistics and the 11th International Joint Conference on Natural Language Processing (Volume 1: Long Papers). Association for Computational Linguistics, August 2021
20. Ridnik, T., et al.: Asymmetric loss for multi-label classification. In: Proceedings of the IEEE/CVF International Conference on Computer Vision (2021)
21. Rousseeuw, P.J.: Silhouettes: a graphical aid to the interpretation and validation of cluster analysis. J. Comput. Appl. Math. **20**, 53–65 (1987)
22. Su, J., Zhu, M., Murtadha, A., Pan, S., Wen, B., Liu, Y.: ZLPR: a novel loss for multi-label classification. arXiv preprint arXiv:2208.02955 (2022)
23. Su, X., Wang, R., Dai, X.: Contrastive learning-enhanced nearest neighbor mechanism for multi-label text classification. In: Proceedings of the 60th Annual Meeting of the Association for Computational Linguistics (Volume 2: Short Papers). Association for Computational Linguistics, Dublin, May 2022
24. Yu, S.C.L., He, J., Gutiérrez-Basulto, V., Pan, J.Z.: Instances and labels: hierarchy-aware joint supervised contrastive learning for hierarchical multi-label text classification (2023)
25. Vu, H.T., Nguyen, M.T., Nguyen, V.C., Pham, M.H., Nguyen, V.Q., Nguyen, V.H.: Label-representative graph convolutional network for multi-label text classification. Appl. Intell. **53**, 14759–14774 (2022)
26. Wang, Z., Wang, P., Huang, L., Sun, X., Wang, H.: Incorporating hierarchy into text encoder: a contrastive learning approach for hierarchical text classification. In: Proceedings of the 60th Annual Meeting of the Association for Computational Linguistics (Volume 1: Long Papers), Dublin, Ireland, May 2022
27. Xiao, L., Huang, X., Chen, B., Jing, L.: Label-specific document representation for multi-label text classification. In: Proceedings of the 2019 Conference on Empirical Methods in Natural Language Processing and the 9th International Joint Conference on Natural Language Processing (EMNLP-IJCNLP). Association for Computational Linguistics, Hong Kong, November 2019
28. Xu, P., Liu, Z., Winata, G.I., Lin, Z., Fung, P.: Emograph: capturing emotion correlations using graph networks (2020)

29. Xu, P., Xiao, L., Liu, B., Lu, S., Jing, L., Yu, J.: Label-specific feature augmentation for long-tailed multi-label text classification. In: Proceedings of the AAAI Conference on Artificial Intelligence, pp. 10602–10610 (2023)
30. Yang, P., Sun, X., Li, W., Ma, S., Wu, W., Wang, H.: SGM: sequence generation model for multi-label classification. In: Proceedings of the 27th International Conference on Computational Linguistics, COLING 2018, Santa Fe, New Mexico, USA, 20–26 August 2018 (2018)
31. Zhang, S., Xu, R., Xiong, C., Ramaiah, C.: Use all the labels: a hierarchical multi-label contrastive learning framework. In: CVPR (2022)
32. Zhang, X., Zhang, Q.W., Yan, Z., Liu, R., Cao, Y.: Enhancing label correlation feedback in multi-label text classification via multi-task learning. arXiv preprint arXiv:2106.03103 (2021)
33. Zhu, J., Wang, Z., Chen, J., Chen, Y.P.P., Jiang, Y.G.: Balanced contrastive learning for long-tailed visual recognition. In: Proceedings of the IEEE/CVF Conference on Computer Vision and Pattern Recognition (2022)

Simultaneous Linear Connectivity of Neural Networks Modulo Permutation

Ekansh Sharma[1,2]([✉]), Devin Kwok[3,4]([✉]), Tom Denton[5], Daniel M. Roy[1,2], David Rolnick[3,4], and Gintare Karolina Dziugaite[6]([✉])

[1] University of Toronto, Toronto, Canada
ekansh@cs.toronto.edu
[2] Vector Institute, Toronto, Canada
[3] McGill University, Montreal, Canada
devin.kwok@mail.mcgill.ca
[4] Mila Quebec AI Institute, Montreal, Canada
[5] Google DeepMind, San Francisco, CA, USA
[6] Google DeepMind, Toronto, Canada
gkdz@google.com

Abstract. Neural networks typically exhibit permutation symmetry, as reordering neurons in each layer does not change the underlying function they compute. These symmetries contribute to the non-convexity of the networks' loss landscapes, since linearly interpolating between two permuted versions of a trained network tends to encounter a high loss barrier. Recent work has argued that permutation symmetries are the *only* sources of non-convexity, meaning there are essentially no such barriers between trained networks if they are permuted appropriately. In this work, we refine these arguments into three distinct claims of increasing strength. We show that existing evidence only supports "weak linear connectivity"—that for each pair of networks belonging to a set of SGD solutions, there exist (multiple) permutations that linearly connect it with the other networks. In contrast, the claim "strong linear connectivity"—that for each network, there exists one permutation that *simultaneously* connects it with the other networks—is both intuitively and practically more desirable. This stronger claim would imply that the loss landscape is convex after accounting for permutation, and enable linear interpolation between three or more independently trained models without increased loss. In this work, we introduce an intermediate claim—that for certain sequences of networks, there exists one permutation that simultaneously aligns matching pairs of networks *from these sequences*. Specifically, we discover that a single permutation aligns sequences of iteratively trained as well as iteratively pruned networks, meaning that two networks exhibit low loss barriers at each step of their

---

E. Sharma and D. Kwok—Equal contribution.

Supplementary Information The online version contains supplementary material available at https://doi.org/10.1007/978-3-031-70368-3_16.

A. Bifet et al. (Eds.): ECML PKDD 2024, LNAI 14947, pp. 262–279, 2024.
https://doi.org/10.1007/978-3-031-70368-3_16

optimization and sparsification trajectories respectively. Finally, we provide the first evidence that strong linear connectivity may be possible under certain conditions, by showing that barriers decrease with increasing network width when interpolating among three networks.

Keywords: Linear mode connectivity · Neural network permutation symmetries · Lottery tickets

1 Introduction

The loss landscape of neural networks is well known to be non-convex. Indeed, linearly interpolating between the weights of two independently trained networks will typically traverse significant loss/error *barriers*, meaning the interpolated networks have much higher loss/0–1 error relative to the two endpoints. There are a number of situations where barriers do not arise, however. Empirically, stochastic gradient descent (SGD) tends to produce training trajectories and basins of attraction that are surprisingly "convex-like". For instance, low error barriers have been observed between certain global minima [8,13,24], and interpolating between points on the same training trajectory leads to monotonically decreasing error [9,12,22].

In this work, we focus on *linear (mode) connectivity* [8,13] where a lack of significant loss/error barriers between networks indicates a certain degree of flatness or even convexity in a region of the loss/error landscape. While linear connectivity has previously been observed in the specific situations described above, recent work conjectures that all *SGD solutions* are linearly connected *modulo permutation*—meaning that there exists permutations (function-preserving per-layer reshuffling of neurons) that enable any networks trained with the same dataset and SGD procedure to be linearly interpolated with low barriers [1,5,6].

What is not yet clear from previous work is which networks can be aligned by a particular permutation. At least two competing claims have emerged: one, where a single permutation is sufficient to mutually connect all network pairs in a set, and two, where a single permutation is only guaranteed to connect one pair of networks. Prior work has described and modelled the former claim, which we call *strong* linear connectivity (Definition 4.2), with the goal of showing that the loss landscape is convex for the set of SGD solutions after accounting for permutations [6]. However, experimental work has thus far focused on the latter claim, which we call *weak* linear connectivity (Definition 4.1), in which individual pairs of networks can be aligned after training [1,6,17] or at initialization [5].

In this work, we make the first distinction between different claims of linear mode connectivity modulo permutation in order to disambiguate between existing results. We define and give evidence for an intermediary claim called *simultaneous weak* linear connectivity (Definition 4.3), where a single permutation connects matching pairs of networks along two sequences of iteratively transformed networks. We verify simultaneous connectivity for sequences of networks resulting from SGD training trajectories, as well as sequences of networks resulting from iterative magnitude pruning (IMP). Finally, we provide the first support for the existence of strong connectivity, by showing that barriers among permuted network triplets reduce as network width increases. Based on this evidence, we conjecture that strong connectivity is possible given sufficiently wide networks.

Contributions.

1. We define three different versions of linear mode connectivity modulo permutation, situating prior work into a more precise framework. We argue that prior evidence only exists for *weak linear connectivity modulo permutation* (see Sect. 4, and Appendix B for a detailed discussion contrasting our definitions and prior work).
2. We demonstrate the existence of *simultaneous weak linear connectivity modulo permutation* between training trajectories. Given a pair of networks and their SGD training trajectories, a single permutation of neurons can be found that aligns not merely the final trained networks, but also each corresponding pair of partially trained networks so that they are linearly connected with low error barrier. Moreover, we find that this result holds for the entire linearly connected basin around each network, so that the same permutation aligns all networks from the same linearly connected mode to a target network from a different mode (Sect. 5.1).
3. We show that simultaneous weak connectivity also exists between iteratively pruned networks. The same permutation aligning two dense trained networks can also be used to align successive iterations of sparse subnetworks found via iterative magnitude pruning (IMP) for lottery tickets [7]. We also find that same alignment allows the sparse mask (i.e., "winning ticket") found via IMP on one network to be reused on another network, meaning that applying the permuted mask at initialization allows the latter network to be trained to the same accuracy as the original dense network (Sect. 5.2).
4. We provide the first evidence towards *strong linear connectivity modulo permutation*, by showing that the barrier between two independently trained networks can be reduced by alignment with a third, unrelated network. As this effect scales with width, we conjecture that sufficiently wide networks exhibit strong connectivity (Sect. 5.3).
5. We identify specific limitations of weight matching algorithms: namely that they are equivalent in performance to activation matching only on trained networks, they rely on larger magnitude weights which appear later in training, they align lower layers earlier in training, and their success or failure is unrelated to a network's stability to SGD noise.

2   Methods

2.1   Preliminaries

Let A and B denote two parameterizations of a given neural network architecture, which are optimized over the loss function ℓ. We take $\ell(X)$ to mean evaluating the network X over a fixed data distribution.

Linear Mode Connectivity and Loss Barrier. The *loss barrier* is defined as the maximum increase in loss on the linear path between A and B:

$$\sup_{\alpha \in (0,1)} \ell(\alpha A + (1 - \alpha)B) - \alpha\ell(A) - (1 - \alpha)\ell(B). \tag{1}$$

Barriers are typically maximized at $\alpha = 0.5$. Our experiments measure barriers in terms of both cross entropy and 0–1 classification error. We refer to barriers in the latter case as *error barriers*. A pair of trained networks are said to be *linearly connected* if the error barrier between them is below a threshold, which is typically set to the expected Monte Carlo noise of the empirical loss estimator. As our experiments mainly compare the relative size of barriers, we informally say that networks are *essentially linearly connected* if the error barrier between them is significantly lowered after applying a permutation (i.e., the loss basin of the network is determined up to some small error).

Sparsity and Iterative Magnitude Pruning. We study sparse subnetworks obtained by an iterative magnitude pruning (IMP) procedure which is used to find "winning tickets" in the lottery ticket hypothesis [7,8]. For a full description of the pruning algorithm, see Appendix A. IMP has been shown to identify non-trivially sparse networks (also called "winning tickets") which are *matching*, meaning they have the same accuracy as their dense counterpart [8]. We refer to such networks as *IMP subnetworks*. Sparse networks pruned for L iterations of IMP are also linearly connected to the next iteration $L + 1$, provided that both networks are matching [15].

2.2   Aligning Networks via Permutation

The intermediate neurons of a neural network can be relabelled without changing its output. Formally, we define a fully-connected feedforward network

$$f(\mathbf{x}_0) = \mathbf{x}_K, \qquad\qquad \mathbf{x}_i = \sigma(\mathbf{W}_i\mathbf{x}_{i-1} + \mathbf{b}_i),$$

with input x_0, layers $i \in \{1, 2, \ldots, K\}$, weights W_i, biases b_i, and pointwise activation function σ. Each $\mathbf{x}_i$ with $i \in \{1, 2, \ldots, K - 1\}$ can be permuted as

$$\mathbf{P}_i\mathbf{x}_i = \sigma\left(\mathbf{P}_i\mathbf{W}_i\mathbf{P}_{i-1}^\top\mathbf{x}_{i-1} + \mathbf{P}_i\mathbf{b}_i\right),$$

where $\mathbf{P}_i$ are permutation matrices in the set of permutation symmetries $\mathcal{S}_{d_i}$, d_i being the dimensions of $\mathbf{x}_i$. If $\mathbf{P}_0$ and $\mathbf{P}_K$ are the identity, f can be rewritten:[1]

$$\mathbf{W}_i' = \mathbf{P}_i\mathbf{W}_i\mathbf{P}_{i-1}^\top, \qquad \mathbf{b}_i' = \mathbf{P}_i\mathbf{b}_i, \quad \text{where } \mathbf{P}_0 = \mathbb{I}_{d_0}, \ \mathbf{P}_K = \mathbb{I}_{d_K}, \text{ and } \mathbf{P}_i \in \mathcal{S}_{d_i}.$$

For convenience, we identify the set of permutations $\{\mathbf{P}_0, \mathbf{P}_1, \ldots, \mathbf{P}_K\}$ as P, and use $P[B]$ to indicate a permutation of B via P (i.e. $\mathbf{P}_i\mathbf{W}_i\mathbf{P}_{i-1}^\top$ and $\mathbf{P}_i\mathbf{b}_i$).

---

[1] See Appendix A for details on handling other types of layers.

To determine the linear connectivity of neural networks *modulo permutation*, we wish to *align* networks by finding a permutation that minimizes

$$P^{\star} = \inf_{P \in \mathcal{S}} d(A, P[B]),$$

where A and B are networks, P is a permutation from $\mathcal{S}$, and d is a function measuring the dissimilarity between two networks. To minimize error barrier with $P^{\star}$, d should ideally measure the error barrier between A and B. However, to simplify the problem we instead define d using two heuristics *weight matching* and *activation matching* developed by prior works [1,20,23].

Weight matching takes d to be the L^2 distance between A and $P[B]$. We use the Greedy-SOBLAP algorithm [1], based on methods from optimal transport, to approximately minimize this distance.[2] Activation matching takes d to be the L^2 distance between each network's intermediate outputs on training data [1,20]. We use essentially the same method as weight matching to minimize this distance. Full details on both matching algorithms can be found in Appendix A.

3   Related Work

Linear Mode Connectivity and Sparsity. Linear mode connectivity arose from investigations on the conditions necessary for obtaining sparse IMP subnetworks ("winning tickets") satisfying the lottery ticket hypothesis [8]. It was found that IMP could identify winning tickets as long as the network to be pruned remained sufficiently "stable" to SGD noise (e.g. random batch order and augmentation). Linear mode connectivity was defined to characterize stability in terms of the loss barrier between independent trained *child networks* spawned from the same parent state. Our experiments use the same approach of spawning child networks from a parent at time t to isolate the effect of SGD noise (Fig. 2 right).

Although the linearly connected mode to which dense networks converge is determined early in training (e.g. 3% of training for ResNet-20 on CIFAR-10, 20% for ResNet-50 on ImageNet), child networks spawned prior to this point will exhibit large error barriers for all networks larger than MNIST-scale [8]. Our experiments refine these observations, teasing apart the relative contributions of initialization and minibatches on feature emergence, and revealing that entire SGD trajectories can be aligned to yield small barriers. Pretraining in the context of lottery tickets has since received considerable attention [15,16]. In particular, winning ticket masks encode information about the final linearly connected basin, suggesting that they are symmetry-dependent [16].

Linear Mode Connectivity Modulo Permutation. The permutation symmetries of neural networks create copies of the global minima in the loss landscape. For overparameterized networks, it is know that these permutation-induced minima are connected by a non-convex, piece-wise linear manifold [19]. In this vein, one

---

[2] Weight matching is approximate, as finding the actual optimum is NP-hard [1,3].

may also consider if distinct minima found by SGD are actually permutations of the same global minimum.

Building on observations of linear mode connectivity in identically initialized networks [8], a conjecture was posed that independently initialized SGD solutions have no error barrier if the symmetries arising from permutation invariance are accounted for [6]. Preliminary evidence for this conjecture investigated the orbit of a single SGD solution when acted upon by permutations. The permuted versions of a network were found to fall into distinct linearly connected basins, with a similar distribution of loss barriers as the barriers between independently initialized and trained networks [6].

Following this conjecture, numerous algorithms have been proposed to directly account for permutation invariance by aligning networks and maximizing their similarity [1,5,10,17,21,25]. In many natural settings, these algorithms have found permutations that considerably reduce or eliminates the loss barrier between two SGD solutions.

Aligning Neural Networks via Permutation. An early method for finding permutations used the correlation or mutual information of activations to align neurons in a one-to-one, bipartite, or few-to-one fashion, discovering that the majority of representations between aligned networks were equivalent [11]. Activation matching was also used to discover mode connectivity, wherein networks are joined by non-linear paths of low loss after permutation [21].

An optimal transport method for aligning network weights or activations has since been developed [1,20,23]. These methods find permutations that maximize similarity between two networks with either the same architecture, or differences only in layer width. Applications include matching the weights of local (client-side) and global models for federated learning [23], merging models that were fine-tuned on different datasets [1,20], and compressing models by combining layers [14]. The connection between weight matching and optimal transport has been deepened to enable weight matching for recurrent networks [2].

The effect of weight and activation matching on linear mode connectivity was overlooked until the recent discovery of empirical evidence for linear mode connectivity modulo permutation on networks satisfying certain conditions [1].[3] Although most works consider trained networks [1,10], networks at initialization have also been observed to be linearly connected by a permutation found after training [5]. Our work differs in that we observe linear connectivity along the entire SGD trajectory, and not only at initialization and convergence. Additionally, we use larger networks which, unlike small fully-connected networks, do not converge to the same linearly connected basin even when starting from the same initialization [8]. Contrary to prior work on smaller networks [5], we do not find that permutations computed at initialization reduce barriers after training.

Alongside these investigations, the optimal transport method for aligning networks has also been improved by iterating over randomized layer orders [1], and resetting normalization statistics to enable the use of batch normalization

---

[3] These conditions include sufficient width, and the use of layer normalization.

[10]. Departing from weight matching, a permutation-finding method has also been developed that directly minimizes barriers using implicit differentiation and the Sinkhorn operator [17].

4   Notions of Linear Connectivity Modulo Permutation

As informally stated in [6], the property of linear mode connectivity modulo permutation (LC mod P) occurs when network parameterizations from a set $\mathcal{F}$ *can be permuted in such a way that there is no barrier on the linear interpolation between any two permuted elements in $\mathcal{F}$.*

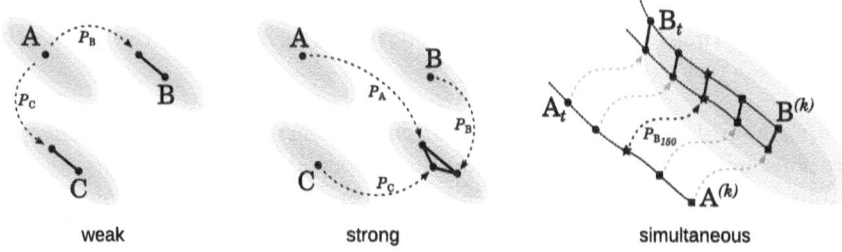

weak   strong   simultaneous

Fig. 1. Notions of linear connectivity modulo permutation. (**Left**) Definition 4.1: for any pair of networks there exists a permutation P that linearly connects the two networks. (**Center**) Definition 4.2: for each network there exists a permutation such that all networks are *mutually* linearly connected after permutation. (**Right**) Definition 4.3: for any pair of training or iterative magnitude pruning trajectories (A_t and $A^{(k)}$ respectively) there exists a permutation P such that the two trajectories are pairwise *simultaneously* linearly connected.

However, this informal notion of LC mod P does not make clear which permutations are shared between different parameterizations, which is important for many downstream applications. For instance, in order to apply gradient updates from one parameterization to another, as may be needed in federated learning [23], a single permutation must be shared throughout the training trajectories of two parameterizations. In order to do model merging between three or more parameterizations without additional retraining [20], a single permutation must simultaneously linearly connect each parameterization to all of the other parameterizations.

To disambiguate these different versions of LC mod P, we start with the weakest interpretation, where each pair of networks is potentially aligned with a different permutation depending on that specific pair (Fig. 1 (Left)).

Definition 4.1 (Weak linear connectivity modulo permutation). *Let $\mathcal{F}$ be a set of networks. We say that* weak linear connectivity modulo permutation (WLC mod P) *holds on $\mathcal{F}$ if there exists a map $Q : \mathcal{F} \times \mathcal{F} \to \mathcal{S}$ such that, for all $A, B \in \mathcal{F}$, the networks A and $Q(A, B)[B]$ are linearly connected.*

In contrast to this weak notion, there is a natural strong notion, where, for a given class of networks, there is a permutation of every network such that all pairs of permuted networks are linearly connected (Fig. 1 (Center)):

Definition 4.2 (Strong linear connectivity modulo permutation). *Let $\mathcal{F}$ be a set of networks. We say that* strong linear connectivity modulo permutation (SLC mod P) *holds on $\mathcal{F}$ if there exists a map $Q' : \mathcal{F} \to \mathcal{S}$ such that, for all $A, B \in \mathcal{F}$, the networks $Q'(A)[A]$ and $Q'(B)[B]$ are* linearly connected.

Note that SLC mod P implies WLC mod P. To see this, let Q' witness SLC mod P for some class $\mathcal{F}$. Then one can verify that Q given by $Q(A, B) = Q'(A)^{-1}Q'(B)$ witnesses WLC mod P on $\mathcal{F}$, because two networks A, B are linearly connected if and only if $P[A], P[B]$ are linearly connected. for every $P \in \mathcal{S}$.

Using the above language, we can summarize existing work precisely. In [6], the authors *prove* that a random pair of sufficiently wide, two-layer networks possess WLC mod P at initialization with high probability [6, Theorem 3.1]. Note that a formal conjecture in [6, Conjecture 1] is often misquoted as stating that WLC mod P holds on a high-probability subset of SGD-trained networks.[4] Instead, this conjecture, when combined with an additional assumption of SGD solutions being closed under permutation, implies SLC mod P for the same subset of networks (but not vice versa). A visualization in the same paper [6, Figure 1] matches neither their proof nor their formal conjecture, but does represent SLC mod P. See Appendix B for an in-depth discussion with proofs.

Empirical evidence supports WLC mod P holding on sufficiently wide, high-probability sets of SGD trained networks, although evidence is still lacking in other settings such as narrow or deep models [1,5,17]. In contrast, there is scant empirical evidence for SLC mod P, at least for standard parametrizations.

Both WLC mod P and SLC mod P are properties that hold for sets of networks, but one may also want to consider properties that hold for trajectories (sequences) of networks (Fig. 1 (Right)).

Definition 4.3 (Simultaneous weak linear connectivity modulo permutation). *Let $\mathcal{C} \subset \mathcal{F}^T$ be a set of sequences of networks. We say that* simultaneous weak linear connectivity modulo permutation *holds on $\mathcal{C}$ if there exists a map $Q : \mathcal{C} \times \mathcal{C} \to \mathcal{S}$ such that, for all $A, B \in \mathcal{C}$ and all $t \in T$, the networks A_t and $Q(A, B)[B_t]$ are linearly connected.*

The property of simultaneous WLC mod P highlights the possibility of finding a single permutation that linearly connects two sequences of networks. In Sect. 5.1, we provide empirical evidence showing that sequences of networks obtained via running SGD are simultaneously WLC mod P. Also, in Sect. 5.2, we

---

[4] In fact, a strictly stronger claim is made: for a certain class of networks $\mathcal{F}$, for all $\theta_1 \in \mathcal{F}$, there is a single permutation that can be applied to θ_1 removing the error barrier between the permuted θ_1 and any other network in the class $\mathcal{F}$. Note that this also means that the networks in $\mathcal{F}$ are piece-wise linearly connected *before* permuting.

empirically show that sequences of iteratively sparsified networks are simultaneously WLC mod P. Simultaneous WLC mod P raises the prospect of interesting applications in federated learning and model fusion.

5   Empirical Findings

In this section, we present our empirical findings regarding *simultaneous weak linear connectivity modulo permutation* (simultaneous WLC mod P). Unless stated otherwise, these results are independent of any algorithmic concerns. In particular, to demonstrate different notions of LC mod P, we simply need to identify *one* permutation that essentially eliminates the error barrier between two networks. Logically, the existence of such a permutation does not depend on the method by which it is found. On the contrary, if we cannot find a permutation that makes two networks linearly connected modulo permutation, this does not rule out the possibility that such a permutation exists and can be found by some other method. In this work, we therefore do not interpret the inability to find a permutation as evidence for or against linear connectivity. In Sect. 6, we discuss results that are algorithm-dependent.

Our experimental setting mimics that of [1] (see Appendix C for details such as hyperparameters, architectures, etc.). The results presented in the main paper are for VGG-16 with layer normalization [4] trained on CIFAR-10. All barriers are plotted using lines to indicate the mean, and shaded regions to indicate the standard deviation over 5 runs. Additional results with different datasets (MNIST, SVHN, CIFAR-10, CIFAR-100) and model architectures (MLP, VGG-16, ResNet-20) are shown in Appendices D and E.

5.1   Training Trajectories Are Simultaneously Weak Linearly Connected Modulo Permutation

We first demonstrate empirically that simultaneous WLC mod P holds for sequences of networks obtained by independent SGD training runs. To show this, we apply weight or activation matching after training, and observe that the resulting loss barrier between independent SGD iterates is highly reduced at all times t. Formally, we consider sequences of dense networks A_t and B_t trained on the same data with independent initializations, minibatches, and other SGD noise, where t denotes the training time. We compute a permutation P_{end} on the networks after training using weight or activation matching

In Fig. 2 (left), we plot the loss barrier between $P_{end}[B_t]$ and A_t As the figure shows, the error barrier after aligning with P_{end} drops from around 70% (as measured between two differently initialized networks after convergence) to 3–8% at any time t in training. This implies that the two networks remain essentially linearly connected modulo permutation throughout training. Previously, [1] only showed that the final weights learned by SGD were linearly connected modulo permutation. Our new observations provide evidence for the strictly stronger conjecture *that SGD trajectories are simultaneously weak linearly connected modulo permutation.*

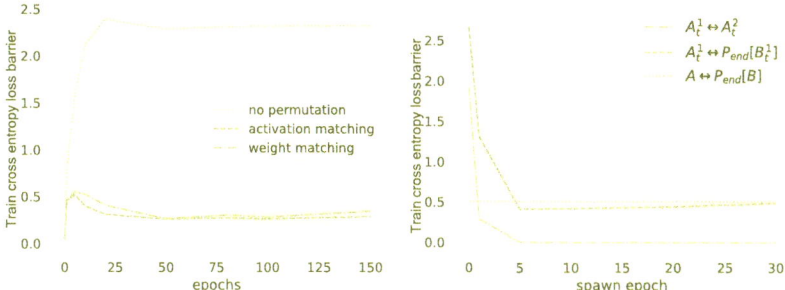

Fig. 2. (Left) loss barrier (y-axis) between networks A_t and permuted B_t at various training times t (x-axis). Permutation is computed at the end of training. The green line corresponds to applying permutations found via weight matching, blue corresponds to permutations found via activation matching, and gray corresponds to no permutation. **(Right)** loss barrier (y-axis) between k "child networks" A^k spawned from A and permuted children $P_{\mathrm{end}}[B_t^k]$ of B spawned at time t (x-axis) from their respective "parent" networks (dashed brown line, $A_t^1 \leftrightarrow P_{\mathrm{end}}[B_t^1]$). P_{end} is computed by aligning parent networks A and B at the end of training (dotted grey line marks the barrier between the permuted parents, $A \leftrightarrow P_{\mathrm{end}}[B]$). These loss barriers are compared against the average loss barrier between independent child networks with the same parent (dot-dashed purple line, $A_t^1 \leftrightarrow A_t^2$). Each child is trained with a different minibatch order starting from the parent's weights at time t. (Color figure online)

In Appendix G.1, we present the results of using permutations P_t computed from earlier in training. Section 6 discusses algorithm-specific takeaways.

The Same Permutation Can Align Multiple Networks from the Same Basin. We next show that a single permutation that linearly connects two SGD solutions after training also connects *other* networks from the same linearly connected modes as each SGD solution. We show this by computing a permutation from the end of two independent training runs, and comparing loss barriers modulo this fixed permutation between networks obtained via an *instability analysis* procedure [8].

[8] defines *instability* to SGD noise at time t as the error barrier after training between two child networks, A_t^1, A_t^2, obtained from a single parent network A at time t, and trained independently to convergence. The first time t from which pairs of children, on average, have no error barrier between them is referred to as the *onset of linear mode connectivity*. Such child networks are not only linearly connected at the end of training, but are also linearly connected throughout their entire training trajectories.

In Fig. 2 (right), we plot the error barrier between the child networks A_t^1 and $P_{\mathrm{end}}[B_t^1]$, where the permutation P_{end} brings the parent networks A and B to the same linearly connected mode. We also plot instability to SGD noise for children spawned at time t (x-axis). The figure shows the onset of linear mode connectivity at around 5 epochs of training. *From the onset of linear mode connectivity onward, we observe that the permutation P_{end} is able to align* *the other child networks throughout training.* This is quite surprising, as the permutation P_{end} was computed on a single pair of parent runs, and is not specifically adapted to align the child networks.

5.2 Iteratively Sparsified Networks Are Simultaneously Weak Linearly Connected Modulo Permutation

We next look at simultaneous weak LC mod P of IMP sparsified networks. We show that a permutation that linearly connects independent dense networks also linearly connects the sequences of sparse networks obtained from the same dense networks via IMP.

Formally, let $A^{(0)}, A^{(1)}, \ldots, A^{(k)}$ be a sequence of increasingly sparse, trained networks obtained by IMP, with $A^{(0)}$ being the original dense, trained network, and $A^{(k)}$ being the sparsest network (assumed to match the accuracy of the dense network). Let $B^{(0)}, \ldots, B^{(k)}$ be another such sequence obtained with the same architecture and data, but using a different initialization and minibatch order.

We identify permutations under which $A^{(i)}$ and $B^{(i)}$ are essentially linearly connected for all i. Specifically, both weight matching and activation matching applied to the dense networks $A^{(0)}$ and $B^{(0)}$ gives a permutation that simultaneously aligns *all* of the sparse IMP subnetworks $A^{(k)}$ and $B^{(k)}$ with lower error barrier than when naively weight matching $A^{(k)}$ to $B^{(k)}$ (Fig. 3, left).

Our findings are complementary to recent work that characterizes the loss landscape of lottery tickets. [15] showed that for any pruning level k, a sparse subnetwork $A^{(k)}$ is essentially linearly connected to its dense counterpart $A^{(0)}$. We further demonstrate that this observation holds, modulo permutation, for networks from different initializations.

Masks Can Be Transported Modulo Permutation. Obtaining an IMP mask is a computationally expensive procedure, as each pruning iteration rewinds and retrains the network's weights from near initialization. Here, we ask whether an IMP mask obtained on one network can be transported to another network via permutation. Indeed, we find this is possible, as the transported mask outperforms one-shot pruning up to a certain sparsity level.

Formally, let $A^{(0)}, A^{(1)}, \ldots, A^{(k)}$ be a sequence of IMP networks of increasing sparsity, where each $A^{(i)}$ is obtained by pruning the 20% smallest weights from $A^{(i-1)}$, and then rewinding and retraining the remaining weights from near initialization. Let $m_A^{(0)}, m_A^{(1)}, \ldots, m_A^{(k)}$ be the corresponding sparsity masks (0 for pruned weights, 1 otherwise) of each IMP network, and let B be an independent run of the dense model which is aligned to $A^{(0)}$ at the end of training by permutation P.

At each pruning level i, we consider the test accuracy of the network obtained after applying $m_A^{(i)}$ to $P[B]$. We obtain this by (1) rewinding the weights of $P[B]$, (2) pruning using the mask $m_A^{(i)}$, and (3) retraining the pruned network to convergence. Since we no longer run IMP on the network B, we compare this

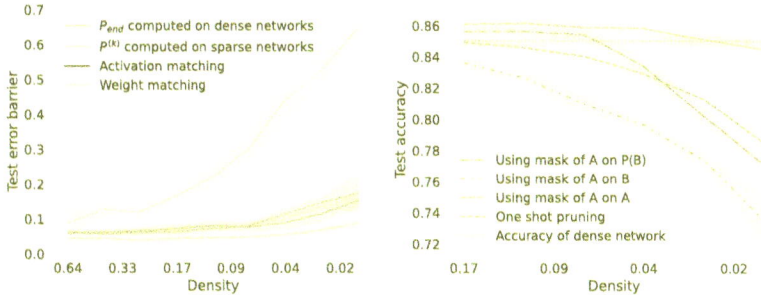

Fig. 3. (Left) error barrier (y-axis, shaded region indicates standard deviation over 5 runs) between permuted sparse IMP subnetworks derived from dense networks trained with different initializations and SGD noise. At each sparsity level (x-axis), pairs of sparse networks are aligned either (1) using a permutation $P^{(k)}$ computed directly on the sparse subnetworks at sparsity level k (orange), and (2) using a permutation computed from the corresponding dense networks (blue). **(Right)** test accuracy of sparse networks using transported masks, at increasing levels of sparsity (x-axis indicates fraction of remaining weights). We use weight matching to compute the permutation. (Color figure online)

procedure to *one-shot pruning*, where we obtain the sparse network by (1) pruning an equivalent fraction $(1 - 0.8^i)$ of the smallest weights of B; (2) rewinding the remaining weights; and (3) retraining to convergence.

We refer to the accuracy of the original dense model as the *matching accuracy*. Figure 3 (right) shows that IMP masks for network A can indeed be transported to B modulo permutation, while preserving matching accuracy. Specifically, we find that applying $m_A^{(i)}$ to $P[B]$ achieves matching accuracy for IMP levels $i \leq 12$, or for sparsities ≥ 0.068. We also show that applying $m_A^{(i)}$ to $P[B]$ achieves matching accuracy up to a higher sparsity level than the one-shot pruning procedure. As one would expect, naively applying $m_A^{(i)}$ to B without permutation fails to achieve matching accuracy at all sparsity levels. Our results suggest that networks within the same linearly connected mode modulo permutation have similar "winning ticket" structures for a range of sparsity levels, and that any mask from this range achieves near-SOTA performance when applied (with appropriate permutations) to other, independently trained models. (See Appendix E for other related experiments.)

5.3 Evidence for Strong Linear Connectivity Modulo Permutation

To search for *strong linear connectivity modulo permutation* (Definition 4.2) between SGD trained networks, we indirectly align two networks relative to a fixed reference network with the following procedure:

1. Independently train 3 networks A, B, C.
2. Align A and B with C to get $P_{A \to C}[A]$ and $P_{B \to C}[B]$.

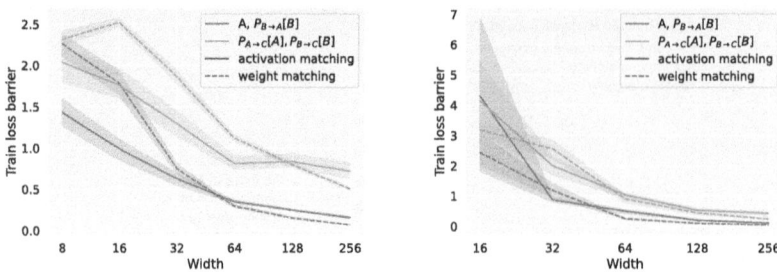

Fig. 4. Test of strong linear connectivity, comparing barriers (y-axis) of networks aligned directly or relative to a reference network (colors) via activation and weight matching (line styles). **(Left)** VGG-16 models of increasing width (x-axis). **(Right)** ResNet-20 models of increasing width (x-axis).

3. Compute the barriers between $P_{A \to C}[A]$ and $P_{B \to C}[B]$.

If strong connectivity holds, we would expect that a single permutation of each of A and B would be sufficient to make them linearly connected with both C and one another.

The results shown in Fig. 4 show that as the width of the networks increases, the loss barrier obtained by indirect alignment between networks A and B decreases. However, indirect alignment still results in much higher barriers than is achievable by aligning A and B directly, meaning the permutations that connect to C do not put A and B in the same linearly connected basin. This failure to find comparable error barriers to the direct alignment may be due to limitations of the weight or activation matching algorithm, which we discuss in Sect. 6. Nevertheless, Fig. 4 supports the conjecture that with greatly increased width, strong linear connectivity modulo permutation is possible between SGD solutions.

6   Algorithmic Aspects of Network Alignment

As eluded earlier, we only have evidence supporting strong linear connectivity modulo permutation for very wide networks. This limitation can perhaps be attributed to the weight matching and activation matching algorithms that we use to align networks (see Sect. 2.2 for descriptions and references for these two algorithms). In this section, we identify potential weaknesses that may hinder the discovery of SLC mod P in more general settings, and summarize our algorithm-specific observations. Our main observation is: **activation matching typically outperforms weight matching, although neither heuristic is superior in all cases.**

Weight Matching Works As Well As Activation Matching for Wide Networks at Convergence. In Fig. 5, we present experiments on weight-matching between networks at different iteration t during training, showing that the weight matching

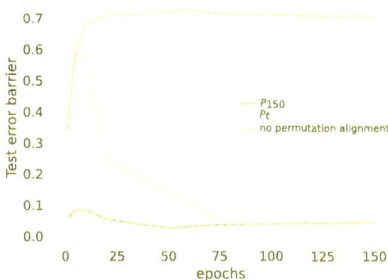

Fig. 5. The error barrier (y-axis) between networks A_t and permuted B_t at different training checkpoints t (x-axis). The brown line corresponds to applying a permutation P_{end} found at the end of training, which successfully eliminates the error barrier between the networks at nearly every epochs. The orange line corresponds to a permutation P_t computed and applied at time t. (Color figure online)

algorithm fails to align networks before approximately 80 epochs (see Appendix G.1 for more discussion). On the other hand, for pruned, sparse networks and networks not trained to convergence, weight matching fails to find permutations that eliminate the error barrier, even when we know that such permutations exist and can be found by aligning the corresponding dense networks. See Sect. 5.2 (Fig. 3) and Appendix E for sparse subnetwork matching results.

Weight Matching Depends on Large Magnitude Weights. In Fig. 6 (left), we experimentally show that weight matching is unaffected when weights are pruned by smallest magnitude, but performs worse when weights are randomly pruned (details in Appendix G.2).

Weight Matching Can Align Lower Layers Earlier in Training. We determine when in training and at which layers the permutations found by weight matching become effective at reducing loss/error barriers, by aligning a subset of layers using the value of their weights from earlier in training (details in Appendix G.3). Figure 6 (right) shows that when this partial alignment is applied in a bottom-up fashion, the lowest layers are aligned sooner than higher layers. This finding is consistent with previous work that has shown that network layers closer to the input converge first [18].

Network "Stability" at Initialization is Neither Necessary nor Sufficient for Permutation Alignment. [8] defined a network to be *stable* to SGD noise if at some time t, two independent runs (with different minibatch randomization) trained from that point forward produce networks that remain linearly connected. It so happens that the experiments of [1], which largely reported the success of weight matching, are between stable networks. However, in Fig. 7 we perform interventions to show that linear connectivity modulo permutation does not require network stability, and vice versa (details in Appendix G.4).

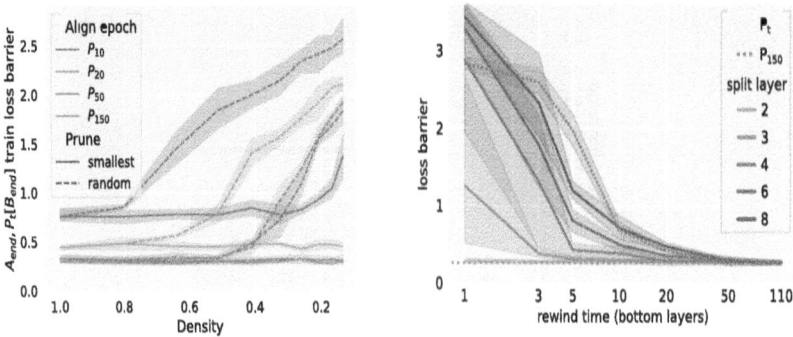

Fig. 6. (Left) effect of magnitude pruning on performance of weight matching algorithm. Loss barrier (y-axis) for networks after training which are aligned with a permutation found via weight matching. Permutations are computed on checkpoints at different training epochs (colors) which are first sparsified (x-axis) via random (dashed) or magnitude (solid) pruning. **(Right)** loss barriers at the end of training between a pair of networks under "bottom-up" partial alignment, which concatenates P_t from input up to layer k with P_{end} for the remaining layers. The rewind time t is the x-axis, and the split point k is indicated by line color. Barriers for P_t computed at time t (x-axis) and at the end of training are included as baselines (dotted lines).

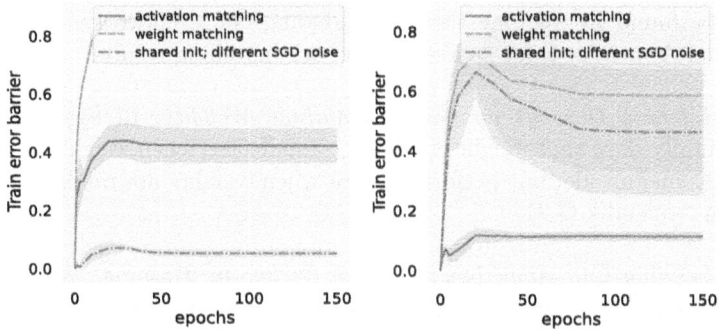

Fig. 7. Stability at initialization is neither sufficient **(left)** nor necessary **(right)** for finding permutation symmetries.

7   Conclusion

In this work, we disambiguate different notions of linear connectivity modulo permutation. We argue that empirical evidence available in the literature only supports *weak linear connectivity modulo permutation* (Definition 4.1).

We show that a permutation aligning two networks up to linear connectivity also simultaneously aligns a large number of related networks. We call this notion of linear connectivity *simultaneous weak linear connectivity* (Definition 4.3). Specifically, we show that the permutation aligning two fully trained networks also aligns the corresponding partially trained networks at any point throughout the training process. We next consider sparse matching subnetworks pruned by iterative magnitude pruning, finding that it is possible to simultaneously align all such subnetworks by applying a permutation that aligns the corresponding dense networks. Furthermore, we show that using the dense-to-dense permutation, sparse masks learned from one network can be reused effectively on another network.

We also define a notion of *strong linear connectivity modulo permutation* which moves beyond the earlier pairwise notion of linear connectivity. Most notably, strong linear connectivity implies that a large set of independently trained networks are all simultaneously and mutually linearly connected after suitable permutations. To this end we experiment with aligning and interpolating between three networks at a time, and demonstrate that strong linear connectivity may be possible with very wide networks.

Finally, our investigations shed light on the behavior of two principal classes of heuristics used in permutation-finding algorithms: weight matching and activation matching. Although neither heuristic is clearly dominant, we observe that activation matching is generally more robust over more settings than weight matching. We find that weight matching depends on large magnitude weights that appear later in training. In keeping with the common wisdom that the initial layers of neural networks are learned earlier, we also find that weight matching identifies the permutation of initial layers earlier.

Acknowledgements. The authors would like to thank Tiffany Vlaar and Utku Evci for feedback on a draft and various ideas, as well as Udbhav Bamba for preliminary implementation work. DMR and DR are supported by Canada CIFAR AI Chairs and NSERC Discovery Grants. The authors also acknowledge material support from NVIDIA in the form of computational resources, and are grateful for technical support from the Mila IDT and Vector teams in maintaining the Mila and Vector Compute Clusters. Resources used to prepare this research were provided, in part, by Mila (mila.quebec), the Vector Institute (vectorinstitute.ai), the Province of Ontario, the Government of Canada through CIFAR, and companies sponsoring the Vector Institute.

Disclosure of Interests. The authors have no competing interests to declare that are relevant to the content of this article.

References

1. Ainsworth, S., Hayase, J., Srinivasa, S.: Git re-basin: merging models modulo permutation symmetries. In: The Eleventh International Conference on Learning Representations (2023)
2. Akash, A.K., Li, S., Trillos, N.G.: Wasserstein barycenter-based model fusion and linear mode connectivity of neural networks (2022)
3. Altschuler, J.M., Boix-Adserà, E.: Wasserstein barycenters are NP-hard to compute. SIAM J. Math. Data Sci. **4**(1), 179–203 (2022)
4. Ba, J.L., Kiros, J.R., Hinton, G.E.: Layer normalization (2016)

5. Benzing, F., et al.: Random initialisations performing above chance and how to find them. In: OPT 2022: Optimization for Machine Learning (NeurIPS 2022 Workshop) (2022)

6. Entezari, R., Sedghi, H., Saukh, O., Neyshabur, B.: The role of permutation invariance in linear mode connectivity of neural networks. In: International Conference on Learning Representations (2022)

7. Frankle, J., Carbin, M.: The lottery ticket hypothesis: finding sparse, trainable neural networks. In: International Conference on Learning Representations (2019)

8. Frankle, J., Dziugaite, G.K., Roy, D., Carbin, M.: Linear mode connectivity and the lottery ticket hypothesis. In: Proceedings of the 37th International Conference on Machine Learning, vol. 119, pp. 3259–3269. PMLR (2020)

9. Goodfellow, I.J., Vinyals, O., Saxe, A.M.: Qualitatively characterizing neural network optimization problems (2015)

10. Jordan, K., Sedghi, H., Saukh, O., Entezari, R., Neyshabur, B.: REPAIR: REnormalizing permuted activations for interpolation repair. In: The Eleventh International Conference on Learning Representations (2023)

11. Li, Y., Yosinski, J., Clune, J., Lipson, H., Hopcroft, J.: Convergent learning: Do different neural networks learn the same representations? In: Proceedings of the 1st International Workshop on Feature Extraction: Modern Questions and Challenges at NIPS 2015, vol. 44, pp. 196–212. PMLR (2015)

12. Lucas, J.R., Bae, J., Zhang, M.R., Fort, S., Zemel, R., Grosse, R.B.: On monotonic linear interpolation of neural network parameters. In: Proceedings of the 38th International Conference on Machine Learning, vol. 139, pp. 7168–7179. PMLR (2021)

13. Nagarajan, V., Kolter, J.Z.: Uniform convergence may be unable to explain generalization in deep learning. In: Advances in Neural Information Processing Systems, vol. 32, pp. 11615–11626 (2019)

14. O'Neill, J., V. Steeg, G., Galstyan, A.: Layer-wise neural network compression via layer fusion. In: Proceedings of the 13th Asian Conference on Machine Learning, vol. 157, pp. 1381–1396. PMLR (2021)

15. Paul, M., Chen, F., Larsen, B.W., Frankle, J., Ganguli, S., Dziugaite, G.K.: Unmasking the lottery ticket hypothesis: What's encoded in a winning ticket's mask? In: The Eleventh International Conference on Learning Representations (2023)

16. Paul, M., Larsen, B., Ganguli, S., Frankle, J., Dziugaite, G.K.: Lottery tickets on a data diet: Finding initializations with sparse trainable networks. In: Advances in Neural Information Processing Systems, vol. 35, pp. 18916–18928 (2022)

17. Peña, F.A.G., Medeiros, H.R., Dubail, T., Aminbeidokhti, M., Granger, E., Pedersoli, M.: Re-basin via implicit Sinkhorn differentiation. In: Proceedings of the IEEE/CVF Conference on Computer Vision and Pattern Recognition, pp. 20237–20246, June 2023

18. Raghu, M., Gilmer, J., Yosinski, J., Sohl-Dickstein, J.: SVCCA: singular vector canonical correlation analysis for deep learning dynamics and interpretability. In: Advances in Neural Information Processing Systems, vol. 30, pp. 6076–6085 (2017)

19. Simsek, B., Ged, F., Jacot, A., Spadaro, F., Hongler, C., Gerstner, W., Brea, J.: Geometry of the loss landscape in overparameterized neural networks: symmetries and invariances. In: Proceedings of the 38th International Conference on Machine Learning, vol. 139, pp. 9722–9732. PMLR (2021)

20. Singh, S.P., Jaggi, M.: Model fusion via optimal transport. In: Advances in Neural Information Processing Systems, vol. 33, pp. 22045–22055 (2020)

21. Tatro, N., Chen, P.Y., Das, P., Melnyk, I., Sattigeri, P., Lai, R.: Optimizing mode connectivity via neuron alignment. In: Advances in Neural Information Processing Systems, vol. 33, pp. 15300–15311 (2020)
22. Vlaar, T.J., Frankle, J.: What can linear interpolation of neural network loss landscapes tell us? In: Proceedings of the 39th International Conference on Machine Learning, vol. 162, pp. 22325–22341. PMLR (2022)
23. Wang, H., Yurochkin, M., Sun, Y., Papailiopoulos, D., Khazaeni, Y.: Federated learning with matched averaging. In: International Conference on Learning Representations (2020)
24. Wortsman, M., Horton, M.C., Guestrin, C., Farhadi, A., Rastegari, M.: Learning neural network subspaces. In: Proceedings of the 38th International Conference on Machine Learning, vol. 139, pp. 11217–11227. PMLR (2021)
25. Yurochkin, M., Agarwal, M., Ghosh, S., Greenewald, K., Hoang, N., Khazaeni, Y.: Bayesian nonparametric federated learning of neural networks. In: Proceedings of the 36th International Conference on Machine Learning, vol. 97, pp. 7252–7261. PMLR (2019)

Fast Fishing: Approximating BAIT for Efficient and Scalable Deep Active Image Classification

Denis Huseljic[(✉)], Paul Hahn, Marek Herde, Lukas Rauch, and Bernhard Sick

Intelligent Embedded Systems, University of Kassel, Kassel, Germany
{dhuseljic,paul.hahn,marek.herde,lukas.rauch,bsick}@uni-kassel.de

Abstract. Deep active learning (AL) seeks to minimize the annotation costs for training deep neural networks. BAIT, a recently proposed AL strategy based on the Fisher Information, has demonstrated impressive performance across various datasets. However, BAIT's high computational and memory requirements hinder its applicability on large-scale classification tasks, resulting in current research neglecting BAIT in their evaluation. This paper introduces two methods to enhance BAIT's computational efficiency and scalability. Notably, we significantly reduce its time complexity by approximating the Fisher Information. In particular, we adapt the original formulation by i) taking the expectation over the most probable classes, and ii) constructing a binary classification task, leading to an alternative likelihood for gradient computations. Consequently, this allows the efficient use of BAIT on large-scale datasets, including ImageNet. Our unified and comprehensive evaluation across a variety of datasets demonstrates that our approximations achieve strong performance with considerably reduced time complexity. Furthermore, we provide an extensive open-source toolbox that implements recent state-of-the-art AL strategies, available at https://github.com/dhuseljic/dal-toolbox.

Keywords: Active Learning · Deep Learning · Fisher Information

1 Introduction

Training deep neural networks (DNNs) requires large amounts of annotated data. In this context, the annotation process represents a significant bottleneck because it is costly and time-consuming. Active learning (AL) is an iterative process that offers a solution by intelligently selecting a subset of informative data points from an unlabeled dataset for annotation. This subset is then employed to train DNNs, reducing the need for extensive annotation processes.

Supplementary Information The online version contains supplementary material available at https://doi.org/10.1007/978-3-031-70368-3_17.

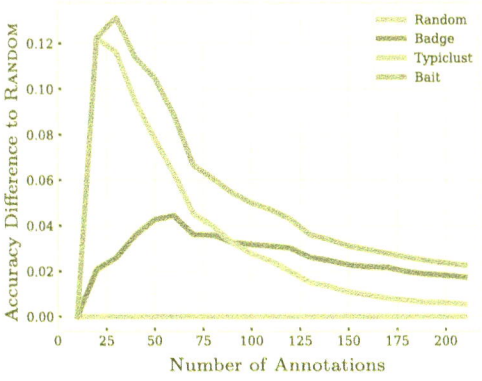
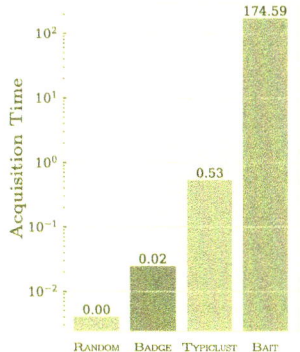

(a) Accuracy improvements compared to a random instance selection for popular strategies.

(b) Average acquisition time per cycle in seconds.

Fig. 1. Comparison of different AL strategies on CIFAR-10.

In recent years, there has been a lot of progress in deep AL, with many studies focusing on developing AL selection strategies. Generally, these strategies can be categorized into uncertainty-based and diversity-based approaches. Uncertainty-based strategies leverage the uncertainty from a DNN, assuming that instances with high uncertainty are informative [13]. However, deep AL often involves batch acquisitions since retraining DNNs is time-consuming and computationally expensive. This makes uncertainty-based strategies less effective due to the risk of picking similar instances in a batch, leading to redundancy. To address this issue, diversity-based strategies have been proposed in deep AL [35]. These prioritize acquiring diverse instances in a batch, thereby reducing information redundancy. Optimally, however, we aim to construct a batch that is both diverse and informative, ensuring the batch's overall informativeness is maximized. To this end, most recent studies seek to balance informativeness and diversity when selecting batches of instances [3,4,15].

In the plethora of those deep AL strategies, BAIT [3] stands out as an approach that offers superior performance. Central to BAIT is optimizing the Bayes risk through an objective function that requires the calculation of the Fisher Information Matrix (FIM). Since the calculation of the FIM in high-dimensional parameter spaces is infeasible, BAIT uses the parameters of the DNN's last layer. Empirically, it shows impressive results across different datasets [8]. Despite its merits, many recent studies neglect to compare their strategy against BAIT [15,33,38,40]. One reason may be the poor time complexity of $\mathcal{O}(D^2 K^3)$ when computing the FIM for a large number of classes K and dimensions D, or the high time and space requirements for the selection based on an FIM per instance. Additionally, the lack of accessible implementations and the complexity of integrating BAIT into existing frameworks could prevent its broad acceptance.

Figure 1 presents the results of an AL experiment on CIFAR-10 with a total budget of 210 instances and a batch acquisition size of 10 instances per cycle. The detailed experimental setup is available in the Appendix A. We evaluate prominent state-of-the-art AL strategies, focusing on performance (i.e., accuracy) and acquisition times (i.e., required time for selecting instances per cycle). Figure 1a plots the accuracy difference between a model trained with an AL strategy and a model trained with randomly sampled instances. Notably, BAIT outperforms every strategy, demonstrating its superior performance. However, the acquisition times in Fig. 1b indicate that BAIT requires significantly more time than others to acquire a batch of instances. This bottleneck will become even more severe when dealing with larger acquisition sizes or more classes. Consequently, despite BAIT's superior accuracy, this example highlights its computational challenges. These practical constraints limit its feasibility across various applications and real-world settings.

For this reason, we aim to improve BAIT's computational efficiency and scalability across classes by introducing two methods to tackle its main bottlenecks. Our first method focuses on efficient computation of the FIM, improving the time complexity to $\mathcal{O}(D^2 K^2)$ by only focusing on the most probable classes in the expectation. Our second method focuses on reducing the size of the FIM, leading to a time complexity of $\mathcal{O}(D^2)$. BAIT's original formulation requires the calculation of an FIM per instance, posing challenges in computational and memory resources, especially as the number of entries in the FIM increases quadratically with more classes. To overcome this, we propose to approximate the original multi-class likelihood through a binary one that significantly lowers the dimensionality, leading not only to an efficient computation of the FIM but also an efficient computation of BAIT's objective. This way, we enable scalability across many classes while enhancing computational efficiency. While the first method prioritizes adherence to the original FIM formulation, the second approach aims to completely disentangle time and space complexity from the number of classes, enabling BAIT's usage on large-scale datasets such as ImageNet [10]. Our contributions can be summarized as follows:

- We propose two approximation methods for BAIT, enhancing its computational efficiency and scalability to many classes without compromising performance.
- We provide a unified and comprehensive study comparing recent state-of-the-art AL strategies across a multitude of image datasets.
- We provide a toolbox that implements recent state-of-the-art strategies, including our version of BAIT, for a straightforward adoption in future research.

2   Related Work

Pool-based deep AL strategies select batches instead of single instances to avoid retraining an DNN with each new acquired instance [34]. AL strategies are typically categorized into uncertainty-based, diversity-based, and hybrid strategies. Uncertainty-based strategies aim to select instances that are assumed to be difficult for the DNN, often close to the decision boundary. Employed in a batch acquisition setting, uncertainty-based strategies greedily select the highest-scoring instances, which can lead to information redundancy. MARGIN sampling [36], a variant of uncertainty sampling, selects instances where the difference between the two highest predicted class probabilities is largest. Recently, it has been shown to work effective in AL [5], despite information redundancy in a batch. BALD [13] assumes a Bayesian DNN and selects instances that maximize the mutual information between predictions and the DNN's posterior distribution. BatchBALD [19] improves the selection of BALD by reducing information redundancy in a batch. In contrast to uncertainty-based strategies, diversity-based strategies aim to select a batch of diverse instances that represent the dataset. CoreSet [35] selects instances that minimize the average distance between labeled and unlabeled instances. Recently, there has been a rise in hybrid strategies that combine uncertainty and diversity-based selection. BADGE [4] selects via k-means++ [2] on instances represented in a gradient space, which results in a selection of difficult and diverse instances. Bayesian Estimate of Mean Proper Scores [38] estimates the loss reduction in combination with k-means to ensure diversity. TYPICLUST [15] focuses on instances with high density through nearest neighbor density estimation. They employ k-means to ensure diversity.

The Fisher information describes the amount of information an observable random variable carries about an unknown parameter. It is commonly used as an approximation to the Hessian in the field of optimization [1]. In the context of AL, [7] motivate its use by the selection of instances that minimize the expected likelihood given the current set of parameters. They propose a near-optimal strategy consisting of two phases, which require multiple calculations of the FIM. Since the size of the FIM scales quadratically with the number of parameters, employing this scheme with DNNs is infeasible. Therefore, BAIT [3] proposes several adaptions, discussed in Sect. 4, to combine this strategy with DNNs. In the context of second-order optimization, there exist several works focusing on the approximation of the FIM. [23] propose to reduce the FIM to its diagonal elements, effectively decreasing the quadratic scaling in the number of parameters to a linear scaling. [17] assume independency between parameter groups and, therefore, reduce the FIM to a block diagonal matrix. [28] propose the Kronecker Factorized Approximate Curvature, which approximates blocks of the FIM as the Kronecker product of two smaller matrices, leading to computational savings in both computation and inversion. While [37] reformulate the score function of the FIM as averages over parameters, [24] improve the efficiency of KFAC by using dimensionality reduction techniques such as low-rank approximations. For a comprehensive comparison regarding these approximations, we refer to [9]. For a more detailed theoretical discussion about the FIM, we refer to [21].

3   Notation

We focus on a pool-based classification setting with instances $x \in \mathcal{X}$ and associated class labels $y \in \mathcal{Y} = \{1, \ldots, K\}$, where K is the number of classes. A DNN consists of a feature extractor and a classification layer. The former is given by $h_\omega : \mathcal{X} \to \mathbb{R}^D$, with parameters ω, mapping the inputs to a hidden representation. The latter is given by $f_\theta : \mathbb{R}^D \to \mathbb{R}^K$, with parameters θ, mapping the hidden representation to a logit space. Class probabilities are obtained via the softmax function $p_\theta(y|x) = [\text{softmax}(f_\theta(h_\omega(x)))]_y$. An AL cycle starts with a labeled pool $\mathcal{L} = \{(x_n, y_n)\}_{n=1}^N$ and a large unlabeled pool $\mathcal{U} = \{x_m\}_{m=1}^M$, where N and M denote the number of labeled and unlabeled instances, respectively. With this setting, we train a DNN on $\mathcal{L}$, assess the informativeness of instances in $\mathcal{U}$, query an oracle for labels on a batch of $B \in \mathbb{N}$ informative instances, add these annotated instances to $\mathcal{L}$ and remove them from $\mathcal{U}$, and repeat the cycle. Hence, across AL cycles, the cardinality N and M of the sets $\mathcal{L}$ and $\mathcal{U}$ are consistently changing. Assuming instance are independent and identically distributed (i.i.d.), we typically train a DNN maximizing $L(\theta) = \sum_{n=1}^N \ln p(y|x, \theta)$, where $\ln p(y|x, \theta)$ is the likelihood function.

4   Time and Space Complexity of BAIT

The general idea of BAIT is to optimize the Bayes risk by adapting the near-optimal two-phase sampling scheme proposed in [7] for the use with DNNs. For this purpose, they introduce three key extensions: (1) focusing on the DNN's last layer due to the high-dimensional parameter space, (2) recalculating the FIM after each retraining to account for changing representations, and (3) developing a greedy method for batch selection. In principle, BAIT selects instances based on an informativeness score that is defined by

$$\arg \min_{x \in \mathcal{U}} \text{tr} \left((\mathbf{M} + \mathbf{I}(x; \theta))^{-1} \mathbf{I}(\theta) \right), \tag{1}$$

where $\mathbf{M} \in \mathbb{R}^{DK \times DK}$ is the FIM of already selected instances (i.e., labeled pool and instances in an acquisition batch), $\mathbf{I}(\theta) \in \mathbb{R}^{DK \times DK}$ is the FIM of all instances (i.e., labeled and unlabeled pool), and $\mathbf{I}(x; \theta) \in \mathbb{R}^{DK \times DK}$ is the FIM of a particular instance $x \in \mathcal{U}$. Note that the computation of the FIM does not require labels. For a detailed derivation and description, we refer to the original work [3].

We see that the informativeness score in Eq. (1) consists of the three FIMs, and hence, the main computations revolve around *calculating* these three matrices. Generally, the FIM reflects the amount of information that the random variable y carries about the parameters θ given a likelihood function $\ln p(y|x, \theta)$. It is defined as the expected outer product of the likelihood's gradient and given by

$$\mathbf{I}(x; \theta) = \mathbb{E}_y[\nabla_\theta \ln p(y|x, \theta) \nabla_\theta^T \ln p(y|x, \theta)], \tag{2}$$

where the expectation is taken over the categorical distribution $y \sim \text{Cat}(y|p_\theta(\cdot|x))$. The FIMs, $\mathbf{I}(\theta)$ and $\mathbf{M}$, are obtained by averaging $\mathbf{I}(x; \theta)$ over the relevant instances.

We aim to reduce BAIT's computational time and memory footprint by approximating Eq. (2) efficiently. Consequently, this simplifies the calculation of the informativeness score from Eq. (1), as all three matrices are calculated more efficiently. Since the outer product has a time complexity of $(KD)^2$ and we need to calculate it repeatedly for each class due to the expectation, we end up with a time complexity of $\mathcal{O}(K(KD)^2)$. As we see, the complexity of this task stem mainly from the **expectation** over the categorical distribution and the dimensionality of the likelihood's **gradient**.

Considering this expectation, we realize that it leads to a cubic time complexity regarding K, which hinders BAIT from being used in a setting with many classes. Moreover, we have these computational costs for every instance within the unlabeled and labeled pool. In addition, the computation of the expectation not only affects time complexity but also drastically increases the memory requirements, which can lead to problems, especially when working with GPUs. Originally, BAIT avoids storing a separate FIM for each instance to mitigate memory constraints. Instead, it retains only the gradients for each instance and class, thus reducing the space complexity from $\mathcal{O}(M(KD)^2)$ to $\mathcal{O}(MK(KD))$. However, this method does not fully solve memory inefficiencies.

The dimensionality of gradients further complicates these problems. Increasing the number of classes not only increases the amount of gradients to be stored (first K) but also increases the dimensionality of those gradients (KD). This is due to the parameters in the last layer increasing linearly with K as the number of classes grows. As a result, the number of elements that need to be stored grows quadratically with the number of classes (expectation and gradient's dimensionality), which is a major challenge for applying BAIT to large-scale problems. In addition, the outer product in Eq. (2) leads to a high-dimensional FIM, which aggravates subsequent operations such as inversions and thus also affects the overall efficiency.

5   Approximations

In the previous section, we analyzed the time and space complexity of BAIT. We demonstrated that the expectation and dimensionality of the likelihood's gradient pose a significant challenge to BAIT's practical applicability. Here, we present our modifications of BAIT to solve these problems by reducing time and space complexity based on approximations of the FIM. In our first approach, we prioritize adherence to the original objective. Conversely, our second strategy aims to reduce the dependency on K, resulting in a distinctly different reformulation of the FIM itself. We refer to the first approximation as BAIT (Exp) and to the second as BAIT (Binary).

5.1   Expectation

Our first method reduces the complexity by focusing on the expectation. This expectation is taken over the distribution $\mathrm{Cat}(y|p_{\boldsymbol{\theta}}(\cdot|\boldsymbol{x}))$ and, intuitively, it weighs the outer products by the model's predicted probability for a respective class:

$$\mathbf{I}(\boldsymbol{x};\boldsymbol{\theta}) = \sum_{y \in \mathcal{Y}} p_{\boldsymbol{\theta}}(y|\boldsymbol{x}) \left(\nabla_{\boldsymbol{\theta}} \ln p(y|\boldsymbol{x},\boldsymbol{\theta}) \nabla_{\boldsymbol{\theta}}^{\mathrm{T}} \ln p(y|\boldsymbol{x},\boldsymbol{\theta}) \right). \tag{3}$$

Table 1. Summary of time and space complexity of the proposed approximations.

	Time Complexity	Space Complexity
BAIT (Exp)	$\mathcal{O}(c(KD)^2)$	$\mathcal{O}(MDcK)$
BAIT (Binary)	$\mathcal{O}(D^2)$	$\mathcal{O}(MD)$

Our idea is that, instead of taking the expectation over the complete categorical distribution, we approximate it by considering only a subset of classes. The motivation is that for many classes, most of the probability mass will be present in the top predictions of our model, leading to a good approximation of the true FIM. This way, we only have to specify the number of top classes that should be considered in the expectation. Formally, this can be seen as taking the expectation over a new categorical distribution, which now only considers the top predictions instead of all classes. To ensure that this categorical distribution and, hence, the FIM is well defined, we additionally need to normalize the selected probabilities. As a result, the expectation is calculated on

$$\text{Cat}(\hat{y}|\hat{p}_\theta(\cdot|\boldsymbol{x}, c)) \quad \text{with} \quad \hat{p}_\theta(\hat{y}|\boldsymbol{x}, c) = \frac{p_\theta(\hat{y}|\boldsymbol{x})}{\sum_{y' \in \mathcal{Y}_c} p_\theta(y'|\boldsymbol{x})}, \tag{4}$$

where $\hat{y} \in \mathcal{Y}_c$ and $\mathcal{Y}_c \subset \mathcal{Y}$ denotes the $c \in \mathbb{N}_{>0}$ top-predictions of the DNN with parameters $\boldsymbol{\theta}$ for instance $\boldsymbol{x}$. This approximation effectively reduces the time complexity from $\mathcal{O}(K(KD)^2))$ to $\mathcal{O}(c(KD)^2))$, where c is a constant factor unaffected by the overall number of classes. Moreover, it lowers the space complexity from $\mathcal{O}(MDK^2)$ to $\mathcal{O}(MDcK)$, effectively diminishing the quadratic growth of memory requirements.

When considering all classes ($c = K$), we retrieve the original FIM formulation as specified in Eq. (2). In contrast, for a single class ($c = 1$), at first glance, our method yields an approximation seemingly akin to the empirical FIM. The empirical FIM is similar to the true FIM but avoids the calculation of expectations by directly leveraging available labels y. Since it offers better time and space complexity by avoiding the expectation, it is frequently employed as a replacement for the true FIM [14,18,41]. However, many studies criticize its use [21,27,32]. From a theoretical perspective, the empirical FIM is not an approximation and does not possess the properties of the true FIM. Comparing the empirical FIM to our approximation, we employ the DNN's prediction instead of the label. As mentioned by [21], the formulation we employ is not equivalent to the empirical FIM, but a biased estimate of the true FIM. An unbiased estimate can be obtained through sampling from $\text{Cat}(y|p_\theta(\cdot|\boldsymbol{x}))$, albeit at the expense of increased variance [21]. Nonetheless, within the context of AL, we empirically found that adopting a biased estimate of the FIM in favor of reduced variance proved advantageous. For a more detailed discussion of this topic, we refer to [21].

5.2   Gradient

Our second method aims to decouple the time and space complexity of computing the FIM from the number of classes. This requires a class-independent formulation for both the expectation and gradient calculations. Consequently, we explore an alternative representation of the FIM, conceptualized via the Hessian of the likelihood function as follows:

$$\mathbf{I}(\boldsymbol{x}; \boldsymbol{\theta}) = -\mathbb{E}_y[\nabla_{\boldsymbol{\theta}}^2 \ln p(y|\boldsymbol{x}, \boldsymbol{\theta})]. \tag{5}$$

This formulation interprets the FIM as the negative expectation of the Hessian with respect to the model parameters. Similar to the original objective function from Eq. (2), we notice that many classes lead to quadratic growth of the Hessian's dimensions. Our key idea involves approximating the Hessian and, consequently, the FIM by adopting a different likelihood $\ln p(y|\boldsymbol{x}, \boldsymbol{\theta})$. This approach is inspired by [25], where class-specific covariance matrices of a Laplace approximation are approximated by considering an upper bound. This upper bound transforms the multi-class classification setting into a binary one, where the maximum probability is considered the positive class' probability. This way, we assume a shared Hessian matrix across classes as a simplification, which significantly reduces the time complexity to $\mathcal{O}(D^2)$ and space complexity to $\mathcal{O}(MD)$. Formally, we replace the categorical likelihood with a Bernoulli likelihood, which is given by

$$\ln p(y|\boldsymbol{x}, \boldsymbol{\theta}) = y \ln \hat{p} + (1 - y) \ln(1 - \hat{p}) \tag{6}$$

where $\hat{p} = \max_y(p_{\boldsymbol{\theta}}(y|\boldsymbol{x}))$ is the highest predicted probability of our DNN. Intuitively, as we basically assume a binary classification setting, the FIM only considers the influence of a subset of parameters that led to the highest predicted probability. When comparing the first approximation BAIT (Exp) with BAIT (Binary), the main difference lies in the modification of the likelihood. This adaption not only ensures better time complexity of the expectation, but also reduces the dimensionality of the gradient $\nabla_{\boldsymbol{\theta}} \ln p(y|\boldsymbol{x}, \boldsymbol{\theta})$ considerably, making it independent of the number of classes. A summary of time and space complexity of the proposed approximations is given in Table 1.

6   Experimental Results

In this section, we investigate the performance of the proposed approximations by comparing them to the original version of BAIT and applying them to real-world tasks across various image datasets.

6.1   Setup

Datasets: We conduct experiments across a variety of image datasets to assess the effectiveness and robustness of our proposed approximations of BAIT. Our

Table 2. Overview over Image datasets.

Dataset	|Train|	|Test|	|Classes|
CIFAR-10	50k	10k	10
STL-10	5k	8k	10
Snacks	5k	1k	20
CIFAR-100	50k	10k	100
Food-101	75k	25k	101
Flowers102	1k	6k	102
StanfordDogs	18k	2k	120
Tiny ImageNet	100k	10k	200
ImageNet	1.2 m	100k	1000

evaluation encompasses a total of nine image datasets. Aiming to ensure a comprehensive evaluation focusing on scalability, we cover a broad range of class numbers. These datasets range from a small (CIFAR-10 [20], STL-10 [39], and Snacks [29]) to a large amount of classes (CIFAR-100 [20], Food-101 [6], Flowers-102 [30], StanfordDogs [42], TinyImageNet [22], ImageNet [10]). For each dataset, we use a 10% split of the training dataset for development of our approximations. Information about the number of instances and classes is summarized in Table 2. A more comprehensive description can be found in Appendix B.

Model: We employ a Vision Transformer (ViT) [12] with pretrained weights obtained through self-supervised learning together with a randomly initialized fully connected layer. Specifically, we use the DINOv2-ViT-S/14 model [31] with approximately 14 million parameters and a feature dimension of $D = 384$ in its final hidden layer. We intentionally select the worst-performing DINO model to simulate real-world deep AL scenarios where optimal pre-trained models may not be readily available. This setting aligns with recent recommendations from the literature [15] that highlight the importance of proper feature representation in deep AL for images. In each AL cycle, we train the DNN's randomly initialized last layer for 200 epochs, employing the Rectified Adam optimizer [26]. We use a training batch size of 128, a learning rate of 0.2, and weight decay of 0.0001. In addition, we utilize a cosine annealing learning rate scheduler. We determined these hyperparameters empirically to be effective across all datasets by evaluating the convergence of a DNN on randomly sampled instances.

AL Setting: We determine the acquisition size, total budget, and the number of initial instances according to the dataset characteristics to guarantee convergence of the AL process. Additionally, the initial labeled pool is chosen randomly. Details are provided in Appendix D. Our evaluation comprises various AL selection strategies, including RANDOM, which selects instances randomly, MARGIN [5], focusing on the uncertainty in the top-2 predicted probabilities, BADGE [4], which selects instances according to the gradient norm of the like-

Table 3. Comparison of BAIT (Exp) to the original formulation of BAIT with varying values for c over multiple datasets.

	RANDOM	BAIT (Exp) $c = 1$	BAIT (Exp) $c = 2$	BAIT (Exp) $c = 5$	BAIT
CIFAR-10 (10 Classes)					
Accuracy (AUC)	82.05	+4.63	+5.39	**+5.50**	+5.37
Acquisition Time (CPU)	00:00	18:07	18:16	22:44	32:07
Acquisition Time (GPU)	00:00	00:41	00:45	01:00	01:23
STL10 (10 Classes)					
Accuracy (AUC)	85.79	+5.61	+6.19	+6.45	**+6.54**
Acquisition Time (CPU)	00:00	01:47	01:49	02:15	03:10
Acquisition Time (GPU)	00:00	00:04	00:04	00:06	00:08
Snacks (20 Classes)					
Accuracy (AUC)	71.93	+7.50	**+8.44**	+8.39	+8.07
Acquisition Time (CPU)	00:00	09:30	10:40	12:40	21:50
Acquisition Time (GPU)	00:00	00:15	00:15	00:20	00:40

lihood, and TYPICLUST [15], employing clustering and density estimation for selection. These strategies are considered state-of-the-art and were chosen based on their robustness and efficiency across different domains [5,15,33]. Additionally, we employ BAIT with our expectation approximation, called BAIT (Exp), and our binary approximation, called BAIT (Binary).

Metrics: To evaluate the effectiveness of a strategy, we consider its learning curve. Therefore, in each cycle, we train a DNN on a labeled pool selected by a given strategy and compute the accuracy on the test dataset. Then, we examine the accuracy improvement over RANDOM by considering the difference of learning curves. Due to space limitations, we only report a selection of learning curves. All remaining curves of all strategies and datasets can be found in Appendix E. As an alternative, we report the area under the learning curve (AUC), which summarizes the learning curve into a single numerical value. All learning curves and numerical values are averaged over ten repetitions to ensure comparability.

6.2 Assessment of Approximations

We compare our approximations to the original BAIT strategy across different image datasets. Due to the memory limitations of the original BAIT formulation (cf. Table 1), it was infeasible to include datasets with more than 20 classes in these experiments. Therefore, we use CIFAR-10, STL-10 and Snacks to assess whether our approximations perform similarly to the original version. We report acquisition times per AL cycle for a batch size of 10 instances, measured on both CPU and GPU. These times were determined using a workstation equipped with an NVIDIA RTX 4090 GPU and an AMD Ryzen 9 7950X CPU.

Table 4. Comparison of BAIT (Binary) and BAIT (Diagonal) to the original formulation of BAIT over multiple datasets.

	RANDOM	BAIT (Diagonal)	BAIT (Binary)	BAIT
CIFAR-10 (10 Classes)				
Accuracy (AUC)	82.05	+1.25	**+5.74**	+5.23
Acquisition Time (CPU)	00:00	01:23	00:13	32:07
Acquisition Time (GPU)	00:00	00:32	00:09	01:23
STL10 (10 Classes)				
Accuracy (AUC)	85.79	+3.43	+6.45	**+6.54**
Acquisition Time (CPU)	0.00	00:08	00:01	03:10
Acquisition Time (GPU)	0.00	00:03	00:01	00:08
Snacks (20 Classes)				
Accuracy (AUC)	71.93	+0.77	**+8.60**	+8.07
Acquisition Time (CPU)	00:00	00:32	00:01	21:50
Acquisition Time (GPU)	00:00	00:11	00:01	00:41

BAIT (Exp): First, we examine the initial approximation discussed in Sect. 5.1, which estimates the expectation by focusing on the most probable classes. Intuitively, increasing the number of considered classes should yield an estimate closer to the true FIM. According to Table 3, most of our approximations ($c > 1$) closely match the performance of the original BAIT formulation across various datasets while reducing acquisition time. Most importantly, this time reduction becomes more substantial when dealing with a larger number of classes, as seen by the acquisition times on the Snacks dataset. In this case, our approximation with $c = 2$ is able to cut BAIT's acquisition time down to a half while providing a better accuracy than the original formulation. Furthermore, in two out of three data sets, our approximations manage to outperform the original BAIT, demonstrating that focusing on the most probable class can even be beneficial for the selection of BAIT.

When considering a single class only in our approximation ($c = 1$), the performance decreases slightly, indicating that a single class may not be enough for an accurate approximation of the FIM. Despite that, this approximation still outperforms RANDOM in terms of accuracy while improving BAIT's time complexity. Furthermore, we already achieve comparable performances to the original BAIT formulation with $c = 2$, suggesting that this number may be sufficient for an accurate approximation. Thus, in the remainder of the experiments, we will fix $c = 2$ for BAIT (Exp).

BAIT (Binary): We examine the effectiveness of our second approximation discussed in Sect. 5.2, which reformulates the multi-class classification task into a binary one. Since this approximation is intended to work on a large number of classes, we also compare it to the diagonal approximation of the FIM. This

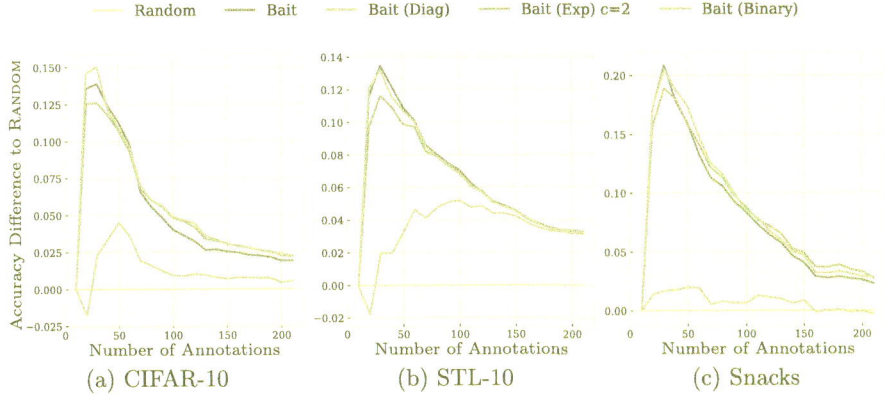

Fig. 2. Accuracy improvement curves of BAIT and its approximations.

approximation only calculates the main diagonal, omitting off-diagonal elements. It is frequently employed to scale the FIM to high dimensions [23], and is therefore suitable for this comparison. We refer to Appendix C for a more detailed explanation. Considering Table 4, we notice that the binary approximation yield similar (or better) accuracies compared to the original version of BAIT across all datasets. In contrast, the diagonal FIM approximation is not able to achieve comparable performances. Regarding the time complexity, we see that both the diagonal and binary approximation of BAIT significantly reduce the time complexity. While the diagonal approximation is slightly slower, our approximation is able to acquire instances quicker and independent of the class number. This is noticeable due to (almost) constant acquisition times between Snacks and STL-10, which both have an identical number of instances but a different number of classes. To ensure that not only the AUC values but also the behavior of the approximations are the similar, we show accuracy improvement curves in Fig. 2 for BAIT and its approximations. They show that our approximations BAIT (Exp) and BAIT (Binary) closely resemble the original formulation, while the diagonal approximation shows strong divergence.

6.3   Benchmark Experiments

Next, we present the main results on all datasets to compare the approximations of BAIT to other state-of-the-art AL strategies. Table 5 shows the average AUC improvement over RANDOM of all strategies across all datasets. In the case of BAIT (Exp), memory limitations make scaling beyond 50 classes infeasible, thus we only report results on the three datasets with less than 100 classes. The learning curves and accuracy improvement curves for each dataset can be found in Appendix E. The results demonstrate that BAIT (Binary) outperforms all state-of-the-art AL strategies on almost all datasets, except for StanfordDogs. This superiority underlines the importance of incorporating BAIT into the evaluation of works proposing novel AL strategies. Our publicly available implementations make this incorporation easy.

Table 5. Benchmark results on image datasets. The best performing strategy is marked as bold.

	RANDOM	MARGIN	BADGE	TYPICLUST	BAIT (Exp)	BAIT (Binary)
CIFAR-10	82.05	+2.38	+2.55	+3.57	+5.43	**+5.74**
STL-10	85.79	+4.42	+3.91	+3.70	+6.19	**+6.45**
Snacks	71.93	+3.32	+3.08	+7.29	+7.50	**+8.60**
CIFAR-100	67.20	+1.49	+1.48	+3.09	N/A	**+4.73**
Food-101	68.05	+0.62	+0.99	+3.42	N/A	**+3.54**
Flowers102	56.00	+12.05	+10.08	+17.64	N/A	**+18.23**
StanfordDogs	65.79	+2.39	+2.59	**+4.22**	N/A	+3.46
Tiny ImageNet	63.71	+0.67	+1.21	+2.52	N/A	**+3.46**
ImageNet	60.74	+0.88	+1.34	−5.76	N/A	**+1.48**

In Fig. 3, we exemplary show accuracy improvement curves for CIFAR-100 and ImageNet. For CIFAR-100, BAIT and TYPICLUST exhibit superior performance in the early stages of AL. This is also corroborated when examining the learning curves of all other datasets. Most likely, this is due to a more effective diverse selection, which have shown to be more beneficial in early AL cycles [15]. As the process progresses, we see that BADGE and MARGIN catch up. At the cycle, the accuracy of TYPICLUST becomes worse than the accuracy of BADGE, MARGIN, and even RANDOM. This demonstrates that TYPICLUST's extensive focus on a diverse selection can harm the accuracy in AL. As studied by [16], there exists a transition where a stronger focus on difficult instances is more beneficial than forcing diversity. Considering our approximation, we see that BAIT performs superior to the other strategies, suggesting that its selection is efficient during the entire AL process. For ImageNet, we see that BAIT provides the strongest improvement compared to the other strategies. Especially, the results demonstrate that TYPICLUST is not suitable for a scenario with a large number of classes and acquisition sizes. By examining the other learning curves, we found that BAIT occasionally leads to a worse final accuracy than BADGE. We suppose that a more accurate approximation of the FIM becomes more important in later AL cycles, and therefore BAIT (Binary) does not exploit the full potential in approximating BAIT. We leave further investigation of this for future work.

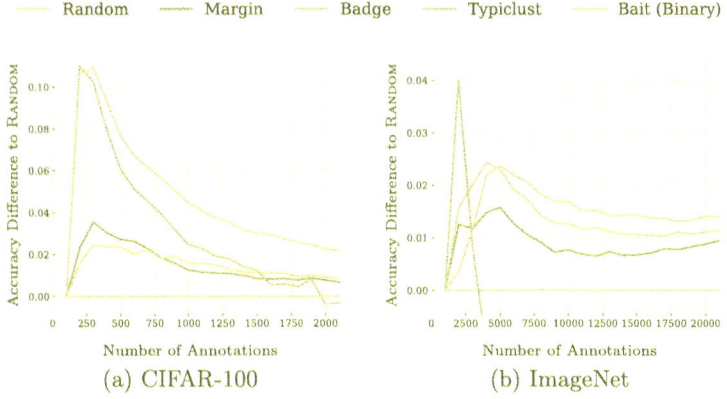

Fig. 3. Accuracy improvement curves of state-of-the-art strategies.

7   Conclusion

In this article, we explored the time and space complexity of BAIT and addressed its scalability issues when applied to tasks with a large number of classes. We presented two methods to reduce its time and space complexity by approximating the FIM. Our first method, BAIT (Exp), modifies BAIT's original formulation by taking the expectation over the most probable classes, reducing the time complexity from $\mathcal{O}(K^3 D^2)$ to $\mathcal{O}(c K^2 D^2)$. Our second method, BAIT (Binary), considers a binary classification task, leading to an alternative likelihood for gradient computations, considerably reducing the time complexity to $\mathcal{O}(D^2)$. This adaptation enables BAIT's usage on large-scale datasets, such as ImageNet. An extensive evaluation across nine image datasets demonstrates that our approximations perform similarly to the original formulation of BAIT, and outperform existing state-of-the-art AL strategies in terms of accuracy.

For practitioners working with image datasets, we suggest using the BAIT (Binary) method. Our research mainly focused on image data and demonstrated that the approximation is effective in this context. For other data modalities, such as text or tabular data, we recommend using BAIT (Exp), as it uses a biased estimator of the FIM and is therefore closer to BAIT's original design. Further, we suggest setting $c = 2$ since focusing on the two most probable classes yields well working approximations. Our implementation is publicly available and provides an easy way to integrate BAIT into existing frameworks.

For future work, we plan to further validate the effectiveness of our approximations by evaluating them on different data modalities. For text data in particular, we are interested in the effectiveness of combining BAIT with models like Bert [11]. For image data, we plan to evaluate our methods even more extensively. We want to ensure that our approximations behave similarly to the original version of BAIT. For this purpose, we will employ more datasets and perform statistical tests for verification.

Acknowledgments. This study was funded by the ALDeep project at the University of Kassel.

Disclosure of Interests. The authors have no competing interests to declare that are relevant to the content of this article.

References

1. Amari, S.I.: Natural gradient works efficiently in learning. Neural Comput. **10**(2), 251–276 (1998)
2. Arthur, D., Vassilvitskii, S., et al.: K-means++: the advantages of careful seeding. In: ACM-SIAM Symposium on Discrete algorithms (2007)
3. Ash, J.T., Goel, S., Krishnamurthy, A., Kakade, S.: Gone fishing: neural active learning with fisher embeddings. In: Advances in Neural Information Processing Systems (2021)
4. Ash, J.T., Zhang, C., Krishnamurthy, A., Langford, J., Agarwal, A.: Deep batch active learning by diverse, uncertain gradient lower bounds. In: International Conference on Learning Representations (2020)
5. Bahri, D., Jiang, H., Schuster, T., Rostamizadeh, A.: Is margin all you need? An extensive empirical study of active learning on tabular data. arXiv preprint arXiv:2210.03822 (2022)
6. Bossard, L., Guillaumin, M., Van Gool, L.: Food-101 – mining discriminative components with random forests. In: European Conference on Machine Learning and Principles and Practice of Knowledge Discovery (2014)
7. Chaudhuri, K., Kakade, S.M., Netrapalli, P., Sanghavi, S.: Convergence rates of active learning for maximum likelihood estimation. In: Advances in Neural Information Processing Systems (2015)
8. Chen, Y., Biros, G.: FIRAL: an active learning algorithm for multinomial logistic regression. In: Advances in Neural Information Processing Systems (2023)
9. Daxberger, E., Kristiadi, A., Immer, A., Eschenhagen, R., Bauer, M., Hennig, P.: Laplace redux-effortless Bayesian deep learning. In: Advances in Neural Information Processing Systems (2021)
10. Deng, J., Dong, W., Socher, R., Li, L.J., Li, K., Fei-Fei, L.: ImageNet: a large-scale hierarchical image database. In: Computer Vision and Pattern Recognition. IEEE (2009)
11. Devlin, J., Chang, M.W., Lee, K., Toutanova, K.: BERT: pre-training of deep bidirectional transformers for language understanding. In: Conference of the North American Chapter of the Association for Computational Linguistics: Human Language Technologies (2019)
12. Dosovitskiy, A., et al.: An image is worth 16x16 words: transformers for image recognition at scale. In: International Conference on Learning Representations (2021)
13. Gal, Y., Islam, R., Ghahramani, Z.: Deep Bayesian active learning with image data. In: International Conference on Machine Learning, pp. 1183–1192. PMLR (2017)
14. George, T., Laurent, C., Bouthillier, X., Ballas, N., Vincent, P.: Fast approximate natural gradient descent in a Kronecker factored eigenbasis. In: Advances in Neural Information Processing Systems (2018)

15. Hacohen, G., Dekel, A., Weinshall, D.: Active learning on a budget: Opposite strategies suit high and low budgets. In: International Conference on Machine Learning (2022)
16. Hacohen, G., Weinshall, D.: How to select which active learning strategy is best suited for your specific problem and budget. In: Advances in Neural Information Processing Systems (2023)
17. Heskes, T.: On "naturalâĂİ learning and pruning in multilayered perceptrons. Neural Comput. **12**(4), 881–901 (2000)
18. Kingma, D.P., Ba, J.: Adam: a method for stochastic optimization. In: International Conference on Learning Representations (2015)
19. Kirsch, A., Van Amersfoort, J., Gal, Y.: BatchBALD: efficient and diverse batch acquisition for deep Bayesian active learning. In: Advances in Neural Information Processing Systems (2019)
20. Krizhevsky, A.: Learning multiple layers of features from tiny images. Master's thesis, University of Toronto (2009)
21. Kunstner, F., Hennig, P., Balles, L.: Limitations of the empirical fisher approximation for natural gradient descent. In: Advances in Neural Information Processing Systems (2019)
22. Le, Y., Yang, X.: Tiny imagenet visual recognition challenge. CS 231N (2015)
23. LeCun, Y., Denker, J., Solla, S.: Optimal brain damage. In: Advances in Neural Information Processing Systems (1989)
24. Lee, J., Humt, M., Feng, J., Triebel, R.: Estimating model uncertainty of neural networks in sparse information form. In: International Conference on Machine Learning (2020)
25. Liu, J.Z., et al.: A simple approach to improve single-model deep uncertainty via distance-awareness. J. Mach. Learn. Res. **24**(42), 1–63 (2023)
26. Liu, L., et al.: On the variance of the adaptive learning rate and beyond. In: International Conference on Learning Representations (2019)
27. Martens, J.: New insights and perspectives on the natural gradient method. J. Mach. Learn. Res. **21**(1), 5776–5851 (2020)
28. Martens, J., Grosse, R.: Optimizing neural networks with Kronecker-factored approximate curvature. In: International Conference on Machine Learning (2015)
29. Matthijs: Snacks dataset (2021). https://huggingface.co/datasets/Matthijs/snacks. Accessed 21 Mar 2023
30. Nilsback, M.E., Zisserman, A.: Automated flower classification over a large number of classes. In: Indian Conference on Computer Vision, Graphics & Image Processing (2008)
31. Oquab, M., et al.: DINOv2: learning robust visual features without supervision. Trans. Mach. Learn. Res. (2023)
32. Pascanu, R., Bengio, Y.: Revisiting natural gradient for deep networks. arXiv preprint arXiv:1301.3584 (2013)
33. Rauch, L., Aßenmacher, M., Huseljic, D., Wirth, M., Bischl, B., Sick, B.: ActiveGLAE: a benchmark for deep active learning with transformers. In: European Conference on Machine Learning and Principles and Practice of Knowledge Discovery, pp. 55–74 (2023)
34. Ren, P., et al.: A survey of deep active learning. ACM Comput. Surv. **54**(9), 1–40 (2021)
35. Sener, O., Savarese, S.: Active learning for convolutional neural networks: a core-set approach. In: International Conference on Learning Representations (2018)
36. Settles, B.: Active learning literature survey. Computer Sciences Technical Report 1648, University of Wisconsin–Madison (2009)

37. Sourati, J., Gholipour, A., Dy, J.G., Tomas-Fernandez, X., Kurugol, S., Warfield, S.K.: Intelligent labeling based on fisher information for medical image segmentation using deep learning. Trans. Med. Imaging **38**(11), 2642–2653 (2019)
38. Tan, W., Du, L., Buntine, W.: Bayesian estimate of mean proper scores for diversity-enhanced active learning. IEEE Trans. Pattern Anal. Mach. Intell. **46**, 3463–3479 (2023)
39. Wang, D., Tan, X.: Unsupervised feature learning with C-SVDDNet. Pattern Recogn. **60**, 473–485 (2016)
40. Yehuda, O., Dekel, A., Hacohen, G., Weinshall, D.: Active learning through a covering lens. In: Advances in Neural Information Processing Systems (2022)
41. Zhang, G., Sun, S., Duvenaud, D., Grosse, R.: Noisy natural gradient as variational inference. In: International Conference on Machine Learning (2018)
42. Zhao, P., Xie, L., Zhang, Y., Tian, Q.: Universal-to-specific framework for complex action recognition. Trans. Multimedia **23**, 3441–3453 (2020)

Understanding Domain-Size Generalization in Markov Logic Networks

Florian Chen[1]([✉]), Felix Weitkämper[2], and Sagar Malhotra[1]

[1] TU Wien, Vienna, Austria
florian.chen@tuwien.ac.at
[2] Ludwig-Maximilians-Universität München, Munich, Germany

Abstract. We study the generalization behavior of Markov Logic Networks (MLNs) across relational structures of different sizes. Multiple works have noticed that MLNs learned on a given domain generalize poorly across domains of different sizes. This behavior emerges from a lack of internal consistency within an MLN when used across different domain sizes. In this paper, we quantify this inconsistency and bound it in terms of the variance of the MLN parameters. The parameter variance also bounds the KL divergence between an MLN's marginal distributions taken from different domain sizes. We use these bounds to show that maximizing the data log-likelihood while simultaneously minimizing the parameter variance corresponds to two natural notions of generalization across domain sizes. Our theoretical results apply to Exponential Random Graphs and other Markov network based relational models. Finally, we observe that solutions that decrease MLN parameter variance, like regularization and Domain-Size Aware MLNs, increase the internal consistency of the MLNs. We empirically verify our results on four different datasets, with different methods to control parameter variance, showing that controlling parameter variance leads to better generalization.

1 Introduction

Given the magnitude and ever-increasing nature of relational data, like social networks and epidemiology data, only a subsample of the data is ever observed. Statistical Relational Learning (SRL) [5,6] methods integrate logic and probability to learn and infer over such data. However, *are parameters estimated from subsampled data a good fit for the model of the larger relational structure?* Shalizi et al. [23] showed that, for most non-trivial probabilistic models on relational structures, it is *probabilistically inconsistent* to apply the same model both to the whole relational structure and to its substructures. Jaeger et al. [10] extend this analysis to a vast array of SRL models. These results show that, unlike independent and identically distributed (iid) data, relational data does not admit consistency of parameter estimation. That is, it is not true that the maximum likelihood (ML) parameter estimate converges to the true model parameters as the size of the observed data grows. In fact, these results show that the notion of a single true parameter, for relational structures of all sizes, is ill-defined for SRL models.

Lack of probabilistic consistency means that using an SRL model learned on a fixed domain for inference on a domain of different size may lead to poor results. The poor generalization behavior of SRL models across domain sizes is indeed observed in multiple empirical studies [11,17,20,30]. Such issues can be ameliorated by using *projective* models—probabilistic models where the same parameters can be used for both the whole relational structure and its subsamples. Formally, projective models capture probability distributions on relational structures (resp. graphs) of size n, where the marginal distribution over substructures (resp. subgraphs) of size $m < n$ does not depend on n. However, Shalizi et al. [23] also show that no projective model can express probability distributions with complex sufficient statistics, like k-cliques for any k larger than two. These results also exclude the possibility of constructing any SRL model with practically desirable First-Order Logic (FOL) features such as transitivity. Given these results, it is unclear what quantitative statements can be made about the generalization behavior of SRL models across domain sizes.

In this paper, we rigorously analyze domain-size generalization for a specific class of SRL models, namely Markov Logic Networks (MLNs). An MLN is a Markov Random Field with features defined in terms of weighted FOL formulas. We first formalize the notion of domain-size generalization of an MLN. We then provide an intuitive argument for what leads to non-projectivity, in terms of the weights that lead to dependence between the smaller and the larger domain. Theorem 1 provides bounds on the difference between the probability distribution induced by an MLN on a subsampled domain and the probability distribution induced by the same MLN on a larger (unseen) domain. We use this analysis to bound the KL divergence between the two distributions in terms of the parameter variance[1] of the MLN. Finally, we show that maximizing the log-likelihood of an MLN on the subsampled domain, while minimizing the parameter variance, corresponds to (i) increasing the log-likelihood for generalization to the larger domain, and to (ii) reducing the KL divergence between the distributions induced by the MLN on the subsampled domain and the larger domain. Finally, we observe that methods like regularization and Domain-Size Aware MLNs [17] minimize the parameter variance, and hence lead to better generalization. We empirically verify these claims on four different datasets, with three different methods for controlling parameter variance. Although the focus of this paper is on MLNs, our results can be generalized to Exponential Random Graph Models (ERGMs) and to any SRL model where template based parameter sharing is used [1,26].

---

[1] We use the term "variance" in a colloquial sense here, as we actually bound the KL divergence in terms of the maximum and the minimum of the weight functions induced by an MLN.

2   Related Work

Lack of probabilistic consistency in probabilistic models on relational structures was first investigated by Shalizi et al. [23]. Jaeger et al. [9] showed that such issues persist in most practically used SRL models. A large array of works have tried to devise new projective models [10,29] or identify and characterize projective fragments of existing SRL models [15,31], which circumvent these issues. However, the proposed new models are currently only of theoretical interest, as no clear way of learning or reasoning with them has been developed. Moreover, existing projective fragments of SRL models are rather restrictive.

These theoretical shortcomings are also reflected in the poor generalization behavior of SRL models in practice [11,17,20,30]. Many works provide heuristic solutions [11,17,20] for better generalization across domain sizes. A particularly relevant family of formalisms adapts the parameter values with the size of the domain [17,28]. In formalisms based on directed graphical models, it has been shown that parameter scaling leads to asymptotically projective models [32], but none of the heuristics developed for MLNs are formally motivated in this way [28]. For MLNs, [14] provides a sound approach for estimating the parameters for a larger unseen domain from a smaller subsample of fixed size. However, the practical applicability of this result is unclear. Furthermore, results provided in [14] rely on learning on a larger domain. Hence, the computational complexity of learning can be prohibitively large in real-world settings.

In comparison to the aforementioned works, we analyze the generalization behavior of an MLN in the most natural setting, i.e., the MLN parameters are learned from a subsample of smaller size, and we analyze the behavior of such a distribution on the larger domain. Our analysis theoretically justifies many of the existing heuristic methods [17]. Our results are also relevant to works in ERGMs [22], investigating the relationship between the sample (resp. the substructure for us) and the population (resp. the larger relational structure for us).

3   Background

3.1   Basic Definitions

The set of integers $\{1, ..., n\}$ is denoted by $[n]$. We use $[m : n]$ to denote the set of integers $\{m, ..., n\}$. Wherever the larger set of integers $[n]$ is clear from context, we will use $[\bar{m}]$ to denote the set $[m + 1 : n]$. For any $d \geq 1$, $\langle n \rangle^d$ represents d-tuples in $[n]^d$, with d distinct elements, appearing in natural order. Hence, $\langle n \rangle^d$ forms a standardized representation of the set of all d-element subsets in $[n]$.

3.2   First-Order Logic

We assume a function-free First-Order Logic (FOL) language $\mathcal{L}$ defined by a finite set of variables $\mathcal{V}$, a finite set of symbols $\mathcal{R}$, and a finite set of domain constants[2] $[n]$. For $a_1, ..., a_k \in \mathcal{V} \cup [n]$ and $R \in \mathcal{R}$ we call $R(a_1, ... a_k)$ an *atom*. If $a_1, ..., a_k \in [n]$, then the atom $R(a_1, ... a_k)$ is called a *ground atom*. A *literal* is an atom or the negation of an atom. We assume *Herbrand Semantics* [7]. Hence, a *world* or an *interpretation* is simply a mapping of each ground atom to a boolean. The set of interpretations over a domain of size n is denoted by $\Omega^{(n)}$. For a subset $I \subset [n]$ we use $\omega \downarrow I$ to denote the partial interpretation induced by I. Thus, $\omega \downarrow I$ is an interpretation over the ground atoms containing only the domain constants in I. For any $\mathbf{c} \in \langle n \rangle^d$ we use $\omega \downarrow \mathbf{c}$ to denote the partial interpretation induced by the domain elements in the tuple $\mathbf{c}$.

Example 1. Consider a formal language comprising only two binary relation symbols, denoted as G and B. We can visualize an interpretation ω as a multi-relational directed graph. In this graph, a directed edge of color green (for G) or blue (for B) connects two nodes x and y if and only if $G(x, y)$ or $B(x, y)$ respectively holds true in ω. For an illustrative interpretation ω on the set $\Delta = [4]$, the graphical representation is as follows:

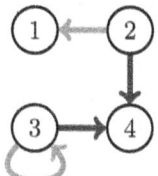

Then, the two subsets $\omega' = \omega \downarrow [2]$ and $\omega'' = \omega \downarrow [\bar{2}]$ can graphically be represented as

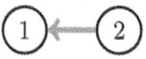   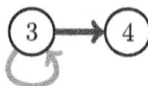

Note that if $\mathbf{c} = \langle 1, 2 \rangle$, then $\omega \downarrow \mathbf{c} = \omega \downarrow [2]$.

Families of Probability Distributions. We will deal with probability distributions on a set of interpretations. A family of probability distributions $\{P^{(n)} : n \in \mathbb{N}\}$ specifies, for each finite domain of size n, a distribution $P^{(n)}$ on the possible n-world set $\Omega^{(n)}$ [10]. We will work with *exchangeable* probability distributions [10]. These are distributions where $P^{(n)}(\omega) = P^{(n)}(\omega')$ if ω and ω' are isomorphic. A distribution $P^{(n)}$ over n-worlds induces a marginal probability distribution over m-worlds $\omega' \in \Omega^{(m)}$, where $m \leq n$, as follows:

$$P^{(n)} \downarrow [m](\omega') = \sum_{\omega \in \Omega^{(n)}: \omega \downarrow [m] = \omega'} P^{(n)}(\omega)$$

Note that due to exchangeability $P^{(n)} \downarrow I$ is the same for all subsets I of size m. Hence, we can always assume any induced m-world to be $\omega \downarrow [m]$. We can now define projectivity as follows:

---

[2] Note that, w.l.o.g., we can assume the domain to be $[n]$ as we can always rename any finite domain of size n with $[n]$.

Definition 1 (Projectivity [10]). *An exchangeable family of probability distributions is called projective if for all* $m < n$:

$$P^{(n)} \downarrow [m] = P^{(m)}$$

4   Learning in Markov Logic

A Markov Logic Network (MLN) Φ is defined by a set of weighted formulas $\{(\phi_i, a_i)\}_i$, where ϕ_i are function-free, quantifier-free, FOL formulas with weights $a_i \in \mathbb{R}$. An MLN Φ induces a probability distribution over the set of interpretations $\Omega^{(n)}$:

$$P_\Phi^{(n)}(\omega) = \frac{1}{Z(n)} \exp\Big(\sum_{(\phi_i, a_i) \in \Phi} a_i N(\phi_i, \omega) \Big) \tag{1}$$

where $N(\phi_i, \omega)$ is the number of true groundings of ϕ_i in ω. The normalization constant $Z(n)$ is called the *partition function* that ensures that $P_\Phi^{(n)}$ is a probability distribution. In the following, we provide an example of an MLN which models the spread of COVID-19 due to contact among different individuals and the impact of vaccines.

Example 2. Let us have a relational language with the unary predicates Covid and Vaccine, and a binary predicate Contact. An MLN can be defined as follows:

$$a_1 \quad \texttt{Vaccine}(x) \rightarrow \neg\texttt{Covid}(x)$$

$$a_2 \quad \texttt{Covid}(x) \wedge \texttt{Contact}(x, y) \rightarrow \texttt{Covid}(y)$$

Like in most SRL models, learning in MLNs is guided by the maximum likelihood (ML) principle. Formally, given an observed relational structure $\omega \in \Omega^{(n)}$, and an MLN Φ, the ML estimate for the weights is given follows:

$$\hat{\mathbf{a}} = \underset{\mathbf{a}}{\arg\max}\, P_\Phi^{(n)}(\omega) \tag{2}$$

where $P_\Phi^{(n)}$ is the probability distribution due to an MLN on the set of interpretations $\Omega^{(n)}$ as defined in Eq. (1). However, in most cases, the observed relational structure ω is a substructure of some larger structure on a larger unobserved domain. For instance, the number of people tested during a pandemic, and the number of contacts reported (say, using a contact-tracing mobile application) are only a subset of the true infection-contact network, which is spread over the entire local or even global population. Hence, our goal is to estimate the parameters for the MLN distribution $P_\Phi^{(n+m)}$ for some (potentially very large) m using only the substructure ω of size n. Formally, we want the following ML estimate:

$$\hat{\mathbf{a}} = \underset{\mathbf{a}}{\arg\max}\, P_\Phi^{(n+m)} \downarrow [n](\omega) \tag{3}$$

However, given that most MLNs are not projective [10,23], the ML estimate in Eq. (3) is not the same as the ML estimate in (2). As m may be very large, it can be computationally prohibitive to make the ML estimate for the distribution $P_{\Phi}^{(n+m)}$. Furthermore, in many cases it may be hard to know or guess the value of m. Hence, our goal would be to analyze the relation between the distributions $P_{\Phi}^{(n)}$ and $P_{\Phi}^{(n+m)} \downarrow [n]$, and use that analysis to subsequently characterize conditions that lead to better ML parameter estimates for $P_{\Phi}^{(n+m)}$, or in other words generalize better to larger domains.

Remark 1. Although projective MLNs can easily be obtained, their expressivity is significantly limited. One projective fragment of MLNs is the σ-determinate MLNs [10,24]. A Markov Logic Network $\Phi := \{\phi_i, a_i\}_i$ is σ-determinate if its formulas ϕ_i satisfy that any two atoms appearing in ϕ_i contain the same variables.

Example 3. Following is an example of a σ-determinate MLN:

$$a_1 \quad \texttt{Covid}(x)$$
$$a_2 \quad \texttt{Contact}(x,y) \wedge \texttt{Contact}(y,x)$$

Even simple MLNs, such as the one presented in Example 2, can not be represented as a σ-determinate MLN. Such shortcomings are true of most projective fragments of MLNs [15]. Hence, our goal in this paper can also be framed as obtaining arbitrarily expressive MLNs that are closer to being projective.

5   Markov Logic Across Domain Sizes

In this section, we analyze how the weights induced by an MLN distribute over different parts of the domain. We present the necessary machinery for our main results in Sect. 6 and create an intuition for what leads to projectivity, and how any MLN can be made *closer* to being projective.

We assume, w.l.o.g., that each k-ary formula in an MLN can be grounded only to k distinct domain constants. This does not restrict the expressivity of an MLN, as an MLN with a formula $\psi(x,y)$ with weight a can be equivalently expressed by replacing $\psi(x,y)$ with two formulas: $\psi(x,x)$ and $\psi(x,y) \wedge (x \neq y)$ with the same weight a. This principle can be generalized to formulas with arbitrary arity. We will use Φ_k to represent the subset of weighted formulas in an MLN Φ with arity k. We now define weight functions for a given MLN Φ.

Definition 2 (weight function). *Given an MLN Φ, we define the weight of an interpretation ω as follows:*

$$w(\omega) = \exp\left(\sum_{(\phi_i, a_i) \in \Phi} a_i N(\phi_i, \omega) \right) \tag{4}$$

We will also need to decompose the weight contribution of different k-tuples to the weight $w(\omega)$. To that end, we define the k-weight functions as follows:

Definition 3 (k-weight function). *Given an MLN Φ, we define the k-weight of an interpretation ω as follows:*

$$w_k(\omega) = \exp\Big(\sum_{(\phi_i,a_i)\in\Phi_k} a_i N(\phi_i,\omega)\Big) \tag{5}$$

where Φ_k is the subset of weighted formulas in Φ with arity k.

In the following two Lemmas, we further decompose the contribution of each k-substructure towards the weight $w(\omega)$.

Lemma 1. *Given an MLN with weight function w and k-weight functions w_k, then:*

$$w(\omega) = \prod_{k\in[d]}\prod_{\mathbf{c}\in\langle n\rangle^k} w_k(\omega\downarrow\mathbf{c}) \tag{6}$$

where d is the largest arity of the formulas in the MLN.

Proof. Let $\phi\in\Phi_k$ be an arbitrary weighted formula with k variables. The weight contribution of ϕ to $\sum_{(\phi_i,a_i)\in\Phi} a_i N(\phi_i,\omega)$ is given by the weighted number of true groundings of ϕ in ω. Since ϕ is always grounded to distinct domain constants and it has arity k, its weight contribution is the sum of its weight contribution to each of the $\omega\downarrow\mathbf{c}$ for $\mathbf{c}\in\langle n\rangle^k$. Repeating the same argument for all arities and all formulas in the MLN, we have that:

$$\sum_{(\phi_i,a_i)\in\Phi} a_i N(\phi_i,\omega) = \sum_{k\in[d]}\sum_{\mathbf{c}\in\langle n\rangle^k}\sum_{(\phi_i,a_i)\in\Phi_k} a_i N(\phi_i,\omega\downarrow\mathbf{c})$$

Hence,

$$\exp\left(\sum_{(\phi_i,a_i)\in\Phi} a_i N(\phi_i,\omega)\right) = \exp\left(\sum_{k\in[d]}\sum_{\mathbf{c}\in\langle n\rangle^k}\sum_{(\phi_i,a_i)\in\Phi_k} a_i N(\phi_i,\omega\downarrow\mathbf{c})\right)$$

$$= \prod_{k\in[d]}\prod_{\mathbf{c}\in\langle n\rangle^k} w_k(\omega\downarrow\mathbf{c})$$

□

Similar weight functions can be constructed for other Markov network based SRL models [1,26] where template based parameter sharing is used.

Lemma 2. *If ω is an interpretation on a domain $[n+m]$, then $w(\omega)$ can be factorized as follows:*

$$w(\omega) = w(\omega\downarrow[n]) \times w(\omega\downarrow[\bar{n}]) \times \prod_{k\in[d]}\prod_{\mathbf{c}\in\langle n+m\rangle^k\setminus\langle n\rangle^k\cup\langle\bar{n}\rangle^k} w_k(\omega\downarrow\mathbf{c}) \tag{7}$$

Proof.

$$w(\omega) = \prod_{k \in [d]} \prod_{\mathbf{c} \in \langle n+m \rangle^k} w_k(\omega \downarrow \mathbf{c})$$

$$= \prod_{k \in [d]} \prod_{\mathbf{c} \in \langle n \rangle^k} w_k(\omega \downarrow \mathbf{c}) \prod_{k \in [d]} \prod_{\mathbf{c} \in \langle \bar{n} \rangle^k} w_k(\omega \downarrow \mathbf{c}) \prod_{k \in [d]} \prod_{\mathbf{c} \in \langle n+m \rangle^k \setminus \langle n \rangle^k \cup \langle \bar{n} \rangle^k} w_k(\omega \downarrow \mathbf{c})$$

$$= w(\omega \downarrow [n]) \times w(\omega \downarrow [\bar{n}]) \times \prod_{k \in [d]} \prod_{\mathbf{c} \in \langle n+m \rangle^k \setminus \langle n \rangle^k \cup \langle \bar{n} \rangle^k} w_k(\omega \downarrow \mathbf{c})$$

□

A key part of our analysis would be understanding the weight contribution to the probability distribution $P_\Phi^{(n+m)} \downarrow [n]$ due to the following term in Lemma 2:

$$\prod_{\mathbf{c} \in \langle n+m \rangle^k \setminus \langle n \rangle^k \cup \langle \bar{n} \rangle^k} w_k(\omega \downarrow \mathbf{c}) \tag{8}$$

Expression (8) captures weight contribution from k-tuples which are strictly not part of the domain $\langle n \rangle^k$, and neither of the domain $\langle \bar{n} \rangle^k$. Intuitively, our goal is to control the weight contributions due to the relations that create dependence between the observed relational structure on the domain $[n]$, and the unobserved relational structure on the domain $[\bar{n}]$.

6   Domain-Size Generalization

In this section, we present the main results of our paper. Let w_k^{max} and w_k^{min} denote the maximum and the minimum of the weight function w_k.

Proposition 1. *Given an interpretation ω on the domain $[n+m]$, then*

$$w(\omega) \le w(\omega \downarrow [n]) \times w(\omega \downarrow [\bar{n}]) \times \prod_{k \in [d]} (w_k^{max})^{\binom{n+m}{k} - \binom{n}{k} - \binom{m}{k}} \tag{9}$$

$$w(\omega) \ge w(\omega \downarrow [n]) \times w(\omega \downarrow [\bar{n}]) \times \prod_{k \in [d]} (w_k^{min})^{\binom{n+m}{k} - \binom{n}{k} - \binom{m}{k}} \tag{10}$$

Proof. The statement follows from Eq. (7) in Lemma 2. The upper bound is obtained by replacing the multiplicative weight contribution of each tuple in $\langle n+m \rangle^k \setminus \langle n \rangle^k \cup \langle \bar{n} \rangle^k$ with w_k^{max}, for all $k \in [d]$. And the lower bound is obtained by replacing the weight contribution of all such tuples with w_k^{min}. □

For ease of notation, we define the following new parameters:

$$M_{max} = \prod_{k \in [d]} (w_k^{max})^{\binom{n+m}{k} - \binom{n}{k} - \binom{m}{k}} \tag{11}$$

$$M_{min} = \prod_{k \in [d]} (w_k^{min})^{\binom{n+m}{k} - \binom{n}{k} - \binom{m}{k}} \tag{12}$$

Proposition 2. *There exists an MLN for which the bounds in Proposition 1 are met for some interpretation ω.*

Proof. Assume an MLN with only the formula $R(x,y) \wedge R(y,z) \wedge R(x,z)$, with weight $a > 0$. It can be checked that the upper bound is met for an $\omega \in \Omega^{(n+m)}$ where all the domain constants are related w.r.t. the relation R. And the lower bound is met by the $\omega' \in \Omega^{(n+m)}$, such that no relation between any of the domain constants exist. □

Proposition 2 shows that bounds in Proposition 1 can not be improved.

Proposition 3. *Given a Markov Logic Network, we have that*

$$M_{min}C_{n,m}Z(n)Z(m) \leq Z(n+m) \leq Z(n)Z(m)C_{n,m}M_{max} \qquad (13)$$

where $C_{n,m}$ is the number of ways in which an interpretation on $[n]$ and an interpretation on $[\bar{n}]$ can be extended to an interpretation on $[n+m]$.

Proof.

$$Z(n+m) = \sum_{\omega} w(\omega)$$

$$\leq \sum_{\omega} w(\omega \downarrow [n]) \times w(\omega \downarrow [\bar{n}]) \times \prod_{d \in [k]} (w_d^{max})^{\binom{n+m}{d} - \binom{n}{d} - \binom{m}{d}}$$

$$= \sum_{\omega} w(\omega \downarrow [n]) \times w(\omega \downarrow [\bar{n}]) \times M_{max}$$

$$= M_{max} \sum_{\substack{\omega' \in \Omega^{(n)} \\ \omega'' \in \Omega^{(m)}}} C_{n,m} \times w(\omega') \times w(\omega'')$$

$$= M_{max}C_{n,m}Z(n)Z(m)$$

□

As the proof of the lower bound follows analogous to the proof of the upper bound, we defer it to the Appendix.

We now present the main result of the paper:

Theorem 1. *Given a Markov Logic Network Φ, then the following inequality holds for all $\omega \in \Omega^{(n)}$:*

$$\frac{M_{min}}{M_{max}} P_\Phi^{(n)}(\omega) \leq P_\Phi^{(n+m)} \downarrow [n](\omega) \leq \frac{M_{max}}{M_{min}} P_\Phi^{(n)}(\omega) \qquad (14)$$

Proof.

$$P_\Phi^{(n+m)} \downarrow [n](\omega') = \sum_{\substack{\omega \in \Omega^{(n+m)} \\ \omega \downarrow [n] = \omega'}} \frac{w(\omega)}{Z(n+m)}$$

Using Proposition 3, we have:

$$P_{\Phi}^{(n+m)} \downarrow [n](\omega') \leq \frac{1}{Z(n)Z(m)M_{min}C_{n,m}} \sum_{\substack{\omega \in \Omega^{(n+m)} \\ \omega \downarrow [n]=\omega'}} w(\omega)$$

Using Proposition 1, we have:

$$P_{\Phi}^{(n+m)} \downarrow [n](\omega') \leq \frac{1}{Z(n)Z(m)M_{min}C_{n,m}} \sum_{\substack{\omega \in \Omega^{(n+m)} \\ \omega \downarrow [n]=\omega'}} w(\omega')w(\omega \downarrow [\bar{n}])M_{max}$$

Hence, we have that

$$P_{\Phi}^{(n+m)} \downarrow [n](\omega') \leq \frac{M_{max}w(\omega')}{Z(n)Z(m)M_{min}C_{n,m}} \sum_{\substack{\omega'' \in \Omega^{(m)} \\ \omega \downarrow [n]=\omega' \\ \omega \downarrow [\bar{n}]=\omega''}} \sum_{\substack{\omega \in \Omega^{(n+m)}}} w(\omega'')$$

$$= \frac{M_{max}w(\omega')}{Z(n)Z(m)M_{min}C_{n,m}} \sum_{\omega'' \in \Omega^{(m)}} C_{n,m}w(\omega'')$$

$$= \frac{M_{max}w(\omega')}{Z(n)Z(m)M_{min}C_{n,m}}C_{n,m}Z(m)$$

$$= \frac{M_{max}}{M_{min}}P_{\Phi}^{(n)}(\omega')$$

□

The proof of the lower bound follows analogous to the proof of the upper bound, we defer it to the Appendix. Let us now denote $\frac{M_{max}}{M_{min}}$ with the symbol Δ.

Corollary 1.

$$-\log P_{\Phi}^{(n+m)} \downarrow [n](\omega) \leq -\log P_{\Phi}^{(n)}(\omega) + \log \Delta$$

Corollary 1 is a simple consequence of Theorem 1 and its proof is therefore deferred to the Appendix.

Corollary 1 shows that minimizing the negative log-likelihood of the observed subsample $\log P_{\Phi}^{(n)}(\omega)$, while simultaneously reducing $\log \Delta$, leads to the upper bound on the negative marginal log-likelihood being reduced. Hence, bringing the parameter estimate closer to the ML estimate as required by Eq. (3). Note that the ML estimate in Eq. (3) takes into account that the observed structure is a subsample of a larger relational structure, and optimizes the weights to get the best estimate for the larger domain size.

As Δ is the quotient of M_{max} and M_{min}, as defined in Eqs. (11) and (12), reducing $\log \Delta$ corresponds to reducing the difference between the largest and the smallest values taken by $\log w_k$. This can be easily achieved by a simple regularization objective on the weights a_i of the MLN.

Theorem 2.
$$KL(P_\Phi^{(n+m)} \downarrow [n] \| P_\Phi^{(n)}) \leq \log \Delta$$

Proof.

$$
\begin{aligned}
KL(P_\Phi^{(n+m)} \downarrow [n] \| P_\Phi^{(n)}) &= \sum_{\omega \in \Omega^{(n)}} P_\Phi^{(n+m)} \downarrow [n](\omega) \times \log \left(\frac{P_\Phi^{(n+m)} \downarrow [n](\omega)}{P_\Phi^{(n)}(\omega)} \right) \\
&\leq \sum_{\omega \in \Omega^{(n)}} P_\Phi^{(n+m)} \downarrow [n](\omega) \times \log \left(\frac{\Delta \times P_\Phi^{(n)}(\omega)}{P_\Phi^{(n)}(\omega)} \right) \\
&= \sum_{\omega \in \Omega^{(n)}} P_\Phi^{(n+m)} \downarrow [n](\omega) \times \log \Delta
\end{aligned}
$$

Note that
$$\sum_{\omega \in \Omega^{(n)}} P_\Phi^{(n+m)} \downarrow [n](\omega) \times \log \Delta$$

is the expectation value of $\log \Delta$ under the distribution $P_\Phi^{(n+m)} \downarrow [n]$. Since the expectation of a constant is the constant itself, we have that:

$$\sum_{\omega \in \Omega^{(n)}} P_\Phi^{(n+m)} \downarrow [n](\omega) \times \log \Delta = \log \Delta$$

$\square$

Theorem 2 gives an easy method of minimizing the upper-bound, on the otherwise intractable, KL-divergence between $P_\Phi^{(n+m)} \downarrow [n]$ and $P_\Phi^{(n)}$. Hence, an MLN learning procedure can be pushed to have smaller $KL(P_\Phi^{(n+m)} \downarrow [n] \| P_\Phi^{(n)})$, and in turn be incentivized towards representing a projective distribution, simply by minimizing the $\log \Delta$ term i.e. the difference between the minimum and the maximum of the k-weight functions.

Corollary 2.

$$- \log P_\Phi^{(n)}(\omega) + KL(P_\Phi^{(n+m)} \downarrow [n] \| P_\Phi^{(n)}) \leq - \log P_\Phi^{(n)}(\omega) + \log \Delta$$

This statement can easily be derived from Theorem 2. We defer its proof to the Appendix.

Corollary 2 characterizes another notion of generalization across varying domain sizes. By minimizing the negative log-likelihood and the difference between M_{min} and M_{max}, we have that the upper-bound on the negative log-likelihood plus the KL divergence between the two distributions is minimized. This minimization can be seen as optimizing a dual objective. On the one hand, the likelihood of the observed substructure is maximized w.r.t. the distribution $P_\Phi^{(n)}$. While on the other hand, $P_\Phi^{(n)}$ is moved closer to $P_\Phi^{(n+m)} \downarrow [n]$ in terms of KL-divergence. This minimization of KL divergence can be seen as incentivizing distributions which are *closer* to being projective.

7   Experiments

In this section, we evaluate the effect of reducing parameter variance on general-ization behavior[3]. Proposition 1 bounds $w(\omega)$ w.r.t. the maxima and minima of w_k. However, for almost all worlds, this bound is loose. This is because, for most worlds, not all k-tuples chosen from across the domains will have the extreme weights. Note that in general, our goal is to minimize the impact of the term presented in Eq. (8). Therefore, it is more effective to reduce the spread between all the weights, rather than merely scaling the upper and the lower bound. Also note that for most MLNs, for some $\omega \in \Omega^k$, we will have that $w_k(\omega) = 1$, i.e., none of the formulas in the MLN will be realized on ω. Thus, in most practical cases, to reduce the spread of the weights a_i, one should reduce their spread around 0.

Multiple approaches discussed in the literature, directly or indirectly, mini-mize the parameter variance [8,17]. We empirically evaluate the effects of three such approaches: L1 regularization, L2 regularization, and Domain-Size Aware Markov Logic Networks (DA-MLNs) [17]. Both L1 and L2 regularization directly work to reduce the spread of the parameters: L1 regularization penalizes the sum of the absolute weight values and L2 regularization penalizes the sum of squared weights. In our setting, we only penalize formulas of arity > 1, because unary formulas do not affect the connecting term discussed in Eq. (8).

A DA-MLN is an adaptation of a regular MLN that reduces the variance of the parameters by down-scaling formula weights depending on the domain size of the dataset it should generalize to. In this section, we will call such datasets *target sets*. A DA-MLN is then given as follows:

$$P_\Phi^{(n)}(\omega) = \frac{1}{Z(n)} \exp\left(\sum_{(\phi_i, a_i) \in \Phi} \frac{a_i}{s_i} N(\phi_i, \omega) \right) \tag{15}$$

The scale-down factor s_i is defined as follows:

$$s_i = \max_{P \in \phi_i}\left(\max\left(1, \prod_{x \in Vars_i(P)^-} |\Delta_x|\right) \right) \tag{16}$$

where $|\Delta_x|$ is the domain size of x in the target set and $Vars_i(P)^-$ is the set of logical variables appearing in ϕ_i but not in the atom P.

To precisely verify our theoretical results, we employ Lifted Inference [12,19] and Lifted Generative Learning [27]. These methods allow us to compute and compare exact dataset likelihoods. In contrast, alternative methods optimize approximate objectives, such as pseudo-likelihood [2], which may interfere with the verification of the theoretical results. However, using lifted methods restricts the expressivity of the MLNs we can test.

---

[3] Our code is available <u>Online</u>.

7.1 Datasets

To provide a thorough analysis of the effects of different methods for generalizing across different domain sizes, we use four datasets commonly used in related literature: Friends & Smokers (FS) [25], IMDB[4] [16], WebKB (see footnote 4) [16] and Nations (see footnote 4) [21].

Friends and Smokers (FS). This synthetic dataset captures information about smoking habits, friendships, and cancer diagnoses of a set of people. The data is created by first randomly selecting 40% of a population to be smokers. Then, 30% of the smokers and 10% of the non-smokers are chosen to suffer from cancer. Lastly, friendships are assigned based on smoking habits, with a 0.8 probability for friendships between people with the same smoking habit, and a 0.1 probability of friendships between people with different smoking habits. For our experiments, we generate a target set of size 500.

IMDB. Taken from the International Movie Database this dataset contains information about movies and, their actors and directors. Also included are certain attributes like gender and work relations of actors and directors. The dataset has a total of 297 constants, of which 268 are of type person, 20 are of type movie, and 9 are of type genre. The dataset contains 3 binary and 3 unary predicates.

WebKB This dataset captures information about web pages from four US universities. For each web page, the original dataset [3] includes a label (e.g. Course, Faculty) as well as textual information about the page contents. Similar to Mihalkova et al. [16], the version we use disregards the textual information and focuses on page classes and relations, for example between courses and teaching assistants. This version of the dataset comprises a total of 989 constants, of which 746 are of type person. The dataset contains 3 binary and 2 unary predicates.

Nations. This dataset contains a set of features of nations and relations between them. Relations include treaties and (economic-)aid, features include governance types and technological advancements. In total, there are 14 nations, 111 features (given as unary predicates), and 56 relations (given as binary predicates).

7.2 Methodology

We compare the generalization behavior of standard generative weight learning to different methods that also reduce parameter variance: L1 regularization, L2 regularization, and DA-MLNs. The structures of the MLNs we use are adopted from Van Haaren et al. [27], who introduced a Lifted Structure Learning (LSL) approach. LSL ensures that the learned structures are liftable and learnable in practice. The Nations dataset, with over 160 predicates, presents an infeasibly

---

[4] Dataset available on the Alchemy website.

large search space of possible clauses for LSL. Hence, we use a hand-crafted MLN of 50 formulas. For weight learning, we employ Lifted Generative Learning [27]. This allows us to compare the exact target set likelihoods, which is the natural evaluation measure for generative learning [4,13,18,27] and is also best suited for validating our theoretical results.

To provide reliable results, we generate 20 training sets for weight learning and 5 target sets for each of the sizes we want to generalize to. For generating a training set, we uniformly sample a subset from a specific type τ of constant: We sample 20 *persons* for FS, 50 *persons* for IMDB and WebKB, and 5 *nations* for the Nations dataset. Now, let I denote the set of the sampled constants. In the training set, we then include all the ground atoms $R(a_1, ...a_k)$ where all the domain constants of type τ in $\{a_1, ..., a_k\}$ are included in I. The process for generating target sets follows a similar approach.

For standard generative weight learning and DA-MLNs, we then learn the weights on each training set and compute the log-likelihood of each target set. For L1 and L2 regularization, to find the best regularization parameter λ, we perform hyperparameter tuning on the values between 10^{-2} and 10^2 on the smallest target sets[5]. As our metric to compare the different approaches, we measure how much the target set log-likelihood improves in comparison to no regularization. This metric measures how well our ML estimate is w.r.t. Eq. (3), i.e., the ML estimate that takes into account the fact that the observed data came from a larger relational structure.

7.3   Results

Figure 1 shows the difference between the average log-likelihood obtained with the regularization approaches and the one obtained without regularization. For each of the four datasets, methods that reduce parameter variance consistently improve target set likelihood by several orders of magnitude (except for L2 regularization on FS). This effect is more pronounced as the target set size grows. Among the methods that reduce parameter variance, L1 and L2 regularization have similar performances. DA-MLNs outperform L1 and L2 on the FS dataset, but underperform on Nations and IMDB, while producing similar results on WebKB. Note that MLNs learned with L1 and L2 regularization are not domain-aware and work with the same parameters across domain sizes. Thus, it is unclear whether domain-aware parameter variance reduction methods are generally preferable to domain-unaware methods in practice. This can be observed in the relative under-performance of DA-MLNs on some of the datasets.

---

[5] Hence, the results on the smallest target sets are slightly biased for L1 and L2. However, for larger target set sizes, no such bias exists.

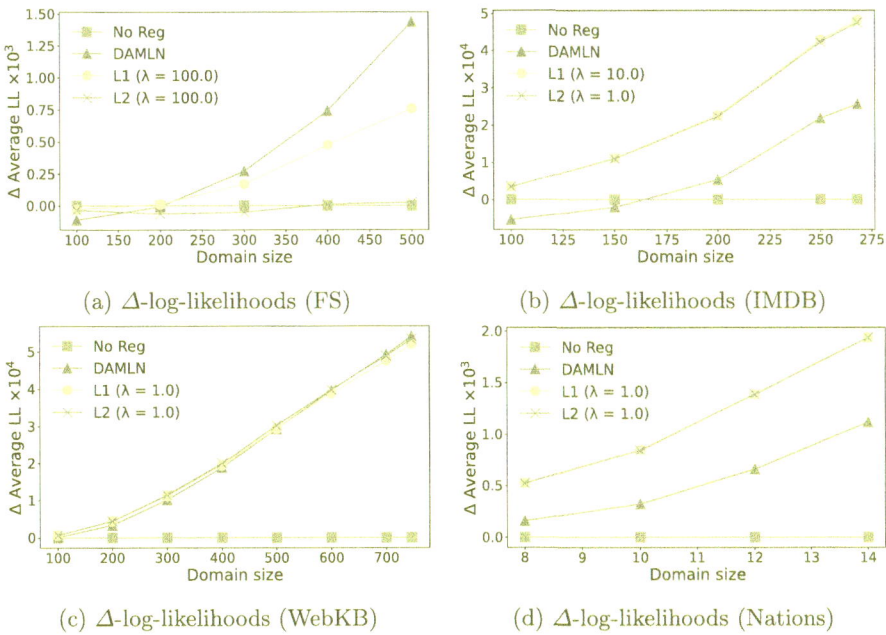

(a) Δ-log-likelihoods (FS)

(b) Δ-log-likelihoods (IMDB)

(c) Δ-log-likelihoods (WebKB)

(d) Δ-log-likelihoods (Nations)

Fig. 1. Results for the Friends & Smokers, IMDB, WebKB, and Nations datasets (Larger values are better)

8 Conclusion

In this paper, we analyze the generalization behavior of Markov Logic Networks when used across domain sizes. We observe that, unlike independent and identically distributed data, relational data does not admit consistency of parameter estimation. We then formalize this inconsistency in terms of the different (and mutually inconsistent) notions of maximum likelihood estimation for the weights of an MLN, when only partial data is observed. In our main theoretical result, we characterize conditions based on the parameter variance of the MLN that minimize this inconsistency. These theoretical conditions motivate and justify weight-learning approaches that decrease parameter variance. To empirically verify these claims we evaluate the generalization performance of three approaches that reduce parameter variance: L1 and L2 regularization, and Domain-Size Aware Markov Logic Networks. Our findings validate that reducing parameter variance consistently improves dataset-likelihoods over larger domains.

Acknowledgments. SM thanks Kilian Rückschloß for pointing towards the problem investigated in this paper.

Appendix

Proof (the lower bound proof for Proposition 3).

$$Z(n+m) = \sum_{\omega} w(\omega)$$

$$\geq \sum_{\omega} w(\omega \downarrow [n]) \times w(\omega \downarrow [\bar{n}]) \times \prod_{d \in [k]} (w_d^{min})^{\binom{n+m}{d} - \binom{n}{d} - \binom{m}{d}}$$

$$= \sum_{\omega} w(\omega \downarrow [n]) \times w(\omega \downarrow [\bar{n}]) \times M_{min}$$

$$= M_{min} \sum_{\substack{\omega' \in \Omega^{(n)} \\ \omega'' \in \Omega^{(m)}}} C_{n,m} \times w(\omega') \times w(\omega'')$$

$$= M_{min} C_{n,m} Z(n) Z(m)$$

□

Proof (the lower bound proof for Theorem 1).

$$P_{\Phi}^{(n+m)} \downarrow [n](\omega') = \sum_{\substack{\omega \in \Omega^{(n+m)} \\ \omega \downarrow [n] = \omega'}} \frac{w(\omega)}{Z(n+m)}$$

Using Proposition 3, we have:

$$P_{\Phi}^{(n+m)} \downarrow [n](\omega') \geq \frac{1}{Z(n)Z(m)M_{max}C_{n,m}} \sum_{\substack{\omega \in \Omega^{(n+m)} \\ \omega \downarrow [n] = \omega'}} w(\omega)$$

Using Proposition 1, we have:

$$P_{\Phi}^{(n+m)} \downarrow [n](\omega') \geq \frac{1}{Z(m)M_{max}C_{n,m}} \sum_{\substack{\omega \in \Omega^{(n+m)} \\ \omega \downarrow [n] = \omega'}} \frac{w(\omega')w(\omega \downarrow [\bar{n}])M_{min}}{Z(n)}$$

$$= \frac{1}{Z(m)M_{max}C_{n,m}} w(\omega') \sum_{\substack{\omega \in \Omega^{(n+m)} \\ \omega \downarrow [n] = \omega'}} \frac{w(\omega \downarrow [\bar{n}])M_{min}}{Z(n)}$$

$$= \frac{1}{Z(m)M_{max}C_{n,m}} w(\omega') \frac{Z(m)C_{n,m}M_{min}}{Z(n)}$$

$$= \frac{M_{min}}{M_{max}} P_{\Phi}^{(n)}(\omega')$$

□

Proof (of Corollary 1).

Using the bound derived in Theorem 1, we have:

$$\Delta^{-1} \times P_\Phi^{(n)}(\omega) \leq P_\Phi^{(n+m)} \downarrow [n](\omega)$$
$$\left(P_\Phi^{(n+m)} \downarrow [n](\omega)\right)^{-1} \leq \Delta \times \left(P_\Phi^{(n)}(\omega)\right)^{-1}$$
$$-\log P_\Phi^{(n+m)} \downarrow [n](\omega) \leq -\log P_\Phi^{(n)}(\omega) + \log \Delta$$

$\square$

Proof (of Corollary 2).

Using Theorem 2, we have:

$$KL(P_\Phi^{(n+m)} \downarrow [n]||P_\Phi^{(n)}) \leq \log \Delta$$
$$-\log P_\Phi^{(n)}(\omega) + KL(P_\Phi^{(n+m)} \downarrow [n]||P_\Phi^{(n)}) \leq -\log P_\Phi^{(n)}(\omega) + \log \Delta$$

$\square$

References

1. Bach, S.H., Broecheler, M., Huang, B., Getoor, L.: Hinge-loss Markov random fields and probabilistic soft logic. J. Mach. Learn. Res. **18**, 109:1–109:67 (2017)
2. Besag, J.: Statistical analysis of non-lattice data. J Roy. Stat. Soc. Ser. D **24**(3), 179–195 (1975)
3. Craven, M., Slattery, S.: Relational learning with statistical predicate invention: better models for hypertext. Mach. Learn. **43**(1/2), 97–119 (2001)
4. Darwiche, A.: Modeling and Reasoning with Bayesian Networks. Cambridge University Press, Cambridge (2009)
5. De Raedt, L., Kersting, K., Natarajan, S., Poole, D.: Statistical Relational Artificial Intelligence: Logic, Probability, and Computation. Morgan & Claypool Publishers, San Rafael (2016)
6. Getoor, L., Taskar, B.: Introduction to Statistical Relational Learning (Adaptive Computation and Machine Learning). The MIT Press, Cambridge (2007)
7. Hinrichs, T., Genesereth, M.: Herbrand logic (2009)
8. Huynh, T.N., Mooney, R.J.: Discriminative structure and parameter learning for Markov logic networks. In: Proceedings of the ICML 2008, pp. 416–423. ACM (2008)
9. Jaeger, M., Schulte, O.: Inference, learning, and population size: projectivity for SRL models. CoRR **abs/1807.00564** (2018)
10. Jaeger, M., Schulte, O.: A complete characterization of projectivity for statistical relational models. In: Proceedings of the IJCAI 2020, pp. 4283–4290. ijcai.org (2020)
11. Jain, D., Barthels, A., Beetz, M.: Adaptive Markov logic networks: learning statistical relational models with dynamic parameters. In: Proceedings of the ECAI 2010, pp. 937–942. IOS Press (2010)

12. Kersting, K.: Lifted probabilistic inference. In: Proceedings of the ECAI 2012, pp. 33–38. IOS Press (2012)
13. Koller, D., Friedman, N.: Probabilistic Graphical Models - Principles and Techniques. MIT Press, Cambridge (2009)
14. Kuzelka, O., Wang, Y., Davis, J., Schockaert, S.: Relational marginal problems: Theory and estimation. In: Proceedings of the AAAI 2018, pp. 6384–6391. AAAI Press (2018)
15. Malhotra, S., Serafini, L.: On projectivity in Markov logic networks. In: Amini, M.R., Canu, S., Fischer, A., Guns, T., Kralj Novak, P., Tsoumakas, G. (eds.) ECML PKDD 2022, Part. LNCS, vol. 13717, pp. 223–238. Springer, Cham (2022). https://doi.org/10.1007/978-3-031-26419-1_14
16. Mihalkova, L., Mooney, R.J.: Bottom-up learning of Markov logic network structure. In: Proceedings of the ICML 2007, pp. 625–632. ACM (2007)
17. Mittal, H., Bhardwaj, A., Gogate, V., Singla, P.: Domain-size aware Markov logic networks. In: Proceedings of the AISTATS 2019, pp. 3216–3224. PMLR (2019)
18. Murphy, K.P.: Machine Learning - A Probabilistic Perspective. MIT Press, Cambridge (2012)
19. Poole, D.: First-order probabilistic inference. In: Proceedings of the IJCAI 2003, pp. 985–991. Morgan Kaufmann (2003)
20. Poole, D., Buchman, D., Kazemi, S.M., Kersting, K., Natarajan, S.: Population size extrapolation in relational probabilistic modelling. In: Straccia, U., Calì, A. (eds.) SUM 2014. LNCS (LNAI), vol. 8720, pp. 292–305. Springer, Cham (2014). https://doi.org/10.1007/978-3-319-11508-5_25
21. Rummel, R.J.: Dimensionality of nations project: attributes of nations and behavior of nation dyads, 1950–1965 (1992)
22. Schweinberger, M., Krivitsky, P.N., Butts, C.T., Stewart, J.R.: Exponential-family models of random graphs: inference in finite, super and infinite population scenarios. Stat. Sci. **35**(4), 627–662 (2020)
23. Shalizi, C.R., Rinaldo, A.: Consistency under sampling of exponential random graph models. Ann. Stat. **41**(2), 508–535 (2013)
24. Singla, P., Domingos, P.M.: Markov logic in infinite domains. In: Proceedings of the UAI 2007, pp. 368–375. AUAI Press (2007)
25. Singla, P., Domingos, P.M.: Lifted first-order belief propagation. In: Proceedings of the AAAI 2008, pp. 1094–1099. AAAI Press (2008)
26. Taskar, B., Abbeel, P., Koller, D.: Discriminative probabilistic models for relational data. In: Proceedings of the UAI 2002, pp. 485–492. Morgan Kaufmann (2002)
27. Van Haaren, J., Van den Broeck, G., Meert, W., Davis, J.: Lifted generative learning of Markov logic networks. Mach. Learn. **103**(1), 27–55 (2016)
28. Weitkämper, F.: Scaling the weight parameters in Markov logic networks and relational logistic regression models. CoRR **abs/2103.15140** (2021)
29. Weitkämper, F.: Projective families of distributions revisited. Int. J. Approx. Reason. **162**, 109031 (2023)
30. Weitkämper, F., Ravdin, D., Fabry, R.: Statistical relational structure learning with scaled weight parameters. In: Bellodi, E., Lisi, F.A., Zese, R. (eds.) ILP 2023. LNCS, vol. 14363, pp. 139–153. Springer, Cham (2023). https://doi.org/10.1007/978-3-031-49299-0_10
31. Weitkämper, F.Q.: An asymptotic analysis of probabilistic logic programming, with implications for expressing projective families of distributions. Theory Pract. Log. Program. **21**(6), 802–817 (2021)
32. Weitkämper, F.: Probabilities of the third type: statistical relational learning and reasoning with relative frequencies. CoRR **abs/2202.10367** (2023)

Retrieval-Augmented Mining of Temporal Logic Specifications from Data

Gaia Saveri[1,2(✉)] and Luca Bortolussi[2]

[1] Department of Computer Science, University of Pisa, Pisa, Italy
`gaia.saveri@phd.unipi.it`
[2] Department of Mathematics, Computer Science and Geoscience,
University of Trieste, Trieste, Italy

Abstract. The integration of cyber-physical systems (CPS) into every-day life raises the critical necessity of ensuring their safety and reliability. An important step in this direction is requirement mining, i.e. inferring formally specified system properties from observed behaviors, in order to discover knowledge about the system. Signal Temporal Logic (STL) offers a concise yet expressive language for specifying requirements, par-ticularly suited for CPS, where behaviors are typically represented as time series data. This work addresses the task of learning STL require-ments from observed behaviors in a data-driven manner, focusing on binary classification, i.e. on inferring properties of the system which are able to discriminate between regular and anomalous behaviour, and that can be used both as classifiers and as monitors of the compliance of the CPS to desirable specifications. We present a novel framework that com-bines Bayesian Optimization (BO) and Information Retrieval (IR) tech-niques to simultaneously learn both the structure and the parameters of STL formulae, without restrictions on the STL grammar. Specifically, we propose a framework that leverages a dense vector database containing semantic-preserving continuous representations of millions of formulae, queried for facilitating the mining of requirements inside a BO loop. We demonstrate the effectiveness of our approach in several signal classifi-cation applications, showing its ability to extract interpretable insights from system executions and advance the state-of-the-art in requirement mining for CPS.

Keywords: Time-series data · Requirement Mining · Temporal Logic

1 Introduction

The adoption of cyber-physical systems (CPS) is pervasive nowadays: wearable devices, internet of things applications and domotics systems are part of our daily lives. However, formal specifications describing their expected behaviour are not always (completely) available. This is extremely serious when CPS are deployed in safety-critical scenarios, such as autonomous vehicles or medical devices, where safety and reliability are key desiderata. Requirement (or specification) mining is the task of inferring system properties from observations of its behaviour [7]. In the context of CPS, such behaviours are often recorded as time series describing the dynamics of the system over time (e.g. readings from a sensor at regular time intervals), hence Signal Temporal Logic (STL) [16,23] is a popular language for expressing requirements [6]. STL is indeed a formalism which allows to reason over real-valued trajectories in a concise yet rich way: for example in STL one can state properties like "the temperature of the room will reach 25 degrees within the next 10 min and will stay above 22° for the next hour". In this work, we tackle the objective of learning STL requirements from observed behaviours of a system, in a data-driven approach: differently from formal methods algorithms, our proposed solution does not require a complete model of the system, but only a fixed dataset of recorded trajectories. Specifically, we frame the STL requirement mining problem in the binary classification scenario, i.e. our goal is to learn a STL formula which is able to discriminate between two classes of labeled trajectories. In this context the mined STL specification can be used both as a classifier and as a monitor for the system at hand, and leveraged for maintenance and bug detection as well as a tool for system modeling and comprehension.

Our contribution consists in proposing a novel framework for STL specification mining, which involves combining Bayesian Optimization (BO) and Information Retrieval (IR) techniques. It allows to simultaneously learn the structure and the parameters of the specifications, without neither the need of a (generative) model of the system under analysis, nor of restrictions on the STL grammar to some specific fragment. Indeed we: (i) leverage the semantics of STL for mapping formulae in a continuous semantic-preserving latent space, using an ad-hoc kernel method [9], see Sect. 2; (ii) build a dense vector database containing the embeddings of millions of formulae, carefully constructed in an incremental way, as detailed in Sect. 3.1. Notably, this step is done once, and the same database is queried in all the application cases; (iii) frame the requirement mining task as an optimization problem in the semantic space of STL formulae, using queries to the database constructed at the previous step for inverting the embeddings, i.e. mapping continuous semantic vectors back to formulae, in Sects. 3. We show the effectiveness of the proposed approach in several binary classification applications, proving that it is able to extract interpretable information from system executions, offering insights on the system under analysis.

2   Background

Signal Temporal Logic (STL) is a linear-time temporal logic which expresses properties on trajectories over dense time intervals [16]. We define as trajectories the functions $\xi : I \to D$, where $I \subseteq \mathbb{R}_{\geq 0}$ is the time domain and $D \subseteq \mathbb{R}^k, k \in \mathbb{N}$ is the state space. The syntax of STL is given by:

$$\varphi := tt \mid \pi \mid \neg\varphi \mid \varphi_1 \wedge \varphi_2 \mid \varphi_1 \mathbf{U}_{[a,b]}\varphi_2$$

where tt is the Boolean *true* constant; π is an *atomic predicate*, i.e. a function over variables $\boldsymbol{x} \in \mathbb{R}^n$ of the form $f_\pi(\boldsymbol{x}) \geq 0$ (we refer to n as the number of variables of a STL formula); $\neg$ and $\wedge$ are the Boolean *negation* and *conjunction*, respectively (from which the *disjunction* $\vee$ follows by De Morgan's law); $\mathbf{U}_{[a,b]}$, with $a, b \in \mathbb{Q}, a < b$, is the *until* operator, from which the *eventually* $\mathbf{F}_{[a,b]}$ and the *always* $\mathbf{G}_{[a,b]}$ temporal operators can be deduced. We can intuitively interpret the temporal operators over $[a, b]$ as follows: a property is *eventually* satisfied if it is satisfied at some point inside the temporal interval, while a property is *globally* satisfied if it is true continuously in $[a, b]$; finally the *until* operator captures the relationship between two conditions φ, ψ in which the first condition φ holds until, at some point in $[a, b]$, the second condition ψ becomes true. We call $\mathcal{P}$ the set of well-formed STL formulae. STL is endowed with both a *qualitative* (or Boolean) semantics, giving the classical notion of satisfaction of a property over a trajectory, i.e. $s(\varphi, \xi, t) = 1$ if the trajectory ξ at time t satisfies the STL formula φ, and a *quantitative* semantics, denoted by $\rho(\varphi, \xi, t)$. The latter, also called *robustness*, is a measure of how robust is the satisfaction of φ w.r.t. perturbations of the signals. Robustness is recursively defined as:

$$\rho(\pi, \xi, t) = f_\pi(\xi(t)) \quad \text{for } \pi(\boldsymbol{x}) = (f_\pi(\boldsymbol{x}) \geq 0)$$
$$\rho(\neg\varphi, \xi, t) = -\rho(\varphi, \xi, t)$$
$$\rho(\varphi_1 \wedge \varphi_2, \xi, t) = \min(\rho(\varphi_1, \xi, t), \rho(\varphi_2, \xi, t))$$
$$\rho(\varphi_1 \mathbf{U}_{[a,b]} \varphi_2, \xi, t) = \max_{t' \in [t+a, t+b]} \left(\min\left(\rho(\varphi_2, \xi, t'), \min_{t'' \in [t,t']} \rho(\varphi_1, \xi, t'')\right) \right)$$

Since they will be used later, we report also the definition of robustness of derived temporal operators: eventually $\rho(\mathbf{F}_{[a,b]}\varphi, \xi, t) = \max_{t' \in [t+a, t+b]} \rho(\varphi, \xi, t)$ and globally $\rho(\mathbf{G}_{[a,b]}\varphi, \xi, t) = \min_{t' \in [t+a, t+b]} \rho(\varphi, \xi, t)$. Robustness is compatible with satisfaction via the following *soundness* property: if $\rho(\varphi, \xi, t) > 0$ then $s(\varphi, \xi, t) = 1$ and if $\rho(\varphi, \xi, t) < 0$ then $s(\varphi, \xi, t) = 0$. When $\rho(\varphi, \xi, t) = 0$ arbitrary small perturbations of the signal might lead to changes in satisfaction value. Intuitively, robustness measures how far is a signal ξ from violating a specification φ, with the sign indicating the satisfaction status. We omit t from the previous notations when properties are evaluated at time $t = 0$.

In this context stochastic processes are probability spaces defined as triplets $\boldsymbol{X} = (\mathcal{T}, \mathcal{A}, \mu)$ of a trajectory space $\mathcal{T}$ and a probability measure μ on a σ-algebra $\mathcal{A}$ over $\mathcal{T}$. Given a STL formula φ with predicates interpreted over state variables of $\boldsymbol{X}$ and a trajectory of the stochastic system $\xi(t)$, its robustness $\rho(\varphi, \xi)$ is a functional $R_\varphi : \mathcal{T} \to \mathbb{R}$, which defines the real-valued random variable $R_\varphi = R_\varphi(\boldsymbol{X})$, following the distribution:

$$\mathbb{P}(R_\varphi(\boldsymbol{X}) \in [a, b]) = \mathbb{P}(\boldsymbol{X} \in \{\xi \in \mathcal{T} | \rho(\varphi, \xi) \in [a, b]\}) \tag{1}$$

The expected value of the previous, i.e. $\mathbb{E}[R_\varphi | \boldsymbol{X}]$, intuitively gives a measure of how strongly a formula is satisfied by trajectories sampled from the process $\boldsymbol{X}$ (the higher the value the more robust the satisfaction) [5]; in the remainder of this paper, this expectation will be approximated by Monte Carlo sampling.

Finding continuous representations of STL formulae is performed with an ad-hoc kernel in [9], by leveraging the quantitative semantics of STL. Indeed, robustness allows formulae to be considered as functionals mapping trajectories into real numbers, i.e. $\rho(\varphi, \cdot) : \mathcal{T} \rightarrow \mathbb{R}$ such that $\xi \mapsto \rho(\varphi, \xi)$. Considering these as feature maps, and fixing a probability measure μ_0 on the space of trajectories $\mathcal{T}$, a kernel function capturing similarity among STL formulae on mentioned feature representations can be defined as:

$$k(\varphi, \psi) = \langle \rho(\varphi, \cdot), \rho(\psi, \cdot) \rangle = \int_{\xi \in \mathcal{T}} \rho(\varphi, \xi) \rho(\psi, \xi) d\mu_0(\xi) \qquad (2)$$

opening the doors to the use of the scalar product in the Hilbert space L^2 as a kernel for $\mathcal{P}$; at a high level, this results in a kernel having high positive value for formulae that behave similarly on high-probability trajectories (w.r.t. μ_0), and viceversa low negative value for formulae that on those trajectories disagree. Intuitively, μ_0 makes *simple* trajectories more probable, considering total variation and number of changes in monotonicity as indicators of complexity of signals. The measure μ_0, operating on piece-wise linear functions over the interval $\mathcal{I} = [a, b]$ (which is a dense subset of the set of continuous functions over $\mathcal{I}$), can be algorithmically defined as in Algorithm 1 (default parameters are set as $a = 0, b = 100, \Delta = 1, m' = m'' = 0, \sigma' = \sigma'' = 1, q = 0.1$). Note that, although the feature space $\mathbb{R}^{\mathcal{T}}$, which we call the *latent semantic space*, into which ρ (and thus Eq. (2)) maps formulae is infinite-dimensional, in practice the kernel trick allows to circumvent this issue by mapping each formula to a vector of dimension equal to the number of formulae which are in the training set used to evaluate the kernel (Gram) matrix. Such embeddings are continuous representations of discrete symbolic objects, namely of STL formulae, which can be used to incorporate them inside gradient-based optimization and learning algorithms.

Bayesian Optimization (BO) is a widely-used framework for black-box optimization problems, e.g. of the form $\boldsymbol{x}^\star = \text{argmax}_{\boldsymbol{x} \in \mathcal{X}} f(\boldsymbol{x})$ being $\mathcal{X} \subset \mathbb{R}^d$, which consists in sequentially at each iteration t observing noisy evaluations of the objective function $y_t = f(\boldsymbol{x}_t) + \varepsilon_t$ where the noise is $\varepsilon_t \sim \mathcal{N}(0, \sigma^2)$, at candidate points $\boldsymbol{x}_t$ chosen by an acquisition function. The Gaussian Process Upper Confidence Bound (GP-UCB) algorithm consists in interpolating the available observations using a Gaussian Process (GP, i.e. a Bayesian non-parametric regression approaches) [24], which is the so-called emulation phase, and adding candidate points to the training set of the GP in an iterative fashion using Upper-Confidence Bound (UCB, [26]) as acquisition function. More specifically it considers f to be a sample path from a GP with 0 mean and stationary kernel function $k : \mathcal{X} \times \mathcal{X} \mapsto \mathbb{R}$, i.e. $f \sim GP(0, k)$, so that at each iteration the posterior mean and variance of $f(\boldsymbol{x})$ are derived, and respectively denoted as $\mu_{t-1}(\boldsymbol{x})$ and $\sigma_{t-1}^2(\boldsymbol{x})$. The UCB acquisition function selects the next candidate point to be evaluated as $\boldsymbol{x}_t = \text{argmax}_{\boldsymbol{x} \in \mathcal{X}} \mu_{t-1}(\boldsymbol{x}) + \beta_t^{\frac{1}{2}} \sigma_{t-1}(\boldsymbol{x})$, with $\beta_t \in \mathbb{R}$.

Algorithm 1. Sampling a trajectory over the interval $[a, b]$ according to μ_0

Input: Δ, a, b, m', m'', σ', σ'', q
Output: ξ

   ▷ sample the starting point
   $\xi_0 \sim \mathcal{N}(m', \sigma')$
   $\xi(t_0) \leftarrow \xi_0$
   ▷ sample the total variation
   $K \sim (\mathcal{N}(m'', \sigma''))^2$
   $y_1, ..., y_{N-1} \sim \mathbb{U}([0, K])$
   $y_0 \leftarrow 0$, $y_n \leftarrow K$
   orderAndRename$(y_0, \ldots, y_n)$
   ▷ now $y_1 \leq y_2 \leq ... \leq y_{N-1}$
   $s_0 \sim \text{Discr}(-1, 1)$
   while $i \leq N$ **do**
      $s \leftarrow \text{Binomial}(q)$ ▷ $P(s = -1) = q$
      $s_{i+1} = s_i \cdot s$
      $\xi(t_{i+1}) = \xi(t_i) + s_{i+1}(y_{i+1} - y_i)$
   end while

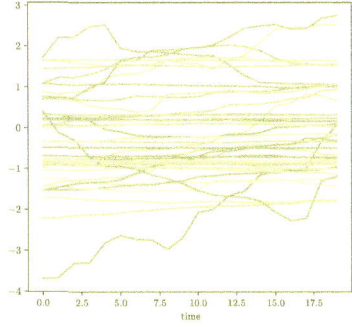

Fig. 1. Signals sampled from μ_0.

3 Retrieval-Augmented STL Requirement Mining

The problem we address with our proposed methodology is that of mining a STL specification from a dataset of labeled trajectories (where labels indicate whether the system complies to some desired behaviour), without the need of inferring or knowing a generative model of the system. This is a supervised two-class classification problem, in which the input is a set of trajectories partitioned in the subset of those labeled as regular (or positive) and those which are anomalous (or negative), denoted respectively as $\mathcal{D}_p$ and $\mathcal{D}_n$, and the output is a (set of) STL formula(e) able to discriminate among the two sets. We also assume that such datasets come from unknown stochastic processes $\boldsymbol{X}_p$ and $\boldsymbol{X}_n$, respectively. We adopt the approach of [19] and optimize the following function in order to mine a STL formula φ with the above desiderata:

$$G(\varphi) = \frac{\mathbb{E}[R_\varphi | \boldsymbol{X}_p] - \mathbb{E}[R_\varphi | \boldsymbol{X}_n]}{\sigma(R_\varphi | \boldsymbol{X}_p) + \sigma(R_\varphi | \boldsymbol{X}_n)} \tag{3}$$

with R_φ defined in Eq. 1. The intuition behind Eq. 3 is that the higher the value of $G(\varphi)$ (i.e. the bigger the difference between the average robustness of the two stochastic processes, relative to the length scale given by their standard deviation σ), the more robustly the inferred φ will separate the sets of positive and negative trajectories. The key insights of our approach are that: (i) we frame the learning problem as the optimization of Eq. 3 in the latent semantic space of formulae, i.e. in the space of embeddings of formulae individuated by the STL kernel of Eq. 2; (ii) we construct a dense (hierarchical) vector database containing the embeddings of millions of STL formulae, that we query anytime we need to invert the embeddings resulting from the previous step. This IR step approximates the inverse of Eq. 2 by finding specification whose latent representation is the nearest to the searched embedding (among the ones in the database), to obtain STL formulae for evaluating $G(\varphi)$. Hence the focus of our procedure is on inferring the semantic features a formula needs to have to discriminate between positive and negative trajectories, without a-priori constraints on its syntactic structure.

Summarizing, the proposed methodology consists in interleaving steps of Bayesian Optimization (BO) and Information Retrieval (IR): the details of these two building blocks are given in Sects. 3.2 and 3.1, respectively.

3.1   Building and Querying the Semantic Vector Database

The STL kernel of Eq. 2 allows to map formulae from the discrete space of their syntactic representation to a continuous space preserving their semantics (i.e. in the individuated latent space semantically similar formulae are mapped to nearby representations), however such map is not invertible. Given the recent advances in indexing structures and search mechanisms for dense vector databases [12], often accompanied by open-source software, we devise a simple yet effective strategy for inverting the STL kernel representations: storing in a reasoned and hierarchical database the embeddings of millions of formulae (more precisely of pairs containing both the formula and the corresponding representation), that we call *semantic vector database*, and using Approximate Nearest Neighbors (ANN) search techniques to obtain STL requirements starting from a candidate embedding.

Selecting STL Formulae. In order to build such database we first need to select the set of formulae to store: we do so by fixing the maximum number M of nodes and the maximum number of variables N (i.e. the maximum dimensionality of the signals) allowed, then we enumerate all templates (i.e. STL formulae where constants are replaced by parameters) satisfying those constraints as detailed in Algorithm 2. Once all the templates have been generated, we instantiate each of them on a set of parameter configurations $\mathcal{P} = \{p_0, \ldots p_{|\mathcal{P}|}\}$ to be filtered by Signature-based Optimization as defined in [18]. Denoting as $\varphi(p)$ the application of the parameter set p to the template formula φ, and given a set of trajectories $\hat{\mathcal{T}} = \{\xi_0, \ldots, \xi_{|\hat{\mathcal{T}}|}\}$ (with $p \in \mathcal{P}$, and $\hat{\mathcal{T}} \subset \mathcal{T}$), it consists in building a matrix $S \in \mathbb{R}^{|\mathcal{P}| \times |\hat{\mathcal{T}}|}$, whose rows are called the *signature* of the formula, s.t. $S(i,j) = \rho(\varphi(p_i), \xi_j)$, i.e. the cell (i,j) of such matrix holds the robustness of the i-th instantiation of φ on the j-th trajectory. In such matrix, the distance between rows can be considered as a proxy for semantic similarity among different parametrizations of the same template, hence a selection on formulae can be done by keeping only those whose signature has a distance higher than a certain threshold $\tau \in \mathbb{R}_{\geq 0}$ from formulae already selected. This helps in removing redundancy in the final database structure.

Algorithm 2. Algorithms for generating STL formulae templates

---

Input: M, N
Output: all_phis   ▷ STL templates of formulae with max N vars and M nodes
   all_phis ← []
   all_phis.append(generateAtomicPropositions())   ▷ $x_i \leq 0$ or $x_i \geq 0$, $\forall i \leq N$
   for $2 \leq m \leq M$ **do**
      prev_phis ←getPhisGivenNodes($m-1$) ▷ retrieve templates with $m-1$ nodes
      unary_ops ←expandByUnaryOperators(prev_phis)   ▷ $F, G, \neg$
      all_phis.append(unary_ops)
      l_list, r_list ←getPairsGivenSum(m)   ▷ all pairs $(l,r) : l+r = m, l \leq r$
      for $(l,r) \in [l\_list, r\_list]$ **do**
         l_phis←getPhisGivenNodes(l)
         r_phis←getPhisGivenNodes(r)
         binary_ops←expandByBinaryOpeators(l_phis, r_phis)   ▷ $\wedge, \vee, U$
         all_phis.append(binary_ops)
      end for
   end for

Computing Formulae Embeddings. Once the list of formulae to be stored in the database is provided, the STL kernel embeddings of such specifications needs to be computed, as for Eq. 2. In order to do so, one needs to (i) fix a probability measure on the space of trajectories and (ii) sample a set of n_{train} formulae against which the kernel is computed. For the first point, we use the default measure μ_0, computed as in Algorithm 1, proved to be quite general in [9] (i.e. even when the test trajectory distribution is not μ_0, the performance of the kernel computed on μ_0 on downstream tasks is comparable to that of the kernel computed on the data generating distribution). Fixing a reference measure on signals allows to compute the embeddings, hence the database, only once and use it in all application cases. For the second point we instead sample from a distribution $\mathcal{F}$ over STL formulae which is algorithmically defined by a syntax-tree random recursive growing scheme, that recursively generates the nodes of a formula given the probability p_{leaf} of each node being an atomic predicate, and a uniform distribution over the other operator nodes. We remark that the n_{train} determines, by construction, the dimension of the vectors stored in the database. Indeed, given the list of formulae to be put in the database, the embedding of each of them is the vector of dimension n_{train} where the i-th entry is the result of Eq. 2 computed on the formula of interest and the i-th training formula.

Index Structure Organization. It is worth noting that the semantic vector database is built once (although it can be further extended) and leveraged in all the applications, hence it is fundamental to organize it in a reasoned way, given the different contexts it can be deployed in. In particular, we hierarchically index the embeddings based on the number of different variables and the number of nodes appearing in the formula. The rationale behind the first choice is that not all signals that we wish to classify with our methodology have the same dimension, the second choice moves instead from the point of view of interpretability:

the smaller the syntax tree of a formula, the higher its human-understandability, hence we incorporate in the retrieval strategy the preference for syntactically simpler formulae.

Semantic Database Implementation. For indexing the embeddings, we use the tools provided by the FAISS Python library [13], which allows us to leverage GPU acceleration for both indexing and querying steps. Since the amount of vectors we need to store is of the order of millions, we opt for the Inverted File Index with Product Quantization (IVFPQ) to accelerate the search procedure (numerical results will follow in the remainder of this Section), which is a composite index that both partitions the embeddings in Voronoi cells and reduces their dimension using Product Quantization [12]. Euclidean (or L_2) distance is used as similarity measure between vectors. We remark that the framework we propose is flexible w.r.t. the choice of the index structure, hence this component can be replaced with any similarity search engine on vector databases without altering the overall structure of the algorithm.

Table 1. Quantiles of $AP@K$, $NDCG@K$ and kernel similarity $k(\varphi, \hat{\varphi})$ between the searched formula φ and the top result $\hat{\varphi}$, across 1000 queries to the semantic database, varying the number of nodes in the syntactic tree of the searched formulae.

$AP@K$				$NDCG@K$				$k(\varphi, \hat{\varphi})$			
n_{nodes}	1quart	median	3quart	99perc	1quart	median	3quart	99perc	1quart	median	3quart	99perc
$(0, 5]$	0.804	1.000	1.000	1.000	1.000	1.000	1.000	1.000	0.810	0.901	0.976	0.992
$(5, 10]$	0.763	1.000	1.000	1.000	1.000	1.000	1.000	1.000	0.774	0.883	0.946	0.979
$(10, 15]$	0.684	0.932	1.000	1.000	1.000	1.000	1.000	1.000	0.728	0.854	0.928	0.972
$(15, 20]$	0.635	0.873	1.000	1.000	1.000	1.000	1.000	1.000	0.742	0.848	0.917	0.976

In order to build the index we enumerate all formulae templates following Algorithm 2 with $M = 5$ and $N = 3$. Given that the measure μ_0 is used for computing the embeddings, we keep into account the typical range of its trajectories (see Fig. 1) for setting the variable thresholds, hence using the grid of 10 equally-spaced points between -4 and 4 as possible parameter instances; for the temporal operators thresholds, we instead leverage the fact that signals for computing the STL kernel representations have 100 points and use the grid of 10 equally-spaced points between 0 and 100 as possible instances for both left and right time-bounds. We select the formulae to store in the index by signature-based optimization with cosine distance and threshold $\tau = 0.9$. Finally, as done in [9], we use $n_{train} = 1000$ STL formulae for computing the embeddings. This leads to a total of $\sim 10\,000\,000$ stored formulae, that we organize in a total of 6 indexes: first splitting them based on the number of variables they contain (namely 1, 2 or 3), then splitting each of them based on the maximum number of nodes allowed (namely 4 or 5). Of course, alongside the embeddings, we store the STL formulae organized in a list structure mirroring the organization of the corresponding representations. Searching for 1000 formulae with 3 variables on such index (hence scanning all the database) requires 1 ± 0.5 s, while computing their embeddings takes 6 ± 2 s exploiting GPU acceleration, finally the whole process of creating the database takes ~ 20 h on a AMD EPYC 7542 machine with 64 GB of RAM and a NVidia A100 with 80 GB of memory. Moreover, storing the index requires ~ 180 MB, which is order of magnitudes less then the ~ 70 GB required in case plain embeddings should be stored (e.g. for exact search).

Semantic Database Effectiveness. In order to assess the effectiveness of our information retrieval system, we use Average Precision at K ($AP@K$) and Normalized Discounted Cumulative Gain at K ($NDCG@K$), to measure both the relevancy of the results and the ranking quality. Defining the precision at K (denoted as $P@K$) as the number of relevant results among the top K retrieved, and rel_k as the indicator function being 1 when the k-th ranked result is relevant and 0 otherwise, then $AP@K = \frac{\sum_{k=1}^{K}(P@K \cdot rel_k)}{\sum_{k=1}^{K} rel_k}$. If we instead define r_k as the *graded relevance* (or gain) of the result at position k, i.e. an integer score indicating its relevance, then the Discounted Cumulative Gain at K is $DCG@K = \frac{\sum_{k=1}^{K} r_k}{\log_2(k+1)}$; the Ideal $DCG@K$ ($IDCG@K$) is the maximum achievable $DCG@K$ given a set of relevance scores and finally $NDCG@K = \frac{DCG@K}{IDCG@K}$. The definitions of $AP@K$ and $NDCG@K$ given above are based on the concept of *relevancy* of the result, which is not immediate to state in our scenario. Intuitively, we consider a retrieved formula $\hat{\varphi}$ semantically similar to a searched formula φ if they behave similarly across an arbitrary number of trajectory, i.e. if they are (un)satisfied on the same subset of signals. Hence, given a similarity threshold $\omega \in [0,1]$ and a set of trajectories $\{\xi_j\}_{j=1}^{n_{\mathrm{traj}}} \subset \mathcal{T}$, we say that $\hat{\varphi}$ is a relevant result w.r.t. the query φ if $\frac{|\{j:s(\varphi,\xi_j)=s(\hat{\varphi},\xi_j)\}|}{n_{\mathrm{traj}}} \geq \omega$. Moreover, we also consider the kernel similarity of Eq. 2 (whose maximum is 1) as an indicator of relevancy of the results, which accounts also for the quantitative semantics of formulae. In our experiments, we used $n_{\mathrm{traj}} = 10\,000$ signals sampled from μ_0, $\omega = 0.9$ and STL formulae sampled from $\mathcal{F}$ with $N = 3$, varying the number n_{nodes} of nodes allowed in the syntax tree of the formula. Results in terms of $AP@K$ and $NDCG@K$ (with $K = 5$) are reported in Table 1, highlighting both the ranking quality of our methodology ($NDCG@K$ always 1) and the relevancy of retrieved results (median $AP@K > 0.87$ in the worst case), as well as resilience to the size of the searched formulae (we recall that the database contains specification with $n_{\mathrm{nodes}} \leq 5$). Moreover, we also observe that the kernel similarity between the query φ and the top retrieved formula $\hat{\varphi}$ is high (median above 0.84 in all tested cases) across all specification dimensions.

3.2 Bayesian Optimization in the Semantic Space of Formulae

As already mentioned, the goal of the methodology we propose is to learn the function mapping semantic embeddings of STL formulae (built as for Eq. 2) to values of $G(\varphi)$ of Eq. 3, given labeled datasets $\mathcal{D}_p$ and $\mathcal{D}_n$ of respectively positive

---

Algorithm 3. STL Requirement Mining Combining BO and IR

Input: $\mathcal{D}_p$, $\mathcal{D}_n$, SemanticDB
Output: $\tilde{\varphi}$
  k_phis, phis, g_phis $\leftarrow$ $getInitialCandidates(\mathcal{D}_p, \mathcal{D}_n)$
  $i \leftarrow 0$
  opt$\leftarrow [0, g\_phis]$
  while $i \leq maxiter$ or $(opt[i] - opt[i-1]) < \varepsilon$ **do**
    new_phis$\leftarrow$optimizeUCB()
    k_phis.append(new_phis)
    phis.append(retrieveFromEmbedding(new_phis, SemanticDB))
    g_phis.append(evaluteObjective(new_phis, $\mathcal{D}_p$, $\mathcal{D}_n$))
    opt.append(max(g_phis))
  end while
  $\tilde{\varphi} \leftarrow$ phis[argmax(g_phis)]

---

and negative trajectories. This implies searching in the latent semantic space for the continuous representation of a formula able to discriminate between signals belonging to $\mathcal{D}_p$ and $\mathcal{D}_n$. Being it a non-linear non-convex optimization problem, we tackle it by means of Bayesian Optimization (BO), and more specifically of the Gaussian Process Upper-Confidence Bound (GP-UCB) algorithm [26]. BO is useful in this scenario also because it allows to efficiently construct a surrogate model of the function, of which we do not know an analytical expression. We recall that the typical BO loop consists in: (i) fit the GP on the available data: pairs $(k(\varphi), G(\varphi))$ in our case, with φ STL formula and $k(\varphi)$ its STL kernel embedding; (ii) optimize the UCB acquisition function to find the best candidate (the kernel embedding $k(\hat{\varphi})$ of a formula $\hat{\varphi}$ improving Eq. 3); (iii) query the objective function, i.e. Equation 3, on the new points to add new data pairs to the training set. In order to obtain an STL formula $\hat{\varphi}$ starting from its embedding $k(\hat{\varphi})$ (i.e. between the steps (ii) and (iii) of the BO loop), we augment the procedure with a retrieval step, querying the semantic vector database on $k(\hat{\varphi})$. An overview of the overall iterative methodology is reported in Algorithm 3, and it consists, at each iteration, in the following:

1. Given the current set of pairs $(k(\varphi), G(\varphi))$ of kernel embeddings and corresponding objective function values, optimize the acquisition function to get a new set of candidate embeddings;
2. In order to evaluate the function $G(\varphi)$ of Eq. 3, we need to invert the embeddings, to obtain STL formulae. This is done with ANN by querying the semantic vector database on the output of the previous step;
3. We evaluate $G(\varphi)$ on the formulae obtained above and record the pair corresponding to the maximum value of $G(\varphi)$.

In our experiments we use GP with Matern kernel (starting with an initial batch of only 10 points) and optimize the UCB acquisition function using Stochastic Gradient Descent (SGD), given the high-dimensionality of the latent semantic space (namely 1000, as detailed in Sect. 3.1); this is done in Python exploiting the BoTorch library [4], which allows to leverage GPU acceleration.

4 Experiments

4.1 Experimental Setting

We test our proposed algorithm across a number of benchmark datasets found in the related literature (see Sect. 5). Unless differently specified, we perform 5-fold cross validation and report mean and standard deviation of the results across the folds. The metrics we use for assessing the quality of the results are: Misclassification Rate (MCR), Precision ($Prec$) and Recall (Rec), defined as follows, given positive and negative trajectory datasets $\mathcal{D}_p$ and $\mathcal{D}_n$, and the mined STL specification $\hat{\varphi}$:

$$MCR = \frac{|\{\xi_j \in \mathcal{D}_p : s(\hat{\varphi}, \xi_j) < 0\}| + |\{\xi_j \in \mathcal{D}_n : s(\hat{\varphi}, \xi_j) > 0\}|}{|\mathcal{D}_p| + |\mathcal{D}_n|}$$

$$Prec = \frac{|\{\xi_j \in \mathcal{D}_p : s(\hat{\varphi}, \xi_j) > 0\}|}{|\{\xi_j \in \mathcal{D}_p : s(\hat{\varphi}, \xi_j) > 0\}| + |\{\xi_j \in \mathcal{D}_n : s(\hat{\varphi}, \xi_j) > 0\}|}$$

$$Rec = \frac{|\{\xi_j \in \mathcal{D}_p : s(\hat{\varphi}, \xi_j) > 0\}|}{|\{\xi_j \in \mathcal{D}_p : s(\hat{\varphi}, \xi_j) > 0\}| + |\{\xi_j \in \mathcal{D}_p : s(\hat{\varphi}, \xi_j) < 0\}|}$$

In the above Equation, using the jargon of binary classification algorithms, the true positives are $TP = \{\xi_j \in \mathcal{D}_p : s(\hat{\varphi}, \xi_j) > 0\}$, the true negatives are $TN = \{\xi_j \in \mathcal{D}_n : s(\hat{\varphi}, \xi_j) < 0\}$, false positives are instead defined as $FP = \{\xi_j \in \mathcal{D}_n : s(\hat{\varphi}, \xi_j) > 0\}$ and finally false negatives $FN = \{\xi_j \in \mathcal{D}_p : s(\hat{\varphi}, \xi_j) < 0\}$.

The retrieval step of Algorithm 3 requires to normalize trajectories of $\mathcal{D}_p$ and $\mathcal{D}_n$ by removing the mean and diving by the standard deviation of each dimension, in order to align their domain with that of μ_0 (i.e. the measure w.r.t. which semantic embeddings are computed). At the end of the procedure, they are scaled back to their original range, and the same inverse normalization is applied to the thresholds of the variables appearing in the mined specification (since all the formulae in the semantic vector database have thresholds compatible with μ_0 by construction). Another aspect to take into account when dealing with the retrieval step of Algorithm 3 is that of time parameters: the sampling rate of μ_0 is 1 per time-step, considering a total of 100 points, but this might vary across trajectory distribution. Assuming that the target distribution $\mathcal{D}_{\text{test}}$ has n_{test} time points, without loss of generality we can consider them to be sampled at rate of 1 per time step, so that a time interval $[a, b]$ referred to a trajectory sampled from μ_0 can be mapped to the corresponding time interval $[a_{\text{test}}, b_{\text{test}}]$ for a trajectory sampled from $\mathcal{D}_{\text{test}}$ as: $[a_{\text{test}}, b_{\text{test}}] = [\lfloor a \cdot (n_{\text{test}}/100) \rfloor, \lfloor a \cdot (n_{\text{test}}/100) \rfloor + \lceil (b - a) \cdot (n_{\text{test}}/100) \rceil]$. Moreover, before querying the semantic vector database, we restrict the search space to only those indexes containing formulae having a number of variables compatible with the dimension of $\mathcal{D}_{\text{test}}$ and, for the sake of interpretability, we first search on indexes containing specifications with at most 4 nodes in the syntactic tree and, if the outcome of the procedure is not satisfactory, we allow also to search for more complex requirements (as we will report in the remainder of this

Section, this has not been necessary in any of test cases). All the experiments are implemented in Python, leveraging GPU acceleration in every sub-procedure thanks to the PyTorch [22] library, and tested on a AMD EPYC 7542 machine with 64 GB of RAM and a NVidia A100 with 80 GB of memory.

4.2 Case Studies

Linear System. In [18] the authors generate trajectories using the following dynamical system: regular trajectories evolve as $\dot{x} = 0.03 \cdot x + w$ while anomalous ones as $\dot{x} = -0.03 \cdot x + w$, being $w(t)$ a white noise with variance 0.04. Following their approach, we generated 100 positive and 100 negative examples, sampling 100 points for each trajectory, a random subset of the obtained signals is reported in Fig. 2(a).

Maritime Surveillance. The *Maritime Surveillance* dataset is a common benchmark for STL anomaly detection, used e.g. in [8,18,19]. It consists in 2-dimensional signals reporting vessels behaviour, divided in 1000 regular and 1000 anomalous examples, each consisting of 61 sampling points. Random signals belonging to the maritime dataset are reported in Fig. 2(d).

Human Activity Recognition (HAR). The HAR dataset created in [1] and publicly available at [14] contains recording of humans performing activities in daily livings. Following the approach [17], we split the 6 possible activities in 3 static postures and 3 dynamic activities, and use these as classification labels (for a total of 36 and 58 trajectories belonging to the mentioned classes, respectively). Signals belonging to this dataset are 1-dimensional, representing the standard

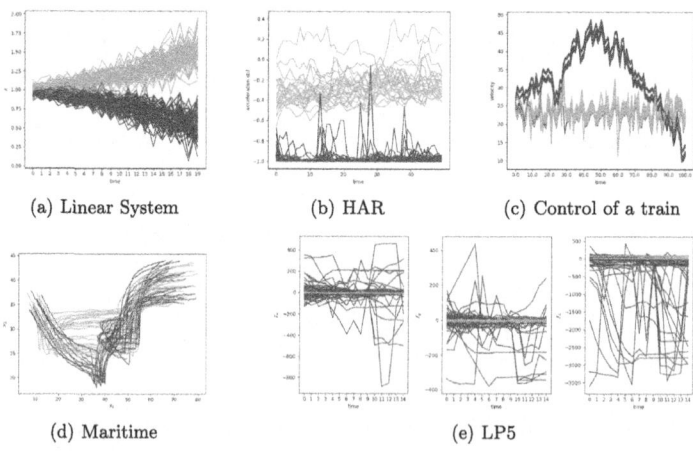

    (a) Linear System   (b) HAR   (c) Control of a train

    (d) Maritime   (e) LP5

Fig. 2. Regular and anomalous trajectories sampled from the testes datasets. (Color figure online)

deviation of the acceleration of the subjects in the x-direction at 50 sampling points. A subset of positive and negative trajectories belonging to the HAR dataset is reported in Fig. 2(b).

Robot Execution Failures in Motion with Part (LP5). The LP5 dataset, available at [14], consists of 3-dimensional signals, representing force measurements along the 3 axes, each recorded at 15 time steps. As in [17], we consider as positive the 44 instances showing normal behaviour and as negative the 26 exhibiting bottom collisions. Trajectories belonging to this dataset are shown in Fig. 2(e).

Cruise Control of Train. Following [18], we test our procedure on the dataset coming from the (noisy) observations of the velocity of a train, simulated in a cruise control scenario. Such signals are 1-dimensional and sampled for 100 points; classes contain 100 trajectories each. A random sample of such trajectories is shown in Fig. 2(c).

Table 2. Best mined formula, mean and std of MCR, Prec and Rec in the test set across the folds for each case study, computational time (in seconds) required to find the reported formula using Algorithm 3.

$\hat{\varphi}$		MCR	$Prec$	Rec	$Time\ (s)$
Linear	$F_{[70,\infty]}(x_0 \leq 1.16)$	0.00 ± 0.00	1.00 ± 0.00	1.00 ± 1.00	7.32 ± 1.45
Maritime	$(x_1 \geq 23.19)U(x_0 \leq 32.56)$	0.07 ± 0.01	0.97 ± 0.01	1.00 ± 0.00	12.4 ± 2.63
HAR	$G(x_0 \geq -0.75)$	0.00 ± 0.00	1.00 ± 0.00	1.00 ± 0.00	8.32 ± 0.25
LP5	$G_{[4,\infty]}(x_2 \geq 0.084)$	0.01 ± 0.01	0.98 ± 0.01	1.00 ± 0.00	11.6 ± 1.44
Train	$G_{[0,36]}(x_0 \leq 37)$	0.03 ± 0.01	0.93 ± 0.02	1.00 ± 0.00	8.21 ± 1.23

Discussion. Results of the retrieval-augmented STL requirement mining procedure are shown in Table 2, where we report the best formula we infer (denoted as $\hat{\varphi}$), along with performance metrics and computational time for each case study. In this context, with *best* formula we denote the mined formula having the lowest MCR in the test-set, using the lower number of nodes as tiebreaker. Notably, we can observe that all mined requirements are interpretable, having a syntax tree of at most 3 nodes, in line with more recent related works (namely [17–19]), and act as almost-perfect discriminators among regular and anomalous trajectories (see MCR in Table 2). Moreover, we can observe that the mined specifications never produce false positives (reporting a recall of 1 in every test case). It is worth noting that even if the LP5 benchmark is 3-dimensional, the mining procedure outputs a formula in which only a variable appears (in line with [17]). We find it interesting to remark the difference between trajectories sampled from the measure μ_0 used to compute STL embeddings (a random sample of signals belonging to this distribution is shown in Fig. 1) and those of the test benchmarks (reported in Fig. 2), to underline the applicability of our methodology across various trajectories distribution, without the need of recomputing the semantic vector database on each data distribution. Finally, another relevant difference between our approach and related works is that we fully exploit GPU acceleration in every step of the mining procedure, resulting in lower computational times. Indeed as reported in Table 2, our method is able to learn a classifier in less than 15s for all the tested cases, while [18] reports a computational time of 39.05 s, 23.17 s, 32.32 s for the Linear, Maritime and Train datasets, respectively; [17] register 22.07 s and 22.83 s for mining a discriminating formula in the HAR and LP5 banchmarks; finally [19] takes 73 s to classify signals belonging to the Maritime dataset.

5   Related Work

Mining STL specifications from data has seen a surge of interest in the last few years, as reported in [7]. At a high level, following [7], contributions in this field can be characterized by: template-based vs template-free procedures (in the first, the mining consists in finding the optimal parameters for a given syntactic skeleton, while in the second both the structure and the parameters of the specification have to be inferred), supervised vs unsupervised approaches (following the availability of labeled trajectory datasets) and online vs offline methodology (the first concerns updating the inferred requirements every time new data are available, the latter considers instead a static set of data). In this scenario our algorithm can be defined as a template-free, supervised, offline approach (besides, it can be cast as an unsupervised learning problem by just changing the objective function of Eq. 3, keeping intact the structure of Algorithm 3). Most of the related works with these characteristics decompose the task in two steps, i.e. as bi-level optimization problem: they first learn the structure of the specification from data, and then instantiate a concrete formula using parameter inference methods. For the first step, a popular extension of STL, called Parametric Signal Temporal Logic (PSTL) has been devised [2], in which parameters replace both threshold constants in numerical predicates and time bounds in temporal operators. In [27] STL specifications are mined using a grid-based discretization of both time and value domain, used to guide a clusterization of the input signals, for assigning to each cluster a descriptive STL formula and obtain the final requirement as disjunction of cluster formulae. In [8] a map between a fragment of PSTL and binary decision trees is established, so that the requirement mining procedure is cast into a classification problem tackled by decision tree, and then translation of the tree to a formula is done to infer the final specification. In [18] STL requirements are mined using systematic enumeration of PSTL templates (in increasing order w.r.t. the syntactic size) and accuracy-driven instantiation of each template (towards optimizing logical classification), guaranteeing that the inferred formula is the smallest one reaching a given accuracy threshold. Another line of works [3,19,25] use instead Genetic Algorithms (GA) to learn the PSTL structure of the specification, and find optimal parameters by Gaussian Process

Upper Confidence Bound (GP-UCB); notably [19] do not put any restrictions on the fragment of PSTL used. Our work differentiate from the cited papers in that it learns simultaneously both the structure and the parameters of the STL requirement, by performing optimization in a continuous space representing the semantics of formulae.

Data Mining for Time Series Data. Typically, anomaly detection and classification in time series data is done by either autoregressive statistical models, distance and density-based clustering methods, or prediction and reconstruction-based deep neural networks [11,21]. Such methods often require to individuate a latent representation of trajectories or to hand-craft features for the analysis, leading to a black-box solution to the problem, which do not offer any insight on the system at hand. Notably, there are works focusing on building time series classifiers which are interpretable, e.g. using Symbolic Fourier Approximation to infer a symbolic representation of the input before learning a linear classifier [10], and using saliency maps to characterize classes [20]. Using STL as formal language for the learned class discriminator allows to describe emergent properties of a system in a rich yet interpretable way, leading to knowledge discovery about the process under analysis [7].

Information Retrieval and Data Mining, as well as their interaction, witnessed a growth of interest recently [15]. In particular, the adoption of the so-called dense vector databases (i.e. specialized storage systems designed for efficient management of dense vectors and supporting advanced similarity search) and retrieval methods has been proven to be effective in injecting factual information into Machine Learning (ML) techniques [28]. Using information retrieval (IR) systems to support ML models access abstractly-represented knowledge stores helps them in generalize beyond training data, without the need of increasing their size, and notably helps for the interpretability and explainability of the outcomes, being inference grounded on retrieved information, often stored in a human-readable format [28].

6   Conclusion

In this work we propose a methodology for mining STL requirements from time-series data, in a purely data-driven setting. In particular, we focus on binary classification of trajectories, such as those coming from observations of Cyber-Physical Systems. The core idea of our algorithm lies in the interaction between Bayesian Optimization and Information Retrieval techniques, by searching from the discriminating STL property directly in a continuous space representing its robust semantics. Doing so, we learn simultaneously the structure and the parameters of the specification, in such a way that it is possible to control its maximum allowed size, enhancing the interpretability of the inferred formula, to foster knowledge discovery about the system. Results of experiments on several benchmarks show that we achieve state-of-art accuracy, drastically reducing

330 G. Saveri and L. Bortolussi

computational time w.r.t. related works, thanks to the exploitation of GPU acceleration in all steps of our methodology. It is also worth mentioning that by changing the optimization objective, our framework can be deployed in an unsupervised classification scenario, as well as in a requirement mining context in which only positive examples are available, extensions that we plan to implement as future work.

References

1. Anguita, D., Ghio, A., Oneto, L., Parra, X., Reyes-Ortiz, J.L.: 21st European Symposium on Artificial Neural Networks, ESANN 2013 (2013)
2. Asarin, E., Donzé, A., Maler, O., Nickovic, D.: Parametric identification of temporal properties. In: Khurshid, S., Sen, K. (eds.) RV 2011. LNCS, vol. 7186, pp. 147–160. Springer, Heidelberg (2012). https://doi.org/10.1007/978-3-642-29860-8_12
3. Aydin, S.K., Gol, E.A.: Synthesis of monitoring rules with STL. J. Circuits Syst. Comput. **29**(11), 2050177:1–2050177:26 (2020)
4. Balandat, M., et al.: BoTorch: a framework for efficient Monte-Carlo Bayesian optimization. In: Advances in Neural Information Processing Systems, vol. 33 (2020)
5. Bartocci, E., Bortolussi, L., Nenzi, L., Sanguinetti, G.: System design of stochastic models using robustness of temporal properties. Theor. Comput. Sci. **587**, 3–25 (2015)
6. Bartocci, E., et al.: Specification-based monitoring of cyber-physical systems: a survey on theory, tools and applications. In: Bartocci, E., Falcone, Y. (eds.) Lectures on Runtime Verification. LNCS, vol. 10457, pp. 135–175. Springer, Cham (2018). https://doi.org/10.1007/978-3-319-75632-5_5
7. Bartocci, E., Mateis, C., Nesterini, E., Nickovic, D.: Survey on mining signal temporal logic specifications. Inf. Comput. **289**(Part), 104957 (2022)
8. Bombara, G., Vasile, C.I., Penedo, F., Yasuoka, H., Belta, C.: A decision tree approach to data classification using signal temporal logic. In: Proceedings of the 19th International Conference on Hybrid Systems: Computation and Control, HSCC 2016, pp. 1–10. ACM (2016)
9. Bortolussi, L., Gallo, G.M., Křetínský, J., Nenzi, L.: Learning model checking and the kernel trick for signal temporal logic on stochastic processes. In: TACAS 2022. LNCS, vol. 13243, pp. 281–300. Springer, Cham (2022). https://doi.org/10.1007/978-3-030-99524-9_15
10. Early, J., Cheung, G.K.C., Cutajar, K., Xie, H., Kandola, J., Twomey, N.: Inherently interpretable time series classification via multiple instance learning. CoRR abs/2311.10049 (2023)
11. Faouzi, J.: Time Series Classification: a review of Algorithms and Implementations. In: Machine Learning (Emerging Trends and Applications). Proud Pen (2022)
12. Han, Y., Liu, C., Wang, P.: A comprehensive survey on vector database: storage and retrieval technique, challenge. CoRR abs/2310.11703 (2023)
13. Johnson, J., Douze, M., Jégou, H.: Billion-scale similarity search with GPUs. IEEE Trans. Big Data **7**(3), 535–547 (2019)
14. Kelly, M., Longjohn, R., Nottingham, K.: The UCI machine learning repository. https://archive.ics.uci.edu
15. Liu, J., et al.: Data mining and information retrieval in the 21st century: a bibliographic review. Comput. Sci. Rev. **34** (2019)

16. Maler, O., Nickovic, D.: Monitoring temporal properties of continuous signals. In: Lakhnech, Y., Yovine, S. (eds.) FORMATS/FTRTFT -2004. LNCS, vol. 3253, pp. 152–166. Springer, Heidelberg (2004). https://doi.org/10.1007/978-3-540-30206-3_12

17. Mohammadinejad, S., Deshmukh, J.V., Puranic, A.G., Vazquez-Chanlatte, M., Donzé, A.: Interpretable classification of time-series data using efficient enumerative techniques. CoRR abs/1907.10265 (2019)

18. Mohammadinejad, S., Deshmukh, J.V., Puranic, A.G., Vazquez-Chanlatte, M., Donzé, A.: Interpretable classification of time-series data using efficient enumerative techniques. In: HSCC 2020: 23rd ACM International Conference on Hybrid Systems: Computation and Control, pp. 9:1–9:10. ACM (2020)

19. Nenzi, L., Silvetti, S., Bartocci, E., Bortolussi, L.: A robust genetic algorithm for learning temporal specifications from data. In: McIver, A., Horvath, A. (eds.) QEST 2018. LNCS, vol. 11024, pp. 323–338. Springer, Cham (2018). https://doi.org/10.1007/978-3-319-99154-2_20

20. Nguyen, T.L., Gsponer, S., Ilie, I., O'Reilly, M., Ifrim, G.: Interpretable time series classification using linear models and multi-resolution multi-domain symbolic representations. CoRR abs/2006.01667 (2020)

21. Oswal, S., Shinde, S., Vijayalakshmi, M.: A survey of statistical, machine learning, and deep learning-based anomaly detection techniques for time series. In: Garg, D., Narayana, V.A., Suganthan, P.N., Anguera, J., Koppula, V.K., Gupta, S.K. (eds.) IACC 2022. CCIS, vol. 1782, pp. 221–234. Springer, Cham (2023). https://doi.org/10.1007/978-3-031-35644-5_17

22. Paszke, A., et al.: Automatic differentiation in PyTorch. In: NIPS-W (2017)

23. Pnueli, A.: The temporal logic of programs. In: 18th Annual Symposium on Foundations of Computer Science (SFCS 1977), pp. 46–57 (1977)

24. Rasmussen, C.E., Williams, C.K.I.: Gaussian Processes for Machine Learning. Adaptive Computation and Machine Learning. MIT Press (2006). https://www.worldcat.org/oclc/61285753

25. Saglam, I., Gol, E.A.: Cause mining and controller synthesis with STL. In: 58th IEEE Conference on Decision and Control, CDC 2019, pp. 4589–4594. IEEE (2019)

26. Srinivas, N., Krause, A., Kakade, S.M., Seeger, M.W.: Information-theoretic regret bounds for Gaussian process optimization in the bandit setting. IEEE Trans. Inf. Theory 58(5), 3250–3265 (2012)

27. Vaidyanathan, P., et al.: Grid-based temporal logic inference. In: 56th IEEE Annual Conference on Decision and Control, CDC 2017, pp. 5354–5359. IEEE (2017)

28. Zamani, H., Diaz, F., Dehghani, M., Metzler, D., Bendersky, M.: Retrieval-enhanced machine learning. In: SIGIR 2022: The 45th International ACM SIGIR Conference on Research and Development in Information Retrieval, pp. 2875–2886. ACM (2022)

CAM-Based Methods Can See Through Walls

Magamed Taimeskhanov[1]([✉]), Ronan Sicre[2], and Damien Garreau[3]

[1] Université Côte d'Azur, Laboratoire J.A. Dieudonné, CNRS, Nice, France
`magamed.taimeskhanov@etu.univ-cotedazur.fr`
[2] Centrale Méditerranée, Aix-Marseille Univ., CNRS, LIS, Marseille, France
`ronan.sicre@lis-lab.fr`
[3] Julius-Maximilians Universität Würzburg, Würzburg, Germany
`damien.garreau@uni-wuerzburg.de`

Abstract. CAM-based methods are widely-used post-hoc interpretability method that produce a saliency map to explain the decision of an image classification model. The saliency map highlights the important areas of the image relevant to the prediction. In this paper, we show that most of these methods can incorrectly attribute an important score to parts of the image that the model cannot see. We show that this phenomenon occurs both theoretically and experimentally. On the theory side, we analyze the behavior of GradCAM on a simple masked CNN model at initialization. Experimentally, we train a VGG-like model constrained to not use the lower part of the image and nevertheless observe positive scores in the unseen part of the image. This behavior is evaluated quantitatively on two new datasets. We believe that this is problematic, potentially leading to mis-interpretation of the model's behavior.

Keywords: Interpretability · Computer Vision · Convolutional Neural Networks · Class Activation Maps

1 Introduction

The recent advances of machine learning pervade all applications, including the most critical. However, deep learning models intrinsically possess many parameters, have complicated architectures, and rely on many non-linear operations, preventing the users to get a good grasp of the rationale behind particular decisions. These models are often called "black boxes" for these reasons [6]. In this respect, there is a growing need for interpretability of the models that are used, which gave birth to the field of eXplainable AI (XAI). When the model to explain is already trained, our main topic of interest, this is often called post-hoc interpretability [25, 26, 35].

Supplementary Information The online version contains supplementary material available at https://doi.org/10.1007/978-3-031-70368-3_20.

In the specific case of image classification, the explanations provided to the user often take the form of a saliency map superimposed to the original image, for instance simply looking at the gradient with respect to the input of the network [30]. The message is simple: the areas highlighted by the saliency maps are used by the network for the prediction. When the first layers of the network are convolutional layers [14], one can take advantage of this and look at the activations of the filters corresponding to the class prediction that we are trying to explain. Indeed, these first layers act like a bank of filters on the input image, and the degree to which they are activated gives us information on the behavior of the network. Thus the first layers possess a certain degree of interpretability, even though it can be challenging to aggregate the information coming from different filters. In any case, the next layers generally consist in a fully-connected neural network, thus suffering from the same caveats as other models. In addition, this second part of the network is equally important for the prediction, but is not taken into account in the explanations we provide if we simply look at activation values.

To solve this problem, a natural idea is to weight each activation map depending on how the second part of the network uses it. In the case of a single additional layer, this is called *class activation maps* [CAM, 36]. The methodology was quickly generalized by [29], using the *average gradient* values of the subsequent layers instead, giving rise to GradCAM, arguably one of the most popular posthoc interpretability method for CNNs. Many extensions are proposed in the following years, we list them in Appendix A and refer to [34] for a recent survey. Without being too technical, for all these methods, the explanations provided consist in a weighted average of the activation maps.

A close inspection of each of these methods reveals that the coefficient associated to each individual map is global, in the sense that the same coefficient is applied to the whole map. The main message of this paper is that this can be problematic, since different parts of the activation map may be used differently by the subsequent layers. Worse, **some parts may even be unused by the subsequent network and still highlighted in the final explanation** (see Fig. 1). Thus we believe that, while giving apparently more-than-satisfying results in practice, CAM-based methods should be used with caution, keeping in mind that some parts of the image may be highlighted whereas they are not even seen by the network.

1.1 Related Work

This paper is inspired by a line of recent works concerned with the reliability of saliency maps claiming that solely relying on the visual explanation provided by a saliency map can be misleading [16,22]. [16] introduces a method for altering the input data with imperceptible perturbations which do not change the predicted label, yet generating different saliency maps. On the other hand, [22] shows that numerous saliency methods generate incorrect scores for the input features when the model prediction is invariant to translation of the input data by a constant.

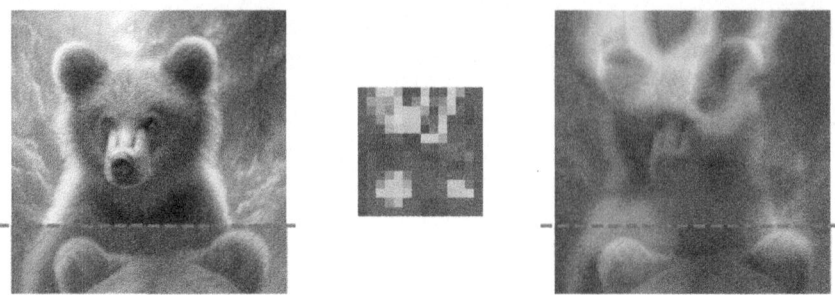

Fig. 1. Example of GradCAM failure on a VGG-like model trained on the ImageNet dataset (masked [**VGG**], see Fig. 4). *Left:* original image; *Middle:* GradCAM explanation before up-sampling; *Right:* original image with GradCAM explanation overlayed as a heatmap. The network does not have access to the red part of the image, **but GradCAM does highlight some pixels in this area.**

It is important to note that neither of these studies specifically challenges the reliability of CAM-based methods.

This perspective on saliency maps is supported by the work of [1], which introduces a randomization-based sanity check indicating that some existing saliency methods are independent of both the model and the data. We note that GradCAM passes the sanity checks proposed by [1]. [11], proposing HiResCAM, are less positive regarding GradCAM pointing out, as we do, that the use of a global coefficient can produce positive explanations where there should not be. Compared to our work, they provide few theoretical explanations and perform experiments on model which are not using parts of the input image. Posthoc interpretability methods in the image realm (not specific to CNN architectures) have been investigated by other works such as [15] which looked into LIME for images [28].

Taking another angle, [20] directly attacks the reliability of GradCAM saliency maps by adversarial model manipulation, *i.e.*, fine-tuning a model with the purpose of making GradCAM saliency maps unreliable. This is achieved by using a specific loss function tailored to this effect. Our approach is different, as we simply force a strong form of sparsity in the model's parameters, not targeting a specific interpretability method.

1.2   Organization of the Paper

We start by looking at GradCAM in Sect. 2. For a given simple CNN architecture described in Sect. 2.1, we derive closed-form expressions for its explanations in Sect. 2.2. Leveraging these expressions, we prove in Sect. 2.3 that GradCAM explanations are positive at initialization, even though a large part of the weights are set to zero.

In Sect. 3, we demonstrate experimentally that this phenomenon remains true after training. To this extent, we proceed in two steps. First, we train to a reasonable accuracy a VGG-like model on ImageNet [9] **which does not see the lower part of input images**, described in Sect. 3.1. Then, we create two datasets (Sect. 3.2) consisting in superposition of images of the same class. We show experimentally in Sect. 3.3 that **CAM-based methods applied to this model wrongly highlights a large portion of the lower part of the images**, misleading the user by showing that the lower part is used for the prediction whereas, by construction it is not. The code for training our model as well as the datasets are provided as supplementary material. Additionally, the code for all experiments is available online.[1] We conclude in Sect. 4.

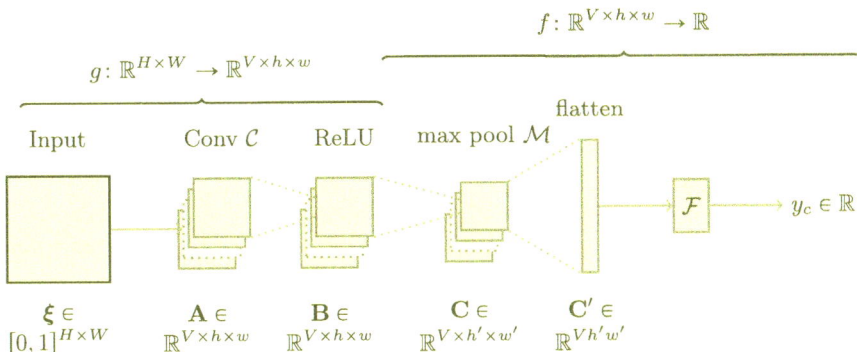

Fig. 2. The model used for the derivation of feature importance scores, [**CNN**]. The number of filters in the convolutional layer $\mathcal{C}$ is $V \in \mathbb{N}^*$. The size of the max pooling filters $k' \in \mathbb{N}^*$ is implicitly defined such that $(h', w') = \frac{1}{k'}(h, w)$ in $\mathbb{N}^*$. The fully-connected neural network $\mathcal{F}(\cdot)$ takes $\mathbf{C}'$ as input and processes it through L layers with ReLU activation functions to produce a raw score y_c, without converting this score into a "probability."

2   Mathematical Description

The model used for the theoretical analysis done in Sect. 2.3 is described in Sect. 2.1, the derivation of GradCAM coefficients in Sect. 2.2.

2.1   A Simple CNN

Let us describe mathematically the model we consider, denoted by [**CNN**] and depicted in Fig. 2. On a high-level, [**CNN**] is a $(L+1)$-layers network, consisting in a single convolution/max pooling layer, followed by a L-layers fully-connected neural network with ReLU activations. Thus the case $L = 1$ corresponds to a single convolutional/max pooling layer followed by a linear transformation.

---

[1] https://github.com/MagamedT/cam-can-see-through-walls.

More precisely, we consider a grayscale image $\boldsymbol{\xi} \in [0,1]^{H \times W}$ as input. For instance, if we consider the MNIST dataset [23], $(H, W) = (28, 28)$. We note that our analysis can be easily extended to RBG images. The convolutional layer $\mathcal{C}(\cdot)$ consists of V filters $\mathbf{F} = (\mathbf{F}_1, \ldots, \mathbf{F}_V)$, represented as a collection of V matrices of shape $k \times k$.

Formally, the output of the convolution step, $\mathbf{A} := \mathcal{C}(\boldsymbol{\xi}) \in \mathbb{R}^{V \times h \times w}$, is given by:

$$\forall v \in [V], \ \forall (i,j) \in [h] \times [w], \qquad \mathbf{A}_{i,j}^{(v)} = \sum_{p,q=1}^{k} \boldsymbol{\xi}_{i+p-1,j+q-1} \mathbf{F}_{p,q}^{(v)}, \tag{1}$$

where $(h, w) = (H - k + 1, W - k + 1)$.

In practice, the filter weights are initialized randomly, typically i.i.d. uniform or Gaussian with proper scaling. There are two main trends on how to scale the variance, either Glorot [17] (also called Xavier), or He [18]. The later with uniform distribution is default for the CNN layer used in PyTorch. However, we assume from now on i.i.d. Gaussian $\mathcal{N}(0, \tau^2)$ initialization in our analysis for mathematical convenience.

After the convolution step, we apply a ReLU non-linearity, denoted by $\sigma :=$ $\max(0, \cdot)$. We define the *rectified activation maps* $\mathbf{B} := \sigma(\mathbf{A}) \in \mathbb{R}^{V \times h \times w}$, where σ is applied coordinate-wise. Next, we consider a down-sampling layer, here a $(k' \times k')$ max pooling $\mathcal{M}(\cdot)$. One can see that the output of the max pooling, $\mathbf{C} := \mathcal{M}(\mathbf{B}) \in \mathbb{R}^{V \times h' \times w'}$, is given by:

$$\forall v \in [V], \ \forall (i',j') \in [h'] \times [w'], \ \mathbf{C}_{i',j'}^{(v)} = \max\left(\mathbf{B}_{k'(i'-1)+1:k'i', k'(j'-1)+1:k'j'}^{(v)}\right), \tag{2}$$

where $(h', w') = \frac{1}{k'}(h, w)$. Note that we assume k' to divide h and w for simplicity.

Finally, let us describe recursively the fully-connected part of [**CNN**], denoted by $\mathcal{F}(\cdot)$:

$$\begin{aligned} \mathcal{F}: \mathbb{R}^{Vh'w'} &\to \mathbb{R} \\ \mathbf{C}' &\mapsto h_L(\mathbf{C}') \end{aligned} \quad \text{with} \quad \begin{cases} h_0(\mathbf{x}) = \mathbf{x}, \\ h_\ell(\mathbf{x}) = \mathbf{W}^{(\ell)} \sigma(h_{\ell-1}(\mathbf{x})) & \text{for } \ell \in [L], \end{cases} \tag{3}$$

where $\mathbf{W}^{(\ell)} \in \mathbb{R}^{d_\ell \times d_{\ell-1}}$ is a weight matrix connecting layer $(\ell - 1)$ and ℓ with $\ell \in [L]$ and d_ℓ the size of layer ℓ. Note that we set $d_0 = Vh'w'$, and $d_L = 1$, since we see the output of our model as the un-normalized logit associated to a given class of a prediction problem. We also, denote by $\mathbf{a}^{(\ell)} := h_\ell(\mathbf{C}')$ the non-rectified activation of layer ℓ and $\mathbf{r}^{(\ell)} := \sigma(\mathbf{a}^{(\ell)})$ its rectified counterpart.

Summary. The model we consider can be described concisely as $\mathcal{F}(\mathcal{M}(\sigma(\mathcal{C}(\cdot))))$. As explained in introduction, given the nature of CAM-based explanations, **it is convenient to split [CNN] in two functions f and g** for the computations of the next Sect. 2.2. More precisely, we write

$$\begin{aligned} [\mathbf{CNN}]: [0,1]^{H \times W} &\to \mathbb{R} \\ \boldsymbol{\xi} &\mapsto f \circ g(\boldsymbol{\xi}) \end{aligned} \quad \text{with} \quad \begin{cases} g(\boldsymbol{\xi}) := \sigma(\mathcal{C}(\boldsymbol{\xi})) \\ f(\mathbf{C}) := \mathcal{F}(\mathcal{M}(\mathbf{C})). \end{cases} \tag{4}$$

Recall that we refer to Fig. 2 for an illustration.

2.2   Closed-Form Expression

The original idea of CAM [36] was limited to computing the saliency map as a linear combination of the feature maps in the last convolutional layer when $L = 1$. Later, GradCAM [29] removed the architecture constraints by computing the average gradient of each feature map with respect to $\mathbf{B}$. In our notation, we have:

Definition 1 (GradCAM). *For an input ξ and model* [**CNN**], *the GradCAM feature scores are given by*

$$[\mathbf{GC}] := \sigma \left(\sum_{v=1}^{V} \alpha_v \mathbf{B}^{(v)} \right) \in \mathbb{R}_+^{h \times w} \,,$$

where each $\alpha_v := \mathrm{GAP}\left(\nabla_{\mathbf{B}^{(v)}} f(\mathbf{B}) \right) \in \mathbb{R}$. Here, GAP denotes the global average pooling, that is, the average of all values, and σ the ReLU as before.

Definition 1 is of course to be taken coordinate-wise. We note that, in practice, [**GC**] is up-sampled and normalized to produce a saliency map with the *same shape* as the input image. To be more precise, what we define as [**GC**] is the middle panel of Fig. 1, whereas the final user will nearly always visualize the right panel. The most important thing to notice in Definition 1 is that α is a **global** coefficient.

We now show why this can be an issue. Looking at Definition 1, whenever the underlying model is not too complicated, one can actually hope to derive a closed-form expression for the feature importance scores of [**GC**] as a function of the model's parameters. This is achieved by:

Proposition 1 (α coefficients for GradCAM, $V = 1$). *Recall that the* **a** *vectors denote the non-rectified activation and* **W** *the weights of the linear part of* [**CNN**]. *Then, for input ξ, the* [**GC**] *coefficient α is given by*

$$\alpha = \frac{1}{hw} \sum_{i,j=1}^{h',w'} \sum_{i_1,\ldots,i_{L-1}=1}^{d_1,\ldots,d_{L-1}} \mathbb{1} a_{i_1}^{(1)}, \ldots, a_{i_{L-1}}^{(L-1)} > 0 \prod_{p=1}^{L} (\mathbf{W}_{i_p, i_{p-1}}^{(p)})^{\top} \,,$$

where we set $i_0 := (i,j)$ and $i_L = 1$.

From Proposition 1, we immediately deduce a closed-form expression for GradCAM explanations. We note that Proposition 1 can be readily extended to an arbitrary number of filters $V > 1$, in which case the **a** and **W** should be interpreted as corresponding to the relevant $v \in [V]$.

The proof of Proposition 1 can be found in Appendix C. In Appendix A, we describe mathematically several other CAM-based methods in the setting of [**CNN**]: XGradCAM [13], GradCAM++ [8], HiResCAM [11], ScoreCAM [32], Opti-CAM [34] and AblationCAM [10]. A close inspection of these definitions reveals that they also use global weighting coefficients applied to the corresponding activation maps, with the notable exception of HiResCAM.

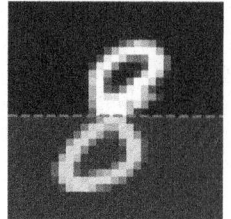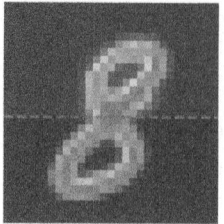

Fig. 3. Illustration of Theorem 1 on an MNIST [23] digit (*left panel*). We set to zero the lower part of $\mathbf{W}$ for [**CNN**], initialize the filter values and remaining weights to i.i.d. $\mathcal{N}(0,1)$, and run GradCAM to get a saliency map (*right panel*). Even though our network does not see the red part of the image, **GradCAM does highlight some pixels in this area**, as predicted by Theorem 1.

2.3 Theoretical Analysis

Leveraging the results of Sect. 2.2, we are able to describe precisely the behavior of GradCAM at initialization for [**CNN**], specifically when the classifier part of our model comprises a single layer ($L = 1$). This analysis is justified by existing works [2,3,12,24,37] showing that, in certain regimes, neural networks stay "near initialization" during training. As announced, we conduct this analysis when the network does not have access to the lower part of the image. Our main result is:

Theorem 1 (Expected GradCAM scores, $L = 1$, masked [CNN]). *Let $\boldsymbol{\xi} \in [0,1]^{H \times W}$ be an input image. Let $\mathbf{m} := \boldsymbol{\xi}_{i:i+k-1,j:j+k-1}$ be the patch of $\boldsymbol{\xi}$ corresponding to index $(i,j) \in [h] \times [w]$. Assume that h' is even, and $\mathbf{W}_{:,-\frac{h'}{2}:,:} = 0$. Assume that the filter values and the non-zero weights are initialized i.i.d. $\mathcal{N}\left(0,\tau^2\right)$. Then, if the number of filters V is greater than 20, we have the following expected lower bound on the GradCAM explanation for pixel (i,j):*

$$\mathbb{E}\left[[\mathbf{GC}]_{i,j}\right] = \mathbb{E}\left[\sigma\left(\sum_{v=1}^{V} \alpha_v \mathbf{B}_{i,j}^{(v)}\right)\right] \geq \frac{V-20}{\sqrt{V}}\sqrt{\frac{h'w'}{16\pi}\frac{\tau^2}{hw}}\,\|\mathbf{m}\|_2\,, \qquad (5)$$

where the expectation in the previous inequality is taken with respect to initialization of the filters and the remaining weights of the linear layer.

Setting $\mathbf{W}_{:,-\frac{h'}{2}:,:}$ to 0 disables the weights within $\mathbf{W}$ that are connected to the lower half part of the activation map $\mathbf{C}_{:,-\frac{h'}{2}:,:}$, effectively preventing [**CNN**] from accessing the lower half of $\mathbf{C}$. In turn, [**CNN**] does not see the lower half of $\boldsymbol{\xi}$, up to side effects. The main consequence of Theorem 1 is that, when the number of filters associated to the class to explain is large enough, $[\mathbf{GC}]_{i,j}$ is positive in expectation if some pixels are activated in the receptive field associated to (i,j). Thus **GradCAM highlights all parts of the image where there is some "activity," even though this information is not used by the network in the end.** We illustrate Theorem 1 in Fig. 3. The main limitation of this analysis is its focus on the behavior at initialization: we investigate in the following whether this behavior also happens after training. Another limitation is the restriction to a single linear layer, but we note that taking $L = 1$ in the fully connected part of [**CNN**] is a dominant approach since ResNet [19].

The proof of Theorem 1 can be found in Appendix D. The key ingredient of the proof is obtaining a probabilistic control of $\sum_q \alpha_q \mathbf{B}^{(q)}$. We note that a similar analysis is possible for other expressions of α, thus other CAM-based methods.

3   Experiments

We know ask the following question: are the consequences of Theorem 1 true after training, and for a more realistic model? To this extent, we train a CNN-based model which by construction cannot access some specified part of the input which we call the *dead zone* (see Fig. 4, details in Sect. 3.1). Clearly, since the dead zone does not influence the output, it should not contain positive model explanations. To test whether this is true, we create two datasets (Sect. 3.2). Each item of the first one is composed of two images from ImageNet with the same label in both the seen and the unseen part of the image. The second dataset is built using generative models on the same categories with two objects in each image located in the seen and unseen part as well. We then check whether CAM-based methods wrongly highlight areas in the dead zone in Sect. 3.3.

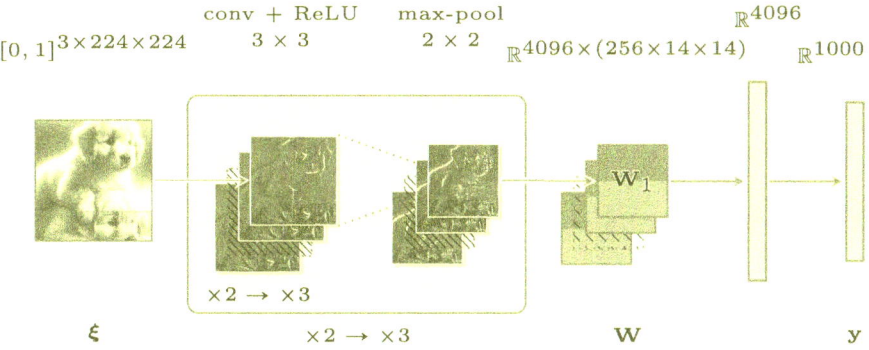

Fig. 4. Our masked VGG16-based model trained on ImageNet with 87.0% top-5 accuracy. The down weights $\mathbf{W}$ are set to 0 and not updated during training. Only the up weights $\mathbf{W}_{:,:,:5,:}$ and the other parameters undergo training. This setting implies that every red part in the channels does not impact the prediction scores, meaning that they are not used. Symbol $\times 2 \rightarrow \times 3$ means the model first uses the green block twice, with each time having 2 consecutive convolutions. Then, it uses the green block three times, with each time having 3 consecutive convolutions. There is no max pooling after the last convolution. (Color figure online)

3.1 Model

Model Definition. The CNN used in our experiments is a modification of a classical VGG16 architecture [31] which we call [**VGG**]. Whereas the original VGG16 model is composed of 5 convolutional blocks including either 2 or 3 convolutional layers with ReLU and max pooling, followed by 3 dense layers. In [**VGG**] we remove the last max pooling (in the fifth convolutional block) and we further apply a mask on selected weights of the first dense layer so the layer can not see the lower part of the activation maps, see Fig. 4 for more details.

Masking. We forbid the network from seeing the dead zone in a very simple way: in the first dense layer $\mathbf{W}$, which has size $4096 \times (256 \times 14 \times 14)$, we permanently set to 0 a band of height 9 corresponding to the lower weights. Formally, this means setting $\mathbf{W}_{:,:,-9:,:} = 0$, which is denoted in red above $\mathbf{W}$ in Fig. 4. Effectively, we are building a wall that stops all information flowing from the last convolutional layer to the remainder of the network. Since the weights $\mathbf{W}$ are directly connected to the final activation map $\mathbf{B} \in \mathbb{R}_{+}^{256 \times 14 \times 14}$, this masking effectively zeroes out the lower sections in each channel denoted by $\mathbf{B}_{:,-9:,:}$. We can trace back the zeroed activations in $\mathbf{B}$ to the preceding activation map $\mathbf{C}$, pinpointing the exact patches in $\mathbf{C}$ that correspond (after convolution) to the features observed in the zeroed activation of $\mathbf{B}$. Because of the side effects in the computation of convolutions, this area of $\mathbf{C}$ is slightly smaller: some pixel activation will still play a role in the model's prediction. Repeating this process until we reach the original image yields a dead zone of height 54 pixels, highlighted in red above $\boldsymbol{\xi}$ in Fig. 4, which covers 24% of the image area. As we mentioned earlier, the other main difference with VGG16 is the removing of the final max pooling layer. This leads to a larger activation layer, allowing us to set weights to zero without hindering too much the network's ability, see Fig. 5. We note that [**VGG**] bears a strong resemblance to [**CNN**]. The main difference is that the convolutional layer of [**CNN**] is replaced by several convolutional blocks in [**VGG**], see Fig. 4.

Training. We train [**VGG**] on Imagenet-1k [9] using classical data augmentation recipe, *i.e.*, random flip and random crop. As optimization algorithm, we use stochastic gradient descent with momentum, weight decay, and a learning rate scheduler. To observe the slight accuracy drop induced by masking a significant part of images, we train a baseline model without masking. We report the train loss and the validation accuracy across training in Fig. 5.

Comparison to SOTA. We also compare the validation top-1 and top-5 accuracy of the VGG16 model found in the PyTorch repository. Our [**VGG**] without max pooling and no masking offers the same performance: 71.5% top-1 and 90.4% top-5 accuracy on the validation set. As we mention in Fig. 5, our model [**VGG**] with masking has lower performance, which is expected as a fourth of the input image, $\boldsymbol{\xi}_{:,171:224,:}$, is unseen by the model. We obtain 66.5%, resp. 71.5%, top-1 and 87.0%, resp. 90.4%, top-5 accuracy on the validation set for our masked

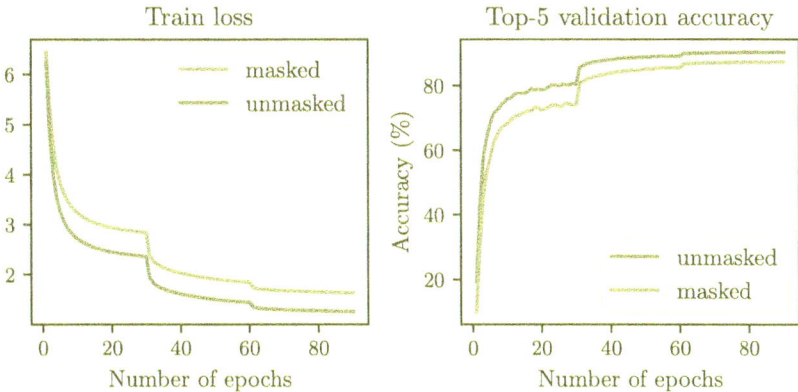

Fig. 5. Plots of the training loss and the top-5 validation accuracy of our unmasked and masked VGG16-like models on Imagenet-1k (2012) [9]. The unmasked [**VGG**] (baseline) and our masked [**VGG**] yield 87.0%, resp. 90.4% top-5 accuracy and 66.5% resp. 71.5% top-1 accuracy on the validation set.

[**VGG**], resp. unmasked [**VGG**]. Nevertheless, we see that [**VGG**] is a **realistic network able to predict ImageNet classes with reasonable accuracy**. We believe that the drop in accuracy is only minor because ImageNet images are centered, and there is enough information in the upper part of the image to achieve near-perfect prediction.

3.2   Proposed Datasets

Objective. To assess how much CAM-based saliency maps emphasize irrelevant areas of an image, we introduce two new datasets in which we control the positions of the image elements using two techniques: cutmix [33] and generative model. More precisely, we produce two datasets, called STACK-MIX and STACK-GEN. Where each image contains two objects, one in the bottom part of the image which is the dead zone for [**VGG**], and the second subject at the top of the image. Therefore, the subject at the center of the image will be mainly responsible for the top-1 predicted score by our masked [**VGG**].

STACK-MIX. We first generate labels for our datasets by randomly sampling 100 classes from the 398 first labels of Imagenet, which corresponds to animals. The first dataset, called STACK-MIX, consists of 100 images featuring one image from each of the 100 classes. Each example ξ is created by mixing, in a cutmix [33] fashion, two images (ξ_1, ξ_2) with same label and sampled randomly in the **validation set** of Imagenet as follows:

$$\xi := \begin{pmatrix} (\xi_1)_{:,\,:170,\,:} \\ (\xi_2)_{:,\,171:224,\,:} \end{pmatrix} \in [0,1]^{3 \times 224 \times 224}, \tag{6}$$

meaning that we create a composite image ξ by superposing an upper vertical slice, taken from the top region of ξ_1 with size $3 \times 170 \times 224$, with a lower

Fig. 6. Sampled images from both of our datasets, *i.e.*, STACK-MIX and STACK-GEN.

vertical slice, taken from the bottom region of ξ_2 with size $3 \times 54 \times 224$. Finally, the quality of the generated images is verified through manual inspection. This dataset lacks realism due to the distinct separation between the two subjects. We address this issue with the help of generative models.

STACK-GEN. The second dataset, called STACK-GEN, consists of 100 images featuring one image from each of the same 100 classes. It was generated using ChatGPT + DALL·E 3 [7,27] by sampling prompts of the following form: "A photo of {animal name} stacked on top of {same animal name}". The word "stacked" determines the positions of the subjects in the generated image, which proceeds as follows: first, ChatGPT refines the original prompt to enhance its suitability for DALL·E 3, then the image is generated. We then preprocess the generated images by selectively editing them to minimize the background and centering the focus on the two animals. This editing involves cropping the images to a 1:1 ratio, ensuring one animal is predominantly within the dead zone as defined by our [**VGG**], while the other is positioned in the upper part of the new image. Figure 6 shows examples of the created images. Note that both datasets are provided in the supplementary material and online.[2]

3.3   Results

For our [**VGG**], we generate saliency maps from various CAM-based methods on our two datasets, STACK-MIX and STACK-GEN, using the predicted category for each example. We used publicly available implementations whenever possible. Regarding Opti-CAM, since our model differs from the one described by [34], we have adjusted the learning rate and number of epochs of the optimization step to achieve a low average drop, as described in the original paper. For each method, we measure how much of the CAM-based saliency maps emphasize the

---

[2] https://github.com/MagamedT/cam-can-see-through-walls.

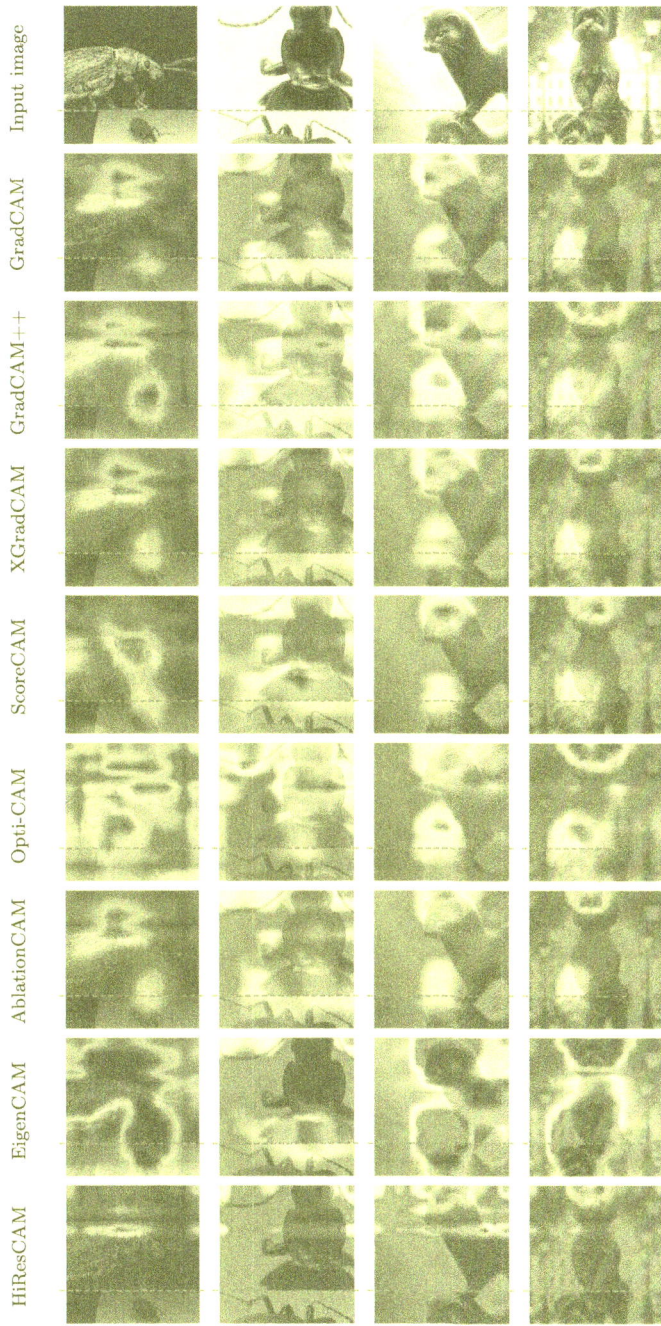

Fig. 7. Saliency maps given by the considered CAM-based methods for [**VGG**]. With the notable exception of HiResCAM, all method highlight parts of images from STACK-GEN and STACK-MIX which are unseen by the network. The lower part in red is unseen by the model.

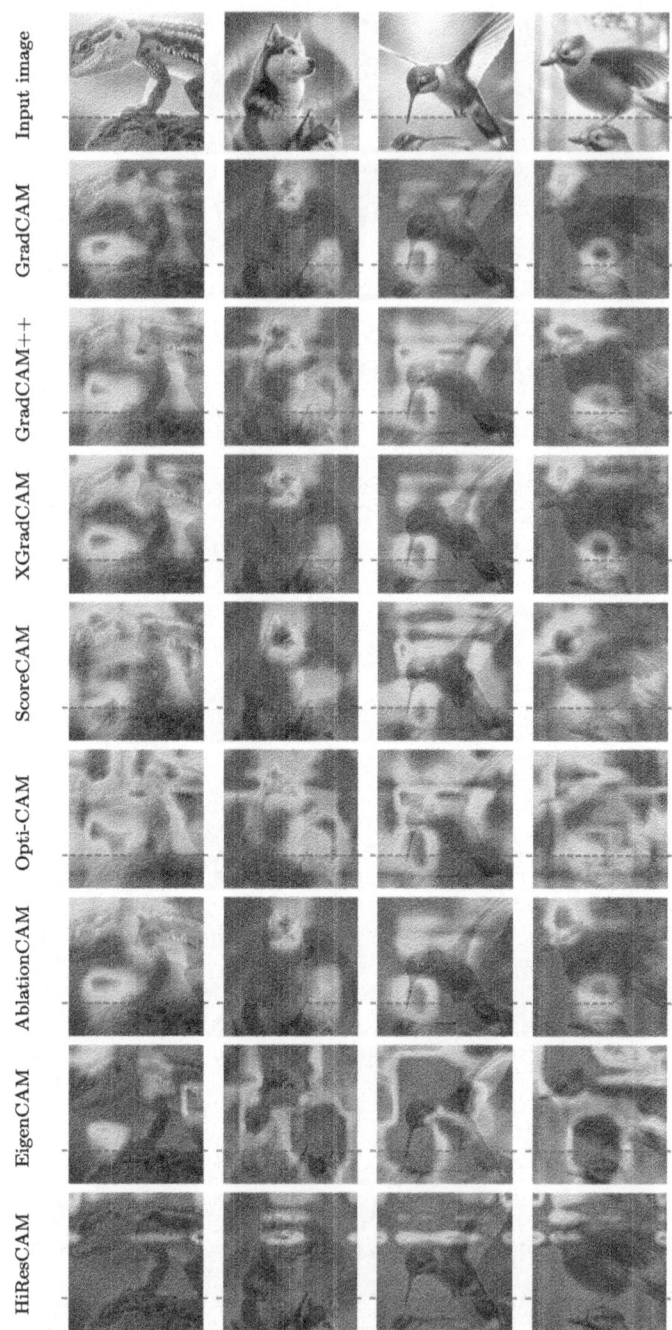

Fig. 8. Saliency maps given by the considered CAM-based methods for [**VGG**]. With the notable exception of HiresCAM, they all highlight parts of images from STACK-GEN which are unseen by the network (this is denoted by the red, rectangular shape in the lower part of the image).

Table 1. Activity in the unseen part of the image, measured by $\mu(\cdot) \times 100$ for several CAM-based methods on both proposed datasets (only images in the validation set are considered).

methods	STACK-MIX ↓	STACK-GEN ↓
GradCAM [29]	22.7 ± 13.4	21.6 ± 11.6
GradCAM++ [8]	28.8 ± 8.1	28.5 ± 7.9
XGradCAM [13]	23.8 ± 9.0	22.8 ± 9.0
ScoreCAM [32]	19.9 ± 10.3	18.5 ± 10.6
Opti-CAM [34]	32.7 ± 7.9	32.0 ± 7.8
AblationCAM [10]	21.0 ± 9.9	20.8 ± 9.6
EigenCAM [4]	51.7 ± 19.7	55.8 ± 21.6
HiResCAM [11]	0.0 ± 0.0	0.0 ± 0.0

unseen part, *i.e.*, the dead zone. We use the metric $\mu(\cdot)$ defined for a upscaled saliency map $\mathbf{\Lambda} \in \mathbb{R}_+^{224 \times 224}$ as follows:

$$\mu(\mathbf{\Lambda}) := \frac{\|\mathbf{\Lambda}_{171:224, :}\|_2}{\|\mathbf{\Lambda}\|_2}, \tag{7}$$

where $\|\cdot\|_2$ is the ℓ^2-norm and the lower part of the image $\boldsymbol{\xi}_{:, 171:224, :}$ is unseen by our [**VGG**]. We note that for a saliency map $\mathbf{\Lambda}$, the lower $\mu(\mathbf{\Lambda})$, the better.

The results can be found in Table 1, Fig. 7 and 8. We observe that every CAM-based method, except HiResCAM, highlights unseen parts of an image to some extent. Moreover, the observations are consistent over both datasets.

We believe that HiResCAM avoids this problem because of how its weighting coefficients are computed. Following the notation of Sect. 2.1, these can be written $\alpha_v^{(4)} := \nabla_{\mathbf{B}^{(v)}} f(\mathbf{B}) \in \mathbb{R}^{14 \times 14}$, and are applied *globally* to $\mathbf{B}^{(v)}$ (see Definition 4 in the Appendix for more details). Because of the masking used in our model, the lower part of $\alpha_v^{(4)}$ is zeroed out, and therefore HiResCAM does not show activity in the lower part of the image.

We notice that, while at first glance HiResCAM appears to perform well in our setting, it has another issue: since $(\alpha_v^{(4)})_{-9:,:} = 0$, the upscaled HiResCAM's saliency map $\mathbf{\Lambda} \in \mathbb{R}_+^{224 \times 224}$ will be zero out in a larger area than the dead zone. Namely, $\mathbf{\Lambda}_{89:224,:} = 0$ which represents 61% of the input image area compared to the 24% of the deadzone. This issue can be observed in Fig. 7 and 8.

4   Conclusion

In this paper, we looked into several CAM-based methods, with a particular focus on GradCAM. We showed that they can highlight parts of the input image that are provably not used by the network. This was also showed theoretically, looking at the behavior of GradCAM for a simple, masked CNN at initialization:

the saliency map is positive in expectation, even in areas which are unseen by the network. Experimentally, this phenomenon appears to remain true, even on a realistic network trained to a good accuracy on ImageNet.

As future work, we would like to extend the theory to a ResNet-like architecture and other CAM-based methods, such as LayerCAM [21]. We also would like to multiply the number of images in our two new datasets, with the hope that this framework can become a standard check for saliency maps explanations.

Acknowledgements. This work was funded in part by the French Agence Nationale de la Recherche (grant number ANR-19-CE23-0009-01 and ANR-21-CE23-0005-01). Most of this work was realized while DG was employed at Université Côte d'Azur. We thank Jenny Benois-Pineau for her valuable insights.

References

1. Adebayo, J., Gilmer, J., Muelly, M., Goodfellow, I., Hardt, M., Kim, B.: Sanity checks for saliency maps. In: Advances in Neural Information Processing Systems (2018)
2. Allen-Zhu, Z., Li, Y., Song, Z.: A convergence theory for deep learning via over-parameterization. In: Proceedings of the 36th International Conference on Machine Learning (2019)
3. Allen-Zhu, Z., Li, Y., Song, Z.: On the convergence rate of training recurrent neural networks. In: Advances in Neural Information Processing Systems (2019)
4. Bany Muhammad, M., Yeasin, M.: Eigen-CAM: visual explanations for deep convolutional neural networks. SN Comput. Sci. **2**(1), 47 (2021)
5. Beauchamp, M.: On numerical computation for the distribution of the convolution of N independent rectified Gaussian variables. Journal de la Société Française de Statistique (2018)
6. Benítez, J.M., Castro, J.L., Requena, I.: Are artificial neural networks black boxes? IEEE Trans. Neural Netw. **8**(5), 1156–1164 (1997)
7. Brown, T., et al.: Language models are few-shot learners. In: Advances in Neural Information Processing Systems (2020)
8. Chattopadhay, A., Sarkar, A., Howlader, P., Balasubramanian, V.N.: Grad-CAM++: generalized gradient-based visual explanations for deep convolutional networks. In: IEEE Winter Conference on Applications of Computer Vision (2018)
9. Deng, J., Dong, W., Socher, R., Li, L.J., Li, K., Fei-Fei, L.: ImageNet: a large-scale hierarchical image database. In: IEEE Conference on Computer Vision and Pattern Recognition (2009)
10. Desai, S., Ramaswamy, H.G.: Ablation-CAM: visual explanations for deep convolutional network via gradient-free localization. In: IEEE Winter Conference on Applications of Computer Vision (WACV) (2020)
11. Draelos, R.L., Carin, L.: Use HiResCAM instead of Grad-CAM for faithful explanations of convolutional neural networks. arxiv preprint arxiv:2011.08891 (2021)
12. Du, S., Lee, J., Li, H., Wang, L., Zhai, X.: Gradient descent finds global minima of deep neural networks. In: Proceedings of the 36th International Conference on Machine Learning (2019)
13. Fu, R., Hu, Q., Dong, X., Guo, Y., Gao, Y., Li, B.: Axiom-based grad-CAM: towards accurate visualization and explanation of CNNs. In: 31st British Machine Vision Conference (2020)

14. Fukushima, K.: Neocognitron: a self-organizing neural network model for a mechanism of pattern recognition unaffected by shift in position. Biol. Cybern. **36**(4), 193–202 (1980)
15. Garreau, D., Mardaoui, D.: What does LIME really see in images? In: International Conference on Machine Learning. PMLR (2021)
16. Ghorbani, A., Abid, A., Zou, J.: Interpretation of neural networks is fragile. In: Proceedings of the AAAI Conference on Artificial Intelligence (2019)
17. Glorot, X., Bengio, Y.: Understanding the difficulty of training deep feedforward neural networks. In: Proceedings of the 13th International Conference on Artificial Intelligence and Statistics (2010)
18. He, K., Zhang, X., Ren, S., Sun, J.: Delving deep into rectifiers: surpassing human-level performance on ImageNet classification. In: Proceedings of the IEEE International Conference on Computer Vision (2015)
19. He, K., Zhang, X., Ren, S., Sun, J.: Deep residual learning for image recognition. In: Proceedings of the IEEE Conference on Computer Vision and Pattern Recognition (2016)
20. Heo, J., Joo, S., Moon, T.: Fooling neural network interpretations via adversarial model manipulation. In: Advances in Neural Information Processing Systems (2019)
21. Jiang, P.T., Zhang, C.B., Hou, Q., Cheng, M.M., Wei, Y.: LayerCAM: exploring hierarchical class activation maps for localization. IEEE Trans. Image Process. **30**, 5875–5888 (2021)
22. Kindermans, P.J., et al.: The (un)reliability of saliency methods. In: Explainable AI: Interpreting, Explaining and Visualizing Deep Learning (2019)
23. LeCun, Y., Bottou, L., Bengio, Y., Haffner, P.: Gradient-based learning applied to document recognition. Proc. IEEE **86**(11), 2278–2324 (1998)
24. Lee, J., et al.: Wide neural networks of any depth evolve as linear models under gradient descent. In: Advances in Neural Information Processing Systems (2019)
25. Linardatos, P., Papastefanopoulos, V., Kotsiantis, S.: Explainable AI: a review of machine learning interpretability methods. Entropy **23**(1), 18 (2021)
26. Lipton, Z.C.: The Mythos of Model Interpretability: In Machine Learning, the Concept of Interpretability is Both Important and Slippery (2018)
27. Ramesh, A., et al.: Zero-shot text-to-image generation. In: International Conference on Machine Learning (2021)
28. Ribeiro, M.T., Singh, S., Guestrin, C.: "Why should i trust you?": explaining the predictions of any classifier. In: Proceedings of the 22nd ACM SIGKDD International Conference on Knowledge Discovery and Data mining (2016)
29. Selvaraju, R.R., Cogswell, M., Das, A., Vedantam, R., Parikh, D., Batra, D.: Grad-CAM: visual explanations from deep networks via gradient-based localization. In: 2017 IEEE International Conference on Computer Vision (ICCV) (2017)
30. Simonyan, K., Vedaldi, A., Zisserman, A.: Deep Inside Convolutional Networks: Visualising Image Classification Models and Saliency Maps. arXiv preprint arXiv:1312.6034 (2013)
31. Simonyan, K., Zisserman, A.: Very deep convolutional networks for large-scale image recognition. In: ICLR (2015)
32. Wang, H., et al.: Score-CAM: score-weighted visual explanations for convolutional neural networks. In: 2020 IEEE/CVF Conference on Computer Vision and Pattern Recognition Workshops (2020)
33. Yun, S., Han, D., Chun, S., Oh, S., Yoo, Y., Choe, J.: CutMix: regularization strategy to train strong classifiers with localizable features. In: IEEE/CVF International Conference on Computer Vision (2019)

34. Zhang, H., Torres, F., Sicre, R., Avrithis, Y., Ayache, S.: Opti-CAM: optimizing saliency maps for interpretability. arXiv preprint arXiv:2301.07002 (2023)
35. Zhang, Y., Tiňo, P., Leonardis, A., Tang, K.: A Survey on Neural Network Interpretability. IEEE Trans. Emerg. Top. Comput. Intell. **5**(5), 726–742 (2021)
36. Zhou, B., Khosla, A., Lapedriza, A., Oliva, A., Torralba, A.: Learning deep features for discriminative localization. In: Proceedings of the IEEE Conference on Computer Vision and Pattern Recognition (2016)
37. Zou, D., Cao, Y., Zhou, D., Gu, Q.: Gradient descent optimizes over-parameterized deep ReLU networks. Mach. Learn. **109**, 467–492 (2020)

Making Alice Appear Like Bob: A Probabilistic Preference Obfuscation Method For Implicit Feedback Recommendation Models

Gustavo Escobedo[1]([✉])[iD], Marta Moscati[1][iD], Peter Muellner[4][iD],
Simone Kopeinik[4][iD], Dominik Kowald[4][iD], Elisabeth Lex[3][iD],
and Markus Schedl[1,2][iD]

[1] Johannes Kepler University Linz, Linz, Austria
{gustavo.escobedo,marta.moscati,markus.schedl}@jku.at
[2] Linz Institute of Technology, Linz, Austria
[3] Graz University of Technology, Graz, Austria
elisabeth.lex@tugraz.at
[4] Know-Center GmbH, Graz, Austria
{pmuellner,skopeinik,dkowald}@know-center.at

Abstract. Users' interaction or preference data used in recommender systems carry the risk of unintentionally revealing users' private attributes (e.g., gender or race). This risk becomes particularly concerning when the training data contains user preferences that can be used to infer these attributes, especially if they align with common stereotypes. This major privacy issue allows malicious attackers or other third parties to infer users' protected attributes. Previous efforts to address this issue have added or removed parts of users' preferences prior to or during model training to improve privacy, which often leads to decreases in recommendation accuracy. In this work, we introduce SBO, a novel probabilistic obfuscation method for user preference data designed to improve the accuracy–privacy trade-off for such recommendation scenarios. We apply SBO to three state-of-the-art recommendation models (i.e., BPR, MultVAE, and LightGCN) and two popular datasets (i.e., MovieLens-1M and LFM-2B). Our experiments reveal that SBO outperforms comparable approaches with respect to the accuracy–privacy trade-off. Specifically, we can reduce the leakage of users' protected attributes while maintaining on-par recommendation accuracy.

Keywords: Recommender Systems · Privacy · Obfuscation · Debiasing · Implicit Feedback

1 Introduction

Recommender systems (RSs) provide relevant content to their users, commonly based on large collections of users' historical interaction data with items, using

A. Bifet et al. (Eds.): ECML PKDD 2024, LNAI 14947, pp. 349–365, 2024.
https://doi.org/10.1007/978-3-031-70368-3_21

collaborative filtering techniques. The historical data used for training of and inference in recommendation models consists of interactions of users with several items and hence represent the preference of each user. While such user-item interaction data is necessary to create an accurate recommendation model, it may also reflect inherent biases in user behavior, which are subsequently encoded or even amplified during model training. For instance, users of music recommender systems from different countries and of different genders tend to prefer different artists and genres [15,17,29], leading to a correlation between users' sensitive attributes and behavioral patterns encoded in their interaction data.

As a consequence, this leads to two important risks: possible privacy breaches and stereotypical or even unfair recommendations. As for *privacy* issues, users' protected information can be leaked when untrusted third parties get access to the users' interaction data [15] or internal user representation of the model [9,33]. For instance, for a group of users that is highly correlated with a list of stereotypical items, private attributes (e. g., gender, occupation, or country) can be unveiled through malicious attacks on the model or the data [2]. Concerning *unfairness*, recommendation models trained on interaction data that is correlated with sensitive user attributes have been shown to impact the quality of recommendations across different user groups distinguished by these attributes [23].

Both problems (privacy concerns and fairness issues) are intertwined because they originate from the correlations between users' interaction behaviors and their sensitive attributes. To mitigate them, several privacy-enhancing methods have been introduced, targeting different stages of the recommendation model's training process (pre-, in-, and post-processing) [27]. Among the pre-processing methods, user preference obfuscation approaches have been proposed to impede malicious attacks that aim at the leakage of private user attributes before training. These approaches primarily consist of adding or removing carefully selected items from users' preference data and have specifically been applied to user-item matrices containing ratings [31,32].

In the work at hand, we introduce *Stereotypicality-Based Obfuscation* (SBO), a probabilistic user preference obfuscation method to counteract inference attacks against private user attributes. Unlike existing methods, SBO selects users and items to obfuscate in a probabilistic fashion, using novel stereotypicality metrics. This limits the number of users whose items require obfuscation and adjusts the selection probability of non-stereotypical items in the sampling process. We demonstrate SBO's performance in terms of recommendation utility and accuracy of an attacker that aims to unveil the users' gender. Experiments with three common recommendation algorithms—BPR-MF, LightGCN, and MultVAE— on two standard recommendation datasets from the movie and music domains— ML-1M (MovieLens) [10] and LFM-2B-100K (Last.fm) [23,30]—showed a favorable accuracy–privacy trade-off of our method.

In the remainder of the paper, we review relevant previous work (Sect. 2), detail the proposed SBO method (Sect. 3), present the setup of our evaluation experiments (Sect. 4), and discuss results (Sect. 5). Ultimately, we summarize our findings and provide an outlook (Sect. 6).

2   Related Work

Related work belongs to two strands of research: privacy-aware RSs (Sect. 2.1) and fairness in RSs through adversarial training (Sect. 2.2). Both can be addressed by altering the user's input data to the RS or the model's latent user representations.

2.1   Privacy-Aware Recommender Systems

RSs typically expose their users to several privacy risks. For example, the disclosure of information that is used to train the recommendation model (e.g., interaction data) [11,40] to third parties, or the inference of information that is not used during model training but correlated with the training data (e.g., gender or age) [36,43].

Various technologies have been employed to address users' privacy concerns, such as homomorphic encryption [14], federated learning [22,24], and differential privacy [25,26]. Homomorphic encryption aims to generate privacy-aware recommendations by utilizing encrypted user data [42]. Federated learning operates under the principle that sensitive user data should remain on the user's device [1]. Lastly, differential privacy (DP) is used to counter privacy risks by incorporating a carefully tuned level of random perturbation into the recommender system [5]. Many works apply DP to protect a user's sensitive attribute. However, malicious parties can still scrutinize the generated recommendations to infer protected attributes [8]. This is the case if non-sensitive interaction data correlates with the user's sensitive attributes and forms distinct patterns that can be uncoded.

For this reason, Weinsberg et al. [36] suggest an approach that detects rating data that is indicative of gender and adds ratings for items indicative of the opposite gender to obfuscate a user's real gender. However, the authors regard the set of items in a user profile as the source of the privacy risk (i.e., the correlation with gender), and their approach leads to a severe drop in recommendation accuracy. In contrast, in the work at hand, we regard the *conjunction* of items in the user profile as the source of the privacy risk, i.e., the correlation of the user's behavioral pattern with gender stereotypes. Additionally, we address the accuracy drop by applying our perturbation mechanism only to users whose behavioral patterns coincide with gender stereotypes.

2.2   Fairness Through Adversarial Training in Recommendation

In the context of RSs, protecting users' privacy often relates to concepts of user fairness [2,4,6,39,41]—a topic of lively interest in research and public communities [3,7,35]. A particular overlap of the two strands exists with so-called fairness through unawareness or fairness through blindness approaches, where "unfair" bias is mitigated by hiding the users' sensitive attributes in the model training process [34]. Thus, privacy and fairness can potentially be ensured if the users' data on protected/sensitive attributes is not encoded in the model.

In RS research, several works use adversarial learning as an in-processing technique [13] to generate feature-independent user embeddings. For instance, to achieve counterfactual fairness, Li et al. [18] apply an adversarial learning module to enforce user embeddings to be independent of the protected attributes. Ganhör et al. [9] and Vassøy et al. [33] add adversarial training to autoencoder-based RSs (e.g., [16]) to remove the implicit information of protected attributes from latent representations of users. Wu et al. [37] use adversarial learning to develop a RS based on two representations of the user: a representation that carries the biased information through sensitive attributes and a bias-free representation that only encodes user interests. Wu et al. [38] develop a graph-based adversarial learning module to increase the fairness of recommendations. More similar to our work, Weinsberg et al. [36] and Strucks et al. [32] use obfuscation to achieve privacy; Slokom et al. [31] show that obfuscation also impacts the fairness of recommendations, while Lin et al. [21] use obfuscation to debias gender from RSs. In contrast to prior works, the work at hand introduces the usage of the user's attribute-specific stereotypicality of items for the probabilistic selection of the data to obfuscate.

3   Methodology

The core idea of the proposed SBO method is to reduce the *stereotypicality* of the users' preferences by applying item obfuscation (imputation and/or removal) at the user level. For this purpose, we first define an item stereotypicality score (I_{Ster}) based on the item's group inclination (IGI). The IGI value indicates how likely it is that a user of a given group consumes a certain item. Then, we use the I_{Ster} values to establish the user's stereotypicality (U_{Ster}) from the interaction data, which enables us to determine each user's degree of stereotypicality concerning the group to which the user belongs. For instance, a male user who predominantly listens to male-associated music tracks will obtain a high user stereotypicality score. U_{Ster} is then used to identify suitable candidates for obfuscation according to a given threshold. For each candidate user selected for obfuscation, we sample a number of items proportional to a fixed percentage of the number of items the user interacted with and apply obfuscation operations on the sample.

We formally present our method in the subsequent sections, focusing on obfuscating gender[1] information because it is a common target for attacks. Note that our method can be easily adapted for other protected attributes. We start by defining the different stereotypicality scores for users and items. Then, we formulate SBO with the supported sub-sampling and obfuscation strategies.

3.1   Item's Group Inclination

We split the set of unique users $U = \{u_1, \ldots, u_{|U|}\}$ in k groups $\{U_g\}_{g=1}^k$, where $U_g \subset U$ and $\bigcap_{g=1}^k U_g = \emptyset$, based on the $k \geq 2$ mutually exclusive values of

---

[1] In this work, we consider only two possible values of gender. However, we acknowledge that the assumption of binary gender is an over-simplification.

the categorical protected attribute p associated with each user. In this work, we split the original set of users by their associated gender. Therefore, we define two groups, U_m and U_f, corresponding to the male and female users, respectively.

Items present different degrees of association to different user groups. Therefore, for each element in the set of unique items $V = \{v_1, \ldots, v_{|V|}\}$, we define the item inclination towards the user group U_g as the fraction between the number of users in U_g that interacted with item v, and the total number of users in U_g. Therefore, given the set of observed interactions $L_{obs} \subset U \times V$, $IGI(v, U_g)$ is given by:

$$IGI(v, U_g) = \frac{|\{u : (u, v) \in L_{obs}\}|}{|U_g|} \tag{1}$$

3.2 Item Stereotypicality

In order to determine if an item is a good candidate for obfuscation, we introduce the item stereotypicality (I_{Ster}), which relates the IGI values of the same item v (Eq. 1) across two user groups. The closer the values of inclination across groups, the closer to zero the value of I_{Ster}. This also indicates that the items closer to the extremes are the most stereotypical ones. The definition of I_{Ster} and its dependence on U_g and $U_{g'}$ is given by:

$$I_{Ster}(v, U_g, U_{g'}) = \frac{IGI(v, U_g) - IGI(v, U_{g'})}{\max\{IGI(v, U_g), IGI(v, U_{g'})\}} \tag{2}$$

Therefore, $I_{Ster}(v, U_g, U_{g'}) = -I_{Ster}(v, U_{g'}, U_g,)$.

Figure 1 shows the distribution of I_{Ster} values over items for the LFM-2B-100K and ML-1M datasets, and for the users in the U_m group, i.e., setting $U_g = U_m$ and $U_{g'} = U_f$ in Eq. 1. Whenever considering a user, we gather the corresponding I_{Ster} values that match the value of the user-protected attribute. In addition, these values are calculated only for items that were consumed by at least one user in each user group.

3.3 User Group Stereotypicality

Next, we introduce a measure of the target user's strength of preference towards group-biased or stereotypical items as defined in Subsect. 3.1. Given a user u and the items in their profile $v \in X_u$, the user's preference towards stereotypical items is measured as the mean U_{Ster}^{mean} or median U_{Ster}^{median} of the distribution of values of I_{Ster}^u over the items in X_u. Throughout this paper, for simplicity, we refer to these scores as U_{Ster} for both definitions (mean and median), but separately explore the effects of both in our results.

The U_{Ster} values are used to determine whether a user is to be considered *highly stereotypical*. Therefore, we define the threshold of user stereotypicality γ as the mean value of all users' U_{Ster} scores. Users with $U_{Ster} \geq \gamma$ are considered *highly stereotypical* and hence selected as targets for obfuscation. Figure 2 shows the values of U_{Ster} of users from LFM-2B-100K and ML-1M in order of descending stereotypicality, as well as the thresholds γ.

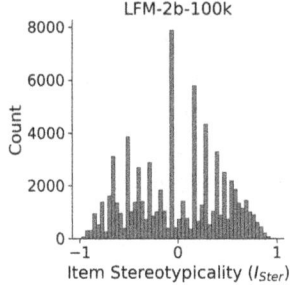 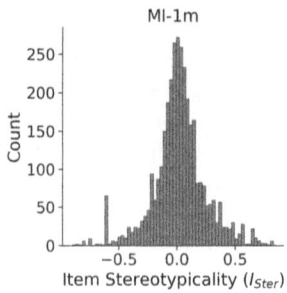

Fig. 1. Distribution of item stereotypicality I_{Ster} $(v, U_g, U_{g'})$ with $U_g = U_m$ and $U_{g'} = U_f$ over the items of the **LFM-2B-100K** (left) and **ML-1M** (right) datasets.

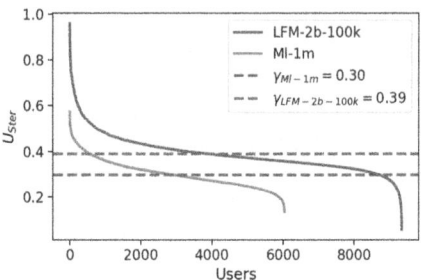

Fig. 2. User group stereotypicality of users from the **LFM-2B-100K** and **ML-1M** datasets, with users in order of descending stereotypicality. The red dotted and green dotted lines indicate the selection threshold $U_{\text{Ster}}^{\text{mean}}$ used for **LFM-2B-100K** and **ML-1M**, respectively. (Color figure online)

3.4 Stereotypicality-Based Obfuscation

Our method SBO consists of three main steps: 1) filtering users according to their U_{Ster} score; 2) sub-sampling candidate items; and 3) obfuscating the users' profiles. Below, we describe each step of the method separately and summarize them in Algorithm 1. First, we compute the list M_u of values of I_{Ster} according to the user's gender label g (each entry of M_u representing a different item). Then, we compute the U_{Ster} values for each user and filter the users with scores higher than the threshold γ, which is considered as a hyper-parameter. Given an obfuscation ratio ρ, the item sub-sampling consists of selecting a set of obfuscation candidates X_u^ρ for the user, containing at most $\rho \cdot |X_u|$ items. For this purpose, we define different sampling pools for the three different obfuscation strategies: *imputation*, *removal*, and *weighted*. Specifically, the sampling pool for *imputation* is $V - X_u$, and the sampling pool for *removal* is X_u; additionally, a weighted combination of these two is the sampling pool for *weighted*. The weight $\omega \leq 1$ decides on the number of items to select for imputation ω and for removal $1 - \omega$ and is treated as a hyper-parameter.[2]

---

[2] We report results for $\omega = 0.5$ only.

Sterotypicality-Based Sampling. To sample the items to select for obfuscation, SBO first selects the items with the highest I_{Ster} scores from the set of obfuscation candidates X_u^ρ. Then, SBO decides on the items to obfuscate by performing a Bernoulli trial on each of the selected items, with a success rate equal to the item's absolute I_{Ster} values. Therefore, items that have high I_{Ster} values are more likely to be obfuscated. The candidate items in X_u^ρ in which Bernoulli trials were successful are inserted in the obfuscation items list C_u, and then obfuscated. The Bernoulli trials are performed on each item independently to use the same I_{Ster} values across all user profiles.

Our aim is to obfuscate the items that are highly stereotypical in the user profile, therefore, when imputing unseen items, we choose items with the most negative I_{Ster} scores (most counter-stereotypical for u's gender). On the contrary, when removing items, we select the items with the most positive I_{Ster} scores (most stereotypical for u's gender). Following the same reasoning, we also define an additional baseline sampling strategy for comparison, *TopStereo*, where the items with the highest I_{Ster} scores in the user profile are selected for removal and the most negative for imputation. In addition, we include the *Random* strategy, which selects items uniformly at random from the user profile for removal and from the set of unexplored items for imputation. After having sub-sampled the list of candidate items X_u^ρ, we perform the selected obfuscation method using the Obfuscate on the user profile X_u and the obfuscation strategy m.

3.5   Attacker Network

As common in literature [9,19], we use a simple feed-forward network as an attacker network. The network is trained on vector representations of the users' interaction data in a supervised manner to predict the private attributes from these representations. The successful prediction of the attribute implies that the current interaction data can reveal the values of the attributes. In our case, this network takes the user preference vectors as input and aims to predict the user's gender.

Algorithm 1: Stereotypicality-based Obfuscation

---

input : List of items the user u interacted with X_u,
     User u's gender label g,
     List of unique items V,
     User groups defined by gender $\{U_m, U_f\}$,
     User stereotypicality threshold γ,
     Obfuscation sampling ratio ρ,
     Obfuscation strategy m
output: Obfuscated list of user u's interactions $\tilde{X}_u$

// Assigning user's stereotypicality
$S_u \leftarrow U_{\text{Ster}}(X_u)$
// User's obfuscation candidate items
$C_u \leftarrow \{\}$
$\tilde{X}_u \leftarrow \{\}$
// Defining the list of item stereotypicality values according to the user's gender label
if $g ==$ male **then**
    | $M_u \leftarrow \{I_{\text{Ster}}(v, U_m, U_f) : v \in V\}$
else
    | $M_u \leftarrow \{I_{\text{Ster}}(v, U_f, U_m) : v \in V\}$
end
// Evaluating the user for high stereotypicality
if $S_u \geq \gamma$ **then**
    // Sub-sampling of candidate items to obfuscate
    $X_u^\rho \leftarrow \text{SubSample}(V, X_u, \rho, m)$
    for $v \in X_u^\rho$ **do**
        $p \leftarrow |M_u(v)|$
        $c \leftarrow \text{BernoulliTrial}(p)$
        if $c ==$ True **then**
            | $C_u \leftarrow C_u \cup \{v\}$
        end
    end
    // Performing obfuscation of the user profile X_u
    $\tilde{X}_u \leftarrow \text{Obfuscate}(X_u, C_u, m)$
else
    | $\tilde{X}_u \leftarrow X_u$
end

---

4 Experimental Setup

4.1 Datasets

We run evaluation experiments on two popular datasets: ML-1M [10][3] and LFM-2B-100K,[4] covering the movie and music domain, respectively (Table 1).

---

[3] https://grouplens.org/datasets/movielens/1m/.
[4] A subset of LFM-2B [23,30], derived by first selecting users with valid gender information, then randomly select $\sim 100k$ unique items that adhere to 5-core filtering.

For the training of recommendation models, we apply 5-core filtering to each dataset, randomly select 20% of each user's interactions, and leave them out as *test* set. We apply the same split procedure on the remaining 80% of interactions to generate the *training* and *validation* sets. For the attackers' training, we perform 5-fold cross-validation over the set of unique users, leaving 20% of them as test users in each fold, and report the average value of the evaluation metrics computed over the folds.

In order to perform obfuscation, we use the concatenation of the *train* and *validation* slices of the original datasets, then we slice the resultant set of interactions following the previously introduced procedure for both recommendation models and attacker networks.

Table 1. Statistical description of datasets

Dataset	Users (Male/Female)	Items	Interactions	Density
ML-1M	6,040 (4,331/1,709)	3,416	999,611	0.0484
LFM-2B-100K	9,364 (7,580/1,784)	99,965	1,820,903	0.0019

4.2 Dataset Obfuscation

The generation of obfuscated datasets is done before training the models with the following hyper-parameters: the user stereotypicality threshold γ is defined as the mean or median as described in Sect. 3.3, the obfuscation ratio parameter is set to $\rho = 0.1$.[5] We perform experiments for all the obfuscation strategies and sampling methods defined in Sect. 3.4. We evaluate SBO against a state-of-the-art obfuscation approach, Perblur, proposed by Slokom et al. [31]. Where available, we used the code provided by the authors[6] and implemented the missing pieces of code. Specifically, we set Perblur's number of user neighbors to 50 for LFM-2B-100K and to 100 for ML-1M. From these neighbors, we collect the 200 and 500 most frequent recommended items for LFM-2B-100K and ML-1M, and used them as personalized lists. Then, we follow the procedure described by Slokom et al. [32] for selecting the 50 most indicative items for each gender. We include in our results both the performance of Perblur with the imputation and with the removal method.

4.3 Algorithms

Recommendation Models. Since the proposed method SBO is largely independent of the recommendation algorithm as long as those are trained on implicit feedback, we carry out our experiments on a selection of well-established recommendation algorithms from different categories: matrix factorization (BPR-MF [28]),

---

[5] We also used $\rho = 0.05$, obtaining similar results, for which we refer the reader to our supplementary material (Appendix A).

[6] https://github.com/SlokomManel/PerBlur.

neural network-based (MULTVAE [20]), and graph-based (LIGHTGCN [12]), hence demonstrating its performance across different types of RSs. We train the RSs for 100 epochs with a learning rate of 0.001 using the Adam optimizer with 512 as batch size. We apply early-stopping with a patience of 10 epochs, using NDCG as validation metric, computed for the top 10 predicted (i. e., recommended) items. The embedding size of all models is set to 64. We use the implementation of the RS models provided by the RecBole[7] framework. Each model is evaluated with each of the dataset obfuscation parameters defined in Sect. 4.2.

Attacker Networks. For the attacker networks, we define the architecture $A = [|V|, l, 2]$ setting the number of nodes of the intermediate layer to $l = 128$ for ML-1M and to $l = 256$ for LFM-2B-100K. Each of the attackers is trained for 50 epochs using the Adam optimizer with 64 as batch size and 0.001 as learning rate with a Cross-Entropy (CE) minimization objective. In order to mitigate the imbalanced distribution of gender, we set proportional weights to each gender category in the CE objective. These networks are applied to all the configurations of parameters defined in Subsect. 4.2.

Evaluation. To assess the recommendation performance, we report the Normalized Discounted Cumulative Gain (NDCG) for the top 10 recommended items. Additionally, we report the Balanced Accuracy (BAcc) to assess the performance of the attacker networks. To ensure the reproducibility of our research, the implementation and complete configuration of our experiments can be found in our publicly available repository.[8]

5   Results and Discussion

In this section, we describe our results, focusing first on the effect on the accuracy–privacy trade-off. We then delve into the effect of SBO's different parameter configurations. Table 2 shows the user's gender obfuscation capabilities of SBO in terms on BAcc for both datasets. Given that both SBO and the baseline Perblur are independent of the recommendation algorithm, the same values of BAcc are valid for the analysis of the performance of the different recommendation algorithms. We also report the results on the dataset without obfuscations, which we refer to as *original.* The BAcc values reported correspond to the best values of the average test results over 5-folds for each obfuscation parameter configuration, with the corresponding NDCG values for each recommendation algorithm, in which at most 10% of the user profiles were obfuscated ($\rho = 0.1$).

We observe that SBO in its variant with *removal* and *SBsampling*, consistently yields the best results in terms of BAcc for both datasets, proving SBO's effectiveness in preventing the attacker's ability to infer user's protected attributes, at the cost of a slight decrease in NDCG. With *removal* and *SBsampling*, SBO delivers ~7% and ~9% of improvement in BAcc with respect

---

[7] https://github.com/RUCAIBox/RecBole.
[8] https://github.com/hcai-mms/SBO.

Table 2. Experimental results on the two datasets ML-1M and LFM-2B-100K. The scores in **bold** indicate the best scores across all models.

Dataset	Obf. Strat.	Sampling	BAcc↓	BPR-MF	LIGHTGCN	MULTVAE
				NDCG ↑		
LFM-2B-100K	orignal	original	0.5501	0.1135	**0.1773**	0.1483
	imputate	PerBlur	0.5522	0.1042	0.1561	0.1402
		Random	0.5427	0.0990	0.1543	**0.1607**
		SBSampling	0.5528	0.1209	0.1764	0.1513
		TopStereo	0.5528	0.1209	0.1764	0.1513
	remove	PerBlur	0.5471	0.1155	0.1764	0.1507
		Random	0.5414	0.1070	0.1564	0.1324
		SBSampling	**0.5136**	0.1138	0.1731	0.1441
		TopStereo	0.5445	**0.1224**	0.1759	0.1518
	weighted	Random	0.5417	0.1055	0.1584	0.1504
		SBSampling	0.5528	0.1209	0.1764	0.1513
		TopStereo	0.5528	0.1209	0.1764	0.1513
ML-1M	original	original	0.6182	0.3445	0.3655	0.3650
	imputate	PerBlur	0.6156	0.3344	0.3581	0.3580
		Random	0.5973	0.3389	0.3592	**0.3718**
		SBSampling	0.8329	0.2866	0.3174	0.3154
		TopStereo	0.8751	0.3111	0.3468	0.3499
	remove	PerBlur	0.6597	0.3437	0.3656	0.3657
		Random	0.6076	0.2904	0.3116	0.3161
		SBSampling	**0.5664**	0.3400	0.3608	0.3586
		TopStereo	0.6124	0.3396	**0.3679**	0.3650
	weighted	Random	0.6001	0.3155	0.3347	0.3441
		SBSampling	0.7255	0.3114	0.3421	0.3383
		TopStereo	0.7335	0.3243	0.3560	0.3578

to the original LFM-2B-100K and the original ML-1M dataset, respectively, at the cost of ∼2% decrease in NDCG across all RSs. Furthermore, when compared with Perblur, ∼8% and ∼6% in improvement in BAcc is achieved on LFM-2B-100K and ML-1M, respectively, which translates into a decrease of ∼4% in NDCG on LFM-2B-100K, and a ∼1% decrease in NDCG on ML-1M.

We observe that when imputing items, SBO can have a negative impact on BAcc for most obfuscation configurations; this may be due to the size of the sampling pool. In this regard, Perblur shows more robustness, which might be attributed to filtering items using the user-based KNN recommendation algorithm. This emphasizes the substantial influence of the selection of obfuscation candidates for imputation of user preferences.

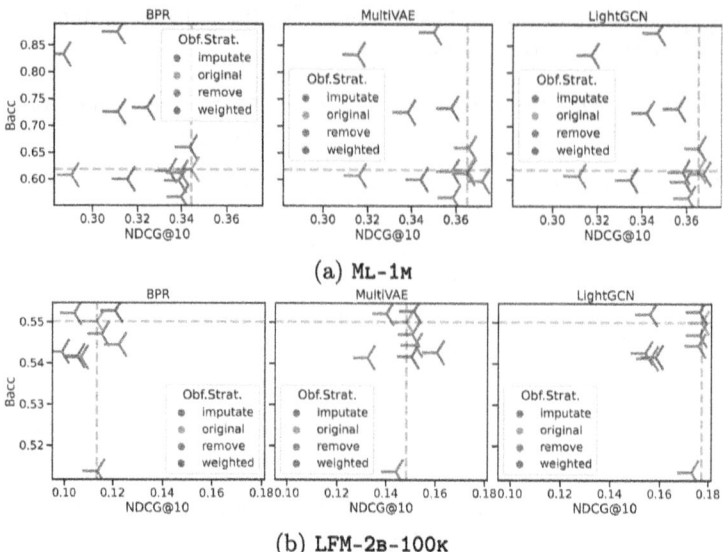

Fig. 3. Performance of the RSs and attacker (NDCG@10 and BAcc) with different obfuscation strategies on (a) ML-1M and (b) LFM-2B-100K. The dotted lines indicate the performances on the datasets without any obfuscation in place.

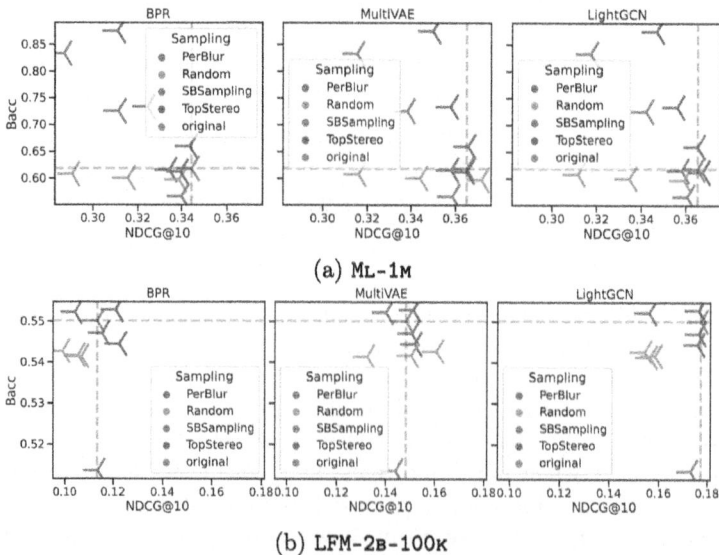

Fig. 4. Performance of the RSs and attacker (NDCG@10 and BAcc) with different sampling methods on (a) ML-1M and (b) LFM-2B-100K. The dotted lines indicate the performances on the datasets without any obfuscation in place.

From Table 2, we can also speculate that on the original LFM-2B-100K it is already hard to infer the users' gender attribute from their preferences, given the low BAcc values reported. In comparison, ML-1M is more exposed to adversarial attacks inferring users' gender (higher BAcc on original dataset), and also more sensitive to the obfuscation methods applied, given the fluctuation in the values of BAcc when different obfuscation strategies are used.

Figure 3 and Fig. 4 show the results of the obfuscation strategy and sampling method obfuscation parameters from Table 2 in terms of two-dimensional plots with NDCG on the x-axis and BAcc on the y-axis, and for each recommendation algorithm. In each subplot, the points closer to the bottom-right corner provide better accuracy–privacy trade-off (higher NDCG and lower BAcc).

In Fig. 3, we see that for both datasets, *removal* is usually below the original dataset BAcc values (below the dotted line), indicating the effectiveness of *removal* in preventing adversarial attacks on protected attributes. Other points clearly show improvements in NDCG, although with a lesser impact on BAcc compared to *removal*. The effect of *removal* is larger on LFM-2B-100K. Furthermore, for the *weighted* strategy, we observe that the performance of SBO mostly falls in the central regions of the plots. Since varying $\omega \in [0, 1]$ allows adjusting the level of *imputation* and *removal*, we speculate that the parameters of *weighted* could be optimized to target better privacy-oriented results. Figure 4 compares the performance of SBO with different sampling methods. We observe that on ML-1M, *SBsampling* and *TopStereo* have decreasing behavior in terms of BAcc while increasing in NDCG values. On the other hand, Perblur has an ascending tendency. On the LFM-2B-100K dataset, the results are more diverse and only partially resemble the trends observed on ML-1M. More importantly, the behavior of *SBsampling* is similar across different recommendation algorithms.

6   Conclusion and Future Work

In this work, we introduced SBO, a novel probabilistic user preference obfuscation method that selects the items to obfuscate based on stereotypicality measures for users and items. Our experiments show that SBO can reach a better accuracy–privacy trade-off than the baselines used for comparison on two recommendation domains (music and movies) by removing highly stereotypical items from the users' profiles. In addition, we show that the different configurations of SBO (obfuscation and sampling strategy) have similar behavior across different recommendation algorithms.

In this work, we limited the analysis to gender as the protected attribute and oversimplified its definition, reducing it to a binary attribute. Therefore, we plan to extend the current work by including an analysis of the effect of SBO with user attributes beyond binary categories, such as age groups or ethnicities. Additionally, our experiments hinted that *imputation* has the potential to achieve a better accuracy–privacy trade-off, a hypothesis that we leave for future work. Finally, further analyses can target the mitigation of other user privacy objectives, such as membership inference.

Acknowledgments. This research was funded in whole or in part by the FFG COMET center program, by the Austrian Science Fund (FWF): P36413, P33526, and DFH-23, and by the State of Upper Austria and the Federal Ministry of Education, Science, and Research, through grants LIT-2021-YOU-215 and LIT-2020-9-SEE-113.

Disclosure of Interests. The authors have no competing interests to declare that are relevant to the content of this article.

References

1. Anelli, V.W., Deldjoo, Y., Di Noia, T., Ferrara, A., Narducci, F.: FedeRank: user controlled feedback with federated recommender systems. In: Hiemstra, D., Moens, M.-F., Mothe, J., Perego, R., Potthast, M., Sebastiani, F. (eds.) ECIR 2021. LNCS, vol. 12656, pp. 32–47. Springer, Cham (2021). https://doi.org/10.1007/978-3-030-72113-8_3
2. Anelli, V.W., Deldjoo, Y., Noia, T.D., Merra, F.A.: Adversarial recommender systems: attack, defense, and advances. In: Ricci, F., Rokach, L., Shapira, B. (eds.) Recommender Systems Handbook, pp. 335–380. Springer, New York (2022). https://doi.org/10.1007/978-1-0716-2197-4_9
3. Deldjoo, Y., Jannach, D., Bellogin, A., Difonzo, A., Zanzonelli, D.: Fairness in recommender systems: research landscape and future directions. User Model. User-Adapted Interact. **34**(1) (2024)
4. Deldjoo, Y., Noia, T.D., Merra, F.A.: A survey on adversarial recommender systems: from attack/defense strategies to generative adversarial networks. ACM Comput. Surv. **54**(2) (2021). https://doi.org/10.1145/3439729
5. Dwork, C.: Differential privacy: a survey of results. In: Agrawal, M., Du, D., Duan, Z., Li, A. (eds.) TAMC 2008. LNCS, vol. 4978, pp. 1–19. Springer, Heidelberg (2008). https://doi.org/10.1007/978-3-540-79228-4_1
6. Dwork, C., Hardt, M., Pitassi, T., Reingold, O., Zemel, R.: Fairness through awareness. In: Proceedings of the 3rd Innovations in Theoretical Computer Science Conference (ITCS), pp. 214–226 (2012)
7. Ekstrand, M.D., Das, A., Burke, R., Diaz, F.: Fairness in recommender systems, pp. 603–646. Springer, New York (2022)
8. Ekstrand, M.D., Joshaghani, R., Mehrpouyan, H.: Privacy for all: ensuring fair and equitable privacy protections. In: Conference on Fairness, Accountability and Transparency, pp. 35–47. PMLR (2018)
9. Ganhör, C., Penz, D., Rekabsaz, N., Lesota, O., Schedl, M.: Unlearning protected user attributes in recommendations with adversarial training. In: Proceedings of the 45th International ACM SIGIR Conference, SIGIR 2022, pp. 2142–2147. ACM, New York (2022). https://doi.org/10.1145/3477495.3531820
10. Harper, F.M., Konstan, J.A.: The movielens datasets: history and context. ACM Trans. Interact. Intell. Syst. **5**(4), 19:1–19:19 (2016). https://doi.org/10.1145/2827872
11. Hashemi, H., et al.: Data leakage via access patterns of sparse features in deep learning-based recommendation systems. In: Workshop on Trustworthy and Socially Responsible Machine Learning (TSRML), in conjunction with the 36th Conference on Neural Information Processing Systems (NeurIPS) (2022)

12. He, X., Deng, K., Wang, X., Li, Y., Zhang, Y., Wang, M.: Lightgcn: simplifying and powering graph convolution network for recommendation. In: Huang, J.X., et al. (eds.) Proceedings of the 43rd International ACM SIGIR Conference on Research and Development in Information Retrieval, SIGIR 2020, Virtual Event, China, 25–30 July 2020, pp. 639–648. ACM (2020)

13. Jin, D., et al.: A survey on fairness-aware recommender systems. Inf. Fusion **100**, 101906 (2023)

14. Kim, S., Kim, J., Koo, D., Kim, Y., Yoon, H., Shin, J.: Efficient privacy-preserving matrix factorization via fully homomorphic encryption. In: Proceedings of the 11th ACM on Asia Conference on Computer and Communications Security (ASIACCS), pp. 617–628 (2016)

15. Krismayer, T., Schedl, M., Knees, P., Rabiser, R.: Predicting user demographics from music listening information. Multim. Tools Appl. **78**(3), 2897–2920 (2019). https://doi.org/10.1007/S11042-018-5980-Y

16. Lacic, E., Reiter-Haas, M., Kowald, D., Reddy Dareddy, M., Cho, J., Lex, E.: Using autoencoders for session-based job recommendations. User Model. User-Adap. Inter. **30**, 617–658 (2020)

17. Lex, E., Kowald, D., Schedl, M.: Modeling popularity and temporal drift of music genre preferences. Trans. Int. Soc. Music Inf. Retrieval **3**(1), 17–31 (2020)

18. Li, Y., Chen, H., Xu, S., Ge, Y., Zhang, Y.: Towards personalized fairness based on causal notion. In: Proceedings of the 44th International ACM SIGIR Conference on Research and Development in Information Retrieval, SIGIR 2021, pp. 1054–1063. ACM, New York (2021)

19. Li, Y., et al.: Making users indistinguishable: attribute-wise unlearning in recommender systems. In: Proceedings of the 31st ACM International Conference on Multimedia, MM 2023, pp. 984–994. ACM, New York (2023)

20. Liang, D., Krishnan, R.G., Hoffman, M.D., Jebara, T.: Variational autoencoders for collaborative filtering. In: Proceedings of the 2018 World Wide Web Conference, WWW 2018, pp. 689–698. International World Wide Web Conferences Steering Committee, Republic and Canton of Geneva, CHE (2018)

21. Lin, C., Liu, B., Zhang, X., Wang, Z., Hu, C., Luo, L.: Privacy-preserving recommendation with debiased obfuscaiton. In: IEEE International Conference on Trust, Security and Privacy in Computing and Communications, TrustCom 2022, Wuhan, China, 9–11 December 2022, pp. 590–597. IEEE (2022)

22. Lin, Y., et al.: Meta matrix factorization for federated rating predictions. In: Proceedings of the 43rd International ACM SIGIR Conference on Research and Development in Information Retrieval (SIGIR), pp. 981–990. Springer, Cham (2020)

23. Melchiorre, A.B., Rekabsaz, N., Parada-Cabaleiro, E., Brandl, S., Lesota, O., Schedl, M.: Investigating gender fairness of recommendation algorithms in the music domain. Inf. Process. Manag. **58**(5), 102666 (2021)

24. Muellner, P., Kowald, D., Lex, E.: Robustness of meta matrix factorization against strict privacy constraints. In: European Conference on Information Retrieval, pp. 107–119 (2021)

25. Müllner, P., Lex, E., Schedl, M., Kowald, D.: ReuseKNN: neighborhood reuse for differentially-private KNN-based recommendations. ACM Trans. Intell. Syst. Technol. **14**(5), 1–29 (2023)

26. Müllner, P., Lex, E., Schedl, M., Kowald, D.: The impact of differential privacy on recommendation accuracy and popularity bias. In: Goharian, N., et al. (eds.) ECIR 2024. LNCS, vol. 14611, pp. 466–482. Springer, Cham (2024). https://doi.org/10.1007/978-3-031-56066-8_33

27. Müllner, P., Lex, E., Schedl, M., Kowald, D.: Differential privacy in collaborative filtering recommender systems: a review. Front. Big Data **6** (2023)

28. Rendle, S., Freudenthaler, C., Gantner, Z., Schmidt-Thieme, L.: BPR: Bayesian personalized ranking from implicit feedback. In: Proceedings of UAI, pp. 452–461 (2009)

29. Schedl, M.: Investigating country-specific music preferences and music recommendation algorithms with the LFM-1B dataset. Int. J. Multim. Inf. Retr. **6**(1), 71–84 (2017)

30. Schedl, M., Brandl, S., Lesota, O., Parada-Cabaleiro, E., Penz, D., Rekabsaz, N.: LFM-2B: a dataset of enriched music listening events for recommender systems research and fairness analysis. In: Proceedings of the 2022 Conference on Human Information Interaction and Retrieval, CHIIR 2022, pp. 337–341. ACM, New York (2022)

31. Slokom, M., Hanjalic, A., Larson, M.A.: Towards user-oriented privacy for recommender system data: a personalization-based approach to gender obfuscation for user profiles. Inf. Process. Manag. **58**(6), 102722 (2021)

32. Strucks, C., Slokom, M., Larson, M.A.: Blurm(or)e: revisiting gender obfuscation in the user-item matrix. In: Burke, R., Abdollahpouri, H., Malthouse, E.C., Thai, K.P., Zhang, Y. (eds.) Proceedings of the Workshop on Recommendation in Multistakeholder Environments co-located with the 13th ACM Conference on Recommender Systems (RecSys 2019), Copenhagen, Denmark, 20 September 2019. CEUR Workshop Proceedings, vol. 2440. CEUR-WS.org (2019)

33. Vassøy, B., Langseth, H., Kille, B.: Providing previously unseen users fair recommendations using variational autoencoders. In: Zhang, J., et al. (eds.) Proceedings of the 17th ACM Conference on Recommender Systems, RecSys 2023, Singapore, Singapore, 18–22 September 2023, pp. 871–876. ACM (2023)

34. Verma, S., Rubin, J.: Fairness definitions explained. In: Proceedings of the International Workshop on Software Fairness, pp. 1–7 (2018)

35. Wang, Y., Ma, W., Zhang, M., Liu, Y., Ma, S.: A survey on the fairness of recommender systems. ACM Trans. Inf. Syst. **41**(3), 1–43 (2023)

36. Weinsberg, U., Bhagat, S., Ioannidis, S., Taft, N.: Blurme: inferring and obfuscating user gender based on ratings. In: Proceedings of the Sixth ACM Conference on Recommender Systems, pp. 195–202 (2012)

37. Wu, C., Wu, F., Wang, X., Huang, Y., , Xie, X.: Fairness-aware news recommendation with decomposed adversarial learning. In: Proceedings of AAAI Conference on Artificial Intelligence, pp. 4462–4469 (2021)

38. Wu, L., Chen, L., Shao, P., Hong, R., Wang, X., Wang, M.: Learning fair representations for recommendation: a graph-based perspective. In: Proceedings of the Web Conference 2021, WWW 2021, pp. 2198–2208. ACM, New York (2021)

39. Wu, Y., Cao, J., Xu, G.: Fairness in recommender systems: evaluation approaches and assurance strategies. ACM Trans. Knowl. Discov. Data **18**(1), 1–37 (2023)

40. Xin, X., et al.: On the user behavior leakage from recommender system exposure. ACM Trans. Inf. Syst. (TOIS) **41**(3), 1–25 (2023)

41. Zemel, R., Wu, Y., Swersky, K., Pitassi, T., Dwork, C.: Learning fair representations. In: International Conference on Machine Learning (ICML), pp. 325–333 (2013)

42. Zhang, M., Chen, Y., Lin, J.: A privacy-preserving optimization of neighborhood-based recommendation for medical-aided diagnosis and treatment. IEEE Internet Things J. **8**(13), 10830–10842 (2021)
43. Zhang, S., Yin, H.: Comprehensive privacy analysis on federated recommender system against attribute inference attacks. IEEE Trans. Knowl. Data Eng. (TKDE) (2023)

Leiden-Fusion Partitioning Method for Effective Distributed Training of Graph Embeddings

Yuhe Bai, Camelia Constantin, and Hubert Naacke[✉]

LIP6, Sorbonne University, Paris, France
{yuhe.bai,camelia.constantin,hubert.naacke}@lip6.fr

Abstract. In the area of large-scale training of graph embeddings, effective training frameworks and partitioning methods are critical for handling large networks. However, they face two major challenges: 1) existing synchronized distributed frameworks require continuous communication to access information from other machines, and 2) the inability of current partitioning methods to ensure that subgraphs remain connected components without isolated nodes, which is essential for effective training of GNNs since training relies on information aggregation from neighboring nodes. To address these issues, we introduce a novel partitioning method, named Leiden-Fusion, designed for large-scale training of graphs with minimal communication. Our method extends the Leiden community detection algorithm with a greedy algorithm that merges the smallest communities with highly connected neighboring communities. Our method guarantees that, for an initially connected graph, each partition is a densely connected subgraph with no isolated nodes. After obtaining the partitions, we train a GNN for each partition independently, and finally integrate all embeddings for node classification tasks, which significantly reduces the need for network communication and enhances the efficiency of distributed graph training. We demonstrate the effectiveness of our method through extensive evaluations on several benchmark datasets, achieving high efficiency while preserving the quality of the graph embeddings for node classification tasks.

Keywords: Distributed Training · Graph Embeddings · Graph Partitioning

1 Introduction

Graph embeddings have become a fundamental technique in machine learning, providing a powerful means of dealing with complex structured data. By transforming nodes, edges, and their interactions within a graph into a compact, lower-dimensional vector space, graph embeddings allow machine learning techniques to be applied to graph data with increased efficiency.

To compute graph embeddings, Graph Neural Networks (GNNs) have gained prominence due to their ability to exploit the inherent structure of graph data.

A. Bifet et al. (Eds.): ECML PKDD 2024, LNAI 14947, pp. 366–382, 2024.
https://doi.org/10.1007/978-3-031-70368-3_22

Among them, the most popular are Graph Convolutional Networks (GCN) [10] and GraphSAGE [6]. Using graph convolution operations, GNNs iteratively aggregate and transform the embeddings of neighboring nodes, culminating in a representation that captures both local and global graph structures.

However, the scalability of GNNs to very large graphs presents a significant challenge. While parallel processing can enhance the efficiency of GNNs by allocating computations across multiple processors or GPUs, for extremely large graphs that exceed the capacity of a single machine, it is crucial to partition the graph and distribute the computational load across multiple machines. While traditional partitioning approaches facilitate distributed learning, they often fail to preserve the structural coherence of the original graph. They typically generate subgraphs that contain multiple connected components and isolated nodes, undermining the performance of GNNs. A connected component is a subgraph in which every pair of nodes is connected by a path, and an isolated node represents a vertex of a graph with no edges and thus of a degree zero. This is because the effectiveness of GNNs depends on the premise that a node's embedding is enriched by the embeddings of its neighbors; if these neighbors lie outside the subgraph, not only will there be more communication, but the quality of the embeddings will also decrease.

To address these challenges, our work introduces a novel partitioning method designed to preserve the structural integrity of subgraphs in a distributed learning framework, followed by a local training strategy. Specifically, we ensure that for any given graph that initially consists of a single connected component, each partition remains a connected component with no isolated nodes. This not only preserves the contextual relevance of node embeddings but also allows local training and eliminates the need for inter-subgraph communication, thereby increasing the efficiency of distributed GNN training. Our contributions are as follows:

1. For an initially connected graph, we proposed a novel partitioning method that guarantees the structural integrity of subgraphs by ensuring that each subgraph remains a single connected component with no isolated nodes.
2. By using single connected components as partitions, we demonstrate the feasibility of achieving high training efficiency for GNNs without sacrificing much accuracy, paving the way for more scalable and efficient distributed learning on very large graphs.

The paper is organized as follows: Sect. 2 presents background knowledge about GNNs and graph embeddings, related work is presented in Sect. 3. Section 4 presents our novel Leiden-Fusion algorithm, and experimental results are discussed in Sect. 5.

2   Background on Graph Embeddings

Graph Neural Networks (GNNs) extend neural network methods to graph data. A typical GNN layer updates the representation of a node based on its neighbors. Graph Convolutional Networks (GCN) [10] and GraphSAGE [6] represent two major advances in the field of GNNs, each introducing unique strategies for aggregating neighborhood information to improve node embeddings. The resulting embeddings are critical in a variety of applications, including but not limited to node classification [19], question answering [7], and recommender systems [21].

GCN [10]: The key idea behind GCN is to update the representation of a node by aggregating the representations of its neighbors. This approach captures the local graph topology in the node embeddings. The formula given for GCN is:

$$\mathbf{h}_v^l = \sigma \left(\frac{1}{|N(v)|} \sum_{u \in N(v)} \mathbf{W}^l \mathbf{h}_u^{l-1} \right) \tag{1}$$

This formula represents how the representation $\mathbf{h}_v^l$ of a node v at layer l is updated. It does this by applying a nonlinear activation function σ (*e.g.*, ReLU function) to the normalized sum of the representations of its u ($u \in N(v)$) neighbors from the previous layer $\mathbf{h}_u^{l-1}$. $\mathbf{W}^l$ is the weight matrix for the layer l.

GraphSAGE [6]: GraphSAGE extends the idea of GCN by incorporating the node's own features along with its neighbors, and by using a sampling strategy that selects a fixed subset of neighbors to aggregate information from, allowing scalability in large graph settings. The formula for GraphSAGE is:

$$\mathbf{h}_v^l = \sigma \left(\mathbf{W}^l \cdot \text{CONCAT} \left(\mathbf{h}_v^{l-1}, \text{AGG} \left(\{ \mathbf{h}_u^{l-1}, \forall u \in N(v) \} \right) \right) \right) \tag{2}$$

In this equation, the new representation of a node v at layer l is obtained by first concatenating the representation of its previous layer $\mathbf{h}_v^{l-1}$ with an aggregated representation of its sampled neighbors' features $\mathbf{h}_u^{l-1}$. The aggregation (AGG) can be a mean, sum, or max operation. This method allows for efficient computation on large-scale graphs and enriches the node embeddings with both central node and sampled neighborhood information.

Thus, the effectiveness of these models relies heavily on their ability to aggregate information from neighboring nodes, underscoring the importance of a partitioning method in a distributed setting that computes partitions as connected graph components. Our partitioning method ensures that the structural integrity of the graph is maintained within each partition, which is crucial for effective local model training.

Figure 1 visually illustrates the process of neighbor aggregation for nodes A and B, contrasting two scenarios based on the partitioning of the graph into subgraphs colored blue (first partition) and gray (second partition).

On the left, both partitions contain a single connected component, ensuring that full neighbor information is available for aggregation. On the right, however, the presence of multiple components and isolated nodes within each partition severely limits the information that nodes A and B can aggregate. Nodes A and B can only aggregate two neighbors instead of four.

More specifically, in a distributed framework with no communication, a lot of neighbor information is lost; with synchronization, a lot of communication

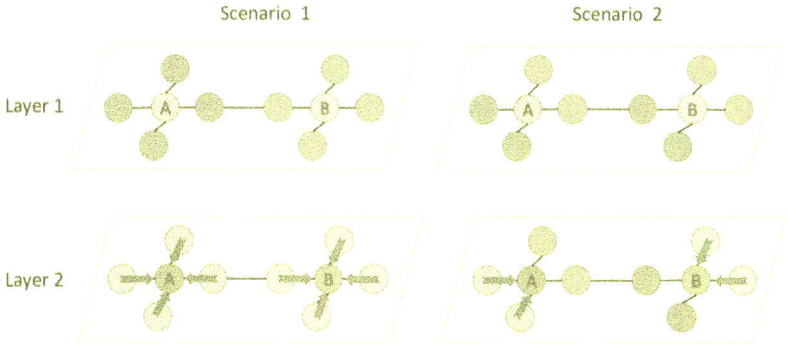

Fig. 1. Aggregation of nodes A and B in GNNs with different partitioning strategies

occurs and there is a delay in information transformation. This illustrates the impact of graph connectivity on the update process in GNN layers.

This illustration highlights the importance of ensuring that each subgraph not only remains a connected component but also avoids isolated nodes to maximize the effectiveness of distributed GNN training. Our partitioning algorithm is specifically designed with this goal in mind and aims to improve the efficiency and effectiveness of GNNs in distributed environments.

3   Related Work

3.1   Partitioning Methods

The goal of most partitioning methods is to reduce edge cuts and ensure load balance, to reduce the communication of synchronized distributed frameworks. We will introduce some of the SOTA partitioning methods.

METIS [9] is one of the most popular algorithms and is used by most SOTA distributed frameworks. For each machine, it aims to form a diagonal-like block in the adjacency matrix, so that when a trainer processes samples in the local partition, most of the embeddings accessed by the batch fall in the local partition, and thus there is little network communication for accessing entity embeddings. METIS focuses mainly on balancing the node size of the partition and minimizing edge cuts.

However, this approach does not directly focus on the component structure within the partitions, which means that it may split a component into multiple partitions, resulting in many isolated nodes. This is problematic for GNN models, which, as discussed earlier, rely heavily on the integrity of the graph structure for effective training.

LPA: The Label Propagation Algorithm [4,13,14] (LPA) was originally designed to detect communities in graphs, using the network structure to determine the communities. In LPA, each node in the graph is initially assigned a unique label. At each iteration of the algorithm, nodes adopt the label that most of their neighbors currently have. This update rule can be written as:

$$\text{label}(v) \leftarrow \text{mode}(\{\text{label}(u) : u \in \mathcal{N}(v)\}) \tag{3}$$

where $\text{label}(v)$ is the label of the node v, and $\mathcal{N}(v)$ is the set of neighbors of v. The mode function selects the most frequent label among the neighbors of a node. The algorithm runs iteratively until convergence or a certain number of epochs is reached, at which point nodes with the same label are considered to be in the same community. One of the main advantages of LPA is its ability to scale naturally to large networks due to its simplicity. To use it for graph partitioning in distributed learning, each node is initially assigned a label ranging from 0 to the number of partitions K.

However, the algorithm has several limitations. It can be sensitive to the initial label assignment and can produce different results on different runs. It can also converge to a trivial solution where all nodes end up with the same label in highly connected graphs. In LPA, each node is initially randomly assigned a label from 0 to n. This means that, for example, for label 0, there may initially be some nodes with label 0 at different positions in the graph. They then propagate separately, forming many small components centered on themselves, resulting in partition 0 having many components quite far apart from each other.

Many other partitioning methods are also used to address specific needs, such as random partitioning, a simple approach where nodes (or edges) are randomly assigned to partitions. It can provide load balancing and high diversity within a partition, but in synchronized frameworks, the communication overhead can be very high; in unsynchronized frameworks, it can lead to poor quality embeddings because each node loses most of its neighbors' information.

3.2 Distributed Training Frameworks

Many frameworks have emerged to facilitate efficient and scalable distributed training of graph embeddings. The key to optimizing distributed training is to reduce the communication required to retrieve and update embeddings. However, no matter how it is reduced, most existing techniques for distributed graph embedding, such as Deep Graph Library (DGL) [18] and PyTorch BigGraph (PBG) [11], require continuous communication.

Spark Local [4] is one of the first frameworks to perform local training of subgraphs to avoid continuous communication. It partitions a graph into subgraphs using LPA (Label Propagation Algorithm) while considering a "landmark graph" which is a small subset of the graph based on node degrees, then they put the landmarks into each subgraph, learn their embeddings locally, and reconcile the embedding spaces using SVD (Singular Value Decomposition) based on the landmark embeddings. However, the quality of the embeddings is degraded because the LPA algorithm can lead to poor-quality partitions. In addition, it is very time-consuming to find the landmarks and add the edges connecting them to each partition.

To address the shortcomings of the current partitioning and distributed training methods, we will introduce our Leiden-Fusion method in the next section.

4   Leiden-Fusion Method

In this section, we outline the main contributions of our work. First, we define the essential features of partitions that allow high-quality embeddings to be computed independently on each partition. We then present a detailed description of our two-step approach.

4.1   Essential Features for Graph Partitioning

As we discussed earlier, we assume that for local training of GNN on subgraphs to be effective, the following conditions must be met:

1. **Each partition should contain one densely connected component.** By ensuring this, most nodes can retain all neighbor information. Only for boundary nodes, a small amount of neighbor information will be lost.
2. **There should be no isolated nodes.** Similar to ensuring one densely connected component, if there are isolated nodes in the subgraphs, these nodes will have no neighbors to aggregate with to update their information, leading to poor training results.

Existing partitioning methods cannot meet these two requirements as we discussed in Sect. 3. Our partitioning method is designed to meet these requirements. The main idea of our approach is to rely on a community detection algorithm and then merge communities in a way that results in densely connected partitions free of isolated nodes.

4.2   Leiden Community Detection

The first step is to obtain densely connected communities using the Leiden algorithm:

The Leiden algorithm [16] is an iterative community detection method that improves on the well-known Louvain algorithm [2], with improvements in terms of quality and speed. The primary goal of the Leiden algorithm is to optimize a modularity function:

$$Q = \frac{1}{2\,m} \sum_c \left(e_c - \gamma \frac{K_c^2}{2\,m} \right) \tag{4}$$

where e_c is the actual number of edges in the community c. The expected number of edges is $\frac{K_c^2}{2m}$, where K_c is the sum of the degrees of the nodes in community c and m is the total number of edges in the network. This modularity is a scalar value that measures the density of links inside communities compared to links between communities. By maximizing the modularity function, Leiden ensures that the resulting communities are densely connected. We abstract the Leiden community detection in Definition 1.

Definition 1 (Leiden communities). *Let $G = (V, E)$ be a graph and $C = \{C_1, \ldots, C_n\}$ be a partition of V which implies $C_i \cap C_j = \emptyset$ for $i \neq j$. Let G_i be the projection of G onto C_i. Let S be the maximum expected size of a community.* **Leiden** : $G \mapsto C$ *associates G with C communities such that it maximizes the modularity of the communities, and each community has less than S vertices, i.e. $\forall C_i \in C, |G_i| \leq S$.*

4.3 Community Fusion

Since the number of communities obtained by Leiden is, in most cases, much larger than the expected number of partitions k, which typically corresponds to the number of machines in a distributed training environment. To address this issue, we propose a novel fusion method to merge these communities.

Our solution is based on the notions of edge cut, defined in Definition 2, and community neighborhood, defined in Definition 3.

Definition 2 (Edge cut). *Let $G = (V, E)$ be a graph. Let V_i, V_j be two disjoint subsets of V. Let G_i (resp. G_j) be the projection of G on V_i (resp. V_j). We define $Cut(G_i, G_j)$ as the set of edges connecting G_i with G_j. We have: $Cut(G_i, G_j) = \{(v, v') \in E | v \in G_i \wedge v' \in G_j\}$*

Definition 3 (Neighbor communities). *Let C be a set of communities in the graph G. The neighboring communities, denoted $Neighbors(C_i)$, are the set of communities that are adjacent to C_i, i.e. $Neighbors(C_i) = \{C_j \in C | Cut(C_i, C_j) \neq \emptyset\}$*

Starting from the initial partitions computed by the Leiden algorithm, for a given partitioning number k, our method iteratively computes k balanced partitions by merging existing partitions with their neighbors. The intuition of the Leiden-Fusion algorithm is shown on Zachary's karate club network [20] in Fig. 2. The goal is to partition the Karate graph into two partitions. First, we get 4 communities through the Leiden community detection algorithm, and then we start from the smallest community, which is the yellow community. We find its most connected neighbor, which is the green community, and merge them. Then the blue community becomes the smallest one to merge with the red one, and finally we get 2 partitions.

The Leiden-Fusion algorithm is described in Algorithm 1. The parameters α and β are used to control the number of nodes assigned to each partition and the maximum size of the initial communities computed by Leiden. We aim to compute balanced partitions whose size is controlled by the variable *max_part_size*, with a tolerance threshold given by α (line 3). We first apply the Leiden community detection algorithm to identify numerous small communities C within the graph G (line 4). Communities are iteratively merged to form larger and larger communities, with each iteration selecting the smallest community in terms of number of nodes (c_{min}) and gradually merging it with its largest edge-cut neighbor community (c_{max_cut}) (lines 5–10). The fusion process ends when $|C|$ equals the desired number of partitions k.

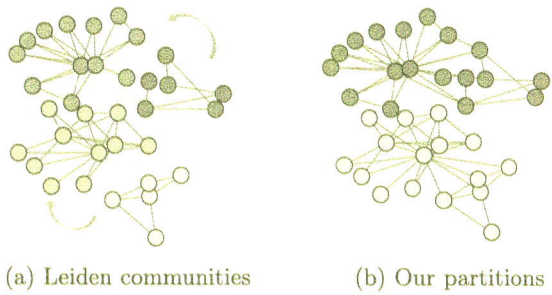

(a) Leiden communities   (b) Our partitions

Fig. 2. Visualization of Leiden community detection and fusion process

---

Algorithm 1 Leiden-Fusion Partitioning Algorithm

1: **Input:** G: graph, k: number of partitions, α, β
2: **Output:** C composed of k subgraphs
3: $max\_part\_size \leftarrow \frac{size(G)}{k} \times (1+\alpha)$
4: $C \leftarrow \text{Leiden}(G, \beta \times max\_part\_size)$ // C is a set of subgraphs
5: **while** $—C— > k$ **do**
6:    $c_{min} \leftarrow \arg\min_{c \in C} size(c)$ // get the smallest community
7:    $c_{max\_cut} \leftarrow \text{LargestEdgeCutNeighbor}(c_{min}, max\_part\_size)$
8:    $c_{merged} \leftarrow c_{max\_cut} \cup c_{min}$ // merge graph c_{min} with graph c_{max_cut}
9:    $C \leftarrow (C \setminus \{c_{min}, c_{max\_cut}\}) \cup \{c_{merged}\}$ // update communities
10: **end while**
11: **return** C

The largest edge-cut neighboring community is computed by Algorithm 2. For each community v to be merged, it finds the most connected community c (given by $|\text{Cut}(v, c)|$, which is the number of edges between v and c) within the size limit given by max_part_size (lines 3–5). If for every neighbor community c the merge exceeds the size limit max_part_size, v will be merged with its smallest neighbor to ensure load balance (lines 6–8).

Algorithm 2 LargestEdgeCutNeighbor

1: **Input:** v, max_part_size
2: **Output:** u
3: $N \leftarrow \{c \in \text{Neighbors}(v) |\, size(c) + size(v) < max\_part\_size\}$
4: **if** $N \neq \emptyset$ **then**
5:    $u \leftarrow \arg\max_{c \in N} |\text{Cut}(v, c)|$ // get the most connected neighbor among N
6: **else**
7:    $u \leftarrow \arg\min_{c \in \text{Neighbors}(v)} size(c)$ // get the smallest neighbor
8: **end if**
9: **return** u

Each partition obtained by this method consists of a single unified component since the initial graph is a connected component and each community computed by the Leiden algorithm is densely connected without isolated nodes.

4.4 Partition Visualization on Karate Dataset

To prove the effectiveness of our algorithm, we compared METIS, LPA, Random and our LF on this Karate dataset, the results are shown in Fig. 3 and Table 1. We can see that our algorithm outperforms on both criteria in the toy example. From Fig. 3 we can see that the LPA method can lead to poor quality partitions because it is sensitive to the initial label assignment. If two nodes at different positions in the graph are assigned the same label (partition 0 in this example), they may propagate to form many components at different positions in the graph, as shown in the figure. Similarly for METIS, we can see that there are many isolated nodes and many components in the partitions.

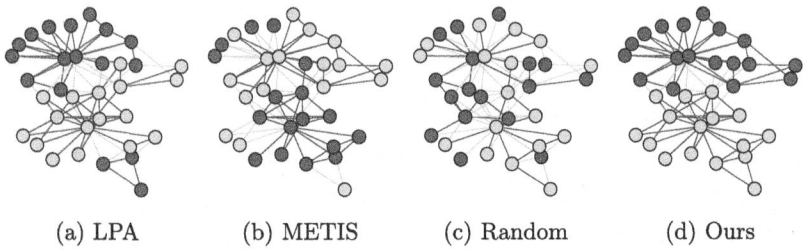

(a) LPA   (b) METIS   (c) Random   (d) Ours

Fig. 3. Comparison of partitioning methods on Karate dataset. ● Partition 0 ● Partition 1 (Color figure online)

As can be seen from Table 1, in this toy example, the partitions obtained by our LF method have zero isolated nodes, each partition has only one component and minimal edge cuts.

Table 1. Evaluation of Partitioning Methods on Karate Dataset

Method	Isolated Nodes		Components		Edge Cuts
	Part 0	Part 1	Part 0	Part 1	Part 0 & 1
LPA	0	0	2	1	17
METIS	4	3	5	4	25
Random	4	1	5	2	45
Ours	0	0	1	1	**10**

Advantages of the Proposed Two-Step Method: Our fusion method can be applied to any graph partitioning technique, but we chose the Leiden community detection method because of its ability to produce well-connected communities. However, Leiden communities vary in size and do not allow specifying the desired number of communities. Our fusion method addresses these limitations by allowing the generation of a specified number of balanced communities. Other graph partitioning methods, such as METIS and LPA, are designed to achieve a given number of partitions. However, they often produce multiple components and isolated nodes, making graph structure reconstruction time-consuming, as shown in the experimental section. This process involves identifying each component within a partition and treating them as separate partitions for fusion.

5   Experimental Results

Setup: We first perform the partitioning methods on one CPU in a centralized way. For METIS, we used the library provided by DGL [18]. For LPA, we reproduced the method of Spark Local [4], and then we implemented our Leiden-Fusion method.

Due to resource limitations, we ran the training process sequentially on a single machine for each partition, which is equivalent to a fully distributed implementation since there is no communication during the training process. The hardware used includes a DELL PowerEdge R650xs with 125 GB of memory and an Intel Xeon Silver 4310 processor with 24 cores/48 threads @ 2.10 GHz, and a DELL PowerEdge R750xa with 2 TB of memory equipped with two Intel Xeon Gold 6330 CPUs, each with 56 cores/112 threads @ 2.00 GHz, and four NVIDIA A100 80 GB PCIe GPUs. The code is available at https://github.com/YuheBAI/leiden-fusion.

Datasets: The datasets we used are the Arxiv and Proteins datasets for node prediction tasks from the Open Graph Benchmark (OGB) [8]. The Arxiv dataset is a directed graph, representing the citation network between all Computer Science (CS) Arxiv papers indexed by MAG [17]. The graph contains 169 343 nodes and 1166 243 edges. The task is to predict the label of each node from 40 subject areas of Arxiv CS papers, which is a multi-classification task. The proteins [15] dataset is an undirected, weighted, and typed (by species) graph. Nodes represent proteins, and edges indicate different types of biologically meaningful associations between proteins, such as physical interactions, co-expression, or homology [3,15]. The graph contains 132 534 nodes and 39 561 252 edges. The task is to predict the presence of protein functions in a multi-label binary classification setup, where there are a total of 112 types of labels to predict. Performance is measured by the average of the ROC-AUC values over the 112 tasks.

Hyperparameter Settings: In the experiments conducted for this paper, specific hyperparameters were set for different parts of the process. During the graph partitioning phase, α, which controls the partition size, was set to 0.05, and β, which controls the size of the Leiden community, was set to 0.5. For the GNN training phase, we used the same hyperparameters as recommended by OGB [8], with the number of epochs reduced to 80 for the Arxiv dataset to avoid overfitting, since training was performed on smaller subgraphs.

5.1   Analysis of Partitions

To evaluate the effectiveness of our partitioning method in terms of subgraph quality, according to the literatures [1,5,12], we adopted the following metrics to measure subgraphs:

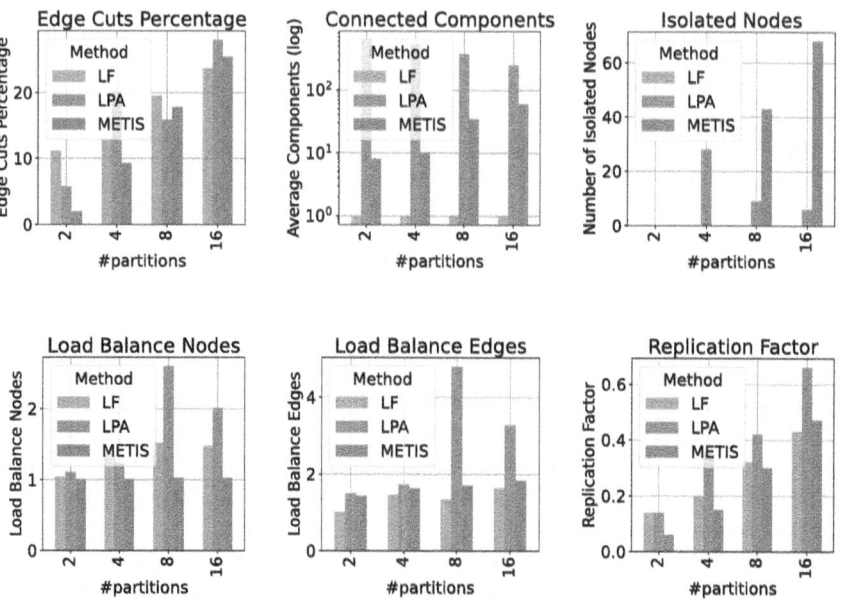

Fig. 4. Comparison of subgraph quality on Arxiv dataset

1. Edge cuts percentage:

$$\tau = \frac{\sum_{i=1}^{k} \Gamma\left(V_i, \bar{V}_i\right)}{m} \tag{5}$$

which is the sum of edge cuts between each partition i and other partitions $\Gamma\left(V_i, \bar{V}_i\right)$ divided by total number of edges m in the graph. Lower edge cuts represent better partition quality.
2. Number of connected components for each partition, which is the number of subgraphs of each partition in which each pair of nodes is connected by a path.
3. Number of isolated nodes for each partition, which is the number of nodes that are not connected to any other nodes.

4. Load balance of nodes:

$$\rho = \frac{\max_{i=1,\dots,k} |P_i|}{|P_{average}|} \tag{6}$$

where $|P_{average}| = \frac{n}{k}$ is the expected number of nodes for each partition in the ideal situation, and $\max_{i=1,\dots,k} |P_i|$ is the maximum number of nodes from k partitions. A lower load balance of nodes represents better partition quality.

5. Load balance of edges: The same formula as for load balance of nodes where $|P_{average}| = \frac{m}{k}$ is the expected number of edges for each partition in the ideal situation, and $\max_{i=1,\dots,k} |P_i|$ is the maximum number of edges from k partitions. A lower load balance of edges represents better partition quality.

6. Replication factor:

$$\mathrm{RF} = \frac{1}{n} \sum_{i \in k} |P_i(v)| \tag{7}$$

where n is the total number of nodes in the graph, and $P_i(v)$ is the total number of replicas of vertices in each partition.

Figure 4 shows the evaluation results of the metrics on the Arxiv dataset, comparing different partitioning methods over different numbers of partitions. The results show that our method excels in minimizing the number of connected components and isolated nodes, ensuring that each partition contains only one connected component and no isolated nodes. In contrast, both LPA and METIS result in multiple connected components and numerous isolated nodes.

In terms of edge cuts and replication factor, our method does not show a significant improvement over other methods when considering 2 to 8 partitions. This is to be expected since the primary goal of our method is not to reduce these factors. However, at 16 partitions, our method performs better than others. This improvement can be attributed to the increase in the number of connected components and isolated nodes in other methods, which negatively affects these factors.

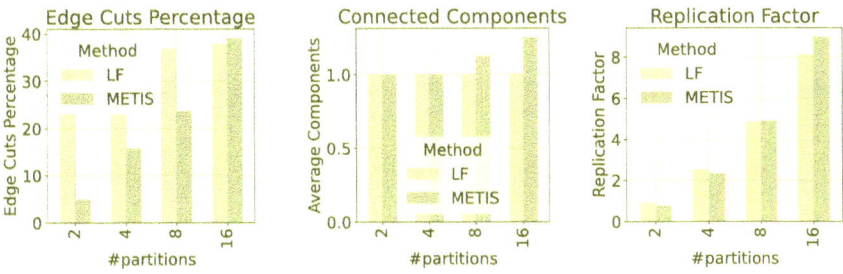

Fig. 5. Comparison on Proteins dataset

Figure 5 shows the results of some metrics on the Proteins dataset. Unlike Arxiv, the Proteins graph is extremely dense with an average node degree of

597, which is 43 times higher than in Arxiv. Therefore, the edge-cut percentage and the replication factor are relatively high, but LF performs relatively better on 16 partitions and wins METIS. Regarding the number of components, once the number of partitions exceeds 4, METIS fails to achieve a single component per partition, while LF remains successful up to 16 partitions.

5.2   Quality Comparison

We evaluated the overall quality of our solution when applied to downstream tasks by following a specific procedure. When creating subgraphs based on partition information, we considered two methods: one that ignores edges between partitions (i.e., inner nodes only) and another that preserves these edges by replicating nodes. These methods will be referred to as *Inner* and *Repli*, respectively. Our goal is to compare the quality of these two approaches.

We train a GCN or GraphSAGE model separately for each partition and obtain the embeddings that are finally combined to train an MLP classifier for the classification task.

Figure 6a shows the accuracy comparison of GCN on the Arxiv dataset with *Inner* and *Repli*, for multi-class prediction, from 2 to 16 partitions, compared to the LPA [4] and METIS [9] partitioning methods, and Fig. 6b shows the corresponding results for GraphSAGE.

Method	Accuracy (%)			
	2	4	8	16
LPA Inner	68.99	66.38	63.07	59.61
LPA Repli	69.60	69.57	67.97	65.62
METIS Inner	**69.59**	68.46	65.68	60.90
METIS Repli	70.32	69.86	68.95	66.70
Our LF Inner	69.33	**69.09**	**66.73**	**65.11**
Our LF Repli	**70.34**	**70.05**	**69.22**	**68.19**

Method	Accuracy (%)			
	2	4	8	16
LPA Inner	69.33	67.86	64.45	62.11
LPA Repli	69.86	68.52	67.37	62.63
METIS Inner	69.90	68.14	67.41	62.98
METIS Repli	70.22	68.54	67.29	64.25
Our LF Inner	**70.63**	**70.90**	68.57	**67.58**
Our LF Repli	70.48	70.46	**69.42**	**68.36**

(a) Accuracy Comparison of GCN on Arxiv Dataset

(b) Accuracy Comparison of SAGE on Arxiv Dataset

Fig. 6. Accuracy Comparison of different methods on Arxiv Dataset

In particular, our LF partitioning method significantly improves the quality compared to the METIS and LPA partitioning methods, for both GCN and SAGE algorithms. For GCN on 16 partitions, LF improves METIS by 6.9% for the *Inner* method and by 2.2% for the *Repli* method. It is important to note that LF achieves almost the highest quality possible which is an accuracy of 71% in a centralized environment. For 16 partitions, the accuracy of the LF method is only 4% lower than that of the centralized solution, while training remains fully localized with low communication costs.

Our method also outperforms for both *Inner* and *Repli*. It should be noted that for all methods, the accuracy of *Repli* is higher than that of *Inner*, which is obvious. In addition, compared to the significant accuracy improvement that GCN brings to *Repli* (for example, for LF 16 partitions, the accuracy is improved by 3%), the improvement for GraphSAGE is not so much (about 1%). The reason may be that GraphSAGE uses a neighbor sampling strategy, so the loss of boundary neighbors has less impact on the model.

We now report quality results for the denser Proteins dataset. Due to its very high density, *Repli* method would replicate too many nodes and increase the training time beyond acceptable limits, thus we only consider the *Inner* method. Table 2 shows the ROC-AUC results of SAGE model.

Table 2. Accuracy Comparison of SAGE on Proteins Dataset

Method	ROC-AUC (%)			
	2	4	8	16
METIS Inner	75.48	67.53	46.45	44.80
Our LF Inner	75.21	65.13	**52.94**	**49.38**

We can see that for 8 and 16 partitions LF's accuracy is more than 10% higher than METIS. This may be because METIS partitions have more than one component. In addition, compared to Arxiv, the accuracy of Proteins drops more when the number of partitions is higher (compared to 76% in centralized training). This may be because we lose more cut edges since the Proteins graph is extremely dense.

5.3 Speed Analysis

Table 3 shows the partitioning time of different partitioning methods on the Arxiv dataset. Note that for our LF, there is 11.5 s of preprocessing time to find communities using Leiden's library [16]. Once we obtain the communities, they can be stored and loaded for further partitioning. Another point is that LF is faster when the number of partitions is larger. This is because LF is an iterative greedy algorithm. For example, two partitions can be considered as obtained by continuing to merge from four partitions. Figure 7 shows the longest training time of all subgraphs obtained by our LF algorithm using the GCN model on the Arxiv dataset. It can be seen that increasing the number of partitions dramatically reduces the training time, while for synchronized distributed frameworks such as DGL [18] and PBG [11], the training time does not decrease much due to numerous communications as discussed in Spark Local [4]. Also, for each partition with *Repli*, the training time increases only a little compared to *Inner*, while the accuracy is much higher as shown in Sect. 5.2.

Table 3. Partitioning time comparison on Arxiv dataset across different methods and partitioning numbers.

Method	Partitioning time (s)			
	2	4	8	16
LPA	71.0	104.5	173.2	327.6
METIS	3.0	3.1	3.1	3.6
Ours (LF)	2.1	2.0	1.8	1.7

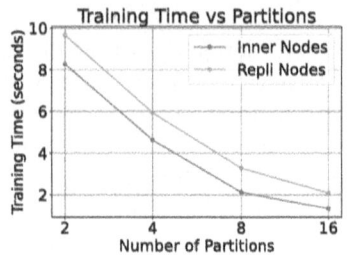

Fig. 7. Training time of LF on Arxiv using GCN

5.4   Impact of Our Fusion Method on Other Partitioning Methods

To further evaluate the benefits of our fusion method, we compared its performance on different partitioning methods. We report the results of our fusion method applied to METIS, LPA, and Leiden for 16 partitions on the GCN model, focusing on partitioning time, edge cuts percentage, and accuracy on the Arxiv dataset. On Table 4 we named each method with a "+F" suffix, which stands for "fusion".

Table 4. Partitioning time(s) and Edge Cuts(%) for 16 partitions on Arxiv

Method	Time(s)	Edge cuts before F(%)	Edge cuts after F(%)
METIS+F	4.8	25.4	25.1
LPA+F	6.6	28.0	27.0
Leiden+F	1.7	–	23.7

We observe that our fusion method reduces the percentage of edge cuts for both METIS and LPA partitioning methods, resulting in improved partition quality. Regarding the fusion time, we note that the fusion process is 2.2 times faster when applied to Leiden compared to METIS (and 3.9 times faster compared to LPA). This is because Leiden inherently guarantees connected communities, whereas for METIS and LPA, we need to additionally identify each connected component.

Table 5 shows the accuracy results for GCN model on Arxiv dataset. Comparing to Fig. 6a, we can observe that our fusion method highly improved the accuracy results for both METIS and LPA partitioning methods, *Inner* results is comparable to the Leiden Fusion, while LF yieds better results for *Repli*. The combination of our Leiden + Fusion method proves its efficiency and effectiveness.

Table 5. Accuracy results (%) for GCN 16 partitions

Method	METIS	METIS+F	LPA	LPA+F	Leiden+F
Inner	60.90	**65.75**	59.61	64.51	65.11
Repli	66.70	67.60	65.62	66.85	**68.19**

6 Conclusion

Current partitioning methods and distributed frameworks face two major challenges in effectively training GNNs that hinder the handling of large networks: 1) the need for continuous communication in synchronized distributed frameworks to access information from other machines, and 2) the inability to ensure that subgraphs remain connected components without isolated nodes. To address these issues, we introduce Leiden-Fusion, a novel partitioning method designed for large-scale graph training with minimal communication. We made the following contributions: (i) For any initially connected graph, our novel partitioning method ensures that each partition is a single densely connected component with no isolated nodes, facilitating effective GNN training. (ii) By adopting a local training strategy without communication, we significantly reduced the training time while maintaining most of the embedding quality. This approach demonstrates that high training efficiency is achievable for GNNs without sacrificing accuracy, enabling more scalable and efficient distributed learning on very large graphs. In future work, we plan to extend our method to handle graphs with multiple components and isolated nodes, and to evaluate its accuracy and efficiency on graphs with different size densities.

Acknowledgement. This work is funded by the SCAI (Sorbonne Center for Artificial Intelligence) at Sorbonne University, France.

References

1. Ayall, T.A., et al.: Graph computing systems and partitioning techniques: a survey. IEEE Access **10**, 118523–118550 (2022)
2. Blondel, V.D., Guillaume, J.L., Lambiotte, R., Lefebvre, E.: Fast unfolding of communities in large networks. J. Stat. Mech: Theory Exp. **2008**(10), P10008 (2008)
3. Consortium, G.O.: The gene ontology resource: 20 years and still going strong. Nucleic acids Res. **47**(D1), D330–D338 (2019)
4. Duong, C.T., Hoang, T.D., Yin, H., Weidlich, M., Nguyen, Q.V.H., Aberer, K.: Scalable robust graph embedding with spark. Proc. VLDB Endowment **15**(4), 914–922 (2021)
5. Gonzalez, E.A.: {PowerGraph}: distributed {Graph-Parallel} computation on natural graphs. In: 10th USENIX Symposium on Operating Systems Design and Implementation (OSDI 12), pp. 17–30 (2012)
6. Hamilton, W., Ying, Z., Leskovec, J.: Inductive representation learning on large graphs. In: Advances in Neural Information Processing Systems, vol. 30 (2017)

7. Hao, et al.: An end-to-end model for question answering over knowledge base with cross-attention combining global knowledge. In: Proceedings of the 55th Annual Meeting of the Association for Computational Linguistics, pp. 221–231 (2017)
8. Hu, W., et al.: Open graph benchmark: datasets for machine learning on graphs. Adv. Neural. Inf. Process. Syst. **33**, 22118–22133 (2020)
9. Karypis, G., Kumar, V.: Metis: A software package for partitioning unstructured graphs, partitioning meshes, and computing fill-reducing orderings of sparse matrices (1997)
10. Kipf, T.N., Welling, M.: Semi-supervised classification with graph convolutional networks. arXiv preprint arXiv:1609.02907 (2016)
11. Lerer, A., et al.: Pytorch-biggraph: a large scale graph embedding system. Proc. Mach. Learn. Syst. **1**, 120–131 (2019)
12. LowY, B., et al.: DistributedGraphLab: aframeworkformachinelearninganddata mininginthecloud. ProceedingsoftheVLDBEndowment **5**(8), 716 (2012)
13. Malewicz et al.: Pregel: a system for large-scale graph processing. In: Proceedings of the 2010 ACM SIGMOD International Conference on Management of Data, pp. 135–146 (2010)
14. Martella, C., Logothetis, D., Loukas, A., Siganos, G.: Spinner: Scalable graph partitioning in the cloud. In: 2017 IEEE 33rd international conference on data engineering (ICDE), pp. 1083–1094. IEEE (2017)
15. Szklarczyk et al.: String v11: protein–protein association networks with increased coverage, supporting functional discovery in genome-wide experimental datasets. Nucleic Acids Res. **47**(D1), D607–D613 (2019)
16. Traag, V.A., Waltman, L., Van Eck, N.J.: From louvain to leiden: guaranteeing well-connected communities. Sci. Rep. **9**(1), 5233 (2019)
17. Wang, K., Shen, Z., Huang, C., Wu, C.H., Dong, Y., Kanakia, A.: Microsoft academic graph: when experts are not enough. Quant. Sci. Stud. **1**(1), 396–413 (2020)
18. Wang, M.Y.: Deep graph library: towards efficient and scalable deep learning on graphs. In: ICLR Workshop on Representation Learning on Graphs and Manifolds (2019)
19. Wang, X., Cui, P., Wang, J., Pei, J., Zhu, W., Yang, S.: Community preserving network embedding. In: Proceedings of the AAAI Conference on Artificial Intelligence, vol. 31 (2017)
20. Zachary, W.W.: An information flow model for conflict and fission in small groups. J. Anthropol. Res. **33**(4), 452–473 (1977)
21. Zhang, E.A.: Collaborative knowledge base embedding for recommender systems. In: Proceedings of the 22nd ACM SIGKDD International Conference on Kowledge Discovery and Data Mining, pp. 353–362 (2016)

Automated Design of Linear Bounding Functions for Sigmoidal Nonlinearities in Neural Networks

Matthias König[1], Xiyue Zhang[2(✉)], Holger H. Hoos[1,3], Marta Kwiatkowska[2], and Jan N. van Rijn[1]

[1] Leiden University, Leiden, The Netherlands
{h.m.t.konig,j.n.van.rijn}@liacs.leidenuniv.nl, hh@aim.rwth-aachen.de
[2] University of Oxford, Oxford, UK
{xiyue.zhang,marta.kwiatkowska}@cs.ox.ac.uk
[3] RWTH Aachen University, Aachen, Germany

Abstract. The ubiquity of deep learning algorithms in various applications has amplified the need for assuring their robustness against small input perturbations such as those occurring in adversarial attacks. Existing *complete* verification techniques offer provable guarantees for all robustness queries but struggle to scale beyond small neural networks. To overcome this computational intractability, *incomplete* verification methods often rely on convex relaxation to over-approximate the nonlinearities in neural networks.

Progress in tighter approximations has been achieved for piecewise linear functions. However, robustness verification of neural networks for general activation functions (*e.g.*, Sigmoid, Tanh) remains under-explored and poses new challenges. Typically, these networks are verified using convex relaxation techniques, which involve computing linear upper and lower bounds of the nonlinear activation functions.

In this work, we propose a novel parameter search method to improve the quality of these linear approximations. Specifically, we show that using a simple search method, carefully adapted to the given verification problem through state-of-the-art algorithm configuration techniques, improves the average global lower bound by 25% on average over the current state of the art on several commonly used local robustness verification benchmarks.

Keywords: Neural Network Verification · Automated Algorithm Configuration · Convex Relaxation

1 Introduction

Over the last decade, deep learning algorithms have gained increasing significance as essential tools across diverse application domains and usage scenarios.

M. König—Work done while at University of Oxford.

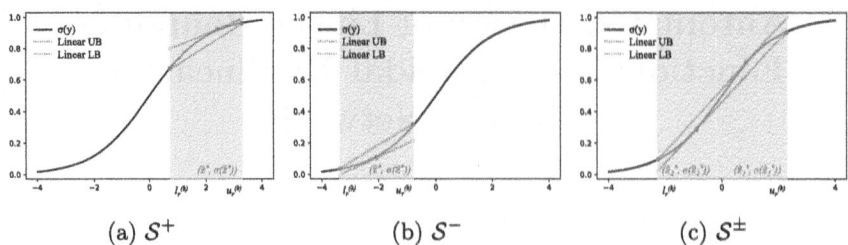

Fig. 1. Linear bounding rules for different cases of the Sigmoid activation function. The x-axis shows the pre-activation bounds, while the y-axis indicates the output of the activation function.

Their applications range from manoeuvre advisory systems in unmanned aircraft to face recognition in mobile phones (see, *e.g.*, [18]). Simultaneously, it is now widely acknowledged that neural networks are susceptible to adversarial attacks, where a given input is manipulated to cause misclassification [31]. Remarkably, in image recognition tasks, the perturbation required can be so subtle that it remains virtually imperceptible to the human eye.

Numerous methods have been proposed to evaluate the robustness of neural networks against adversarial attacks. Early methods involve empirical attacks [7,12,22]; however, these approaches do not provide a comprehensive assessment of neural network robustness, as a defence mechanism against one type of attack might still be vulnerable to another, potentially novel, class of attacks. Consequently, there have been efforts to develop approaches that formally verify the robustness of neural networks against adversarial attacks [3,6,11,13,19,33,37]. These formal verification methods enable a principled assessment of neural network robustness, providing provable guarantees on desirable properties, typically in the form of a pair of input-output specifications. However, *complete* verification of neural networks, where the verifier is theoretically guaranteed to provide a definite answer to the property under verification, is a challenging NP-complete problem [19]. Solving this problem usually requires computationally expensive methods, *e.g.*, SMT solvers [19] or mixed integer linear programming systems [34], limiting scalability and efficiency.

The aforementioned computational complexity is mainly due to the non-linear activation functions in a neural network, which results in the neural network verification problem becoming a non-convex optimisation problem. In light of this, *incomplete* methods have been proposed that exploit *convex relaxation* techniques, which approximate the nonlinearities using linear symbolic bounds, to provide sound and efficient verification [1,11,28,30,39]; completeness can be achieved through branch and bound (see, *e.g.* [5]). Most formal verification methods are limited to ReLU-based networks (see [23]), which satisfy the piecewise linear property. However, convex relaxation techniques are applicable to *more general* commonly-used activation functions, such as Sigmoid or Tanh, by approximating them in terms of piecewise linear functions. This enables an extension of formal verification methods based on convex relaxation to general activation functions, though these remain under-explored.

An important application of neural network verification tools is *robustness certification*, which computes guarantees that the prediction of the network is stable (invariant) around a given input point. CROWN [39] is the first generic framework leveraging adaptive linear bounds to efficiently certify robustness for general activation functions. Another series of techniques, *e.g.*, DeepZ [28], DeepPoly [30] employs abstract interpretation and abstract transformers for commonly-used activation functions, such as Sigmoid and Tanh. CROWN computes linear upper and lower bounds of nonlinear functions, while DeepZ and DeepPoly incorporate convex relaxation into the abstract transformer.

Convex relaxation methods typically over-approximate the activation functions in all nonlinear neurons in the neural network, which inevitably introduces imprecision. Consequently, the corresponding verification algorithms may fail to prove the robustness of a neural network when the original network satisfies the specification and thus weaken certification guarantees. A globally, *i.e.*, network-wise, tighter over-approximation is crucial for strong certification guarantees. However, for Sigmoidal activations, it remains unclear how to design or configure the linear approximation of nonlinear neurons to achieve globally tight output bounds, which are directly used to determine the robustness of neural networks with respect to a specific property under investigation.

Motivated by the need to reduce imprecision of the bounding functions (as an over-approximation of the activation function), in this work we introduce an automated and systematic method to compute tighter bounding functions to improve certification guarantees. These bounding functions are defined by the tangent point, where they touch the activation function. To this end, we propose a novel parameter search method for identifying the tangent points to find better-suited linear bounding functions by considering different cases of Sigmoidal activation functions. To tackle the infinite search space (of tangent points) for the bounding functions, our approach leverages state-of-the-art algorithm configuration techniques.

Concretely, we use automated algorithm configuration techniques to find optimal hyper-parameters of the search method used for obtaining the tangent points of the linear bounding functions, which has previously been done using binary search [39]. These hyper-parameters control the initial tangent point per neuron as well as the rate at which these initial points are updated (we will refer to the latter as a *multiplier* for the remainder of this work) until a feasible bounding function has been obtained. Notice that these hyper-parameters are set for the entire network, *i.e.*, all neurons share the same starting point and multiplier; however, they eventually result in different tangent points, as the search process only ends once a feasible bounding function has been found.

Moreover, we show that, by using our proposed method, we improve the average lower bound on the network output by 25% on average across several verification benchmarks, and can certify robustness for instances that were previously unsolved.

2   Related Work

In the following, we give some background on using convex relaxation for neural network robustness verification and on automated algorithm configuration.

2.1   Convex Relaxation for Neural Network Verification

We use $f : \mathbb{R}^n \to \mathbb{R}^m$ to denote a neural network trained to make predictions on an m-class classification problem. Let $\mathbf{W}^{(i)}$ denote the weight matrix and $\mathbf{b}^{(i)}$ denote the bias for the i-th layer. We use $\hat{z}^{(i)}$ to denote the pre-activation neuron values, and $z^{(i)}$ to denote the post-activation ones, such that we have $\hat{z}^{(i)} = \mathbf{W}^{(i)} z^{(i-1)} + \mathbf{b}^{(i)}$ (*i.e.*, the post-activation values of layer $i-1$ are linearly weighted to become the pre-activation values of layer i). We use σ to denote the activation functions in intermediate layers, such that $z = \sigma(\hat{z})$. In this work, we focus on neural networks with sigmoidal activation functions, including Sigmoid $\sigma(x) = 1/(1 + e^{-x})$ and Tanh $\sigma(x) = (e^x - e^{-x})/(e^x + e^{-x})$. We use $f(x)$ to denote the neural network output for all classes and $f_j(x)$ to denote the output associated with class j (with $1 \leq j \leq m$). The final decision is computed by taking the label with the greatest output value, *i.e.*, arg $\max_j f_j(x)$.

Neural Network Verification. The robustness to adversarial perturbations is one of the most important properties of neural networks, requiring that the predictions of a neural network should be preserved for a local input region $\mathcal{C}$, typically defined as a l_p norm ball with a radius of ϵ around the original input x_0, in the following way:

$$\forall x \in \mathcal{C}, \arg\max_j f_j(x) = \arg\max_j f_j(x_0). \tag{1}$$

For an input instance x_0 with ground-truth label y_0, verifying the robustness property can then be transformed into proving that $\forall x \in \mathcal{C}, f_{y_0}(x) - f_j(x) \geq 0$ for all $j \neq y_0$ where $j \in [1, m]$.

Given a verification problem, for example, to check whether the output constraint $f_{y_0}(x) - f_j(x) \geq 0$ is satisfied, we can append an additional layer at the end of the neural network, such that the output property can be merged into the new neural network function $g : \mathbb{R}^n \to \mathbb{R}$ where $g(x) = f_{y_0}(x) - f_j(x)$. While this should formally be done for each possible network output $j \neq y_0$, to simplify the notation we use a single g.

In this way, the verification problem can be formulated canonically as follows, i.e., to prove or falsify:

$$\forall x \in \mathcal{C}, g(x) \geq 0 \tag{2}$$

One way to verify the property is to solve the optimisation problem $\min_{x \in \mathcal{C}} g(x)$. However, due to the nonlinearity of the activation function σ, the neural network verification problem is NP-complete [19]. To address such intractability, state-of-the-art verification algorithms leverage convex relaxation to transform the verification problem into a convex optimisation problem.

Definition 1 (Convex relaxation of activation functions). *For a non-linear activation function $\sigma(\hat{z})$ with pre-activation bounds $\hat{z} \in [l, u]$, convex relaxation relaxes the non-convex equality constraint $z = \sigma(\hat{z})$ to two convex inequality constraints by computing the linear lower and upper bounding functions $h_L(\hat{z}) = \alpha_L\hat{z} + \beta_L$ and $h_U(\hat{z}) = \alpha_U\hat{z} + \beta_U$, such that $h_L(\hat{z}) \leq \sigma(\hat{z}) \leq h_U(\hat{z})$.*

In Fig. 1, the pre-activation bounds are indicated by the grey-shaded area, h_L is illustrated by the red dotted line, whereas h_U is indicated by the green dotted line. Notice that the parameters $\alpha_L, \beta_L, \alpha_U, \beta_U$ of the linear functions depend on the pre-activation bounds l and u. With such convex inequality constraints, we can propagate the relaxations through layers using an efficient back-substitution procedure to compute the linear lower bounds (denoted by $\underline{g}$) and upper bounds (denoted by $\overline{g}$) for the neural network g.

Definition 2 (Convex relaxation of neural networks). *A convex relaxation of the neural network $g : \mathbb{R}^n \to \mathbb{R}$ over an input region $\mathcal{C}$, are two linear functions $\underline{g}$ and $\overline{g}$ such that $\underline{g}(x) \leq g(x) \leq \overline{g}(x)$ for all $x \in \mathcal{C}$.*

Using convex relaxation, the verification problem is then reduced to the following convex optimisation problem:

$$g^* = \min_{x \in \mathcal{C}} \underline{g}(x) \tag{3}$$

The robustness property is proved if the optimal solution $g^* \geq 0$, as we have $\forall x \in \mathcal{C}, g(x) \geq \underline{g}(x)$. Meanwhile, convex relaxation of the nonlinear constraints inevitably introduces approximation. As a consequence, the computed minimum might fail to satisfy $\min_{x \in \mathcal{C}} \underline{g}(x) \geq 0$ even in cases in which the network is robust, i.e., $\min_{x \in \mathcal{C}} g(x) \geq 0$. Instead, the true verification remains unknown. Therefore, improving the quality of the approximation is crucial to reducing false failures in robustness verification, and thus strengthening certification guarantees, which is the aim of our contribution.

2.2   Automated Algorithm Configuration

In general, the algorithm configuration problem can be described as follows: Given an algorithm A (also referred to as the *target algorithm*) with parameter configuration space Θ, a set of problem instances Π, and a cost metric $c : \Theta \times \Pi \to \mathbb{R}$, find a configuration $\theta^* \in \Theta$ that minimises cost c across the instances in Π:

$$\theta^* \in \arg\min_{\theta \in \Theta} \sum_{\pi \in \Pi} c(\theta, \pi) \tag{4}$$

The general workflow of the algorithm configuration procedure starts with selecting a configuration $\theta \in \Theta$ and an instance $\pi \in \Pi$. Next, the configurator initialises a run of algorithm A with configuration θ on instance π, and measures the resulting cost $c(\theta, \pi)$. The configurator uses this information about the target algorithm's performance to find a configuration that performs well on the training instances. This is enabled by a surrogate model, which provides a posterior probability distribution that characterises the potential cost for $c(\theta, \pi)$ at a configuration θ. At every time point t, we have a new observation of the cost c_t at a new configuration point θ_t. The posterior distribution is updated based on the augmented observation set $S_t = S_{t-1} \cup \{(\theta_t, c_t)\}$, and the next configuration point is selected by maximising the acquisition function in the following form:

$$\theta_{t+1} = \arg\max_{\theta \in \Theta} a_t(\theta, S_t) \qquad (5)$$

where a_t denotes the acquisition function. A typical choice for the acquisition functions is the expected improvement [16].

Once the configuration budget (e.g., time budget or number of trials) is exhausted, the procedure returns the current incumbent θ^*, which represents the best configuration found so far. Finally, when running the target algorithm with configuration θ^*, it should result in lower cost (such as average running time) or improved solution quality across the benchmark set.

Automated algorithm configuration has already been shown to work effectively in the context of formal neural network verification [21], but also in other, related domains, such as SAT solving [14,17], scheduling [8], mixed-integer programming [15,24], evolutionary algorithms [2], answer set solving [10], AI planning [35], and machine learning [9,32].

3   Method

We consider the task of local robustness verification with Sigmoidal activation functions. To this end, we focus on convex relaxation based perturbation analysis as employed in the CROWN framework [39], and use automated algorithm configuration techniques to improve the linear bounds of the nonlinear activation functions. Notice that, by default, CROWN employs binary search to obtain the points at which the linear bounding functions are tangent to the activation function [39].

In this study, on the other hand, we used SMAC [16] to guide the search for suitable tangent points. SMAC is a widely known, freely available, state-of-the-art configurator based on sequential model-based optimisation (also known as Bayesian optimisation). The main idea of SMAC is to construct and iteratively update a statistical model of target algorithm performance to guide the search for promising configurations; SMAC uses a Random Forest regressor [4].

3.1   Configuration Objective

The objective of the configuration procedure is to maximise the minimum of the global lower bound g^* as defined in Eq. (3) for each instance under verification. Notice that the optimisation procedure that solves Eq. (3) is performed separately for each instance, i.e., $|\Pi| = 1$. Recall that an instance is verified to be robust when $g^* > 0$, as outlined in Sect. 2.1.

By design, SMAC solves a minimisation problem; see Eq. (4). Since we are interested in maximising the lower bound of the network output, we apply appropriate sign changes and define the cost metric c as the negative of the global lower bound.

3.2 Configuration Space

The CROWN framework [39] proposed a general certification solution for neural networks with nonlinear activation functions, which relies on convex relaxation to compute the output bounds. Since Sigmoidal activation functions (Sigmoid/Tanh/Arctan) share the same features, that is, convex on the negative side ($x < 0$) and concave on the other side ($x > 0$), the authors of [39] leveraged this feature and proposed a general method to compute the parameters (i.e., the tangent points) of linear upper and lower functions h_U, h_L.

Based on the curvature of Sigmoidal activation functions, each nonlinear neuron of the i-th layer (denoted as $[n_i]$) is categorised into one of the three cases: $\mathcal{S}^+$, $\mathcal{S}^-$, and $\mathcal{S}^\pm$ where $\mathcal{S}^+ = \{j \in [n_i] \mid 0 \le l_j^{(i)} \le u_j^{(i)}\}$, $\mathcal{S}^- = \{j \in [n_i] \mid l_j^{(i)} \le u_j^{(i)} \le 0\}$, and $\mathcal{S}^\pm = \{j \in [n_i] \mid l_j^{(i)} \le 0 \le u_j^{(i)}\}$. Intuitively, $\mathcal{S}^+$ represents the case in which the pre-activation bounds are both positive, $\mathcal{S}^-$ represents the case in which they are both negative and $\mathcal{S}^\pm$ represents the case in which l is negative and u is positive. Different bounding rules are then proposed for these three cases; these are illustrated in Fig. 1.

Bounding rules for S^+ domain. For any node $j \in \mathcal{S}^+$, the activation function $\sigma(\hat{z}_j)$ is concave; hence, a tangent line of $\sigma(\hat{z}_j)$ (represented in green in Fig. 1) at any tangent point $\hat{z}_j^* \in [l_j^{(i)}, u_j^{(i)}]$ is a valid upper bounding function $h_{U,j}^{(i)}$, and the linear function passing the two endpoints, $(l_j^{(i)}, \sigma(l_j^{(i)}))$ and $(u_j^{(i)}, \sigma(u_j^{(i)}))$, serves as a valid lower bounding function $h_{L,j}^{(i)}$.

To select good values of $\hat{z}_j^*$, we propose a search method whose behaviour depends on two hyper-parameters, which are optimised by SMAC:

– starting point s at which each tangent point $\hat{z}_j^*$ is initialised;
– multiplier ψ to change the value of $\hat{z}_j^*$ if the linear bound is found to be invalid.

This search method is illustrated in Algorithm 1. The algorithm initiates and maintains the best configuration $\theta^* = (s^*, \psi^*)$ found so far, which leads to a tighter bounding function (defined by the tangent points $\hat{z}_j^*$ for each node j) and, thus, a tighter global lower bound g^*. To find optimal settings for the hyper-parameters s and ψ, SMAC samples values of s and ψ from a pre-defined configuration space Θ. Configurations θ, as introduced in Eq. (4), are sampled from $s \in [0.01, 2]$ and $\psi \in [1.01, 3]$. Note that these ranges are based on empirical observations; values outside this range typically result in extremely loose bounds. In principle, other ranges could be used to sample values of s and ψ, respectively, as long as $s > 0$ and $\psi > 1$. Also note that the search method will be the same for the entire network; however, it will result in different tangent points per node.

Algorithm 1. Our proposed tangent point search method for sigmoidal functions. It details only the search method in the S^+ domain (for space reasons). The hyper-parameter s determines the initial tangent point and the hyper-parameter ψ determines the rate by which s is changed. The values of s and the update rate ψ are not exposed as hyper-parameters in vanilla CROWN. In our case, these are optimised by SMAC, where SMAC uses a fixed budget of $n_{\max}$ evaluation calls. Notice that in vanilla CROWN, binary search is performed once initial bounds have been obtained to move tangent points closer to 0; this step is not performed in our method.

1: **procedure** OPTIMISESEARCHPARAMETERS(x, s, ψ)
2:      Initialize cost $c_0 = \inf$ ▷ first iteration of the optimisation loop can overwrite it
3:      Initialise observation set $S_0 = \{\emptyset\}$
4:      **for** $n = 1, \cdots, n_{\max}$ **do**
5:          Select starting point and multiplier rate based on observation set
6:          $(s_n, \psi_n) \leftarrow \underset{(s,\psi) \in \Theta}{\arg\max}\, a_n((s,\psi), S_{n-1})$
7:          **for all** node j in network f **do**
8:              **if** $j \in S^+$ **then**
9:                  Initialise upper bound tangent point for current node $\hat{z}_j^* \leftarrow s_n$
10:                 **while** Tangent point $\hat{z}_j^*$ leads to invalid upper bound $h_{U,j}^{(i)}$ **do**
11:                     Update $\hat{z}_j^*$: $\hat{z}_j^* \leftarrow \hat{z}_j^* \cdot \psi_n$
12:                 **end while**
13:             **else if** $j \in S^-$ **then**
14:                 Determine tangent point for lower bounding function, in a similar fashion as above
15:             **else**   ▷ In this case, $j \in S^\pm$;
16:                 Determine tangent point for both lower bounding and upper bounding function, in a similar fashion as above
17:             **end if**
18:         **end for**
19:         Evaluate cost $c_n = -1 \cdot g^*(x)$ ▷ Linear bounding layer-by-layer via convex relaxation as per Def. 2 and Eq. 3
20:         Augment observation set $S_n = S_{n-1} \cup \{((s_n, \psi_n), c_n)\}$
21:         Train surrogate model using S_n
22:         **if** $c_n < c^*$ **then**
23:             $s^* \leftarrow s_n$, $\psi^* \leftarrow \psi_n$, $c^* \leftarrow c_n$
24:         **end if**
25:     **end for**
26:     **return** c^*, (s^*, ψ^*)
27: **end procedure**

Based on a given hyperparameter configuration, we can identify the tangent points $\hat{z}_j^*$ for each node j, and compute the global lower bound g^* as well as the cost value $c = -1 \cdot g^*(x)$. At each iteration, we can acquire the parameter configuration based on the updated surrogate model (trained with the new observation set S_n). With the selected candidate configuration, we perform a sanity check to ensure the validity of the linear upper bounding function incurred by the tangent points $\hat{z}_j^*$ at each node. An upper bound is considered invalid if, at any given point $\hat{z}_j^*$, the value of $h_{U,j}^{(i)}(\hat{z}_j^*)$ is smaller than the value of the nonlinear sigmoidal $\sigma(\hat{z}_j^*)$. The global lower bound g_n (also the cost value c_n) is then updated with the new upper bounding function. The loop terminates and returns the best-achieved cost value c^* as well as the corresponding configuration (s^*, ψ^*) when the maximum iteration limit $n_{\texttt{max}}$ is reached. Notice that SMAC has several additional options that can slightly deviate from the description [16].

While Algorithm 1 elaborates on tangent point search for the upper bounding function, it can be similarly configured to search for lower bounding functions with two key modifications. Firstly, the parameters are initialised to different value ranges. Secondly, the validity evaluation is to check the bounding function will always return a lower value than the sigmoidal function. We introduce the details for S^- and $S^\pm$ domains in the following.

Bounding Rules for S^- Domain. Symmetrically, for any node $j \in \mathcal{S}^-$, $h_{U,j}^{(i)}$ is defined as the linear function passing the two endpoints and $h_{L,j}^{(i)}$ can be a tangent line of $\sigma(\hat{z}_j)$ at any point $\hat{z}_j^* \in [l_j^{(i)}, u_j^{(i)}]$. In this case, we employ a similar search method, except that a configuration θ^* is now sampled from $s \in [-0.01, -2]$ and $\psi \in [-1.01, -3]$. Again, other ranges could be used to sample values of s and ψ, respectively, as long as $s < 0$ and $\psi < -1$. Furthermore, a bound is considered invalid if, at any given point $\hat{z}_j^*$, the value of $h_{L,j}^{(i)}(\hat{z}_j^*)$ is larger than the value of $\sigma(\hat{z}_j^*)$.

Bounding Rules for $S^\pm$ Domain. Lastly, for any node $j \in \mathcal{S}^\pm$, $h_{U,j}^{(i)}$ is a tangent line passing $(l_j^{(i)}, \sigma(l_j^{(i)}))$ and a tangent point $(\hat{z}_1^*, \sigma(\hat{z}_1^*))$, where $\hat{z}_1^* \geq 0$, and $h_{L,j}^{(i)}$ is a tangent line that passes $(u_j^{(i)}, \sigma(u_j^{(i)}))$ and a tangent point $(\hat{z}_2^*, \sigma(\hat{z}_2^*))$, where $\hat{z}_2^* \leq 0$. Again, we employ the same search method to obtain $\hat{z}_j^*$ for each bound. Moreover, for any $h_{U,j}^{(i)}$, a configuration θ^* is sampled from $s \in [0.01, 2]$ and $\psi \in [1.01, 3]$, while for any $h_{L,j}^{(i)}$, a configuration θ^* is sampled from $s \in [-0.01, -2]$ and $\psi \in [-1.01, -3]$. Notice that we can use similar rules for bounding Tanh functions, which share the Sigmoidal shape but range from –1 to 1 instead of 0 to 1.

4   Setup for Empirical Evaluation

We will empirically investigate the effectiveness of the proposed search procedure. We consider two types of neural network architecture: convolutional neural networks (CNNs) and fully connected neural networks (FNNs). We evaluate the effectiveness of our approach on the ERAN benchmark [25–28,30]. Following the naming conventions, we refer to these networks as ConvMed and FNN_6x500. These networks are adversarially trained on the CIFAR-10 dataset with $\epsilon = 0.0313$. In addition, to investigate whether our findings hold for additional datasets, we consider a CNN with similar architecture that is adversarially trained on the MNIST dataset with $\epsilon = 0.3$. Furthermore, to demonstrate the

Table 1. Experimental results obtained from our automated bound configuration method, compared to the vanilla CROWN algorithm, which relies on binary search to obtain the tangent points (with α-optimisation for Sigmoid networks). Boldfaced values indicate superior performance.

Dataset	Network	Activation	Epsilon	Avg. Global Lower Bound $\mathbf{g}^*$			# Certified Instances	
				Baseline	Configured	Improvement	Baseline	Configured
CIFAR10	ConvMed	Sigmoid	0.0313	−15.611	**−5.506**	184%	0	**1**
CIFAR10	ConvMed	Sigmoid	0.0157	−1.727	**−1.633**	6%	13	13
CIFAR10	ConvMed	Sigmoid	0.0078	−0.709	**−0.702**	1%	29	**31**
CIFAR10	ConvMed	Sigmoid	0.0039	−0.440	−0.440	−	39	39
CIFAR10	FNN_6x500	Sigmoid	0.0313	−18.448	**−16.433**	12%	0	0
CIFAR10	FNN_6x500	Sigmoid	0.0157	−17.819	**−16.234**	10%	0	0
CIFAR10	FNN_6x500	Sigmoid	0.0078	−15.075	**−13.478**	12%	0	0
CIFAR10	FNN_6x500	Sigmoid	0.0039	−6.731	**−5.915**	14%	5	**8**
MNIST	ConvMed	Sigmoid	0.3	−5.153	**−4.175**	23%	3	**4**
MNIST	ConvMed	Sigmoid	0.12	−9.484	**−8.563**	10%	0	0
MNIST	ConvMed	Sigmoid	0.06	−0.110	**−0.070**	57%	54	54
MNIST	ConvMed	Sigmoid	0.03	2.753	2.756	−	92	92
CIFAR10	ConvMed	Tanh	0.0313	−65.849	**−55.565**	19%	0	0
CIFAR10	ConvMed	Tanh	0.0157	−44.504	**−33.723**	32%	0	0
CIFAR10	ConvMed	Tanh	0.0078	−10.835	**−9.348**	16%	2	2
CIFAR10	ConvMed	Tanh	0.0039	−2.299	**−2.259**	2%	15	**16**

generality of our method to other Sigmoidal activation functions, we perform the evaluation on a neural network trained on CIFAR-10 with similar architecture, adversarially trained with $\epsilon = 0.0313$, but using Tanh activation functions. Further details about the considered networks can be found in the ERAN repository.

We verify each network for local robustness with respect to the first 100 instances in the test set of the MNIST and CIFAR-10 datasets, respectively, initially with perturbation radii equivalent to the values used during training. We also evaluate our method on a broader range of perturbation radii; specifically, we considered $\epsilon \in \{0.0157, 0.0078, 0.0039\}$ for CIFAR-10 and $\epsilon \in \{0.12, 0.06, 0.03\}$ for MNIST. Notice that these values of ϵ are in line with commonly chosen values from the verification literature [20,29,36].

For verification, we employ CROWN with default settings as provided by the authors. For Sigmoid-based networks, CROWN also performs α-optimisation [38]; however, this is not implemented for Tanh activation functions, where we run CROWN without further optimisation. As we are purely interested in the global lower bound on the network output, we skip the PGD attack used for upper bound computation. For the configuration procedure, we set the number of evaluation calls performed by SMAC to 150, *i.e.*, the configuration process

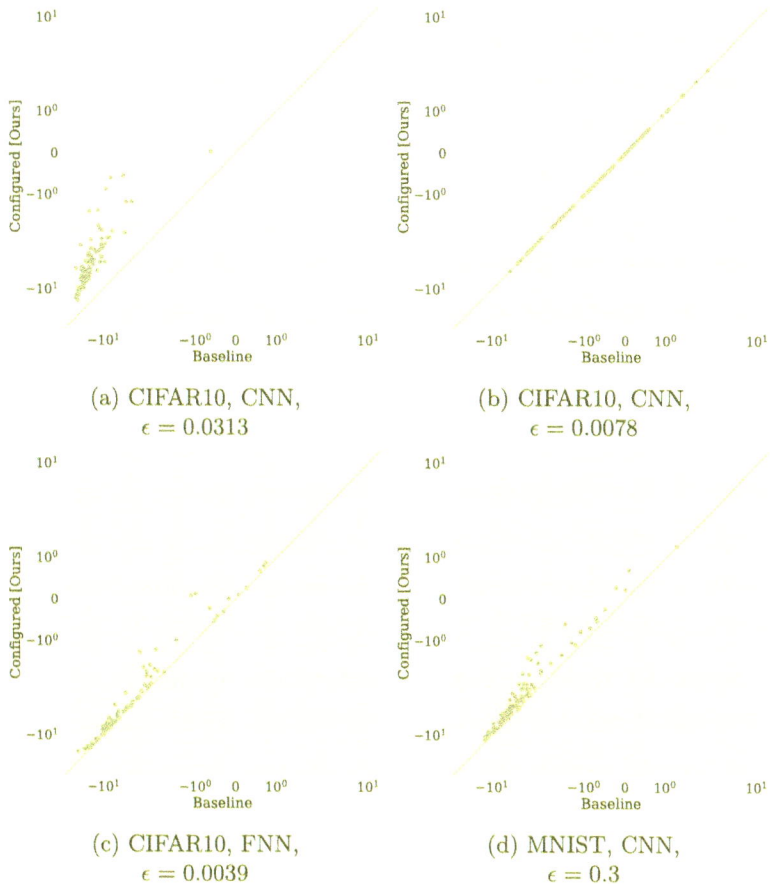

(a) CIFAR10, CNN,
$\epsilon = 0.0313$

(b) CIFAR10, CNN,
$\epsilon = 0.0078$

(c) CIFAR10, FNN,
$\epsilon = 0.0039$

(d) MNIST, CNN,
$\epsilon = 0.3$

Fig. 2. Experimental results obtained for Sigmoid-based networks. Each dot represents a problem instance and the global lower bound, *i.e.*, the value of g^*, for that instance achieved by the baseline approach (x-axis) *vs* our method (y-axis).

terminates after 150 trials. Experiments are performed on a cluster of machines equipped with NVIDIA GeForce GTX 1080 Ti GPUs with 11 GB video memory.[1]

5   Experimental Results and Discussion

Table 1 shows the average global lower bound over the given test instances achieved by the vanilla CROWN algorithm, which represents our baseline, and those achieved by CROWN in combination with our configured bounding method. In addition, we report the absolute number of instances for which robustness certification could be obtained by each approach.

---

[1] Code is available at https://github.com/ADA-research/nnv-bound-configuration.

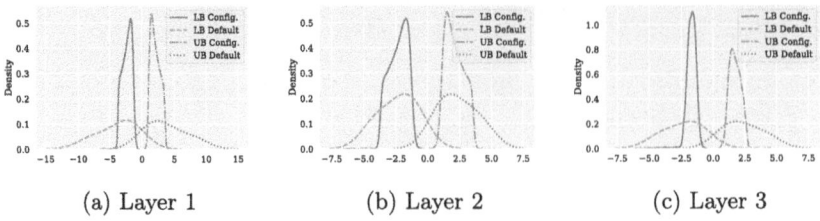

(a) Layer 1   (b) Layer 2   (c) Layer 3

Fig. 3. Probability density functions of the values of $\hat{z}_j^*$ obtained by our method as well as vanilla CROWN for lower and upper bounding functions per activation layer. Remember that $\hat{z}_j^*$ determines the tangent point of the bounding function of a given node j.

5.1 Sigmoid-Based Networks

We first report the results obtained for the Sigmoid-based CNN trained on the CIFAR-10 dataset. These can be found in Table 1. When certifying this network with $\epsilon = 0.03$, we obtained an improvement in the average global lower bound of 184% (-15.611 *vs* -5.506). This is also visualised in Fig. 2a. As shown there, the global lower bound is improved by almost an order of magnitude for some of the instances. Furthermore, using our configured bounds, we could certify robustness for an instance that could not be solved by the baseline approach.

For $\epsilon = 0.0157$ and $\epsilon = 0.0078$, we achieved improvements of 6% and 1%, respectively. Interestingly, although the latter is a rather small improvement in the average global lower bound, our method could verify two additional instances, which the baseline method was unable to solve. These results are visualised in Fig. 2b. Although instances generally lie very close to the equality line, our method could improve on instances with lower bounds very close to 0, for which even a very small increase can lead to certified robustness.

Lastly, when $\epsilon = 0.0039$, we achieved a similar performance as the baseline method. This indicates that the effectiveness of configured bounds decreases as the perturbation radius becomes very small.

Next, we investigate whether our approach extends to fully connected neural networks; experimental results are shown in Table 1. In general, bounds obtained for this network type are looser than those obtained for the CNN, irrespective of the perturbation radius. Nonetheless, our method achieved improvements in the average global lower bound between 10 and 14 per cent across all considered perturbation radii.

For FNNs, we achieved the greatest improvement when the perturbation radius is smallest, *i.e.* $\epsilon = 0.0039$. In this scenario, we were able to certify robustness for 3 additional instances, which were previously unsolved; see also Fig. 2c for a visualisation of these results.

Table 1 also shows the results from our experiments on the CNN trained on the MNIST dataset. When verifying this network with $\epsilon = 0.3$, we achieved an improvement in the average global lower bound of 23%. Furthermore, using our configured bounds, we could again certify robustness for an instance that could not be solved previously. This is also visualised in Fig. 2d.

For $\epsilon = 0.12$ and $\epsilon = 0.06$, we achieved improvements in the average global lower bound of 10% and 57%, respectively. Lastly, when $\epsilon = 0.03$, our method did not improve over the baseline. This again shows that, for CNNs, configuring the bounds is less effective if the perturbation radius is minimal.

5.2   Tanh-Based Networks

Next, we report the results obtained for the Tanh-based CNN trained on the CIFAR-10 dataset; these are also presented in Table 1. Notably, we achieved consistent improvement for any given value of ϵ. Furthermore, we found that the global lower bounds are generally much lower than those obtained for the Sigmoid-based CNN, although verifying the same properties. Nevertheless, when $\epsilon = 0.0039$, our method enables the verification of an additional instance, which was previously unsolved. Overall, our results demonstrate the strength of our approach, and its potential to improve verification performance on networks with non-piecewise linear activation functions in general.

5.3   Distribution of Tangent Points

To gain a better understanding of the difference between bounding functions obtained by our method and vanilla CROWN, we perform an empirical analysis of tangent point distributions obtained for the linear bounding functions of the Sigmoid-based CNN network trained on CIFAR-10 when verified for local robustness with $\epsilon = 0.0313$. Notice that this benchmark shows the greatest improvement in the average global lower bound.

Figure 3 shows probability density functions of tangent points for lower and upper bounding functions per activation layer of the considered network. Notably, they show that the distribution of tangent points found by the configured bounding algorithm is much more centred around a specific value than those obtained by the baseline approach. Furthermore, the difference between the empirical distribution functions of tangent points obtained by our search method and the baseline approach was determined as statistically significant by means of a Kolmogorov-Smirnov test with a standard significance threshold of 0.05. Moreover, these empirical observations hint towards the existence of an optimal region for bounding parameters of Sigmoidal activation functions, which might be difficult to identify using the baseline approach without automated configuration of the tangent point search method.

6   Conclusions and Future Work

In this work, we have shown that automated algorithm configuration can provide a systematic and effective way for designing bounding functions of nonlinearities beyond the commonly studied piecewise linear activation functions (*e.g.*, ReLU).

Specifically, our new method achieved consistent improvements in average global lower bound across several benchmarks and perturbation radii.

At the same time, we see several fruitful avenues for future work. First of all, the proposed search method for obtaining the tangent points is controlled by only two hyper-parameters. A more sophisticated method, allowing for a more fine-grained configuration of the search method, could improve performance further. In addition, one could configure the hyper-parameters of the search method as well as those of the α-optimisation method, $e.g.$, the learning rate of the projected gradient algorithm, jointly.

Overall, we see this study as a promising step towards the automated design of versatile and efficient neural network verification algorithms.

Acknowledgment. This research was partially supported by TAILOR, a project funded by EU Horizon 2020 research and innovation program under GA No. 952215. It further received funding from the ERC under the European Union's Horizon 2020 research and innovation program (FUN2MODEL, grant agreement No. 834115) and ELSA: European Lighthouse on Secure and Safe AI project (grant agreement No. 101070617 under UK guarantee).

References

1. Bak, S., Tran, H.D., Hobbs, K., Johnson, T.T.: Improved geometric path enumeration for verifying ReLU neural networks. In: Proceedings of the 32nd International Conference on Computer Aided Verification (CAV 2020), pp. 66–96 (2020)
2. Bezerra, L.C., López-Ibánez, M., Stützle, T.: Automatic component-wise design of multiobjective evolutionary algorithms. IEEE Trans. Evol. Comput. **20**(3), 403–417 (2015)
3. Botoeva, E., Kouvaros, P., Kronqvist, J., Lomuscio, A., Misener, R.: Efficient verification of ReLU-based neural networks via dependency analysis. In: Proceedings of the 34th AAAI Conference on Artificial Intelligence (AAAI-20), pp. 3291–3299 (2020)
4. Breiman, L.: Random forests. Mach. Learn. **45**(1), 5–32 (2001)
5. Bunel, R., Lu, J., Turkaslan, I., Torr, P.H.S., Kohli, P., Kumar, M.P.: Branch and bound for piecewise linear neural network verification. J. Mach. Learn. Res. **21**, 42:1–42:39 (2020)
6. Bunel, R., Turkaslan, I., Torr, P., Kohli, P., Mudigonda, P.K.: A unified view of piecewise linear neural network verification. In: Advances in Neural Information Processing Systems 31 (NeurIPS 2018), pp. 1–10 (2018)
7. Carlini, N., Wagner, D.: towards evaluating the robustness of neural networks. In: Proceedings of the 38th IEEE Symposium on Security and Privacy (IEEE S&P 2017), pp. 39–57 (2017)
8. Chiarandini, M., Fawcett, C., Hoos, H.H.: A modular multiphase heuristic solver for post enrolment course timetabling. In: Proceedings of the 7th International Conference on the Practice and Theory of Automated Timetabling (PATAT 2008) (2008)
9. Feurer, M., Springenberg, J.T., Hutter, F.: initializing bayesian hyperparameter optimization via meta-learning. In: Proceedings of the 29th AAAI Conference on Artificial Intelligence (AAAI-15), pp. 1128–1135 (2015)

10. Gebser, M., Kaminski. R., Kaufmann, B., Schaub, T., Schneider, M.T., Ziller, S.: A portfolio solver for answer set programming: preliminary report. In: Proceedings of the 10th International Conference on Logic Programming and Nonmonotonic Reasoning (LPNMR2019), pp. 1–6 (2011)

11. Gehr, T., Mirman, M., Drachsler-Cohen, D., Tsankov, P., Chaudhuri, S., Vechev, M.: AI2: safety and robustness certification of neural networks with abstract interpretation. In: Proceedings of the 39th IEEE Symposium on Security and Privacy (IEEE S&P 2018), pp. 3–18 (2018)

12. Goodfellow, I.J., Shlens, J., Szegedy, C.: Explaining and harnessing adversarial examples. In: Proceedings of the 3rd International Conference on Learning Representations (ICLR 2015), pp. 1–11 (2015)

13. Henriksen, P., Lomuscio, A.: Efficient neural network verification via adaptive refinement and adversarial search. In: Proceedings of the 24th European Conference on Artificial Intelligence (ECAI 2020), pp. 2513–2520 (2020)

14. Hutter, F., Babic, D., Hoos, H.H., Hu, A.J.: Boosting verification by automatic tuning of decision procedures. In: Proceedings of Formal Methods in Computer Aided Design (FMCAD 2007), pp. 27–34 (2007)

15. Hutter, F., Hoos, H.H., Leyton-Brown, K.: Automated configuration of mixed integer programming solvers. In: Proceedings of the 7th International Conference on Integration of Artificial Intelligence (AI) and Operations Research (OR) Techniques in Constraint Programming (CPAIOR 2010), pp. 186–202 (2010)

16. Hutter, F., Hoos, H.H., Leyton-Brown, K.: Sequential model-based optimization for general algorithm configuration. In: Proceedings of the 5th International Conference on Learning and Intelligent Optimization (LION 5), pp. 507–523 (2011)

17. Hutter, F., Lindauer, M., Balint, A., Bayless, S., Hoos, H., Leyton-Brown, K.: the configurable SAT solver challenge (CSSC). Artif. Intell. **243**, 1–25 (2017)

18. Julian, K.D., Kochenderfer, M.J., Owen, M.P.: Deep neural network compression for aircraft collision avoidance systems. J. Guid. Control. Dyn. **42**(3), 598–608 (2019)

19. Katz, G., Barrett, C., Dill, D.L., Julian, K., Kochenderfer, M.J.: Reluplex: an efficient SMT solver for verifying deep neural networks. In: Proceedings of the 29th International Conference on Computer Aided Verification (CAV 2017), pp. 97–117 (2017)

20. König, M., Bosman, A.W., Hoos, H.H., van Rijn, J.N.: Critically assessing the state of the art in neural network verification. J. Mach. Learn. Res. **25**(12), 1–53 (2024)

21. König, M., Hoos, H.H., Rijn, J.N.v.: Speeding up neural network robustness verification via algorithm configuration and an optimised mixed integer linear programming solver portfolio. Mach. Learn. **111**(12), 4565–4584 (2022)

22. Kurakin, A., Goodfellow, I., Bengio, S.: Adversarial examples in the physical world. arXiv preprint arXiv:1607.02533 (2016)

23. Li, L., Xie, T., Li, B.: Sok: Certified robustness for deep neural networks. In: Proceedings of the 44th IEEE Symposium on Security and Privacy (SP2023), pp. 1289–1310 (2023)

24. Lopez-Ibanez, M., Stützle, T.: Automatically improving the anytime behaviour of optimisation algorithms. Eur. J. Oper. Res. **235**(3), 569–582 (2014)

25. Müller, C., Serre, F., Singh, G., Püschel, M., Vechev, M.: Scaling polyhedral neural network verification on GPUs. In: Proceedings of Machine Learning and Systems 3 (MLSys 2021), pp. 1–14 (2021)

26. Singh, G., Ganvir, R., Püschel, M., Vechev, M.: Beyond the single neuron convex barrier for neural network certification. In: Advances in Neural Information Processing Systems 32 (NeurIPS 2019), pp. 1–12 (2019)
27. Singh, G., Gehr, T.: Boosting robustness certification of neural networks. In: Proceedings of the 7th International Conference on Learning Representations (ICLR 2019), pp. 1–12 (2019)
28. Singh, G., Gehr, T., Mirman, M., Püschel, M., Vechev, M.: Fast and effective robustness certification. In: Advances in Neural Information Processing Systems 31 (NeurIPS 2018), pp. 1–12 (2018)
29. Singh, G., Gehr, T., Püschel, M., Vechev, M.: An abstract domain for certifying neural networks. In: Proceedings of the 3rd ACM on Programming Languages (POPL 2019), pp. 1–30 (2019)
30. Singh, G., Gehr, T., Püschel, M., Vechev, M.: An abstract domain for certifying neural networks. In: Proceedings of the 46th ACM SIGPLAN Symposium on Principles of Programming Languages (ACMPOPL 2019), pp. 1–30 (2019)
31. Szegedy, C., et al.: Intriguing properties of neural networks. In: Proceedings of the 2nd International Conference on Learning Representations (ICLR 2014), pp. 1–10 (2014)
32. Thornton, C., Hutter, F., Hoos, H.H., Leyton-Brown, K.: Auto-WEKA: combined selection and hyperparameter optimization of classification algorithms. In: Proceedings of the 19th ACM SIGKDD International Conference on Knowledge Discovery and Data Mining (KDD2013), pp. 847–855 (2013)
33. Tjeng, V., Xiao, K., Tedrake, R.: Evaluating robustness of neural networks with mixed integer programming. In: Proceedings of the 7th International Conference on Learning Representations (ICLR 2019), pp. 1–21 (2019)
34. Tjeng, V., Xiao, K.Y., Tedrake, R.: Evaluating robustness of neural networks with mixed integer programming. In: 7th International Conference on Learning Representations, ICLR 2019, New Orleans, LA, USA, 6-9 May 2019. OpenReview.net (2019). https://openreview.net/forum?id=HyGIdiRqtm
35. Vallati, M., Fawcett, C., Gerevini, A.E., Hoos, H., Saetti, A.: Automatic generation of efficient domain-specific planners from generic parametrized planners. In: Proceedings of the 6th Annual Symposium on Combinatorial Search (SOCS), pp. 184–192 (2013)
36. Wang, S., et al.: Beta-crown: efficient bound propagation with per-neuron split constraints for neural network robustness verification. In: Advances in Neural Information Processing Systems 34 (NeurIPS 2021), pp. 29909–29921 (2021)
37. Xiang, W., Tran, H.D., Johnson, T.T.: Output reachable set estimation and verification for multilayer neural networks. IEEE Trans. Neural Netw. Learn. Syst. **29**(11), 5777–5783 (2018)
38. Xu, K., et al.: Fast and complete: enabling complete neural network verification with rapid and massively parallel incomplete verifiers. In: Proceedings of the 9th International Conference on Learning Representations (ICLR 2021) (2021)
39. Zhang, H., Weng, T.W., Chen, P.Y., Hsieh, C.J., Daniel, L.: Efficient neural network robustness certification with general activation functions. In: Advances in Neural Information Processing Systems 31 (NeurIPS 2018) vol. 31, pp. 4944–4953 (2018)

Efficiently Predicting Mutational Effect on Homologous Proteins by Evolution Encoding

Zhiqiang Zhong$^{(\boxtimes)}$ and Davide Mottin

Aarhus University, Aarhus, Denmark
{zzhong,davide}@cs.au.dk

Abstract. Predicting protein properties is paramount for biological and medical advancements. Current protein engineering mutates on a typical protein, called the *wild-type*, to construct a family of homologous proteins and study their properties. Yet, existing methods easily neglect subtle mutations, failing to capture the effect on the protein properties. To this end, we propose EvoLMPNN, Evolution-aware Message Passing Neural Network, an efficient model to learn evolution-aware protein embeddings. EvoLMPNN samples sets of anchor proteins, computes evolutionary information by means of residues and employs a differentiable evolution-aware aggregation scheme over these sampled anchors. This way, EvoLMPNN can efficiently utilise a novel message-passing method to capture the mutation effect on proteins with respect to the anchor proteins. Afterwards, the aggregated evolution-aware embeddings are integrated with sequence embeddings to generate final comprehensive protein embeddings. Our model shows up to 6.4% better than state-of-the-art methods and attains 36× inference speedup in comparison with large pre-trained models. Code and models are available at https://github.com/zhiqiangzhongddu/EvolMPNN.

1 Introduction

Can we predict important properties of a protein by directly observing only the effect of a few mutations on such properties? This basic biological question [16] has recently engaged the machine learning community due to the current availability of benchmark data [8,30,44]. Proteins are sequences of amino acids (or residues), which are the cornerstone of life and influence a number of metabolic processes, including diseases [28]. For this reason, protein engineering stands at the forefront of modern biotechnology, offering a remarkable toolkit to manipulate and optimise existing proteins for a wide range of applications, from drug development to personalised therapy [1].

One fundamental process in protein engineering progressively mutates an initial protein, called the *wild-type*, to study the effect on the protein's properties [38]. These mutations form a family of *homologous proteins* as in Fig. 1. This process is appealing due to its cheaper cost compared to other methods and reduced time and risk [41].

A. Bifet et al. (Eds.): ECML PKDD 2024, LNAI 14947, pp. 399–415, 2024.
https://doi.org/10.1007/978-3-031-70368-3_24

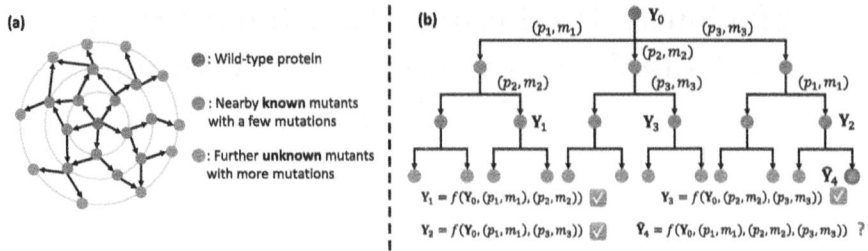

Fig. 1. Protein property prediction on homologous protein family. (a) An example homologous protein family with labelled nearby mutants with few mutations. We aim to predict the label of unknown mutants with more mutations. (b) The evolutionary pattern for (a). For instance, $\mathbf{Y}_0$ is the label vector of the corresponding protein sequence, and (p_1, m_1) indicates mutation m_1 at position p_1 of the protein's amino acid sequence.

Yet, the way mutations affect the protein's properties is not completely understood [5], as it depends on a number of chemical reactions and bonds among residues. For this reason, machine learning offers a viable alternative to model complex interactions among residues. Initial approaches employed *feature engineering* to capture protein's evolution [14,35]; yet, a manual approach is expensive and does not offer enough versatility. Advances in NLP and CV inspired the design of deep *protein sequence encoders* [18,39] and general purpose Protein Language Models (PLMs) that are pre-trained on large-scale datasets of sequences. Notable PLMs include ProtBert [4], AlphaFold [19], TAPE Transformer [30] and ESM [33]. These models mainly rely on Multiple Sequence Alignments (MSAs) [25] to search on large databases of protein evolution. While this process focuses on conserved regions, it is insensitive to subtle yet crucial mutations in less conserved regions and introduces additional computational burdens [6,29].

To overcome the limitations of previous models, we propose EVOLMPNN, Evolution-aware Message Passing Neural Network, to predict the mutational effect on homologous proteins. Our fundamental assumption is that there are inherent correlations between protein properties and the sequence differences among them, as shown in Fig. 1-(b). EVOLMPNN devises a novel message-passing method to integrate both protein sequence and evolutionary information by identifying where and which mutations occur on the target protein sequence, compared with known protein sequences and predicts the mutational effect on the target protein property. To avoid the costly *quadratic* pairwise comparison among proteins, we devise a theoretically grounded (see Sect. 4.6) *linear* sampling strategy to compute differences only among the proteins and a fixed number of anchor proteins (Sect. 4.2). We additionally introduce two extensions of our model, EVOLGNN and EVOLFORMER, to include available data on the relation among proteins (Sect. 4.5). The theoretical computation complexities of proposed methods are provided to guarantee their efficiency and practicality.

We apply the proposed methods to three benchmark homologous protein property prediction datasets with nine splits. Empirical evaluation results (Sect. 5.1) show up to 6.7% Spearman's ρ correlation improvement over the best performing baseline models, reducing the inference time by 36× compared with pre-trained PLMs.

2   Preliminary and Problem

In protein engineering, we first receive a *set of proteins* $\mathcal{M} = \{\mathcal{P}_i\}_{i=1,2,...,M}$ in which each protein can be associated with a label vector $\mathbf{Y}_i \in \mathbb{R}^\theta$ that describes its biomedical properties, *e.g.*, fitness, stability, fluorescence, and solubility. Each protein $\mathcal{P}_i$ is a linear chain of *amino-acids* $\mathcal{P}_i = \{r_j\}_{j=1,2,...,N}$. While a protein sequence folds into specific 3D conformation to perform some biomedical functions, each amino-acid is considered as a *residue*. Residues form peptide bonds and can interact with each other by different chemical bounds [28]. In short, the function of a protein is mainly determined by the chemical interactions between residues. Since the 3D structure is missing in benchmark datasets [8,30,44], we assume no 3D protein information in this paper.

Homologous Protein Family. A set of protein sequences ($\mathcal{M}$) is a *homologous protein family* if there exists an ancestral protein $\mathcal{P}_{\mathrm{WT}}$, called *wild-type*, such that any $\mathcal{P}_i \in \mathcal{M}$ is obtained by mutating $\mathcal{P}_{\mathrm{WT}}$ through substitution, deletion, insertion and truncation of residues [27]. As shown in Fig. 1-(a), a homologous protein family can be organised together by representing their evolutionary relationships and Fig. 1-(b) illustrates the detailed evolutionary patterns.

Research Problem. Protein engineering based on homologous proteins is a promising and essential direction for designing novel proteins of desired properties [16]. Understanding the relation between protein sequence and property is one essential step. Practically, biologists perform experiments in the lab to manually label the property $\mathbf{Y}_{\mathrm{TRAIN}}$ of a set of protein $\mathcal{M}_{\mathrm{TRAIN}} \subset \mathcal{M}$ and the follow-up task is predicting $\hat{\mathbf{Y}}_{\mathrm{TEST}}$ of the rest proteins $\mathcal{M}_{\mathrm{TEST}} \subset \mathcal{M}$. However, due to their shared ancestry, homologous proteins typically have similarities in their amino acid sequences, structures, and functions. *Accurately predicting the homologous protein property by distinguishing these subtle yet crucial differences is still an open challenge.*

3   Related Work

Feature Engineering. Besides conducting manual experiments in labs to measure protein properties, the basic solution is to design different feature engineering methods based on relevant biological knowledge, to extract useful information from protein sequence [14]. [8] introduce using Levenshtein distance [22] and BLOSUM62-score [10] relative to wild-type to design protein sequence features. In another benchmark work, [44] adopt another two typical protein sequence feature descriptors, *i.e.*, Dipeptide Deviation from Expected Mean (DDE) [35] and Moran correlation (Moran) [14]. For more engineering methods, refer to the comprehensive review [21].

Protein Representation Learning. In the last decades, propelled by the outstanding achievements of machine learning and deep learning, protein representation learning has revolutionised protein property prediction research. Early work along this line adopts the idea of word2vec [26] to protein sequences. To increase model capacity, deeper *protein sequence encoders* were proposed by the Computer Vision and Nature Language Processing communities [18,39]. The latest works develop *Protein Language Models*, which focus on employing deep sequence encoder models for protein sequences and are pre-trained on million- or billion-scale sequences. Well-known works include ProtBert [4], AlphaFold [19], TAPE Transformer [30], ESM [33], AlphaFold2 [19] and Ankh [11]. However, most current research on protein sequences defines it as a language problem and does not sufficiently consider the subtle yet crucial evolutionary patterns in homologous proteins. For instance, [19,33] explore protein Multiple Sequence Alignments (MSAs) [25,31] to capture the mutational effect. Nevertheless, the MSA searching process introduces additional computational burden and is insensitive to subtle but crucial sequence differences [29]. [6] indicate the shortcomings of MSAs on easily neglecting the presence of minor mutations, which can propagate errors to downstream protein sequence representation learning tasks. The message-passing idea has been explored in protein folding tasks [9,19]. We build on this direction, by devising a novel model that captures subtleties in the mutations for homologous protein property prediction tasks.

4   Framework

EvoLMPNN is a novel framework that integrates both protein sequence information and evolution information by means of residues. As a result, EvoLMPNN accurately predicts the mutational effect on homologous protein families. First, in Sect. 4.1, we introduce *embedding initialisation* for protein sequence and residues and the update module for residue embedding (Sect. 4.2). The *evolution encoding* in Sect. 4.3 is the cornerstone of the model that ameliorates protein embeddings. We conclude in Sect. 4.4 with the generation of *final proteins embeddings and model optimisation*. We complement our model with a theoretical analysis to motivate our methodology and a discussion of the computation complexity (Sect. 4.6). We additionally propose extended versions of EvoLMPNN that deal with available protein-protein interactions (Sect. 4.5) (Fig. 2).

4.1   Embedding Initialisation

Protein Sequence Embedding. Given a *set of proteins* $\mathcal{M} = \{\mathcal{P}_i\}_{i=1,2,...,M}$, we first adopt a (parameter-frozen) PLM model [25,32][1] as protein sequence

---

[1] We do not fine-tune PLM in this paper for efficiency consideration.

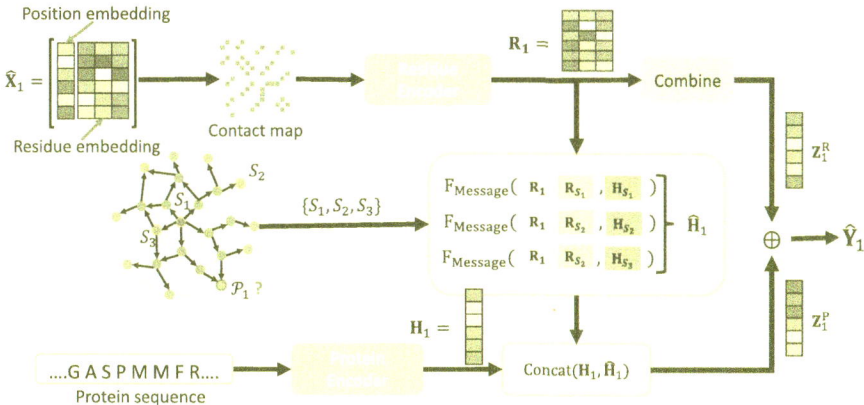

Fig. 2. Our EvolMPNN framework encodes protein mutations via a sapient combination of residue evolution and sequence encoding.

encoder to initialise protein-sequence embeddings (**H**) as one d-dimensional real-valued vector for every protein $\mathcal{P}_i$, which include *macro* (*i.e.*, protein sequence) level information as the primary embedding.

$$\mathbf{H} = \text{PlmEncoder}(\{\mathcal{P}_i\}_{i=1,2,...,M}), \tag{1}$$

where the obtained protein embedding $\mathbf{H} \in \mathbb{R}^{M \times d}$ and $\mathbf{H}_i$ corresponds to each protein $\mathcal{P}_i$. Different encoders can extract information on various aspects, however, existing PLM models that rely on MSAs are not sensitive enough to capture the evolution pattern information in homologous protein families [29]. [6] systematically indicate the shortcomings of MSAs on easily neglecting the presence of minor mutations, which can propagate errors to downstream protein sequence representation learning tasks.

Residue Embedding Initialisation. In order to properly capture the evolution information in homologous proteins, we delve into the residue level for *micro* clues. We adopt two residue embedding initialisation approaches, *i.e.*, one-hot encoding (Φ^{OH}) and pre-trained PLM encoder (Φ^{PLM}), to generate protein's initial residue embeddings $\mathbf{X}_i = \{\mathbf{x}_j^i\}_{j=1,2,...,N}$, where $\mathbf{x}_j^i \in \mathbb{R}^d$. In particular, Φ^{OH} assigns each protein residue[2] with a binary feature vector $\mathbf{x}_j^i$, where $\mathbf{x}_{jb}^i = 1$ indicates the appearance of the b-th residue at $\mathcal{P}_i$'s j-th position. By stacking N residues' feature vectors into a matrix, we can obtain $\mathbf{X}_i \in \mathbb{R}^{N \times d}$. On the other hand, following the benchmark implementations [48], PlmEncoder can export residue embeddings similar to Eq. 1. Formally, Φ^{PLM} initialises protein residue embeddings as $\mathbf{X}_i = \text{PlmEncoder}(\{r_j\}_{j=1,2,...,N})$.

Position Embedding. Another essential component of existing PLM is the positional encoding, which was first proposed by [39]. This positional encoding effectively captures the relative structural information between residues and

---

[2] There are 20 different amino acid residues commonly found in proteins.

integrates it with the model. In our case, correctly recording the position of each residue in the protein sequence plays an essential role in identifying each protein's corresponding mutations. Because a mutation occurring at different positions can have varying effects on protein properties due to its impact on protein structure. Therefore, after initialising residue embeddings, we further apply positional embedding on each protein's residues. Particularly, we randomly initialise a set of d position embeddings $\Phi^{\mathrm{Pos}} \in \mathbb{R}^{N \times d}$, and it will be learned in the training process of the entire framework. We denote the residue embedding empowered by position embedding as $\hat{\mathbf{X}}_i = \mathbf{X}_i \odot \Phi^{\mathrm{Pos}}$, where $\odot$ indicates the element-wise multiplication.

4.2   Residue Embedding Update

To maintain a stable 3D structure, 3D protein folding depends on the strength of different chemical bonds between residues. Previous studies manually designed residue contact maps to model the residue-residue interactions to learn effective residue embeddings [17,32]. In this paper, we adopt the residue-residue interaction to update residue embeddings but eschew the requirement of manually designing the contact map. Instead, we assume the existence of an implicit fully connected residue contact map of each protein $\mathcal{P}_i$ and implement the Transformer model [39,43] to adaptively update residue embeddings. Denote $\mathbf{R}_i^{(\ell)}$ as the input to the $(\ell+1)$-th layer, with the first $\mathbf{R}_i^{(0)} = \hat{\mathbf{X}}_i$ be the input encoding. The $(\ell+1)$-th layer of residue embedding update module can be formally defined as follows:

$$\mathrm{Att}_i^h(\mathbf{R}_i^{(\ell)}) = \mathrm{SOFTMAX}(\frac{\mathbf{R}_i^{(\ell)}\mathbf{W}_Q^{\ell,h}(\mathbf{R}_i^{(\ell)}\mathbf{W}_K^{\ell,h})^{\mathrm{T}}}{\sqrt{d}}),$$

$$\hat{\mathbf{R}}_i^{(\ell)} = \mathbf{R}_i^{(\ell)} + \sum_{h=1}^{H} \mathrm{Att}_i^h(\mathbf{R}_i^{(\ell)})\mathbf{R}_i^{(\ell)}\mathbf{W}_V^{\ell,h}\mathbf{W}_O^{\ell,h}, \qquad (2)$$

$$\mathbf{R}_i^{(\ell+1)} = \hat{\mathbf{R}}_i^{(\ell)} + \mathrm{ELU}(\hat{\mathbf{R}}_i^{(\ell)}\mathbf{W}_1^{\ell})\mathbf{W}_2^{\ell},$$

where $\mathbf{W}_O^{\ell,h} \in \mathbb{R}^{d_H \times d}$, $\mathbf{W}_Q^{l,h}, \mathbf{W}_K^{l,h}, \mathbf{W}_V^{l,h} \in \mathbb{R}^{d \times d_H}$, $\mathbf{W}_1^{\ell} \in \mathbb{R}^{d \times r}$, $\mathbf{W}_2^{\ell} \in \mathbb{R}^{d_t \times d}$ are learnable parameters, H is the number of attention heads, d_H is the dimension of each head, d_t is the dimension of the hidden layer, ELU [7] is an activation function, and $\mathrm{Att}_i^h(\mathbf{R}_i^{(\ell)})$ refers to as the attention matrix. After each Transformer layer, we add a normalisation layer $i.e.$, LayerNorm [2], to reduce the over-fitting problem proposed by [39]. After stacking L_r layers, we obtain the final residue embeddings as $\mathbf{R}_i = \mathbf{R}_i^{(L_r)}$.

4.3   Evolution Encoding

In homologous protein families, all proteins are mutants derived from a common wild-type protein $\mathcal{P}_{\mathrm{WT}}$ with different numbers and types of mutations. In this paper, we propose to capture the evolutionary information via the following assumption.

Assumption 1 (Protein Property Relevance). *Assume there is a homologous protein family* $\mathcal{M}$ *and a function* F_{DIFF} *can accurately distinguish the mutations on mutant* $\mathcal{P}_i$ *compared with any* $\mathcal{P}_j$ *as* $F_{\text{DIFF}}(\mathcal{P}_i, \mathcal{P}_j)$. *For any target protein* $\mathcal{P}_i$, *its property* $\mathbf{Y}_i$ *can be predicted by considering 1) its sequence information* $\mathcal{P}_i$; *2)* $F_{\text{DIFF}}(\mathcal{P}_i, \mathcal{P}_j)$ *and the property of* $\mathcal{P}_j$, *i.e.,* $\mathbf{Y}_j$. *Shortly, we assume there exists a function* f *that maps* $\mathbf{Y}_i \leftarrow f(F_{\text{DIFF}}(\mathcal{P}_i, \mathcal{P}_j), \mathbf{Y}_j)$.

Motivated by Assumption 1, we take both protein sequence and the mutants difference $F_{\text{DIFF}}(\mathcal{P}_i, \mathcal{P}_j)$ to accurately predict the protein property. To encode the protein sequence, we employ established tools described in Sect. 4.1. Here instead, we describe the evolution encoding to realise the function of $F_{\text{DIFF}}(\mathcal{P}_i, \mathcal{P}_j)$.

The naïve solution to extract evolutionary patterns in a homologous family is constructing a complete phylogenetic tree [15] based on the mutation distance between each protein pair. Yet, finding the most parsimonious phylogenetic tree is **NP**-hard [34].

To address the aforementioned problems, instead of constructing the phylogenetic tree, we compute the distance among a few sampled proteins we call *anchor proteins* and all the other proteins. Theoretical analysis to validate this design is discussed in Sect. 4.6. Specifically, denote $\mathbf{H}_i^{(\ell)}$ as the input to the $(\ell + 1)$-th block and define $\mathbf{H}_i^{(0)} = \mathbf{H}_i$. The evolution localisation encoding of the $(\ell + 1)$-th layer contains the following key components: *(i)* k anchor protein $\{\mathcal{P}_{S_i}\}_{i=1,2,\ldots,k}$ selection. *(ii)* Evolutionary information encoding function F_{DIFF} that computes the difference between residues of each protein and those of the anchor protein, and target protein's evolutionary information is generated by summarising the obtained differences $\mathbf{d}_{ij}$ as follows:

$$\mathbf{d}_{ij} = \text{COMBINE}(\mathbf{R}_i - \mathbf{R}_{S_j}), \tag{3}$$

where COMBINE can be implemented as differentiable operators, such as, CONCATENATE, MAX POOL MEAN POOL and SUM POOL; here we use the MEAN POOL to obtain $\mathbf{d}_{ij} \in \mathbb{R}^d$. *(iii)* Message computation function F_{MESSAGE} that integrates protein features and evolutionary information as one message from an anchor protein. Specifically, F_{MESSAGE} combines protein sequence feature information of two proteins with their evolutionary differences. We empirically find that the simple element-wise product between $\mathbf{H}_j^{(\ell)}$ and $\mathbf{d}_{ij}$ attains good results

$$F_{\text{MESSAGE}}(i, j, \mathbf{H}_j^{(\ell)}, \mathbf{d}_{ij}) = \mathbf{H}_j^{(\ell)} \odot \mathbf{d}_{ij}, \tag{4}$$

(iv) Aggregate evolutionary messages from k anchors and combine them with the protein's embedding as the updated protein embedding, which contains the protein sequence and evolutionary information:

$$\hat{\mathbf{H}}_i^{(\ell)} = \text{COMBINE}(\{F_{\text{MESSAGE}}(i, j, \mathbf{H}_j^{(\ell)}, \mathbf{d}_{ij})\}_{j=1,2,\ldots,k}), \tag{5}$$

$$\mathbf{H}_i^{(\ell+1)} = \text{CONCAT}(\mathbf{H}_i^{(\ell)}, \hat{\mathbf{H}}_i^{(\ell)})\mathbf{W}^\ell, \tag{6}$$

where $\mathbf{W}^\ell \in \mathbb{R}^{2d \times d}$ transform concatenated vectors to the hidden dimension. After stacking L_p layers, we obtain the final protein sequence embedding $\mathbf{Z}_i^{\mathrm{P}} = \mathbf{H}_i^{(L_p)}$.

4.4 Final Embedding and Optimisation

After obtaining protein $\mathcal{P}_i$'s residue embeddings $\mathbf{R}_i$ and sequence embedding $\mathbf{Z}_i^{\mathrm{P}}$, we summarise its residue embeddings as a vector $\mathbf{Z}_i^{\mathrm{R}} = \mathrm{MEAN\ POOL}(\mathbf{R}_i)$. The final protein embedding summarises the protein sequence information and evolution information as the comprehensive embedding $\mathbf{Z}_i = \mathrm{CONCAT}(\mathbf{Z}_i^{\mathrm{P}}, \mathbf{Z}_i^{\mathrm{R}})$ and the final prediction is computed as $\hat{\mathbf{Y}}_i = \mathbf{Z}_i \mathbf{W}^{\mathrm{FINAL}}$ where $\mathbf{W}^{\mathrm{FINAL}} \in \mathbb{R}^{d \times \theta}$, θ is the number of properties to predict. Afterwards, we adopt a simple and common strategy, similar to [44], to solve the protein property prediction tasks.

Algorithm 1: The framework of EVOLMPNN

Input: Protein set $\mathcal{M} = \{\mathcal{P}_i\}_{i=1,2,\dots,M}$ and each protein sequence $\mathcal{P}_i$ contains a residue set $\{r_j\}_{j=1,2,\dots,N}$; Message computation function $\mathrm{F}_{\mathrm{MESSAGE}}$ that outputs an d dimensional message; $\mathrm{COMBINE}(\cdot)$ and $\mathrm{CONCAT}(\cdot)$ operators.

Output: Protein embeddings $\{\mathbf{Z}_i\}_{i=1,2,\dots,M}$

1   $\mathbf{H}_i \leftarrow \mathrm{PLMENCODER}(\mathcal{P}_i)$
2   $\mathbf{X}_i \leftarrow \Phi^{\mathrm{OH}}(\{r_j\}_{j=1,2,\dots,N}) \,/\, \Phi^{\mathrm{PLM}}(\{r_j\}_{j=1,2,\dots,N})$
3   $\hat{\mathbf{X}}_i \leftarrow \mathbf{X}_i \odot \Phi^{\mathrm{POS}}$
4   $\mathbf{R}_i^{(0)} \leftarrow \hat{\mathbf{X}}_i$
5   **for** $\ell = 1, 2, \dots, L_r$ **do**
6   **for** $i = 1, 2, \dots, N$ **do**
7         $\mathbf{R}_i^{(\ell)} \leftarrow \mathrm{NODEFORMER}(\mathbf{R}_i^{(\ell-1)})$
8   **end**
9   **end**
10   $\mathbf{R}_i \leftarrow \mathbf{R}_i^{(L_r)}$
11   $\mathbf{H}_i^{(0)} \leftarrow \mathbf{H}_i$
12   **for** $\ell = 1, 2, \dots, L_p$ **do**
13      $\{S_j\}_{j=1,2,\dots,k} \sim \mathcal{M}$
14   **for** $i = 1, 2, \dots, M$ **do**
15   **for** $j = 1, 2, \dots, k$ **do**
16            $d_{ij} = \mathrm{COMBINE}(\mathbf{R}_i - \mathbf{R}_{S_j})$
17   **end**
18         $\hat{\mathbf{H}}_i^{(\ell)} = \mathrm{COMBINE}(\{\mathrm{F}_{\mathrm{MESSAGE}}(i, j, \mathbf{H}_j^{(\ell)}, d_{ij})\}_{j=1,2,\dots,k})$
            $\mathbf{H}_i^{(\ell+1)} = \mathrm{CONCAT}(\mathbf{H}_i^{(\ell)}, \hat{\mathbf{H}}_i^{(\ell)}) \mathbf{W}^\ell$
19   **end**
20   **end**
21   $\mathbf{Z}_i^{\mathrm{P}} = \mathbf{H}_i^{(L_p)}$
22   $\mathbf{Z}_i^{\mathrm{R}} = \mathrm{MEAN\ POOL}(\mathbf{R}_i)$
23   $\mathbf{Z}_i = \mathrm{CONCAT}(\mathbf{Z}_i^{\mathrm{P}}, \mathbf{Z}_i^{\mathrm{R}})$

Specifically, we adopt the MSELoss ($\mathcal{L}$) to measure the correctness of model predictions on training samples against ground truth labels. The objective of learning the target task is to optimise model parameters to minimise the loss $\mathcal{L}$ on this task. The framework of EvolMPNN is summarised in Algorithm 1.

4.5 Extensions on Observed Graph

EvolMPNN does not leverage any information from explicit geometry among proteins, where each protein only communicates with randomly sampled anchors (Sect. 4.3). However, it is often possible to have useful structured data $G = (\mathcal{M}, \mathbf{A})$ that represents the relation between protein-protein by incorporating specific domain knowledge [47].[3] Therefore, here we introduce EvolGNN, an extension of EvolMPNN on the possibly observed protein interactions.

EvolGNN. We compute the evolution information as Eq. 3. The evolution information can be easily integrated into the pipeline of message-passing neural networks, as an additional structural coefficient:

$$\mathbf{m}_a^{(\ell)} = \text{Aggregate}^{\mathcal{N}}(\{\mathbf{A}_{ij}, \underbrace{\mathbf{d}_{ij}}_{\text{Evol. info.}}, \mathbf{H}_j^{(\ell-1)} \mid j \in \mathcal{N}(i)\}),$$

$$\mathbf{m}_i^{(\ell)} = \text{Aggregate}^{\mathcal{I}}(\{\mathbf{A}_{ij}, \underbrace{\mathbf{d}_{ij}}_{\text{Evol. info.}} \mid j \in \mathcal{N}(i)\}) \, \mathbf{H}_i^{(\ell-1)}, \qquad (7)$$

$$\mathbf{H}_i^{(\ell)} = \text{Combine}(\mathbf{m}_a^{(\ell)}, \mathbf{m}_i^{(\ell)}),$$

where $\text{Aggregate}^{\mathcal{N}}(\cdot)$ and $\text{Aggregate}^{\mathcal{I}}(\cdot)$ are two parameterised functions. $\mathbf{m}_a^{(\ell)}$ is a message aggregated from the neighbours $\mathcal{N}(i)$ of protein $\mathcal{P}_i$ and their structure ($\mathbf{A}_{ij}$) and evolution ($\mathbf{d}_{ij}$) coefficients. $\mathbf{m}_i^{(\ell)}$ is an updated message from protein $\mathcal{P}_i$ after performing an element-wise multiplication between $\text{Aggregate}^{\mathcal{I}}(\cdot)$ and $\mathbf{H}_i^{(\ell-1)}$ to account for structural and evolution effects from its neighbours. After, $\mathbf{m}_a^{(\ell)}$ and $\mathbf{m}_i^{(\ell)}$ are combined together to obtain the update embedding $\mathbf{H}_i^{(\ell)}$.

EvolFormer. Another extension relies on pure Transformer structure, which means the evolution information of $\mathcal{M}$ can be captured by every protein. The evolution information can be integrated into the pipeline of Transformer, as additional information to compute the attention matrix:

$$\text{Att}^h(\mathbf{H}^{(\ell)}) = \text{Softmax}(\frac{\mathbf{H}^{(\ell)}\mathbf{W}_Q^{\ell,h}(\mathbf{H}^{(\ell)}\mathbf{W}_K^{\ell,h})^{\text{T}}}{\sqrt{d}} + \underbrace{\text{Mean Pool}(\{\mathbf{R}_i\}_{i=1,2,...,M})}_{\text{Evol. info.}}),$$

$$(8)$$

Other follow-up information aggregation and feature vector update operations are the same as the basic Transformer pipeline, as described in Eq. 2.

---

[3] Available contact map describes residue-residue interactions can be easily integrated as relational bias of Transformer [43] as we used in Sect. 4.2.

4.6   Theoretical Analysis

Anchor Selection. Inspired by [45], we adopt Bourgain's Theorem [3] to guide the random anchor number (k) of the evolution encoding layer. Briefly, support by a constructive proof (Theorem 2 [24]) of Bourgain Theorem (Theorem 1), only $k = O(\log^2 M)$ anchors are needed to ensure the resulting embeddings are guaranteed to have low distortion (Definition 1), in a given metric space $(\mathcal{M}, \mathrm{F_{DIST}})$. EVOLMPNN can be viewed as a generalisation of the embedding method of Theorem 2, where $\mathrm{F_{DIST}}(\cdot)$ is generalised via message passing functions (Eq. 3–Eq. 6). Therefore, Theorem 2 offers a theoretical guide that $O(\log^2 M)$ anchors are needed to guarantee low distortion embedding. Following this principle, EVOLMPNN choose $k = \log^2 M$ random anchors, denoted as $\{S_j\}_{j=1,2,\ldots,\log^2 M}$, and we sample each protein in $\mathcal{M}$ independently with probability $\frac{1}{2^j}$. Detailed discussion and proof refer to Appendix A.

Complexity Analysis. The computation costs of EVOLMPNN, EVOLGNN, and EVOLFORMER come from residue encoding and evolution encoding, since the protein sequence and residue feature initialisation have no trainable parameters. The residue encoder introduces the complexity of $O(MN)$ following an efficient implementation of NodeFormer [43]. In the evolution encoding, EVOLMPNN performs communication between each protein and $\log^2 M$ anchors, which introduces the complexity of $O(M\log^2 M)$; EVOLGNN performs communication between each protein and K neighbours with $O(KM)$ complexity; EVOLFORMER performs communication between all protein pairs, which introduces the complexity of $O(M)$, following the efficient implement, NodeFormer. In the end, we obtain the total computation complexity of EVOLMPNN - $O((N + \log^2 M)M)$, EVOLGNN - $O((N + K)M)$ and EVOLFORMER - $O((N + 1)M)$.

5   Experimental Study

In this section, we empirically study the performance of EVOLMPNN. We validate our model on three benchmark homologous protein family datasets and evaluate the methods on nine data splits to consider comprehensive practical use cases. Our experiments comprise a comprehensive set of state-of-the-art methods from different categories. We additionally demonstrate the effectiveness of two extensions of our model, EVOLGNN and EVOLFORMER, with different input features. We conclude our analysis by studying the influence of some hyper-parameters and investigating the performance of EVOLMPNN on high mutational mutants.

Datasets and Splits. We perform experiments on benchmark datasets of several important protein engineering tasks, including AAV, GB1 and Fluorescence, and generate three splits on each dataset. Data statistics are summarised in Table 1. The split λ-VS-REST indicates that we train models on wild-type protein and mutants of no more than λ mutations, while the rest are assigned to test. The split LOW-VS-HIGH indicates that we train models on sequences with

Table 1. Dataset splits and corresponding statistics; if the split comes from a benchmark paper, we report the corresponding citation.

Landscape	Split	# Total	#Train	#Valid	#Test
AAV [5]	2-VS-REST [8]	82,583	28,626	3,181	50,776
	7-VS-REST [8]	82,583	63,001	7,001	12,581
	LOW-VS-HIGH [8]	82,583	42,791	4,755	35,037
GB1 [42]	2-VS-REST [8]	8,733	381	43	8,309
	3-VS-REST [8]	8,733	2,691	299	5,743
	LOW-VS-HIGH [8]	8,733	4,580	509	3,644
Fluorescence [36]	2-VS-REST	54,025	12,712	1,413	39,900
	3-VS-REST [44]	54,025	21,446	5,362	27,217
	LOW-VS-HIGH	54,025	44,082	4,899	5,044

target value scores equal to or below wild-type, while the rest are assigned to the test.

Baselines. As baseline models, we consider methods in four categories. First, we selected four *feature engineer* methods, *i.e.*, Levenshtein [8], BLOSUM62 [8], DDE [35] and Moran [14]. Second, we select four *protein sequence encoder* models, *i.e.*, LSTM [18], Transformer [30], CNN [30] and ResNet [46]. Third, we select four *pre-trained PLM models*, *i.e.*, ProtBert [12], ESM-1b [33], ESM-1v [25] and ESM-2 [23]. In the end, we select four *GNN-based methods* which can utilise available graph structure, *i.e.*, GCN [20], GAT [40], GraphTransformer [37] and NodeFormer [43].

Implementation. We follow the PEER benchmark settings[4], including train and test pipeline, model optimisation and evaluation method (evaluation is Spearman's ρ metric), adopted in [44] to make sure the comparison fairness. For the baselines, including feature engineer, protein sequence encoder and pre-trained PLM, we adopt the implementation provided by benchmark Torch-drug [48] and the configurations reported in [44]. For the GNN-based baselines, which require predefined graph structure and protein features, we construct K-NN graphs [13], with $K = \{5, 10, 15\}$, and report the best performance. As features, we use the trained sequence encoder, which achieves better performance, used also in our method. In addition, we adopt ESM-1b as the residue encoder on GB1 dataset and adopt One-Hot encoding on AAV and Fluorescence datasets to speed up the training process. All experiments are conducted on two NVIDIA GeForce RTX 3090 GPUs and two NVIDIA RTX A6000 GPUs, and we report the mean performance of three runs with different random seeds. We present more details at https://github.com/zhiqiangzhongddu/EvolMPNN.

---

[4] https://github.com/DeepGraphLearning/PEER_Benchmark.

5.1 Effectiveness

EVOLMPNN Outperforms all Baselines on 9 Splits. Table 2 summarises performance comparison on AAV, GB1 and Fluorescence datasets. EVOLMPNN achieves new state-of-the-art performance on most splits of three datasets, with up to 6.7% improvements to baseline methods. This result vindicates the effectiveness of our proposed design to capture evolution information for homologous protein property prediction.

Table 2. Quality in terms of Spearman's ρ correlation with target value. NA indicates a non-applicable setting. * Used as a feature extractor with pre-trained weights frozen. † Results reported in [8,44]. Top-2 performances of each split are marked as **bold** and underline.

Category	Model	Dataset								
		AAV			GB1			Fluorescence		
		2-vs-R.	7-vs-R.	L.-vs-H.	2-vs-R.	3-vs-R.	L.-vs-H.	2-vs-R.	3-vs-R.	L.-vs-H.
Feature Engineer	Levenshtein	0.578	0.550	0.251	0.156	-0.069	-0.108	0.466	0.054	0.011
	BLOSUM62	NA	NA	NA	0.128	0.005	-0.127	NA	NA	NA
	DDE	0.649†	0.636	0.158	0.445†	0.816	0.306	0.690	0.638†	0.159
	Moran	0.437†	0.398	0.069	0.069†	0.589	0.193	0.445	0.400†	0.046
Protein Seq.Encoder	LSTM	0.125†	0.608	0.308	-0.002†	-0.002	-0.007	0.256	0.494†	0.207
	Transformer	0.681†	0.748	0.304	0.271†	0.877	0.474	0.250	0.643†	0.161
	CNN	0.746†	0.730	<u>0.406</u>	0.502†	0.857	0.515	<u>0.805</u>	<u>0.682</u>†	0.249
	ResNet	0.739†	0.733	0.223	0.133†	0.542	0.396	0.594	0.636†	0.243
Pre-trained PLM	ProtBert	0.794†	0.719	0.322	0.634†	0.866	0.308	0.451	0.679†	0.201
	ProtBert*	0.209†	0.507	0.277	0.123†	0.619	0.164	0.403	0.339†	0.161
	ESM-1b	0.821†	0.735	0.385	0.704†	0.878	0.386	0.804	0.679†	0.221
	ESM-1b*	0.454†	0.573	0.241	0.337†	0.605	0.178	0.528	0.430†	0.091
	ESM-1v	0.826	0.741	0.394	0.721	<u>0.884</u>	0.390	0.804	<u>0.682</u>	<u>0.251</u>
	ESM-1v*	0.533	0.580	0.171	0.359	0.632	0.180	0.562	0.563	0.070
	ESM-2	0.824	0.734	0.390	0.712	0.874	0.372	0.791	0.668	0.201
	ESM-2*	0.475	0.581	0.199	0.422	0.632	0.189	0.501	0.511	0.084
GNN-based Methods	GCN	0.824	0.730	0.361	0.745	0.865	0.466	0.755	0.677	0.198
	GAT	0.821	0.741	0.369	<u>0.757</u>	0.873	0.508	0.768	0.667	0.208
	GraphTransf.	<u>0.827</u>	<u>0.749</u>	0.389	0.753	0.876	<u>0.548</u>	0.780	0.678	0.231
	NodeFormer	<u>0.827</u>	0.741	0.393	<u>0.757</u>	0.877	0.543	0.794	0.677	0.213
Ours	EVOLMPNN	**0.835**	**0.757**	**0.433**	**0.768**	**0.889**	**0.584**	**0.809**	**0.684**	**0.262**

Manual Construction of Homology Graphs Proves to be Less Effective. Notably, GNN-based methods that utilise manually constructed graph structure do not enter top-2 on 8 of 9 splits and two Transformer structure models, *i.e.*, GraphTransformer and NodeFormer, often outperform such methods. It can be understood since homology graph construction is a challenging biomedical task [29], the simple K-NN graph construction is not an effective solution.

Large-Scale PLM Models are Dominated by Simple Models. Surprisingly, we find that smaller models, such as CNN and ResNet, can outperform large ESM variants pre-trained on million- and billion-scale sequences. For instance, ESM-1v has about 650 million parameters and is pre-trained on around 138 million UniRef90 sequences [25]. Yet, CNN outperforms ESM-1v on three splits of `Fluorescence` dataset. This indicates the necessity of designs targeting specifically the crucial homologous protein engineering task.

Table 3. Results on `GB1` datasets (metric: Spearman's ρ) of our proposed methods, with different residue embeddings. Top-2 performances of each split marked as **bold** and underline.

Model	Split		
	2-vs-R.	3-vs-R.	L.-vs-H.
State-of-the-art	0.757	0.878	0.548
EvolMPNN (Φ^{OH})	0.766	0.877	0.553
EvolGNN (Φ^{OH})	0.764	0.866	0.536
EvolFormer (Φ^{OH})	0.764	0.868	0.537
EvolMPNN (Φ^{PLM})	**0.768**	**0.881**	**0.584**
EvolGNN (Φ^{PLM})	<u>0.767</u>	<u>0.879</u>	<u>0.581</u>
EvolFormer (Φ^{PLM})	0.766	0.879	0.575

Our Proposed Extension Models Outperform all Baselines on `GB1` Dataset. We performed additional experiments on `GB1` datasets to investigate the performance of two extended models, *i.e.*, EvolGNN and EvolFormer and study the influence of different residue embedding initialisation methods. The results summarised in Table 3 evince that EvolMPNN outperforms the other two variants in three splits, and all our proposed models outperform the best baseline. This result confirms the effectiveness of encoding evolution information for homologous protein property prediction. Besides, the models adopting the PLM encoding Φ^{PLM} achieve better performance than those using the one-hot encoding Φ^{OH}. From this experiment, we conclude that residue information provided by PLM helps to capture protein's evolution information.

5.2 Analysis of Performance

The Performance of EvolMPNN Comes from its Superior Predictions on High Mutational Mutants. For the Low-vs-High split of `GB1` dataset, we group the test proteins into 4 groups depending on their number of mutations. Next, we compute three models, including EvolMPNN, ESM-1b (fine-tuned PLM model) and CNN (best baseline), prediction performances on each protein group and present the results in Fig. 3. EvolMPNN outperforms two baselines in all 4 protein groups. Notably, by demonstrating EvolMPNN's clear edge in groups of no less than 3 mutations, we confirm the generalisation effectiveness from low mutational mutants to high mutational mutants. **As per**

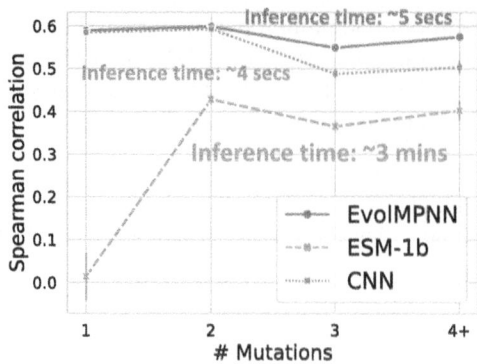

Fig. 3. Performance on protein groups of different numbers of mutations, with the Low-vs-High split and avg. epoch inference time on GB1 dataset.

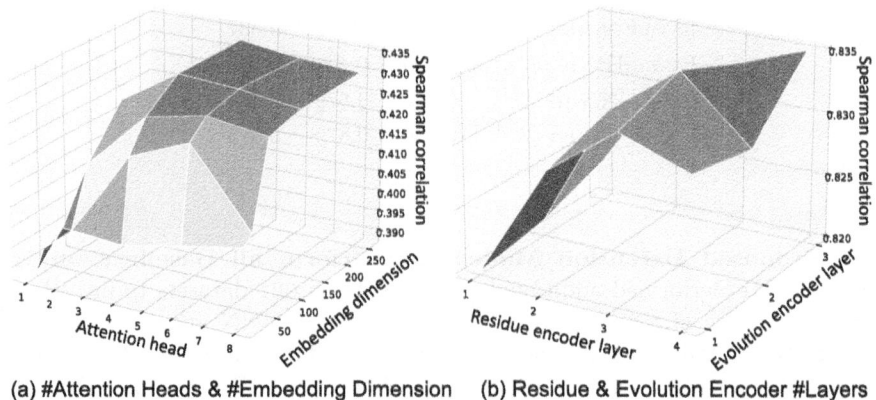

(a) #Attention Heads & #Embedding Dimension   (b) Residue & Evolution Encoder #Layers

Fig. 4. EvolMPNN performance on AAV's Low-vs-High (a) and 2-vs-Rest (b) splits, with different hyper-parameters.

inference time, EvolMPNN and CNN require similar inference time (≈ 5 s), $36\times$ faster than ESM-1b (≈ 3 min).

Influence of Hyper-Parameter Settings on EvolMPNN. We present in Fig. 4 a group of experiments to study the influence of some hyper-parameters on EvolMPNN, including the number of attention heads, embedding dimension and the number of layers of residue encoder and evolution encoder.

6   Conclusion and Future Work

We propose EvolMPNN that integrates both protein sequence information and evolution information by means of residues to predict the mutational effect on homologous proteins. Empirical and theoretical studies show that EvolMPNN and its extended variants (EvolGNN and EvolFormer) achieve outstanding performance on several benchmark datasets while retaining reasonable computation complexity. In future work, we intend to incorporate 3D protein structure information towards general-purpose homologous protein models. In addition, it would be interesting to experiment with different approaches to selecting anchor sets, for example, using central nodes-based selections.

Acknowledgments. This work is supported by the Horizon Europe and Innovation Fund Denmark under the Eureka, Eurostar grant no E115712 - AAVanguard.

References

1. Alley, E.C., Khimulya, G., Biswas, S., AlQuraishi, M., Church, G.M.: Unified rational protein engineering with sequence-based deep representation learning. Nat. Methods **16**(12), 1315–1322 (2019)
2. Ba, J.L., Kiros, J.R., Hinton, G.E.: Layer normalization. CoRR abs/1607.06450 (2016)
3. Bourgain, J.: On lipschitz embedding of finite metric spaces in hilbert space. Israel J. Math. **52**, 46–52 (1985)
4. Brandes, N., Ofer, D., Peleg, Y., Rappoport, N., Linial, M.: Proteinbert: a universal deep-learning model of protein sequence and function. Bioinformatics **38**(8), 2102–2110 (2022)
5. Bryant, D.H., et al.: Deep diversification of an AAV capsid protein by machine learning. Nat. Biotechnol. **39**(6), 691–696 (2021)
6. Chatzou, M., et al.: Multiple sequence alignment modeling: methods and applications. Brief. Bioinform. **17**(6), 1009–1023 (2016)
7. Clevert, D.A., Unterthiner, T., Hochreiter, S.: Fast and accurate deep network learning by exponential linear units (ELUs). CoRR abs/1511.07289 (2015)
8. Dallago, C., et al.: FLIP: benchmark tasks in fitness landscape inference for proteins. In: Proceedings of the 2021 Annual Conference on Neural Information Processing Systems (NeurIPS) (2021)
9. Dauparas, J., et al.: Robust deep learning-based protein sequence design using proteinmpnn. Science **378**(6615), 49–56 (2022)
10. Eddy, S.R.: Where did the BLOSUM62 alignment score matrix come from? Nat. Biotechnol. **22**(8), 1035–1036 (2004)
11. Elnaggar, A., et al.: Ankh: optimized protein language model unlocks general-purpose modelling. CoRR abs/2301.06568 (2023)
12. Elnaggar, A., et al.: Prottrans: toward understanding the language of life through self-supervised learning. IEEE Trans. Pattern Anal. Mach. Intell. (TPAMI) **44**(10), 7112–7127 (2022)
13. Eppstein, D., Paterson, M.S., Yao, F.F.: On nearest-neighbor graphs. Discrete Comput. Geom. **17**, 263–282 (1997)
14. Feng, Z.P., Zhang, C.T.: Prediction of membrane protein types based on the hydrophobic index of amino acids. J. Protein Chem. **19**, 269–275 (2000)
15. Fitch, W.M., Margoliash, E.: Construction of phylogenetic trees: a method based on mutation distances as estimated from cytochrome c sequences is of general applicability. Science **155**(3760), 279–284 (1967)
16. Fowler, D.M., Fields, S.: Deep mutational scanning: a new style of protein science. Nat. Methods **11**(8), 801–807 (2014)

17. Gao, Z., et al.: Hierarchical graph learning for protein-protein interaction. Nat. Commun. **14**(1), 1093 (2023)
18. Hochreiter, S., Schmidhuber, J.: Long short-term memory. Neural Comput. (1997)
19. Jumper, J., et al.: Highly accurate protein structure prediction with alphafold. Nature **596**(7873), 583–589 (2021)
20. Kipf, T.N., Welling, M.: Semi-supervised classification with graph convolutional networks. In: Proceedings of the 2017 International Conference on Learning Representations (ICLR) (2017)
21. Lee, D., Redfern, O., Orengo, C.: Predicting protein function from sequence and structure. Nat. Rev. Mol. Cell Biol. **8**(12), 995–1005 (2007)
22. Li, Y., Liu, B.: A normalized levenshtein distance metric. IEEE Trans. Pattern Anal. Mach. Intell. (TPAMI) **29**(6), 1091–1095 (2007)
23. Lin, Z., et al.: Evolutionary-scale prediction of atomic-level protein structure with a language model. Science **379**(6637), 1123–1130 (2023)
24. Linial, N., London, E., Rabinovich, Y.: The geometry of graphs and some of its algorithmic applications. Combinatorica **15**, 215–245 (1995)
25. Meier, J., Rao, R., Verkuil, R., Liu, J., Sercu, T., Rives, A.: Language models enable zero-shot prediction of the effects of mutations on protein function. In: Proceedings of the 2021 Annual Conference on Neural Information Processing Systems (NeurIPS), pp. 29287–29303 (2021)
26. Mikolov, T., Sutskever, I., Chen, K., Corrado, G.S., Dean, J.: Distributed representations of words and phrases and their compositionality. In: Proceedings of the 2013 Annual Conference on Neural Information Processing Systems (NIPS), pp. 3111–3119 (2013)
27. Ochoterena, H., Vrijdaghs, A., Smets, E., Claßen-Bockhoff, R.: The search for common origin: homology revisited. Syst. Biol. **68**(5), 767–780 (2019)
28. Pauling, L., Corey, R.B., Branson, H.R.: The structure of proteins: two hydrogen-bonded helical configurations of the polypeptide chain. Proc. Natl. Acad. Sci. **37**(4), 205–211 (1951)
29. Pearson, W.R.: An introduction to sequence similarity ("homology") searching. Curr. Protoc. Bioinform. **42**(1), 3–1 (2013)
30. Rao, R., et al.: Evaluating protein transfer learning with TAPE. In: Proceedings of the 2019 Annual Conference on Neural Information Processing Systems (NeurIPS), pp. 9686–9698 (2019)
31. Rao, R., et al.: MSA transformer. In: Proceedings of the 2021 International Conference on Machine Learning (ICML), vol. 139, pp. 8844–8856. PMLR (2021)
32. Rao, R., Meier, J., Sercu, T., Ovchinnikov, S., Rives, A.: Transformer protein language models are unsupervised structure learners. In: Proceedings of the 2021 International Conference on Learning Representations (ICLR) (2021)
33. Rives, A., et al.: Biological structure and function emerge from scaling unsupervised learning to 250 million protein sequences. Proc. Natl. Acad. Sci. **118**(15), e2016239118 (2021)
34. Sankoff, D.: Minimal mutation trees of sequences. SIAM J. Appl. Math. **28**(1), 35–42 (1975)
35. Saravanan, V., Gautham, N.: Harnessing computational biology for exact linear b-cell epitope prediction: a novel amino acid composition-based feature descriptor. Omics J. Integr. Biol. **19**(10), 648–658 (2015)
36. Sarkisyan, K.S., et al.: Local fitness landscape of the green fluorescent protein. Nature **533**(7603), 397–401 (2016)

37. Shi, Y., Huang, Z., Feng, S., Zhong, H., Wang, W., Sun, Y.: Masked label prediction: unified message passing model for semi-supervised classification. In: Proceedings of the 2021 International Joint Conferences on Artifical Intelligence (IJCAI), pp. 1548–1554 (2021)

38. Siezen, R.J., de Vos, W.M., Leunissen, J.A., Dijkstra, B.W.: Homology modelling and protein engineering strategy of subtilases, the family of subtilisin-like serine proteinases. Protein Eng. Des. Sel. 4(7), 719–737 (1991)

39. Vaswani, A., et al.: Attention is all you need. In: Proceedings of the 2017 Annual Conference on Neural Information Processing Systems (NIPS), pp. 5998–6008 (2017)

40. Velickovic, P., Cucurull, G., Casanova, A., Romero, A., Lio, P., Bengio, Y.: Graph attention networks. In: Proceedings of the 2018 International Conference on Learning Representations (ICLR) (2018)

41. Wang, M., Si, T., Zhao, H.: Biocatalyst development by directed evolution. Biores. Technol. **115**, 117–125 (2012)

42. Wu, N.C., Dai, L., Olson, C.A., Lloyd-Smith, J.O., Sun, R.: Adaptation in protein fitness landscapes is facilitated by indirect paths. Elife **5**, e16965 (2016)

43. Wu, Q., Zhao, W., Li, Z., Wipf, D.P., Yan, J.: Nodeformer: a scalable graph structure learning transformer for node classification. In: Proceedings of the 2022 Annual Conference on Neural Information Processing Systems (NeurIPS), pp. 27387–27401 (2022)

44. Xu, M., et al.: PEER: a comprehensive and multi-task benchmark for protein sequence understanding. In: Proceedings of the 2022 Annual Conference on Neural Information Processing Systems (NeurIPS) (2022)

45. You, J., Ying, R., Leskovec, J.: Position-aware graph neural networks. In: Proceedings of the 2019 International Conference on Machine Learning (ICML), vol. 97, pp. 7134–7143. PMLR (2019)

46. Yu, F., Koltun, V., Funkhouser, T.A.: Dilated residual networks. In: Proceedings of the 2017 Conference on Computer Vision and Pattern Recognition (CVPR), pp. 636–644. IEEE (2017)

47. Zhong, Z., Barkova, A., Mottin, D.: Knowledge-augmented graph machine learning for drug discovery: a survey from precision to interpretability. CoRR abs/2302.08261 (2023)

48. Zhu, Z., et al.: Torchdrug: a powerful and flexible machine learning platform for drug discovery. CoRR abs/2202.08320 (2022)

Interpretable and Fair Mechanisms for Abstaining Classifiers

Daphne Lenders[1,2]([⊠]), Andrea Pugnana[3], Roberto Pellungrini[4], Toon Calders[1,2], Dino Pedreschi[3], and Fosca Giannotti[4]

[1] Adrem Data Lab, University of Antwerp, Antwerp, Belgium
daphne.lenders@uantwerpen.be
[2] DigiTax, University of Antwerp, Antwerp, Belgium
[3] KDD Lab, University of Pisa, Pisa, Italy
[4] KDD Lab, Scuola Normale Superiore, Pisa, Italy

Abstract. Abstaining classifiers have the option to refrain from providing a prediction for instances that are difficult to classify. The abstention mechanism is designed to trade off the classifier's performance on the accepted data while ensuring a minimum number of predictions. In this setting, often fairness concerns arise when the abstention mechanism solely reduces errors for the majority groups of the data, resulting in increased performance differences across demographic groups. While there exist a bunch of methods that aim to reduce discrimination when abstaining, there is no mechanism that can do so in an explainable way. In this paper, we fill this gap by introducing Interpretable and Fair Abstaining Classifier (IFAC), an algorithm that can reject predictions both based on their uncertainty and their unfairness. By rejecting possibly unfair predictions, our method reduces error and positive decision rate differences across demographic groups of the non-rejected data. Since the unfairness-based rejections are based on an interpretable-by-design method, i.e., rule-based fairness checks and situation testing, we create a transparent process that can empower human decision-makers to review the unfair predictions and make more just decisions for them. This explainable aspect is especially important in light of recent AI regulations, mandating that any high-risk decision task should be overseen by human experts to reduce discrimination risks. (Code and Appendix for this work is available on: https://github.com/calathea21/IFAC).

Keywords: Reject Option · Fair ML · Interpretable ML

1 Introduction

Over the last 15 years, much research has been conducted on creating fairness-aware classification algorithms. While a lot of work has been done on creating automatized solutions based on some mathematical definition of fairness, recently the call for more flexible approaches has been growing. Rather than trying to define or achieve fairness through one numeric measure for the entire

A. Bifet et al. (Eds.): ECML PKDD 2024, LNAI 14947, pp. 416–433, 2024.
https://doi.org/10.1007/978-3-031-70368-3_25

system, there is a growing recognition that we need to understand under which circumstances unfairness occurs, which groups are most affected by it, and which differences in the treatment of demographic groups might be justifiable [12,45]. Because of the delicate and nuanced nature of these questions, there is also an increased consensus that automated algorithms cannot be used alone in the identification and resolution of bias, but instead should actively be overseen and adapted by human experts with sufficient knowledge about a domain and the historic biases in place. This call for human-in-the-loop approaches for algorithmic fairness is now even mandated by AI legislation, such as the EU AI Act [16]. Despite the clear call that human oversight and control are necessary, the legislation says little about how it should take place [16]. A way to put humans in the loop during the deployment of a system is provided by the framework of selective classification. The original idea behind this framework is to build a classifier that abstains from making a prediction when it is not certain about it. In other words, these models *reject* ambiguous instances and pass them to better decision models or human experts, to increase accuracy over all non-rejected instances. Even though this idea originally dates back to the 1970 s [9], it has only barely been explored in the context of increasing the fairness of models, by abstaining from predictions that might be unfair. Ensuring the interpretability of such abstentions, and explaining why instances are seen as unfair can further empower humans to understand whether to override original decisions or not, and increase the overall fairness of the decision process [43].

In this work, we exploit this idea by proposing an Interpretable Fair Abstaining Classifier (IFAC) for building selective classifiers that do not only abstain from making decisions in cases of uncertainty but also in cases of unfairness. We do so by adding an inherently interpretable mechanism for unfairness-based rejections to a selective classifier, thus allowing the user to inspect the unfair decisions of the model and the instances they need to review.

The paper is organized as follows: in Sect. 2 we list the main papers in the literature relevant to our work, in Sects. 3 and 4 we provide, respectively, the necessary mathematical background and formulation of our method, in Sect. 5 we provide a thorough experimental evaluation of our method and finally in Sect. 6 we discuss our results and conclude the paper.

2   Related Literature

Fairness in Classification.   Classifiers exhibiting discriminatory behaviour towards certain demographic groups have been a concern for some time now [36]. Over the years, many metrics have been proposed to measure discrimination in these settings. These include *group metrics*, such as demographic parity and equal odds, that compare how classifiers behave over different population groups in the data. Particularly, demographic parity compares a classifier's output ratios and equal odds its error ratios across demographics [36]. Next to *group metrics*, there are *individual metrics* to identify for one instance at a time whether they are affected by discrimination. These metrics operate on the principle of *treating likes alike* and check if similar individuals receive similar decision outcomes [36]. When it comes to mitigating bias in classification tasks, a common approach is to choose one of the available metrics and build a classifier to satisfy the associated fairness goal while maintaining its predictive accuracy [7,24,48]. Recently, however, the simplicity of these approaches has been criticized: optimizing for group metrics comes with the risk of *cherry-picking*, the practice of arbitrarily changing prediction labels in pursuit of some "superficial" fairness goal, without further attention to whether the decisions make sense on an individual level [18]. Contrarily, only paying attention to individual fairness does not ensure that discrimination does not still happen globally, and certain demographic groups are not systematically excluded from receiving favourable decision outcomes [18]. Hence, researchers have argued that instead of fixating on one fairness goal in an automated manner, any efforts to detect and mitigate discrimination should be guided by domain experts, who can take a more holistic approach to fairness, and make nuanced considerations about the nature of bias and how to address it [25,41,45]. Related to this, researchers have also pointed out the importance of addressing intersectional discrimination [12]. This describes the unique discrimination that people from a combination of marginalized groups (e.g., black women) face, which cannot be solely explained by the "sum" of discrimination faced by each marginalized group in isolation (e.g., being black and being female) [13]. Currently, many works on fair classification only focus on discrimination experienced by demographic groups as defined by a single binary-sensitive feature. Recognizing that algorithmic harms can only be combated when understanding how they uniquely unfold, some studies like [6,19,46] have started incorporating intersectionality in their research.

Prediction with a Reject Option. The idea to allow a machine learning model to abstain in the prediction stage dates back to the 1970 s, when it was introduced for classification tasks [9]. Two main frameworks allow one to learn abstaining models, i.e. ambiguity rejection and novelty rejection [26]. The former focuses on abstaining from instances where mistakes are more likely; the latter builds methods that abstain on instances that are largely dissimilar from the training data distribution [30,35,47]. Within ambiguity rejection, we can further distinguish between Learning to Reject (LtR) [9] and Selective Prediction (SP) [15]. The former (LtR) requires one to define a class-wise cost function that penalizes mispredictions and rejections [10,11]. The latter (SP) requires instead one to either pre-define a target coverage c to achieve and minimize the risk *(bounded-abstention)* [23,28,38,39], or fix a target risk e to guarantee and maximize the coverage *(bounded-improvement)* [21,22].

Fairness and Reject Option. There are a few works that analyze the effects on fairness caused by a reject option. [29] show that even if abstaining can improve the overall accuracy, some demographic groups can be negatively impacted by the reject option. [31] propose a surrogate loss for the classification task considering performance on different subgroups of instances. The proposed loss allows enforcing a sufficiency condition to avoid unfair results. A similar approach for the regression task is proposed by [42]. [40] provide a theoretical analysis of the selective classification framework when introducing a fairness constraint in the bounded-abstention problem.

Explainability and Reject Option. The study of explainable AI (XAI) methods in the context of abstaining classifiers is limited. [17] propose a reject option for natively interpretable models such as prototype-based ones. [2] consider counterfactual techniques to explain reject options of learning vector quantization classifiers. [3] introduce semi-factual explanations for the reject option, yielding a model-agnostic approach at the expense of potentially high complexity. Finally, [4] propose a model-agnostic framework to explain the abstention mechanism, including counterfactual, semi-factual, and factual approaches.

3 Background

3.1 Selective Classification

Consider the triplet $(\mathbf{L}, \mathbf{S}, Y)$: $\mathbf{L}$ represents the legally-grounded features and takes values in $\mathcal{L} \subseteq \mathbb{R}^{d_l}$; $\mathbf{S}$ refers to the sensitive attributes and takes values in $\mathcal{S} \subseteq \mathbb{R}^{d_s}$; Y is the (binary) target variable, whose domain is $\mathcal{Y} = \{0, 1\}$. For example, if Y encodes being rich and our goal is to predict Y given some set of features, $\mathbf{L}$ could include educational level and employment status, while $\mathbf{S}$ could refer to gender or race. We denote with $\mathcal{X} = \mathcal{L} \times \mathcal{S}$ the whole feature space and with $\mathbf{X} = (\mathbf{L}, \mathbf{S})$ the pair of both legally grounded and sensitive features.

Given the hypothesis space $\mathcal{H}$ of functions (classification models) mapping $\mathcal{X}$ to $\mathcal{Y}$, a learning algorithm aims to find a hypothesis $h \in \mathcal{H}$ such that it minimizes some risk measure $R(h) = \mathbb{E}[l(h(\mathbf{X}), Y)]$, where $l : \mathcal{Y} \times \mathcal{Y} \to \mathbb{R}$ is a *loss function* and $\mathbb{E}$ is computed over the joint probability distribution $P(\mathbf{X}, Y)$.

To reduce the classifier's error rates, one can add a selection mechanism that allows the model to abstain from predicting over more difficult-to-classify instances. More formally, we can define a selective classifier[1] as:

$$(h, g)(\mathbf{x}) = \begin{cases} h(\mathbf{x}) & \text{if} \quad g(\mathbf{x}) = 1 \\ \text{abstain} & \text{otherwise}, \end{cases} \tag{1}$$

where $g : \mathcal{X} \to \{0, 1\}$ is the so-called *selection function* or *rejector*[2].

In practice, the selection function is often obtained by setting a threshold τ on a confidence function $\upsilon : \mathcal{X} \to \mathbb{R}$, which determines the portion of the data on which the classifier is more likely to misclassify. In such a case, the selection function can be defined as $g(\mathbf{x}) = \mathbb{1}\{\upsilon(\mathbf{x}) \geq \tau\}$.

To avoid rejecting too many instances, the selective classification framework introduces *the coverage*, i.e. the percentage of instances for which the selective classifier must provide a prediction. Coverage is denoted as $\phi(g) = \mathbb{E}[g(\mathbf{X})]$ and can be traded off for performance improvements. In this case, performance

---

[1] In this work, we use the terms *abstaining* and *selective* interchangeably.
[2] We use the term abstain and reject when $g(\mathbf{x}) = 0$ and accept or selects when $g(\mathbf{x}) = 1$.

is measured through the risk over the accepted region, commonly called the *selective risk* and defined as $R(h, g) = \frac{\mathbb{E}[l(h(\mathbf{X}), Y) g(\mathbf{X})]}{\phi(g)}$.

To find a selective classifier that minimizes selective risk, it is necessary to select a lower bound c as a *target coverage* [23]. Given a target coverage c, an optimal selective predictor (h, g) (parameterized by θ^*, ψ^*) is defined as:

$$\underset{\theta \in \Theta, \psi \in \Psi}{\arg\min} R(h_\theta, g_\psi) \quad \text{s.t.} \quad \phi(g_\psi) \geq c \qquad (2)$$

We learn the optimal parameters using an empirical counterpart of selective risk and coverage, using an i.i.d. dataset $\mathcal{D} = \{(\mathbf{x}_i, y_i)\}_{i=1}^n$ drawn from P.

Finally, we call *coverage-calibration* the post-training procedure of estimating the threshold τ for the target coverage c specified in Eq. 2. This is generally done by estimating the $(1-c) \cdot 100$-th percentile of the confidence function over a held-out calibration dataset.

3.2 Measuring Fairness With Association Rules and Situation Testing

Association Rules: In our methodology, we make use of association rules to identify discriminatory behaviour of a base classifier h, upon which g can decide to reject its predictions. Let us assume we have access to a dataset of realizations $\mathcal{D}$. We recall $\mathbf{x}_i = (\mathbf{l}_i, \mathbf{s}_i) = (l_i^1, \cdots, l_i^{d_l}, s_i^1, \cdots, s_i^{d_s})$, where l_i^j refers to the value taken by the jth legally grounded feature of instance i and s_i^j to the jth sensitive feature of instance i.

We call a specific realization of a single variable within $\mathbf{x}_i$ an *item*, e.g. if we consider the variable `race`, `race=White` is an item. Let $\mathcal{I}$ be the set of all possible items. A subset I of $\mathcal{I}$ is called an *itemset*.

We can decompose I into its legally grounded and sensitive parts, $I = (I_L, I_S)$, where I_L is an itemset containing only legally grounded features and I_S is an itemset that contains only sensitive ones. A transaction T is a subset of I with exactly one item for every feature in $\mathbf{x}$. In other words, a sampled instance's features $\mathbf{x}_i$ can be seen as a transaction T. For a transaction T, we say T *verifies* itemset (I_L, I_S) if $(I_L, I_S) \subseteq T$. The support of itemset (I_L, I_S) with respect to the dataset $\mathcal{D}$ is denoted as $supp_\mathcal{D}((I_L, I_S)) = \frac{|\{T \in \mathcal{D} : (I_L, I_S) \subseteq T\}|}{|\mathcal{D}|}$.

A decision rule is an expression $(I_L, I_S) \rightarrow Y$. The support of a decision rule is $supp_\mathcal{D}((I_L, I_S) \rightarrow Y) = supp_\mathcal{D}((I_L, I_S), Y)$. The confidence of the rule is then defined as $conf_\mathcal{D}((I_L, I_S) \rightarrow Y) = \frac{supp_\mathcal{D}((I_L, I_S), Y)}{supp_\mathcal{D}((I_L, I_S))}$.

To measure the impact of the sensitive features of a decision rule, the Selective Lift (*slift*) measure introduced by [34] can be used. In this paper we use the definition *by difference* of *slift*, which is detailed as follows:

$$slift_\mathcal{D}((I_L, I_S) \rightarrow Y) = conf_\mathcal{D}((I_L, I_S) \rightarrow Y) - conf_\mathcal{D}((I_L, \neg I_S) \rightarrow Y) \qquad (3)$$

Computing $conf_\mathcal{D}(I_L, \neg I_S) \rightarrow Y$ requires one to take the confidence of all the transactions that verify I_L but do not verify I_S.

Example. Consider an association rule `race = Black, education = Masters` $\rightarrow$ `income = low`, with `race` $\subseteq$ **S** and `education` $\subseteq$ **L** and `income` $= Y$. Imagine the confidence of this rule is 0.90 and its slift is 0.50. This means that the confidence of `race` $\neq$ `Black, education = Masters` $\rightarrow$ `income = low` is 0.90-0.50 = 0.40. Because of this high difference `race` $\neq$ `Black, education = Masters` could be seen as a subgroup at risk of discrimination.

As indicated by [33], decision rules can be learned on the original data using algorithms like Apriori [1] and then filtered according to fairness-based policies.

Situation Testing: Since association rules only detect global discrimination patterns, one can use the Situation Testing algorithm to further analyse fairness on a local level [44]: To check whether instance $\mathbf{x}_i$ receives a fair outcome Y, we use a distance function to search $\mathcal{D}$ for $\mathbf{x}_i$'s k-nearest neighbors from a reference group and a non-reference group, meaning we obtain two sets of instances $\mathcal{K}_{tr}^r$ and $\mathcal{K}_{tr}^{nr}$. A reference group is defined by sensitive feature values of those instances from the data we assume to be treated favorably, for instance, `race = White, sex = Male`. All instances not belonging to this group are seen as the non-reference group. To define instance $\mathbf{x}_i$'s individual discrimination score we calculate the ratio of positive decision ratio for $\mathcal{K}_{tr}^r$ and $\mathcal{K}_{tr}^{nr}$: $dec_r = \frac{|\{j \in \mathcal{K}_{tr}^r : y_j = 1\}|}{k}$, $dec_{nr} = \frac{|\{j \in \mathcal{K}_{tr}^{nr} : y_j = 1\}|}{k}$ and take the difference between both $(dec_r - dec_{nr})$. If this score exceeds the user-defined individual discrimination threshold t, it indicates that the treatment reserved to instance i depends on its sensitive characteristics.

4   Methodology

We propose to learn a selective classifier that does not only reject instances based on the uncertainty of their predictions but also their unfairness. In doing so we can decrease unfairness over all non-rejected instances. Further, by providing explanations for why some predictions are marked as unfair, we aid human reviewers in understanding whether the fairness concerns are indeed justified and enable a more informed decision process over them. We call our approach IFAC (Interpretable and Fair Abstaining Classifier). The intuition behind IFAC is visualized in Fig. 1: on top of the base classifier h we have our rejector g, which takes an instance's features $\mathbf{x}_i$ and the classifier h's prediction as its input. The rejector first executes a global fairness analysis on this instance, checking if it falls under any subgroups at risk of discrimination, as identified by discriminatory association rules (Sect. 3.2). If it does, it performs a local fairness check using Situation Testing [44], evaluating how the prediction for $h(\mathbf{x}_i)$ compares to the labels of similar instances in the data. After this, a *certainty assessment* is performed. Depending on the outcome of the assessment and the former fairness analysis there are four possibilities for our rejector: in case the prediction is deemed as fair and it exceeds a dedicated confidence threshold, the prediction is kept. Contrary, fair predictions that fall below this threshold are rejected. If we are dealing with an unfair prediction exceeding a separate confidence threshold for unfair data, it also gets rejected: though the prediction is certain, we have reasons to doubt it, because it is unfair. Finally, on predictions that are both unfair and uncertain, IFAC flips the original classifier $h(\mathbf{x}_i)$ prediction. The reasoning behind these interventions is that predictions that are neither fair nor certain are probably inaccurate, to begin with, and it is safe to alter them. This flipping mechanism is also added in case the user-defined *coverage* for IFAC does not allow to reject *all* unfair predictions. A complete walk-through example of how IFAC makes rejections is provided in Appendix A.

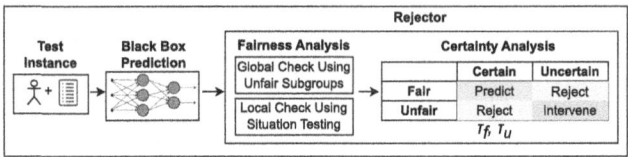

Fig. 1. Intuition behind IFAC

Now that we have described the basic intuition behind how IFAC is applied, we outline how it is learned. Given some data $\mathcal{D}$, we split it into a training set $\mathcal{D}_{tr}$ and two validation sets $\mathcal{D}_{val_1}$, $\mathcal{D}_{val_2}$. Then, given the target coverage c and the unfair reject weight w_u[3], IFAC is devised as follows:

1. **Learn a classifier:** we train classifier h from $\mathcal{D}_{tr}$. We highlight that any off-the-shelf probabilistic classifier can be considered, making our approach model-agnostic;
2. **Learn at-risk subgroups:** we extract association rules from validation set $\mathcal{D}_{val_1}$. The rules allow us to understand if there are correlations between sensitive features $\mathbf{S}$ and predictions of h, and, consequently, identify at-risk subgroups [33];
3. **Situation Testing:** we prepare the hyperparameters and distance function to run Situation Testing.
4. **Calibration:** we use the second validation set $\mathcal{D}_{val_2}$ to calibrate the rejection strategy, considering both *unfairness* and *uncertainty*:
   (i) the learned association rules are applied on $\mathcal{D}_{val_2}$;
   (ii) situation testing is performed for those instances falling under discriminatory patterns. This allows one to split the sample into a *fair* part $\mathcal{D}_{val_2^f}$ and an *unfair* one $\mathcal{D}_{val_2^u}$;
   (iii) depending on c and w_u, we estimate two different rejection thresholds, i.e. τ_f and τ_u. These thresholds are computed following the *coverage-calibration* procedure described in Sect. 3, ranking instances w.r.t. the confidence function over samples $\mathcal{D}_{val_2^f}$ and $\mathcal{D}_{val_2^u}$ respectively.

Fig. 2 summarizes the steps needed to learn IFAC. In the rest of this section, we further detail steps 2, 3, and 4.

---

[3] The unfair reject weight w_u determines how many rejections can be made based on unfairness concerns.

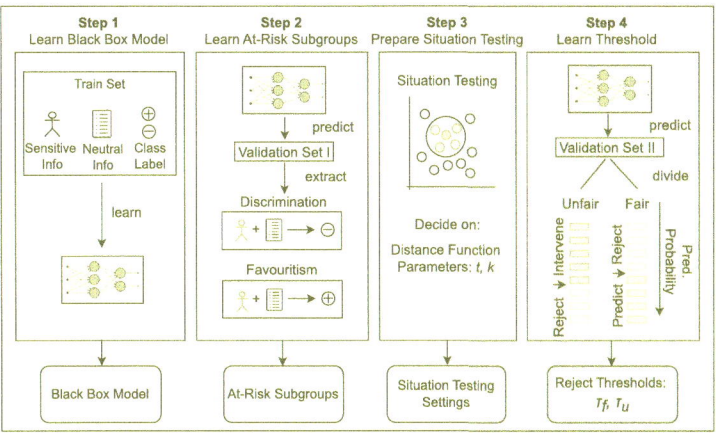

Fig. 2. The four steps for learning IFAC.

4.1 Step 2: Learn At-Risk Subgroups

To learn global patterns of unfairness, we use discriminatory association rules, as described in Sect. 3.2. To do so we apply h on the first validation set $\mathcal{D}_{val_1}$ and extract the association rules for the data and hs predictions with the apriori algorithm. We do so separately for each sensitive feature value and their combination. For example, let us have two sensitive attributes `sex` and `race` with two possible values, `F,M` and `W,B` respectively. We apply apriori and extract rules for each of the itemsets: {`sex=M`}, {`sex=F`}, {`race=W`}, {`race=B`}, {`sex=M ∧ race=B`}, {`sex=M ∧ race=W`}, {`sex=F ∧ race=W`}, {`sex=F ∧ race=B`}. Thus, the number of rules found meeting minimum support is not biased towards the largest demographic groups in the data.

As per our previous notation, we extract rules in the form of $(I_L, I_S) \rightarrow Y$, for some prediction outcome $h(\mathbf{x})$ in a binary classification setting $Y \in \mathcal{Y} = \{0, 1\}$. We say that rules with $Y = 0$ describe potentially discriminated subgroups, while rules with $Y = 1$ describe potentially favored ones. We extract favoring associations only for fixed reference groups defined for our data, e.g. white men (as described in Sect. 3.2). After extracting both favoring and discriminatory associations, we filter out statistically significant rules meeting an *slift* threshold. We calculate statistical significance using Z-test, testing if the proportion of some decision outcome Y is significantly different for the groups (I_L, I_S) and $(I_L, \neg I_S)$ [8]. We only select rules with $p < 0.01$. Further, we filter out *high-slift* rules by checking for which ones the following holds:

$$conf_{\mathcal{D}_{val_1}}((I_L, I_S) \rightarrow Y_v) - slift_{\mathcal{D}_{val_1}}((I_L, I_S) \rightarrow Y_v) < 0.5 \qquad (4)$$

Which in the context of binary classification is true *iff*:

$$conf_{\mathcal{D}_{val_1}}((I_L, \neg I_S) \rightarrow Y_v) < conf_{\mathcal{D}_{val_1}}((I_L, \neg I_S) \rightarrow \neg Y_v) \qquad (5)$$

Intuitively, this means that we only select the groups $\{I_L, I_S\}$ for which negating the sensitive part of the group ($\{I_L, \neg I_S\}$) yields higher confidence for value Y_v w.r.t. the opposite value $\neg Y_v$ (brief proof in Appendix Section B).

4.2   Step 3: Situation Testing

Part of the abstention mechanism of IFAC is based on a local fairness check for instances that are covered by global discrimination patterns. The aim is to use the global check to identify larger subgroups at risk of unfair treatment, while the local check allows us to execute a more fine-grained analysis taking all of an instance's characteristics into account. Our local fairness check is performed via Situation Testing, comparing a prediction $h(\mathbf{x}_i)$ for instance i with the decision labels of similar instances from $\mathcal{D}_{tr}$ (see Sect. 3.2). For the algorithm, a suitable distance function must be chosen e.g. we can consider the one used by Luong et al. [44] or one learned from the data [32]. We follow Luong's suggestion of a context-dependent approach and let an expert choose hyperparameters t and k depending on the decision task [44].

4.3   Step 4: Calibrate Rejection Strategy

Whether the rejector keeps, rejects, or intervenes on the original prediction for $\mathbf{x}$, depends on the (un)certainty of the base classifier. To evaluate the confidence of the classifier, we resort to the softmax response $v(\mathbf{x}) = \max_{y \in \mathcal{Y}} s_y$ [20,22], where $s_y(\mathbf{x}) \approx P(Y = y | \mathbf{X} = \mathbf{x})$ is an estimate of the conditional probability. We then estimate two thresholds τ_f and τ_u to choose between prediction, intervention, and abstention. The final selective classifier is in the form:

$$(h, g)(\mathbf{x}) = \begin{cases} h(\mathbf{x}) & \text{if } Fair(\mathbf{x}) \text{ and } v(\mathbf{x}) => \tau_f \\ \text{abstain} & \text{if } Fair(\mathbf{x}) \text{ and } v(\mathbf{x}) < \tau_f \\ 1 - h(\mathbf{x}) & \text{if } \neg Fair(\mathbf{x}) \text{ and } v(\mathbf{x}) < \tau_u \\ \text{abstain} & \text{if } \neg Fair(\mathbf{x}) \text{ and } v(\mathbf{x}) >= \tau_u \end{cases}$$

To learn τ_f and τ_u, h is applied on our second validation dataset $\mathcal{D}_{val_2}$ and its predictions are extracted. We then first extract those predictions that fall under discriminatory associations as learned in Step 2. After, we apply the Situation Testing algorithm as set up in Step 3 on those instances, and extract all that fail this individual fairness test. We consider those as the unfair fraction of the validation data ($\mathcal{D}_{val_2^u}$) and the remaining ones as the fair fraction $\mathcal{D}_{val_2^f}$. The number of rejections that can be made for both groups is determined by two parameters given by the user, namely the target coverage c and the unfair reject weight w_u. Given that the $\mathcal{D}_{val_2}$ consists of N instances of which N_u belong to $\mathcal{D}_{val_2^u}$ and N_f belong to $\mathcal{D}_{val_2^f}$, we calculate the number of total rejections (N_{rej}), the number of unfairness-based rejections (N_{ufr}) and the number of uncertainty-based rejections (N_{ucr}) as follows:

$$N_{rej} = \lceil (1 - c) \cdot N \rceil; \quad N_{ufr} = min(\lceil N_{rej} \cdot w_u \rceil, N_u); \quad N_{ucr} = N_{rej} - N_{ufr} \quad (6)$$

We then proceed by separately ordering the fair and unfair instances of the validation data according to the confidence function $v(\mathbf{x})$. On the fair instances, we determine the threshold τ_f such that N_{ucr} instances fall below this threshold, and on the unfair sample such that N_{ufr} instances exceed τ_u.

5   Experimental Evaluation

The goal of our experimental section aims to address the following questions:

Q1: Does IFAC achieve comparable results to state-of-the-art selective classifiers in terms of predictive performance and fairness?

Q2: How does IFAC explain the drivers behind unfairness-based rejections, and how could these explanations be utilized?

Q3: How do *coverage c* and the *unfair-reject weight u_w* affect our results?

5.1   Experimental Settings

Data and Baselines.   We run experiments considering two real datasets, namely ACSINCOME [14] and WISCONSINRECIDIVISM [5]. The former is about predicting high or low income based on instances' education, occupation etc. We define sex (male vs. female) and race (white vs. black vs. other) as sensitive attributes and take the group of white men as our reference group. We compare their treatment to each intersectional group based on race and sex.

WISCONSINRECIDIVISM contains information about criminal defendants, like their type of offense, number of prior offenses, etc. The task is to predict if they will not recidivate. We take race as the sensitive attribute (white vs. black vs. other). Because of a base classifiers' lower False Negative and higher False Positive rates on white people, we define this as the reference group[4].

We use different classification algorithms, namely a Random Forest, a Neural Network, and an XGBoost Classifier. We fitted all models with the default parameters of the corresponding Python libraries. Starting from these base classifiers, we compare IFAC with the following model-agnostic methods:

- *Full Coverage* (FC): the classifier itself when predicting on all the instances ($c = 1.00$)
- *Uncertainty Based Abstaining Classifier* (UBAC): The plug-in algorithm by [27]. This is the most well-known model-agnostic method and achieves state-of-the-art performance [37]. As for IFAC, we consider $v(\mathbf{x}) = max_{y \in \mathcal{Y}} s_y(\mathbf{x})$ as the confidence function. The rejection threshold is computed according to the *coverage-calibration* procedure.

---

[4] For full details on the preprocessing steps executed on both datasets we refer to our github repository.

Because we consider discrimination based on non-binary sensitive attributes (and in the case of ACSINCOME even intersectional discrimination), we do not compare with the fair abstention mechanism of Schreuder et al. [40] as a baseline, which only works on a single binary sensitive feature.

Hyperparameters. For **Q1** and **Q2**, we set $c = .80$ for the abstaining classifiers. Further, for IFAC we set the *unfair reject weight* (w_u) equal to 1.0. The intuition behind this is that if the coverage is large enough, IFAC should abstain from predicting any unfair instance, and only if not, fairness interventions should be performed. For the Situation Testing algorithm used by IFAC we set k, i.e. the number of neighbors used for the fairness comparisons to 10, and t to 0.3. For extracting discriminatory association rules we use the apriori algorithm of `apyori` with min. support of 0.01 and min. confidence of 0.85.

Metrics. For **Q1**, we evaluate predictive performance in terms of accuracy, precision, and recall on all non-rejected instances. Concerning fairness measures, we report the False Negative, False Positive, and Positive Decision Rates for the different demographic groups of each dataset. Further, we report the range and the standard deviation across demographic groups over these measures. Note, that we define these measures regarding the desirable label of each dataset. Hence, the positive decision ratio for ACSINCOME is the ratio of *high* income prediction, and for WISCONSINRECIDIVISM it is the ratio of *non-recidivism* predictions.

Experimental Setup. We split each dataset into training, two validation, and a test part (40% for train, 15% for each validation, and 30% for test) and train the classifiers on the former. For IFAC we learn the discriminatory associations on the first validation set. The reject thresholds for both IFAC and UBAC are calibrated based on the second. Finally, we randomly split the test set into 10 samples [32] and compute the final metrics on each of these samples. We provide results as averages and standard errors over these 10 test set samples.

5.2   Results

Q1: Performance and Fairness. We describe the predictive performance on each dataset and each classifier-methodology combination in Table 1. As can be seen, both selective classification methods improve upon the performance of FC, however, for UBAC this improvement is slightly larger, especially for the income prediction task.

In Fig. 3 we can see how the increased performance of UBAC comes at the cost of its fairness. In this Figure, we highlight the results of a Random Forest classifier combined with different selective classification methods, showing the average False Negative -, False Positive, and Positive Decision Rates (FNR, FPR, and PDR) over demographic groups (the results for Neural Networks and XGBoost follow the same patterns and are included in the Appendix). We also highlight the range of these metrics across demographics (i.e. the performance difference between the highest- and lowest performing group) and the standard deviation. Fairer classifiers should score lower on both metrics, to ensure that there are no big performance differences across groups.

Table 1. Performance Results ACSINCOME and WISCONSINRECIDIVISM

		ACSINCOME			WISCONSINRECIDIVISM		
		Acc.	Rec.	Prec.	Acc.	Rec.	Prec.
RF	FC	.78 ± .01	.57 ± .02	.65 ± .03	.62±.01	.77±.01	.65±.01
	UBAC	**.83 ± .01**	**.62 ± .02**	**.69 ± .03**	**.65±.01**	**.83±.01**	**.66±.01**
	IFAC	.80 ± .01	.59 ± .04	.64 ± .03	**.65±.01**	**.83±.01**	**.66±.01**
NN	FC	.80 ± .01	.58 ± .03	.71 ± .03	.63±.01	0.74±.01	.65±.01
	UBAC	**.86 ± .01**	**.62 ± .03**	**.77 ± .03**	**.66±.02**	**.77±.01**	**.68±.02**
	IFAC	.83 ± .01	.58 ± .03	.73 ± .02	**.66±.02**	.76±.01	**.68±.02**
XGB	FC	.81 ± .01	.60 ± .03	.73 ± .03	.63±.01	.77±.01	.65±.01
	UBAC	**.87 ± .01**	**.64 ± .03**	**.78 ± .03**	**.66±.01**	**.83±.01**	**.68±.01**
	IFAC	.84 ± .01	.59 ± .03	.75 ± .03	**.66±.01**	.82±.01	**.68±.01**

Starting with ACSINCOME, we see that for UBAC this is not the case: we observe an especially unequal distribution of FNR across demographic groups, with the highest difference being 0.4 (between white men and black women). This difference is even higher than for the FC classifier, as the UBAC selection mechanism only decreases the FNR for white men while increasing it for others. With using IFAC this effect does not occur: through rejecting predictions that are at high risk of unfairness, FNRs decrease for minority groups like women or black people, and overall the rates become more equal across demographics, bringing the range down to 0.2 and the std. to 0.08. The patterns are slightly less strong when considering the FPR and PDR across demographics, but still hold. Similar patterns occur for WISCONSINRECIDIVISM: the range and standard deviation for FNR, FPR, and PDR across demographics decrease when using IFAC, while they increase with UBAC. We acknowledge that the effect is less strong here, but attribute this to IFACs selection criteria for unfair instances being too strict. In Appendix D we show results with a lower threshold t for situation testing (meaning that more instances can get rejected out of unfairness concern), where IFAC makes FNR, FPR, and PDR nearly equal across groups. Further, we highlight how equalizing error rates across demographics is only the first step towards improving the fairness of the decision task. As we illustrate in the next section, enabling humans to review rejected instances and the explanation behind them, is the most crucial contribution of our method.

Q2: Explaining Unfair Rejections. One of the main advantages of IFAC is that it can explain why rejected predictions are seen as unfair. In Fig. 4 we show some explanations behind rejected instances for both of our datasets, and we use the ACSINCOME case to highlight how a human expert can utilize them. We see two instances that were both rejected based on the same global pattern of unfairness: the classifier predicting "low income" ratios for black women, aged between 30 and 39 working in management, than for people with the same age

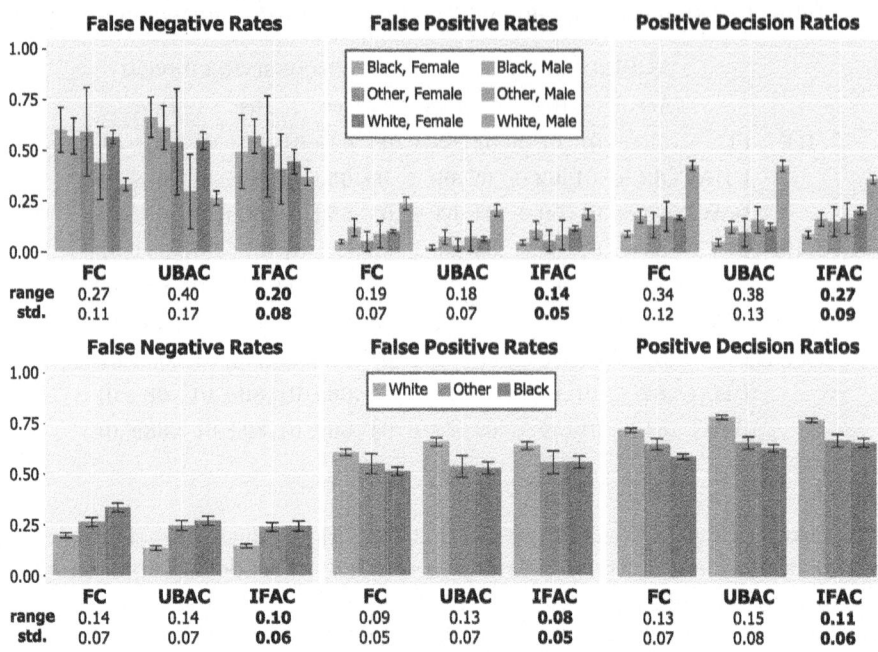

Fig. 3. Performance measures over demographic groups when applying a Random Forest in combination with various selective classifiers on ACSINCOME (above) and WISCONSINRECIDIVISM (below). A regular UBAC increases differences in error- as well as positive decision rates among groups. Using IFAC, and rejecting instances based on unfairness, diminishes these differences.

and occupation, but different demographics. While an algorithm only analyses such patterns statistically, human experts can examine them with sensitivity surrounding their historical context. For instance, it is well known that racism and sexism contribute to hostile work environments for black women. Hence, a human expert can reason how these dynamics may hinder fair compensation in roles like management, that are normally associated with high salaries.

The results of situation testing provide further insight into the unfairness of the classifier: For both instances, a high ratio of the 10 most similar white men have a high income; explaining why their own low income predictions are marked as unfair. However, for the first instance, many of the white men considered for the comparison have a higher education level and amount of working hours than her. Since it makes sense, that people working part-time do not get the same compensation as people working full-time, the low income prediction could be seen as justified and a human reviewer could decide to keep it. For the second case, all similar white men do share the instances' education level, working hours, etc. Hence, there is no justification for why she would be the only one receiving a low income prediction, and a human expert could decide to override this decision.

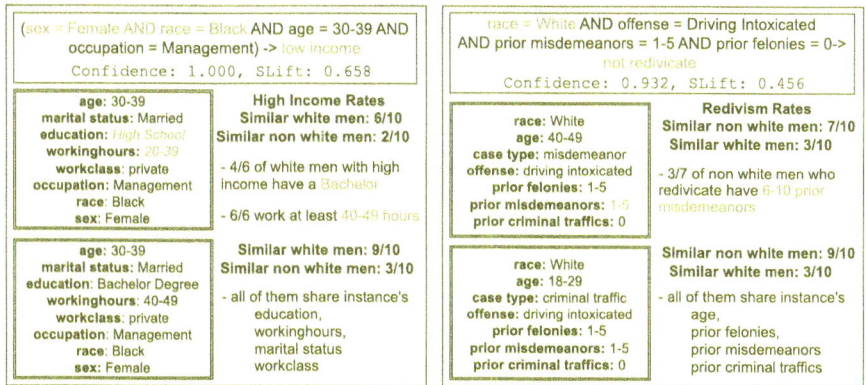

Fig. 4. Examples for ACSINCOME (left) and WISCONSINRECIDIVISM (right) of two rejected instances, and the explanation behind their rejections.

To conclude, these examples show how IFAC's interpretable-by-design rejector can have a large impact in increasing the fairness of a decision process. In particular, our approach goes beyond a rough statistical analysis of discriminatory patterns and allow for the integration of human domain knowledge to achieve a much deeper fairness assessment.

Q3: Effects of c and w_u. In this section, we explore the effect of parameters c and w_u on IFAC's performance. Out of space constraints, we only report the results with a Random Forest as a base-classifier on ACSINCOME. The results for the other classifiers and the other dataset follow the same pattern and are included in the Appendix. In Fig. 5 we visualize how the accuracy, the range in positive decision ratio across demographics, and the standard deviation change as a function of the coverage and the w_u. Unsurprisingly, for both UBAC and IFAC the accuracy drops as the coverage increases. Regardless of the coverage and the w_u UBAC outperforms IFAC. Further, we see that a lower w_u comes at the cost of accuracy, especially when the coverage is high. Intuitively this makes sense: w_u determines how many of the unfair predictions are rejected, and for how many an intervention is performed. With the low weight of 0.25, the majority of unfair prediction labels are simply flipped, and only the ones with very high prediction probability are abstained from. With an increase in coverage, this pattern is more extreme, as the general number of instances that can be abstained from is lower. When observing the effect of differing coverages and w_u on the fairness of the predictions, we observe that performing more interventions (as a result of a lower w_u) has a desirable effect: both the range and standard deviation of positive decision ratios decreases across demographics. The effect is again larger for higher coverages because fewer allowed rejections mean more interventions, which bring the positive decision ratios across demographic groups closer together.

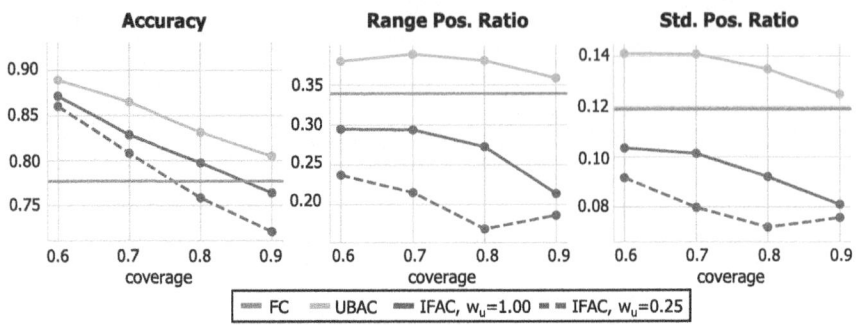

Fig. 5. Effects of c and w_u parameters in our selective classification settings.

6   Discussion and Conclusion

In this paper, we have introduced IFAC, an Interpretable and Fair Abstaining Classifier. This classifier rejects predictions from a base classifier, both in cases of uncertainty and unfairness. Unfairness rejections are based on the interpretable-by-design methods of unfair association patterns and situation testing. Through our experiments, we have shown how using our abstention mechanism yields satisfying overall performance, while improving fairness across demographic groups over all non-rejection instances. This stands in contrast to a regular uncertainty-based abstaining classifier, that does not take the fairness of predictions into account. We have also shown how the explanations behind our abstention mechanism, can empower human decision-makers to review the rejected instances and make fairer decisions for them. This holds immense potential for complying with recent AI regulations, which require automated decision-making processes to be supervised by humans to mitigate the risks of discrimination. By only having to review instances at high risk of unfairness, our framework can make this process more practical and time-efficient. To further empower human users, further research could involve human experts in the selection of *at-risk* subgroups and in choosing distance function and parameters for Situation Testing. Also, user studies can help in understanding how humans engage with such a system. For this, one should consider adding explanations for all non-rejected instances, so that humans can still explore the base classifier in the accepted cases.

Acknowledgments. D. Lenders and T. Calders were funded by Digitax Centre of Excellence UAntwerp and by Research Foundation Flanders under FWO file number: V467123N. A. Pugnana and R. Pellungrini and D. Pedreschi and F. Giannotti have received funding by PNRR - M4C2 - Investimento 1.3, Partenariato Esteso PE00000013 - "FAIR - Future Artificial Intelligence Research" - Spoke 1 "Human-centered AI", funded by the European Commission under the NextGeneration EU programme, ERC-2018-ADG G.A. 834756 "XAI: Science and technology for the eXplanation of AI decision making" and Prot. IR0000013. This work was also funded by the European Union under Grant Agreement no. 101120763 - TANGO. Views and opinions expressed are however those of the author(s) only and do not necessarily reflect those of the European

Union or the European Health and Digital Executive Agency (HaDEA). Neither the European Union nor the granting authority can be held responsible for them. The work has also been realised thanks to NextGenerationEU - National Recovery and Resilience Plan, PNRR) - Project: "SoBigData.it - Strengthening the Italian RI for Social Mining and Big Data Analytics" - Prot. IR000001 3 - Notice n. 3264 of 12/28/2021.

References

1. Agrawal, R., Srikant, R.: Fast algorithms for mining association rules in large databases. In: VLDB, pp. 487–499. Morgan Kaufmann (1994)
2. Artelt, A., Brinkrolf, J., Visser, R., Hammer, B.: Explaining reject options of learning vector quantization classifiers. In: IJCCI, pp. 249–261. SCITEPRESS (2022)
3. Artelt, A., Hammer, B.: "even if ..." - diverse semifactual explanations of reject. In: SSCI, pp. 854–859. IEEE (2022)
4. Artelt, A., Visser, R., Hammer, B.: "i do not know! but why?"- local model-agnostic example-based explanations of reject. Neurocomputing **558**, 126722 (2023)
5. Ash, E., Goel, N., Li, N., Marangon, C., Sun, P.: WCLD: curated large dataset of criminal cases from wisconsin circuit courts (2023)
6. Cabrera, Á.A., Epperson, W., Hohman, F., Kahng, M., Morgenstern, J., Chau, D.H.: Fairvis: visual analytics for discovering intersectional bias in machine learning. In: 2019 IEEE Conference on Visual Analytics Science and Technology (VAST), pp. 46–56. IEEE (2019)
7. Calmon, F., Wei, D., Vinzamuri, B., Natesan Ramamurthy, K., Varshney, K.R.: Optimized pre-processing for discrimination prevention. In: Advances in Neural Information Processing Systems, vol. 30 (2017)
8. Casella, G., Berger, R.L.: Statistical Inference, Duxbury Press. Pacific Grove, CA (2002)
9. Chow, C.K.: On optimum recognition error and reject tradeoff. IEEE Trans. Inf. Theory **16**(1), 41–46 (1970)
10. Condessa, F., Bioucas-Dias, J.M., Castro, C.A., Ozolek, J.A., Kovacevic, J.: Classification with reject option using contextual information. In: ISBI, pp. 1340–1343. IEEE (2013)
11. Cortes, C., DeSalvo, G., Mohri, M.: Theory and algorithms for learning with rejection in binary classification. Ann. Math. Artif. Intell. **92**, 1–39 (2023)
12. Costanza-Chock, S., Raji, I.D., Buolamwini, J.: Who audits the auditors? recommendations from a field scan of the algorithmic auditing ecosystem. In: FAccT, pp. 1571–1583. ACM (2022)
13. Crenshaw, K.: Demarginalizing the intersection of race and sex: a black feminist critique of antidiscrimination doctrine, feminist theory and antiracist politics. In: University of Chicago Legal Forum, vol. 1989 (1989)
14. Ding, F., Hardt, M., Miller, J., Schmidt, L.: Retiring adult: new datasets for fair machine learning, pp. 6478–6490 (2021)
15. El-Yaniv, R., Wiener, Y.: On the foundations of noise-free selective classification. J. Mach. Learn. Res. **11**, 1605–1641 (2010)
16. Enqvist, L.: Human oversight' in the EU artificial intelligence act: what, when and by whom? Law Innov. Technol. **15**(2), 508–535 (2023)
17. Fischer, L., Hammer, B., Wersing, H.: Optimal local rejection for classifiers. Neurocomputing **214**, 445–457 (2016)
18. Fleisher, W.: What's fair about individual fairness? In: Proceedings of the 2021 AAAI/ACM Conference on AI, Ethics, and Society, pp. 480–490 (2021)

19. Foulds, J.R., Islam, R., Keya, K.N., Pan, S.: An intersectional definition of fairness. In: 2020 IEEE 36th International Conference on Data Engineering (ICDE), pp. 1918–1921. IEEE (2020)
20. Franc, V., Prusa, D., Voracek, V.: Optimal strategies for reject option classifiers. J. Mach. Learn. Res. **24**(11), 1–49 (2023)
21. Gangrade, A., Kag, A., Saligrama, V.: Selective classification via one-sided prediction. In: AISTATS, vol. 130, pp. 2179–2187. PMLR (2021)
22. Geifman, Y., El-Yaniv, R.: Selective classification for deep neural networks. In: NIPS, pp. 4878–4887 (2017)
23. Geifman, Y., El-Yaniv, R.: Selectivenet: a deep neural network with an integrated reject option. In: ICML, vol. 97, pp. 2151–2159. PMLR (2019)
24. Goel, N., Yaghini, M., Faltings, B.: Non-discriminatory machine learning through convex fairness criteria. In: Proceedings of the 2018 AAAI/ACM Conference on AI, Ethics, and Society, pp. 116–116 (2018)
25. Goethals, S., Martens, D., Calders, T.: PreCoF: counterfactual explanations for fairness. Mach. Learn. **113**, 1–32 (2023)
26. Hendrickx, K., Perini, L., der Plas, D.V., Meert, W., Davis, J.: Machine learning with a reject option: a survey. ArXiv **abs/2107.11277** (2021). https://api.semanticscholar.org/CorpusID:236318084
27. Herbei, R., Wegkamp, M.H.: Classification with reject option. Can. J. Stat. **34**(4), 709–721 (2006)
28. Huang, L., Zhang, C., Zhang, H.: Self-adaptive training: beyond empirical risk minimization. In: NeurIPS (2020)
29. Jones, E., Sagawa, S., Koh, P.W., Kumar, A., Liang, P.: Selective classification can magnify disparities across groups. In: ICLR (2021)
30. Kühne, J., März, C., et al.: Securing deep learning models with autoencoder based anomaly detection. In: PHM Society European Conference, vol. 6, pp. 13–13 (2021)
31. Lee, J.K., et al.: Fair selective classification via sufficiency. In: ICML. Proceedings of Machine Learning Research, vol. 139, pp. 6076–6086. PMLR (2021)
32. Lenders, D., Calders, T.: Learning a fair distance function for situation testing. In: Kamp, M., et al. Machine Learning and Principles and Practice of Knowledge Discovery in Databases, ECML PKDD 2021, Communications in Computer and Information Science, vol. 1524, pp. 631–646. Springer, Cham (2021). https://doi.org/10.1007/978-3-030-93736-2_45
33. Pedreschi, D., Ruggieri, S., Turini, F.: Discrimination-aware data mining. In: KDD, pp. 560–568. ACM (2008)
34. Pedreschi, D., Ruggieri, S., Turini, F.: Measuring discrimination in socially-sensitive decision records, pp. 581–592. SIAM (2009)
35. Perini, L., Davis, J.: Unsupervised anomaly detection with rejection. In: NeurIPS (2023)
36. Pessach, D., Shmueli, E.: A review on fairness in machine learning. ACM Comput. Surv. (CSUR) **55**(3), 1–44 (2022)
37. Pugnana, A., Perini, L., Davis, J., Ruggieri, S.: Deep neural network benchmarks for selective classification. arXiv preprint arXiv:2401.12708 (2024)
38. Pugnana, A., Ruggieri, S.: AUC-based selective classification. In: AISTATS, vol. 206, pp. 2494–2514. PMLR (2023)
39. Pugnana, A., Ruggieri, S.: A model-agnostic heuristics for selective classification. In: AAAI, pp. 9461–9469. AAAI Press (2023)
40. Schreuder, N., Chzhen, E.: Classification with abstention but without disparities. In: UAI. Proceedings of Machine Learning Research, vol. 161, pp. 1227–1236. AUAI Press (2021)

41. Selbst, A.D., Boyd, D., Friedler, S.A., Venkatasubramanian, S., Vertesi, J.: Fairness and abstraction in sociotechnical systems. In: Proceedings of the Conference on Fairness, Accountability, and Transparency, pp. 59–68 (2019)
42. Shah, A., et al.: Selective regression under fairness criteria. In: ICML. Proceedings of Machine Learning Research, vol. 162, pp. 19598–19615. PMLR (2022)
43. Stevens, A., Deruyck, P., Veldhoven, Z.V., Vanthienen, J.: Explainability and fairness in machine learning: improve fair end-to-end lending for kiva. In: SSCI, pp. 1241–1248. IEEE (2020)
44. Thanh, B.L., Ruggieri, S., Turini, F.: k-nn as an implementation of situation testing for discrimination discovery and prevention. In: KDD, pp. 502–510. ACM (2011)
45. Wachter, S., Mittelstadt, B.D., Russell, C.: Why fairness cannot be automated: bridging the gap between EU non-discrimination law and AI. Comput. Law Secur. Rev. **41**, 105567 (2021)
46. Wang, A., Ramaswamy, V.V., Russakovsky, O.: Towards intersectionality in machine learning: Including more identities, handling underrepresentation, and performing evaluation. In: Proceedings of the 2022 ACM Conference on Fairness, Accountability, and Transparency, pp. 336–349 (2022)
47. Wang, X., Yiu, S.: Classification with rejection: scaling generative classifiers with supervised deep infomax. In: IJCAI, pp. 2980–2986. ijcai.org (2020)
48. Zafar, M.B., Valera, I., Rogriguez, M.G., Gummadi, K.P.: Fairness constraints: mechanisms for fair classification. In: Artificial intelligence and statistics, pp. 962–970. PMLR (2017)

Boosting Long-Tail Data Classification with Sparse Prototypical Networks

Alexei Figueroa[1], Jens-Michalis Papaioannou[1,6](✉), Conor Fallon[1],
Alexandra Bekiaridou[2,3], Keno Bressem[4,5], Stavros Zanos[2,3], Felix Gers[1],
Wolfgang Nejdl[6], and Alexander Löser[1]

[1] DATEXIS, Berliner Hochschule für Technik, Berlin, Germany
{afigueroa,michalis.papaioannou,gers,aloeser}@bht-berlin.de
[2] Elmezzi Graduate School of Molecular Medicine, Manhasset, USA
[3] Feinstein Institutes for Medical Research, Northwell Health, Manhasset, USA
{abekiaridou,szanos}@northwell.edu
[4] Klinikum rechts der Isar, Department of Radiology, Technical University Munich,
Munich, Germany
bressem@dhm.mhn.de
[5] Department of Radiology and Nuclear Medicine, German Heart Center Munich,
Munich, Germany
[6] L3S, Leibniz University Hannover, Hanover, Germany
nejdl@l3s.de

Abstract. Clinical Decision Support Systems (CDSS) have become ubiquitous in healthcare facilities, leveraging the increasing presence of Electronic Health Records (EHR). Predicting clinical outcomes from clinical text, such as identifying diagnoses based on the admission state of patients, is among the core tasks that a CDSS must address. The state-of-the-art for this task has been set by transformer encoder models, recently superseded by encoders enhanced with a prototypical network. This task remains a significant challenge due to the substantial imbalance of the outcome labels, which is characterized by a long-tailed distribution where the majority of diagnoses are under-represented. Motivated by recent biologically inspired findings in deep learning, we propose S-Proto, a novel, efficient, and sparse prototypical layer. Our method achieves state-of-the-art performance in outcome diagnosis prediction, without compromising on the explainability characteristics of prototypical encoders. Quantitative results demonstrate that our approach is robust to the challenges presented by clinical notes, and transfers successfully to a second, unseen dataset. Qualitative evaluation with medical doctors shows that S-Proto is capable of disaggregating the representations of a disease that manifests differently in patient cohorts.

Keywords: Prototypical Networks · Sparsity · Long-Tail · NLP

A. Figueroa and J.-M. Papaioannou—Both authors contributed equally.

A. Bifet et al. (Eds.): ECML PKDD 2024, LNAI 14947, pp. 434–449, 2024.
https://doi.org/10.1007/978-3-031-70368-3_26

1   Introduction

Clinical Decision Support Systems (CDSS) are used to assist medical profession-als not only in their decision-making process, but also in allocating effectively the limited resources of healthcare facilities. CDSS assist clinicians with pre-dictions on diagnoses, medications, and procedures to streamline patient care in hospitals and clinics. An important component of CDSS is the integration of a patient's Electronic Health Record (EHR) data. These records consolidate multimodal information spanning text, tabular data, images, and time series. A set of tasks focusing on the text modality of these records is Clinical Outcome Prediction [1], which involves predicting clinical outcomes at discharge given the admission state of a patient in text form. In this work, we focus on one of these tasks: diagnoses prediction, where a model is required to annotate the presence (and absence) of medical conditions and probable future outcome diagnoses. The target space for this task is the set of codes of the International Classification of Diseases (ICD), which are billing codes used for clinical documentation. This task is an extreme multilabel-classification problem. MIMIC is the largest pub-licly available dataset that contains clinical notes and is annotated with these labels. Hence, we use the latest release MIMIC-IV [16].

Diagnosis prediction from clinical text has been mainly addressed using networks from the BERT [10] family. These approaches have focused on pre-training with a variety of data, tasks, data augmentation, and architecture modifications to improve classification performance. Recently ProtoPatient [2] achieved state-of-the-art (SOTA) performance on diagnoses prediction by adapt-ing methods from prototypical networks in computer vision, adding new prop-erties of explainability to the representations of the clinical notes. In general, these methods struggle to perform classification when there is a large imbalance in annotated labels [2,21]. As a result, they excel at predicting the majority classes, performing strongly in metrics such as AUROC, but show great room for improvement in performance metrics that reflect this imbalance, such as PR-AUC. In MIMIC-IV, 90% of the diagnoses are associated with only 20% of the patients; or alternatively, 80% of the samples relate to 10% of the labels. This is a very pronounced long-tail, typical for clinical data.

S-Proto: To tackle this we propose a novel method for these high class-imbalance multilabel classification problems. Our method boosts prediction per-formance across all parts of the distribution, especially in the long-tail. It achieves SOTA performance in both AUROC and PR-AUC. S-Proto is based on recent advancements in prototypical networks in NLP, and biologically-inspired net-works. It leverages the expansion of the dimensionality of the prototypical layer. We refer to a single dimension in this expansion as a *sub-network*. S-Proto efficiently learns these *sub-networks* through a *winner-takes-all* mechanism, as observed in biological neural circuits. For a given input, this mechanism selects a *winner sub-network* and inhibits the rest. The selection of *winner sub-networks* proves to be crucial to the scores achieved, both in classification metrics and in training efficiency. Our Experiments show the importance of multiple *win-*

ner sub-networks to adapt to the variance of clinical notes in both the presence and absence of a diagnosis. We confirm with medical doctors that the *winner* sub-networks that predict the presence of a diagnosis, learn clinically-relevant phenotypical hierarchies from the data. Additionally, we evaluate quantitatively the explanations provided by S-Proto, showing that our increased complexity does not deteriorate their faithfulness.

We summarise our contributions as follows:

1. To our knowledge, we are the first to propose an efficient, sparse, and multi-dimensional prototypical layer for clinical text based on the winner-takes-all mechanism.
2. We evaluate our approach and all baselines on an extreme-multi-label classification problem at the example of diagnoses prediction. We achieve new state-of-the-art performance addressing the long-tail without compromising on explainability. Additionally, we demonstrate the transferability of the performance gains using a second unseen dataset.
3. A qualitative analysis with medical professionals shows that our approach learns clinically meaningful hierarchies (cohorts of patients).
4. We make the source code to reproduce our experiments and our methods openly available[1].

The remaining of this work is structured as follows: Sect. 2 discusses related work, Sect. 3 describes the task and the data used to train and evaluate the models. In Sect. 4 we present the core of our method, Sect. 5 and Sect. 6 detail our experimental setup and results. We discuss our most relevant findings in Sect. 7 and conclude with further directions of research in Sect. 8.

2   Related Work

Recent studies have applied pre-trained language models (LMs) to perform clinical outcome prediction using unstructured text from EHRs [1]. Architectures have centered on transformers of the BERT family [10], enhancing performance by literature augmentation [22], multi-modal knowledge integration [13,31], and multi-lingual knowledge transfer [23]. Most recently, ProtoPatient leverages an additional prototypical layer [29] on top of PubMedBERT [30], and achieves SOTA performance while simultaneously enhancing interpretability. Going beyond these works, we generalize ProtoPatient to multiple prototypes, applying a sub-network selection strategy based on a winner-takes-all mechanism [8,19]. This improves overall performance, boosting the long-tail significantly while maintaining explainability.

Classification with Multiple Prototypes. Multiple class-prototypes have been widely proposed in computer vision [7,15,18], including applications on clinical imaging data [9]. In the text domain, and close to our work, Proto-lm [32]

---

[1] https://github.com/DATEXIS/sproto.

propose an approach based on multiple class-prototypes, focusing on improving interpretability by using a loss only applicable to *multi-class* classification tasks. In contrast to Proto-lm, we tackle an extreme *multi-label* classification task in the clinical-text domain, with a novel sparse prototypical layer.

Relation to Mixture of Experts (MoE). Recently in NLP, very large sparse MoE networks [11,12,17,27] have resulted from *conditional computation* [4], where only parts of the model are activated depending on the input. We can also think of our approach as a simplified MoE where the expert is gated by the winner-takes-all mechanism. Essentially, we *condition the activation* of the most relevant (*winner*) sub-network on an admission note.

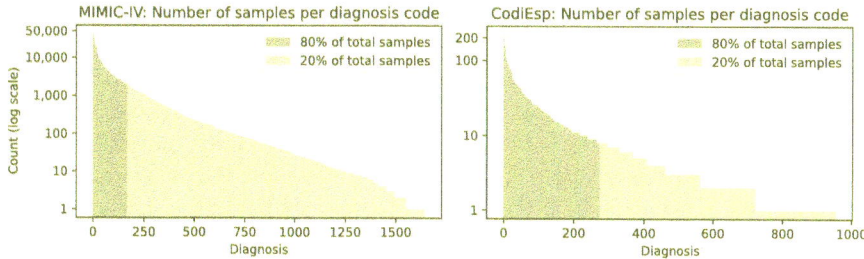

Fig. 1. Power Law Distribution of the outcome diagnoses. The head (purple) consists of 80% of the patients and the tail (yellow) 20 % of the patients. For MIMIC, the tail represents 90% of the labels and the head 10%. For CodieEsp, it is 71% and 29% respectively. The figures show that a large majority of labels are in the long-tail of both distributions. (Color figure online)

3   Task: Diagnoses Prediction

We evaluate our method on the extreme multi-label classification task of diagnoses outcome prediction as presented in [1]. For our experiments, we use the clinical text data from MIMIC-IV. A very challenging aspect of this task is that diagnosis codes (labels) are power-law distributed with a pronounced long-tail (Fig. 1). This is exacerbated by frequent diagnoses being under-annotated. We also evaluate our approach on a second dataset CodiEsp in a zero-shot setting. We discuss details about the dataset and its performance in Sect. 7.

MIMIC-IV is a publicly available database consisting of EHRs, curated for researchers, stemming from the Beth Israel Deaconess Medical Center. A principal difference between this version and older versions of MIMIC is that MIMIC-IV contains Intensive Care Unit (ICU) data as well as EHRs from other divisions

Table 1. Distribution of admission notes and ICD-10 codes per dataset split. Note that not all labels are present in every dataset split because 9% of the labels appear fewer than three times. We show the number of labels and patients for the evaluation of CodiEsp on the right side of the table.

MIMIC-IV					CodiEsp
	Total	Train	Test	Val	Test
# Patients	**119,175**	102,199	7,618	9,358	999
Diagnostic Labels	**1,643**	1,643	1,183	1,214	**955**

of the hospital. This leads to a more pronounced long tail. We update the diagnosis prediction task of [1] with the data from MIMIC-IV [16]. To simulate the state of a patient at admission time from a given discharge note, we use the methodology as in [1] with the distinction that we focus on three-digit ICD-10 codes, instead of ICD-9. Our classification target space consists of 1643 unique labels. We use stratified sampling [26] to create three similarly distributed data splits (train/val/test). Note that not all labels appear in every dataset split, as certain labels occur fewer than three times in MIMIC-IV. We ensure that every label is at least once in the training set and that every patient is present in only one of the splits to avoid overfitting (a patient can have multiple re-admissions). The details of the splits are shown in Table 1.

4   Methods

Both our method and ProtoPatient phrase the task of diagnoses prediction as a latent spatial clustering of patients. For ProtoPatient, two learnable components per label are used: **one** *attention* vector and **one** *prototype* vector. The attention vector enables textual explanations, while the prototype vector is a cluster centroid that defines a *prototypical patient*: the typical patient for a disease.

As we note in Sect. 3 the distribution of diagnoses is long-tailed and thus very challenging. We argue that a way to boost classification and tackle the variance of patients and diagnoses, especially in the long-tail, is to learn **multiple centroids** per diagnosis, i.e. multiple *prototypical patients*. Our method achieves this in two steps:

1. We expand the dimensionality of both the *attention* and *prototype* vectors for every diagnosis, thus learning multiple prototypes and attention vectors for every label (diagnosis). We refer to the combination of these two vectors as a *sub-network*.
2. We select the best single *sub-network* using the winner-takes-all mechanism. This sub-network we refer to as the *winner sub-network*.

These two steps are illustrated in detail in Fig. 2. The winner-takes-all selection mechanism models a biologically-inspired process, where the neuron that activates the most in response to an input inhibits all other competing neurons

[8,19]. Analogously, we select (activate) only one *winner* of the multiple *sub-networks* for every class, depending on the admission note of a patient. This resulting sparse activation makes the model efficient to train since only a portion of the parameters is affected by the loss at every training step. Since the single activated *winner sub-network* is computationally equivalent to ProtoPatient, the explainability properties of ProtoPatient should also be preserved in our approach. We analyze this further in Sect. 7.

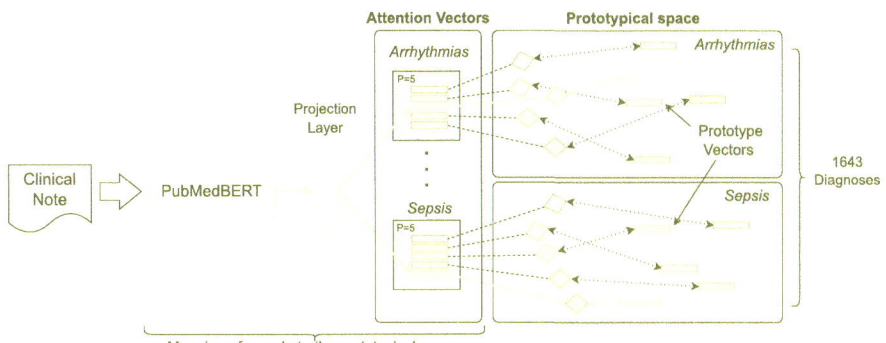

Fig. 2. S-Proto Overview. Starting from the left, a clinical note is encoded by Pub-MedBERT and then projected by a linear layer. Then this representation is mapped to the prototypical space by the attention-vectors (colored rectangles left) for each diagnosis. Note that every admission note is mapped P(=5) times for one diagnosis, these are shown as *rhombuses*, and the distance to each prototype vector (colored rectangles right) is computed. The **winner-takes-all** mechanism chooses the *winner sub-network* (highlighted in orange). Only the chosen *sub-network* parameters are updated at each training step. (Color figure online)

S-Proto. We formally introduce our method. Let $enc(x) \in \mathbb{R}^{T \times d}$ be the Pub-MedBERT encoding for the admission note x, T be the token dimension, and d the hidden dimension of this encoding. This encoding is down-projected with a linear layer $L \in \mathbb{R}^{d \times h}$ where h is the hidden dimension of this projection, thus:

$$\psi_t = \langle enc_t(x), L \rangle \tag{1}$$

where $t \in T$ denotes the token index and $\psi_t \in \mathbb{R}^h$ is a single projected token embedding.

Our approach uses P sub-networks per diagnosis. Let $W \in \mathbb{R}^{C \times P \times h}$ be the label-wise attention vectors and $u \in \mathbb{R}^{C \times P \times h}$ be the prototypical vectors, where $c \in C$ denotes a diagnosis, and $p \in P$ denotes a single *sub-network*. A mapped admission note into the prototypical space $v_{c,p}$ is computed as follows:

$$v_{c,p} = \sum_{t \in T} S_{t,c,p} \odot \psi_t \qquad S_{t,c,p} = \frac{\exp(\phi_{t,c,p})}{\sum_{t \in T} \exp(\phi_{t,c,p})} \qquad \phi_{t,c,p} = W_{c,p} \odot \psi_t \tag{2}$$

The output of a sub-network p and class c is the Euclidean distance: $\epsilon_{c,p} = \|u_{c,p} - v_{c,p}\|_2$, and a prediction $\hat{y}_{c,p}$ is computed from these distances as follows:

$$\hat{y}_{c,p} = \sigma(-\epsilon_{c,p}) \tag{3}$$

where σ is the sigmoid activation function. We use the winner-takes-all selection to choose the single index p that corresponds to the closest prototype vector $u_{c,p}$ to the mapped admission note $v_{c,p}$. This is achieved by masking the predictions with a Kronecker δ function as in Eq. 4.

$$\hat{y} = [y_0, ..., y_c] \text{ where } \hat{y}_c = \sum_{p \in P} \hat{y}_{c,p}\delta_{p,\hat{p}} \text{ and } \hat{p} = \underset{p}{\mathrm{argmax}}(-\epsilon_{c,p}) \tag{4}$$

$\hat{y}$ is the vector containing the predictions produced by the *winner* sub-networks for every diagnosis. The loss is then defined as $BCE(\hat{y}, y)$, where BCE is the binary cross entropy loss function. Since $\hat{y}$ only involves *winner* sub-networks $\hat{p}$, within the prototypical layer only the parameters $W_{c,\hat{p}}$ and $u_{c,\hat{p}}$ are adjusted after each training step.

Sparsity and Efficiency. The consequence of the introduced loss is that it allows us to update a subset of parameters during training. In practice, we accomplish this by running two forward passes on the prototypical layer: first, to identify the *winner sub-network* ($\hat{p}$) for each class; and second, to compute the corresponding loss. Since only the second step involves gradient computation, the subsequent backpropagation step has the same memory footprint as ProtoPatient. We compute the update of the most computationally intensive component of our architecture, PubMedBERT, and the projection layer L only once. Thus, we can keep comparable batch sizes to ProtoPatient, with a prototypical layer that is P times larger, with a marginal additional computational cost. This effect can be seen in the training times of our experiments in Table 2.

5   Experiments

We fine-tune and evaluate the following methods for the diagnosis prediction task to compare with [2]. However, we train all models with the latest version of the MIMIC-IV data, which is roughly three times larger.

PubMedBERT. To assess the effects of the novel prototypical layer we compare against this baseline. This also illustrates a full ablation of our method.

ProtoPatient. [2] This method is a special case of our method when P=1 (see: Sect. 4). We evaluate this model to assess whether the novel competing sub-networks and our selection mechanism result in performance benefits. We are especially interested in the results for diagnoses in the long tail. This method is the strongest baseline we can compare against in the diagnoses prediction task.

S-Proto Variants. The number of prototypes P created per class controls the sparsity of the respective model. In the context of the diagnoses prediction task, this hyperparameter could be viewed as a budget to explain the variance of a diagnosis. We train three variants of our approach for three values of P - namely 3, 5, and 15 - corresponding to the number of competing sub-networks per class. We denote these S-Proto-3P, S-Proto-5P and S-Proto-15P respectively. We limit our experiments to a maximum of $P = 15$, since $1/3$ of the labels occur less frequently than 15 times in the training data.

ProtoPatient-XL We claim that our choice of architecture is the reason for performance improvements. To investigate if the change in performance of S-Proto is due to increased model size, we evaluate a scaled version of ProtoPatient. Therefore, we increase the hidden dimension h of the projection layer L, label-wise attention W, and prototype parameters u to five times the dimensionality of ProtoPatient to compare against S-Proto-5P.

5.1   Finetuning and Hyperparameters

For ProtoPatient, ProtoPatient-XL, and the S-Proto variants, we use the best-performing hyperparameters and the same stopping criteria as in [2]. We randomly initialize the prototypical layer. For the S-Proto variants we use a hidden dimension h of 256 as in [2], for ProtoPatient-XL we set this to $5 \cdot h$ or 1280. For PubMedBERT we do hyper-parameter optimization (HPO) on 50 trials with *ray* [20] using *HyperOpt* [5], resulting in a learning rate and warmup steps in the range of the findings of [1]. For all methods we fine-tune on 7xA100 40GB GPUs.

6   Results

We present the results of the evaluation of the diagnoses prediction task based on MIMIC-IV in Table 2. We group the variants of S-Proto at the bottom of the table. We report the same metrics as [2]: AUROC and PR-AUC. Additionally, we report the training time for each experiment in days.

S-Proto Outperforms Existing Approaches. All the S-Proto variants report higher scores for AUROC and PR-AUC when compared to PubMed-BERT, ProtoPatient, and ProtoPatient-XL. The best performing baseline is ProtoPatient-XL, which demonstrates a PR-AUC improvement of 2.8 p.p. (11.6% relative) and ties on AUROC when compared to ProtoPatient. S-Proto-15P achieves the best performance in AUROC 1.46 p.p. gain (relative 1.62%) when compared to ProtoPatient-XL. The best variant of S-Proto for PR-AUC is 5P and it shows an increase of 3.71 p.p (relative 13%) when compared to ProtoPatient-XL. Given that ProtoPatient-XL and S-Proto-5P have the same number of parameters for the attention vectors and prototype vectors, we demonstrate that our architecture choices regarding sparsity and dimensionality are beneficial. Generally, all S-Proto variants outperform ProtoPatient-XL significantly in PR-AUC, and ProtoPatient by an even larger margin. The gains in PR-AUC of S-Proto variants show that our approach better addresses the challenges of the long-tail, and therefore succeed at a larger number of labels.

Table 2. Performance metrics of all the approaches on MIMIC-IV admission notes. Classification scores are reported in % points. The S-Proto variants outperform all baselines, S-Proto-5P is the best in PR-AUC and S-Proto-15P is the best in AUROC. The S-Proto variants take similar time to train independently of the choice of P. ProtoPatient-XL outperforms ProtoPatient but scales poorly when compared to S-Proto-5P. PubMedBERT performs the worst in the classification metrics but is the fastest model to train.

Model	AUROC(Macro)	AUROC(Micro)	PR-AUC	Training time[Days]
PubmedBERT	88.63	96.86	20.40	**0.25**
ProtoPatient	89.16	97.08	24.25	1.92
ProtoPatient-XL	90.25	97.27	27.07	5.30
S-Proto-3P	91.43	97.51	29.51	2.80
S-Proto-5P	91.66	97.53	**30.78**	2.69
S-Proto-15P	**91.71**	**97.54**	30.53	3.07

Efficiency of S-Proto. S-Proto-5P (and even S-Proto-15P which has 3 times more parameters in the prototypical layer) take a fraction of the time to train compared to ProtoPatient-XL. In the case of S-Proto-5P, it trains in roughly 49% of the time taken to train ProtoPatient-XL. On average all S-Proto variants take roughly the same time to train regardless of the dimensionality P, between 2.7 and 3.1 days, as we show in Table 2. We argue that the variance in this time is mostly determined by the time taken to converge while training and less so due to the dimensionality P.

7 Analysis and Discussion

Boosted Performance in the Long-Tail. To gain further knowledge on where the performance gains of S-Proto come from - we decompose the labels into the *Head* and *Tail* of their distribution depending on how often each label occurs in the training set. As in Fig. 1, *Head* corresponds to the group of labels with 80% of all samples or ≈ 10% of the labels, while *Tail* is the remaining 20% of all the samples or ≈ 90% of the labels. The large majority of the diagnoses are in the long-tail, therefore we focus on PR-AUC since it more informative about this part of the distribution. In addition, we measure F-1, precision, and recall for the *Head* and the *Tail* of the distribution. We present this in Table 3.

All models perform similarly on the *Head* labels. However, S-Proto-5P leads with a significant margin on the *Tail*, where PR-AUC is 4.5 p.p (18.8%) and

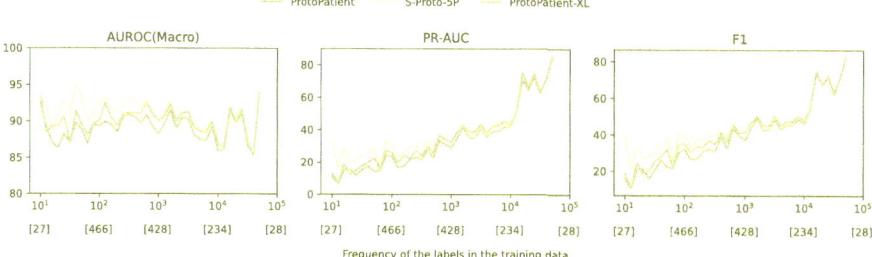

Fig. 3. Performance metrics for the labels in the training data based on their frequency (sorted ascending from left-to right). Numbers in square brackets indicate the number of samples in the respective range. All models present very similar scores on the classes occuring more than 10,000 times (22% of all labels). However, the classes that see the more benefits from S-Proto are the ones that occur less than 1,000 times(78% of all labels)

7.3 p.p (34.6%) higher than ProtoPatient-XL and ProtoPatient respectively. Although ProtoPatient-XL employs attention and prototype vectors with the same number of parameters as S-Proto-5P, its linear layer L is 5 times larger, resulting in a larger model overall. The improvements of S-Proto-5P in the long-tail compared to ProtoPatient-XL reinforce the hypothesis that the enhanced performance of our method is not solely attributed to a larger model size.

We demonstrate in Fig. 3 a more granular view of the scores for all diagnoses based on their frequency in the training data. We notice that after diagnoses appearing roughly more than 10,000 times (22% of all labels) the performance is similar for all models. In contrast, the labels occuring less than 1,000 times (78% of all labels) are the ones that benefit the most of S-Proto. This confirms that not only S-Proto sets the SOTA for clinical outcome prediction, but also it

Table 3. The table breaks down the classification metrics to measure the performance for the *Head* and *Tail* of the diagnoses distribution. All models perform similarly on the *Head* with ProtoPatient-XL leading in PR-AUC, F1, and Precision and S-Proto-5P leading in AUROC and Recall. S-Proto-5P outperforms significantly all methods in the *Tail* for all metrics, showing that our method boosts classification in the long-tail.

Class group	Model	AUROC(Macro)	PR-AUC	F1	Precision	Recall
Head	ProtoPatient	88.81	43.37	49.34	49.11	52.28
	ProtoPatient-XL	89.65	**46.32**	**51.11**	**51.04**	53.73
	S-Proto-5P	**89.88**	45.29	50.99	49.77	**54.43**
Long Tail	ProtoPatient	89.21	21.06	28.90	31.54	50.47
	ProtoPatient-XL	90.35	23.86	32.20	34.39	51.98
	S-Proto-5P	**91.95**	**28.36**	**36.90**	**39.45**	**54.18**

shows the most significant gains in the most challenging part of the distribution of diagnoses, where the other models perform generally worse.

Table 4. Performance Metrics on CodiEsp

Model	AUROC(Macro)	AUROC(Micro)	PR-AUC
ProtoPatient	87.57	86.70	29.03
ProtoPatient-XL	89.95	84.68	32.15
S-Proto-5P	**91.37**	**89.96**	**38.87**

Transfer to a Second Dataset. To evaluate the robustness of our approach we evaluate S-Proto-5P, ProtoPatient, and ProtoPatient-XL on CodiEsp, a second clinical dataset. This consists of 1000 clinical case reports in Spanish manually selected by doctors from different clinical wards, e.g. cardiology and oncology. We follow [23] in the definition of the diagnosis prediction task, but use only the English translation of the original notes. We keep only the labels that are common with MIMIC-IV, which results in 955 unique labels. Details of this split are shown in Table 1 *Right*. They also exhibit a long-tailed distribution, illustrated in Fig. 1 *Right*. We show our results in Table 4.

Similar to our results on MIMIC, we observe that S-Proto-5P outperforms ProtoPatient and ProtoPatient-XL. The largest gains are achieved in PR-AUC. Performance improves by 9.84 p.p ($\approx 34\%$) and 6.72 p.p. ($\approx 23\%$) respectively. We argue that a possible reason for the improved performance in PR-AUC compared to MIMIC is the distribution shift in label frequency. Namely, we observe that 168 ($\approx 18\%$) labels from the *Head* of CodiEsp belong to the *Tail* in MIMIC. In contrast, only 48 ($\approx 5\%$) labels from the *Head* in MIMIC are found in the *Tail* of CodiEsp. The results of this evaluation highlight that our method generalizes well to other unseen clinical data. In practice, this is a very desirable property because clinical data is very heterogenous and transformer encoders are known to learn *annotation artifacts*, i.e. spurious class-conditioned features in text [14,24,33].

Competing Sub-Networks Boost Classification Power. At prediction time S-Proto uses the best classifier (*winner* sub-network) depending on the input. Note that by definition there is only one *winner* sub-network per admission note (sample). Since there are multiple admission notes for a given diagnosis, multiple winner sub-networks may describe the same disease. Intuitively this means that one diagnosis may be characterized by multiple *prototypical patients*. Figure 4 *Left* shows the number of different *winner* sub-networks for every diagnosis in the test set for S-Proto-5P. We observe that only the minority (11%) of the labels have only **one** *winner* sub-network and the majority has multiple. Multiple *winners* indicate that multiple sub-networks are needed to describe

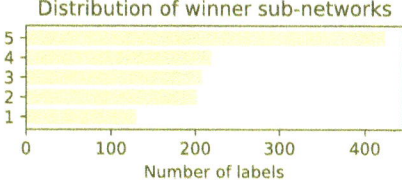
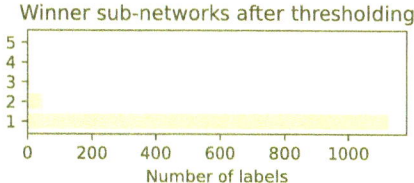

Fig. 4. Left: Number of all *winner* sub-networks that are needed to capture the variance of the patients per diagnosis. **Right**: Number of *winner* sub-networks that only predict the *presence* of a diagnosis (after thresholding predictions). This contrast suggests that S-Proto-5P also learns *negative prototypical patients* to explain the *absence* of a disease.

the high variance in the admission notes of the same diagnosis. This also shows that the performance gains of S-Proto are directly benefiting from the multiple *prototypical patients* learned for a diagnosis.

Most Winner Sub-Networks Define the Absence of a Diagnosis. In practice, assigning one diagnosis to a clinical note is a binary decision, i.e. we need to define a threshold for the predictions. We determine a threshold per diagnosis based on the highest F1 score. We show in Fig. 4 *Right*, that for most of the diagnoses only one winner is able to surpass this threshold, i.e. predict the presence of a disease. This is in sharp contrast with the number of sub-networks that are needed to describe a disease (Left of Fig. 4). Intuitively, this means that S-Proto also learns *negative prototypical patients* focusing on the absence of a diagnosis. This *contrastive* effect is only possible with multiple competing sub-networks describing both the presence and the absence of a diagnosis. We argue this effect determines the *performance boost in the long-tail*.

Qualitative Analysis with Doctors. S-Proto chooses *winner* sub-networks to generate a prediction for every admission note. We examine further the results in Fig. 4 *Right*, namely the group of labels for which two *winner* sub-networks predict the presence of the diagnosis. With the help of medical doctors, we qualitatively evaluate 30 admission notes for 5 diagnoses exhibiting this novel feature. Given two groups of admission notes, each corresponding to a single *winner* sub-network, the doctors are tasked to assess whether the token saliencies(e.g. see Fig. 5 *Middle* and *Right*) describing these sub-networks belong to meaningful *cohorts* of the same disease.

Learned Hierarchies in Disease Cohorts. For all five diagnoses (30 patients), the doctors find that the two groups correspond to different cohorts of patients. For instance, in the case of *Arrhythmias*, the medical doctors confirm that the two *winner* sub-networks are meaningful disease phenotypes, i.e. the severe (*Ventricular*) and more benign (*Supra-ventricular*) arrhythmias. Additionally, for one of these diseases(*Other disorders of the peritoneum* or K66), the

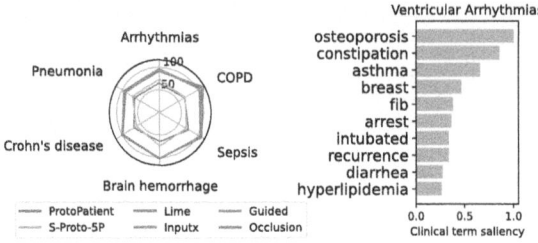
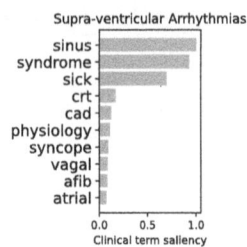

Fig. 5. Left: Faithfulness evaluated for 6 diseases (lower is better); both ProtoPatient and S-Proto perform similarly. **Middle, Right**: Saliencies of the closest patients to the prototypes to arrhythmias, S-Proto learns separate phenotypical descriptors for Ventricular Arrhythmias (Center) and Supra-Ventricular Arrhytmias(Right).)

doctors highlight how in one of the *winner* sub-networks a different diagnosis would be more suitable for the phenotype of these patients, namely *Retroperitoneal hematoma* or K68. Finally, for one disease the cohorts identified do not represent different phenotypes but differ in complications or risk factors of the underlying disease. This suggests that S-Proto is capable of capturing a meaningful hierarchy within diagnoses. This learned hierarchy can be seen as an additional property that allows for more granular explanations e.g. severe vs benign arrhythmias or patients with different risk factors. In practice, this may support clinical professionals in clinical coding, as well as in identifying errors [25].

Faithfull Explanations are Maintained. We evaluate our model's interpretability by examining its performance on the faithfulness metric [3] to compare with the results of [2]. This score measures how much the performance degrades when a portion of the tokens in the input are removed. Iteratively, the input to the model is masked, based on sorted token saliencies (ranked by importance). Faithfulness is measured by the AUC of the resulting performance scores for every level of masking. Note, that a lower score reflects a higher degree of sensitivity to the masked input, i.e. better model faithfulness [3].

Comparisson to Post-hoc XAI Methods. First, we examine faithfulness for a group of 6 diseases, 3 of which were evaluated in [2]. We employ explainability methods from [3], namely *Lime, InputXGradient, Guided-BackPropagation* and *Occlusion*. We limit ourselves to 6 diseases since these are computationally very taxing. We compare them to ProtoPatient and our S-Proto-5P model. In Fig. 5 *Left* we demonstrate these results. We observe that both S-Proto-5P and ProtoPatient present similar performance, and outperform the other token-saliency generation methods.

Faithfulness of S-Proto on all Labels. We evaluate faithfulness for all S-Proto variants, ProtoPatient and ProtoPatient-XL on the test data (1183 labels) since these are more computationally efficient at generating explanations. We normalize the change in performance for all models and levels of masking based on

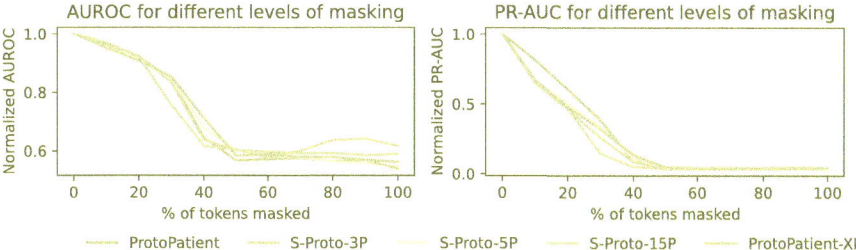

Fig. 6. AUROC (left) and PR-AUC(right) evaluated for different levels of masking of tokens based on saliencies generated by S-Proto, ProtoPatient-XL, and ProtoPatient for all MIMIC diseases in our test set. The saliencies(token importance) are ranked, i.e. 20% corresponds to the top 20% of most important tokens. All variants of S-Proto retain a similar level of faithfulness to ProtoPatient.

the best-achieved score (0% of tokens masked) to make the models comparable. Figure 6 shows the results of this evaluation for 10 levels of masking and all S-Proto and ProtoPatient variants.

We note that in line with our observations in Sect. 4, S-Proto and ProtoPatient are both similarly sensitive to the masking of the most important tokens. Thus, we argue that even though S-Proto is more complex, it preserves the faithful predictions of the original ProtoPatient model.

8   Conclusion

In this work we present S-Proto, a novel architecture leveraging a sparse network of multiple prototypes. We show with an extensive quantitative evaluation that our approach sets state-of-the-art performance for both AUROC and PR-AUC. Our analysis shows that our approach significantly improves performance for the long-tail (majority of classes). This is emphasized by the transferability of these gains to a second unseen dataset. The sparsity of S-Proto allows for efficient scaling, boosting performance significantly with no loss of interpretability. Our qualitative evaluation with medical doctors shows that S-Proto presents a novel interpretability property of finding meaningful cohorts of patients.

Limitations and Future Work. Although we demonstrate performance improvements, power-law distributions like diagnosis codes remain challenging for machine learning models. Future directions of research expanding our methods include the control over the selection of sub-networks beyond the mechanism of winner-takes-all. Additionally, in this work, we train S-Proto only on MIMIC. In practice, healthcare data segregation is a common problem. This could be tackled by our method in continual learning scenarios due to its sparsity [6,28].

Acknowledgments. We would like to thank the reviewers for their helpful suggestions and comments. Our work is funded by the German Federal Ministry of Education and Research (BMBF) under the grant agreements 01|S23013C (More-with-Less).

01|S23015A (AI4SCM) and 16SV8857 (KIP-SDM). This work is also funded by the Deutsche Forschungsgemeinschaft (DFG, German Research Foundation) Project-ID 528483508 - FIP 12, as well as the European Union under the grant project 101079894 (COMFORT - Improving Urologic Cancer Carewith Artificial Intelligence Solutions).

Disclosure of Interests. The authors declare no relevant competing interests.

References

1. van Aken, B., Papaioannou, J., Mayrdorfer, M., Budde, K., Gers, F.A., Löser, A.: Clinical outcome prediction from admission notes using self-supervised knowledge integration. In: EACL, pp. 881–893. Association for Computational Linguistics (2021)
2. van Aken, B., et al.: This patient looks like that patient: Prototypical networks for interpretable diagnosis prediction from clinical text. In: AACL/IJCNLP (1), pp. 172–184. Association for Computational Linguistics (2022)
3. Atanasova, P., Simonsen, J.G., Lioma, C., Augenstein, I.: A diagnostic study of explainability techniques for text classification. In: EMNLP (1), pp. 3256–3274. Association for Computational Linguistics (2020)
4. Bengio, Y., Léonard, N., Courville, A.C.: Estimating or propagating gradients through stochastic neurons for conditional computation. CoRR abs/1308.3432 (2013)
5. Bergstra, J., Yamins, D., Cox, D.D.: Hyperopt: a python library for optimizing the hyperparameters of machine learning algorithms. In: SciPy, pp. 13–19. scipy.org (2013)
6. Bricken, T., Davies, X., Singh, D., Krotov, D., Kreiman, G.: Sparse distributed memory is a continual learner. In: ICLR. OpenReview.net (2023)
7. Chen, C., Li, O., Tao, D., Barnett, A., Rudin, C., Su, J.: This looks like that: deep learning for interpretable image recognition. In: NeurIPS, pp. 8928–8939 (2019)
8. Dasgupta, S., Stevens, C.F., Navlakha, S.: A neural algorithm for a fundamental computing problem. Science **358**(6364), 793–796 (2017)
9. Deuschel, J., et al.: Multi-prototype few-shot learning in histopathology. In: ICCVW, pp. 620–628. IEEE (2021)
10. Devlin, J., Chang, M., Lee, K., Toutanova, K.: BERT: pre-training of deep bidirectional transformers for language understanding. In: NAACL-HLT (1), pp. 4171–4186. Association for Computational Linguistics (2019)
11. Du, N., et al.: Glam: efficient scaling of language models with mixture-of-experts. In: ICML. Proceedings of Machine Learning Research, vol. 162, pp. 5547–5569. PMLR (2022)
12. Fedus, W., Zoph, B., Shazeer, N.: Switch transformers: scaling to trillion parameter models with simple and efficient sparsity. J. Mach. Learn. Res. **23**, 120:1–120:39 (2022)
13. Grundmann, P., Oberhauser, T., Gers, F.A., Löser, A.: Attention networks for augmenting clinical text with support sets for diagnosis prediction. In: COLING, pp. 4765–4775. International Committee on Computational Linguistics (2022)
14. Gururangan, S., Swayamdipta, S., Levy, O., Schwartz, R., Bowman, S.R., Smith, N.A.: Annotation artifacts in natural language inference data. In: NAACL-HLT (2), pp. 107–112. Association for Computational Linguistics (2018)

15. Hase, P., Chen, C., Li, O., Rudin, C.: Interpretable image recognition with hierarchical prototypes. In: HCOMP, pp. 32–40. AAAI Press (2019)
16. Johnson, A., Bulgarelli, L., Pollard, T., Horng, S., Celi, L.A., Mark, R.: MIMIC-IV (2021). https://doi.org/10.13026/s6n6-xd98. https://physionet.org/content/mimiciv/1.0/
17. Lepikhin, D., et al.: Gshard: scaling giant models with conditional computation and automatic sharding. In: ICLR. OpenReview.net (2021)
18. Li, X., Tian, T., Liu, Y., Yu, H., Cao, J., Ma, Z.: Adaptive multi-prototype relation network. In: 2020 Asia-Pacific Signal and Information Processing Association Annual Summit and Conference (APSIPA ASC), pp. 1707–1712 (2020)
19. Liang, Y., et al.: Can a fruit fly learn word embeddings? In: ICLR. OpenReview.net (2021)
20. Liaw, R., Liang, E., Nishihara, R., Moritz, P., Gonzalez, J.E., Stoica, I.: Tune: a research platform for distributed model selection and training. CoRR abs/1807.05118 (2018)
21. Naik, A., Lehman, J., Rosé, C.P.: Adapting to the long tail: a meta-analysis of transfer learning research for language understanding tasks. Trans. Assoc. Comput. Linguist. **10**, 956–980 (2022)
22. Naik, A., Parasa, S., Feldman, S., Wang, L.L., Hope, T.: Literature-augmented clinical outcome prediction. In: NAACL-HLT (Findings), pp. 438–453. Association for Computational Linguistics (2022)
23. Papaioannou, J., et al.: Cross-lingual knowledge transfer for clinical phenotyping. In: LREC, pp. 900–909. European Language Resources Association (2022)
24. Poliak, A., Naradowsky, J., Haldar, A., Rudinger, R., Durme, B.V.: Hypothesis only baselines in natural language inference. In: *SEM@NAACL-HLT, pp. 180–191. Association for Computational Linguistics (2018)
25. Searle, T., Ibrahim, Z.M., Dobson, R.J.B.: Experimental evaluation and development of a silver-standard for the MIMIC-III clinical coding dataset. CoRR abs/2006.07332 (2020)
26. Sechidis, K., Tsoumakas, G., Vlahavas, I.: On the stratification of multi-label data. In: Gunopulos, D., Hofmann, T., Malerba, D., Vazirgiannis, M. (eds.) ECML PKDD 2011. LNCS (LNAI), vol. 6913, pp. 145–158. Springer, Heidelberg (2011). https://doi.org/10.1007/978-3-642-23808-6_10
27. Shazeer, N., et al.: Outrageously large neural networks: The sparsely-gated mixture-of-experts layer. In: ICLR (Poster). OpenReview.net (2017)
28. Shen, Y., Dasgupta, S., Navlakha, S.: Algorithmic insights on continual learning from fruit flies. CoRR abs/2107.07617 (2021)
29. Snell, J., Swersky, K., Zemel, R.S.: Prototypical networks for few-shot learning. In: NIPS, pp. 4077–4087 (2017)
30. Tinn, R., et al.: Fine-tuning large neural language models for biomedical natural language processing. Patterns **4**(4), 100729 (2023)
31. Winter, B., Rosero, A.F., Löser, A., Gers, F.A., Siu, A.: KIMERA: injecting domain knowledge into vacant transformer heads. In: LREC, pp. 363–373. European Language Resources Association (2022)
32. Xie, S., Vosoughi, S., Hassanpour, S.: Proto-LM: a prototypical network-based framework for built-in interpretability in large language models. In: EMNLP (Findings), pp. 3964–3979. Association for Computational Linguistics (2023)
33. Zellers, R., Holtzman, A., Bisk, Y., Farhadi, A., Choi, Y.: Hellaswag: can a machine really finish your sentence? In: ACL (1), pp. 4791–4800. Association for Computational Linguistics (2019)

Author Index

GPSR Compliance

The European Union's (EU) General Product Safety Regulation (GPSR) is a set of rules that requires consumer products to be safe and our obligations to ensure this.

If you have any concerns about our products, you can contact us on ProductSafety@springernature.com

In case Publisher is established outside the EU, the EU authorized representative is:

Springer Nature Customer Service Center GmbH
Europaplatz 3
69115 Heidelberg, Germany

The manufacturer's authorised representative in the EU is Springer
Nature Customer Service Centre GmbH, Europaplatz 3, 69115 Heidelberg,
Germany. If you have any concerns regarding our products, please
contact ProductSafety@springernature.com

Printed and bound by CPI Group (UK) Ltd, Croydon, CR0 4YY
06/05/2026
02103807-0001